Avoiding Questionable Research Practices in Applied Psychology

William O'Donohue • Akihiko Masuda
Scott Lilienfeld
Editors

Avoiding Questionable Research Practices in Applied Psychology

Springer

Editors
William O'Donohue
Department of Psychology
University of Nevada
Reno, NV, USA

Akihiko Masuda
Department of Psychology
University of Hawaii at Manoa
Honolulu, HI, USA

Scott Lilienfeld
Department of Psychology
Emory University
Atlanta, GA, USA

ISBN 978-3-031-04967-5 ISBN 978-3-031-04968-2 (eBook)
https://doi.org/10.1007/978-3-031-04968-2

This Springer imprint is published by the registered company Springer Nature Switzerland AG
The registered company address is: Gewerbestrasse 11, 6330 Cham, Switzerland

Contents

Part I
Introduction

Chapter 1
Questionable Research Practices in Clinical Psychology

William O'Donohue and Akihiko Masuda

Abstract Research into fundamental questions such as psychotherapy outcome and process forms a fundamental component of clinical science. At a broader level, the aim of this volume is to present useful, practical information—for both consuming current research and improving one's own research—for researchers, instructors, and trainees (e.g., doctoral students) in clinical psychology. Simultaneously, at a more specific level, it is also our thesis that an improved understanding of questionable research practices (QRPs) derived from this book offers students and researchers to more accurately and deeply understand psychological science and clinical psychology and to learn to avoid errors in their own research. While taking these aims into consideration, we have organized this book into three major sections. The first section of this volume (i.e., Chaps. 1, 2, 3, and 4) offers a general introduction to the issues of QRPs, setting them into a historical and current landscape in psychological science and clinical psychology. The second section of this volume (Chaps. 5, 6, 7, 8, 9, 10, 11, 12, and 13) introduces some of the notable exemplars of QRPs and QRPs in various research contexts. Finally, in the third and last section of this volume (Chaps. 14, 15, 16, 17, 18, and 19), newly emerging models for minimizing the impact of QRPs in research are presented.

Keywords Questionable research practices · Clinical psychology · Clinical science

Background

Research into fundamental questions such as psychotherapy outcome and process forms a fundamental component of clinical science (McFall, 1991). For a variety of reasons, replications of these studies are also a key component of clinical science. Knowledge is thought to be reliable: A true result ought to appear again in similar

W. O'Donohue (✉)
Department of Psychology, University of Nevada, Reno, NV, USA
e-mail: wto@unr.edu

A. Masuda
University of Hawai'i at Mānoa, Honolulu, HI, USA

W. O'Donohue et al. (eds.), *Avoiding Questionable Research Practices in Applied Psychology*, https://doi.org/10.1007/978-3-031-04968-2_1

testing circumstances. This generalizability question is critical because practitioners essentially depend on the reliability of these results when they consume this research and use it as a basis for their clinical decision-making (e.g., choice of treatments for a particular client). To put it another way, practitioners want the general results to "replicate" with their clients. Therefore, replication studies can provide information about the extent to which scientific results are indeed reproducible.

When a result is found not to be reproducible, then additional key questions are raised that can range broadly as: "Are the original results actually spurious/artifactual/false?" or "Was there some sort of error (intentional or unintentional) in conducting or interpreting the replication study that showed a failure to replicate?" or "Because in any attempt of conducting a replication, is it simply impossible to create identical situations (e.g., use the same participants, history effects given the passage of time create differences in participants, etc), and are some of the unavoidable differences just due to legitimate boundary conditions to the effect?" (e.g., Schmidt, 2009). However, even if this is the case, what exactly are these boundary conditions and is the situation that I want to generalize to within or outside these boundaries? Replication failures or even when it is unknown if a result can replicate can lead to very problematic consequences, such as unexpected treatment failures.

Replication Crisis in Psychology

Social psychology has been the subfield in psychology that in the last few decades paid the most attention to replications and has generally been regarded as having a replication crisis. They should be seen not as unique but as showing leadership regarding these issues. Several key findings in social psychology have simply failed to replicate (see Chap. 4 in this volume). This of course can call into question findings in which replications have not yet been attempted. The question becomes, "Would this finding replicate?" or relatedly, "Does this finding seem robust even though it has not been put to the test of replication?"

There are several types of replications (e.g., direct or literal, partial, or conceptual) and questions can be raised about what it means for a study to be replicated or not to be replicated (see Chaps. 4 and 14 in this volume). It is often the case that replications can be very expensive to conduct in time, effort, and money and this is certainly one reason why there are so few. Replicating grant research may not even be feasible unless one also can obtain a similar-sized grant to obtain the necessary resources, and granting agencies may not value replication studies sufficiently to award such grants. There are views that journal editors will more likely reject replications than original studies and views that conducting replications are not as career enhancing as conducting original research (see Chap. 10 in this volume). These factors can all conspire to create few replications and thus the reproducibility of studies in clinical science can remain an open question.

Replication Studies in Clinical Psychology

These reasons may have a role in the relatively low amount of published replication research in clinical psychology. As a crude index, the APA PsycINFO search (conducted on October 29, 2021) using the following key words revealed the following number of citations:

- "Replication" (in title) and "cognitive behavior therapy" (in any text) = 40
- "Replication" (in title) and "psychodynamic" (in title) = 4
- "Replication" (in title) and "Rogerian" (in title) = 0
- "Replication" (in title) and "eclectic" (in title) = 0
- "Replication" (in title) and "cognitive therapy" (in title) = 84
- "Failure to replicate" (in title) and "behavior therapy" (in title) = 12
- "Replication" (in title) and "acceptance and commitment therapy" (in title) = 3
- "Replication" (in title) and "dialectical behavior therapy" (in title) = 1
- "Questionable research practices" (in title) and "behavior therapy" (in title) = 0.

These are admittedly crude indices and the reader is cautioned that some replication studies could be missed in these procedures, but on the other hand it is also important to note that these values do not mean that each citation was an actual replication study, as a review of the titles and abstracts revealed that some of these citations were commentaries, literature reviews, and types of studies other than replication studies. However, given the decades that this literature search covers, these numbers seem alarmingly low: it seems fair to say that there have been few replication attempts for the vast majority of results in the field of clinical psychology.

In addition, another concern is the extent to which psychologists pay attention to studies that report replication failures. In the social psychology literature, there is evidence that when studies fail to replicate, psychologists still pay more attention to the original study rather than the replication failure. For example, Darley and Gross (1983) initially published a study that showed that social class information biased participants' interpersonal judgments. However, subsequently, Baron et al. (1995) published an article that described two experiments with much larger samples that not only failed to replicate the original findings but interestingly reported findings were in the opposite direction. However, Jussim, Crawford, Anglin, Stevens, and Duarte (2016) in an analysis of citations since 1996, found that the original study (i.e., Darley & Gross, 1983) was cited 852 times, while the failed replication had been cited only 38 times (according to Google Scholar searches conducted on September 11, 2015). It may be the case that psychologists prefer the "good news" of a positive finding and attempt to avoid or minimize a replication failure that casts doubt on a positive finding. Clinical scientists reasonably want to have tools to help individuals, and perhaps for them, replication failures can be seen as undermining the hope associated with these possible tools.

Questionable Research Practices

Questionable research practices (QRPs) have been implicated by many as being responsible for findings that can be created by researchers that produce initially false results (Fiedler & Schwarz, 2016; John et al., 2012; Wicherts, 2011). False results would hopefully not be replicated as this would allow the science to identify and perhaps be cleansed of false information. QRPs can range from: the use of the "file drawer" to hide negative results (Rosenthal, 1979; Rotton et al., 1995), *p*-hacking (e.g., Wicherts et al., 2016), hypothesizing after results are known (HARKing; Kerr, 1998), selective reporting of multiple outcome variables (O'Donohue et al., 2016a), deciding to collect more data when the results are not significant (*p*-hacking), and failing to disclose all experimental conditions (also see this whole volume for a further explication of the different types of QRPs as well as their definitions).

John, Lowenstein, and Prelac (2012) surveyed over 2000 academic psychologists and asked them to self-disclose their usage of QRPs. Their conclusion was that the results indicated "surprisingly" high self-admission rates of a wide variety of QRPs. There was a corresponding low rate of self-admission of outright manufacture of data. Thus, this seems to be consistent that the major problem in the distortion of the scientific research base in psychology is not fabricating data but the use of QRPs to produce favored results. It must also be noted that it is likely that these data are still an underestimate of the use of QRPs in psychology as these values can be affected by the reluctance to disclose any wrongdoing. Some of the key findings are presented in Table 1.1.

Table 1.1 Self-admission rate of various questionable research practice (QRP) reported in John et al. (2012)

	QRP item	Self-admission rate (%)
1.	In a paper, failing to report all of a study's dependent measures	66.5
2.	Deciding whether to collect more data after looking to see whether the results were significant	58.0
3.	In a paper, failing to report all of a study's condition	27.4
4.	Stopping collecting data earlier than planned because one found the result that one had been looking for	22.5
5.	In the paper, "rounding off" a *p*-value (e.g., reporting that a *p*-value of 0.054 is less than 0.05)	23.3
6.	In a paper, selectively reporting studies that "worked"	50.0
7.	Deciding whether to exclude data after looking at the impact of doing so on the results	43.4
8.	In a paper, reporting unexpected finding as having been predicted from the start	35.0
9.	In a paper, claiming that results are unaffected by demographic variables (e.g., gender) when one is actually unsure (or knows that they do)	4.5
10.	Falsifying data	1.7

Note: Self-admission rates reported in the table are ones endorsed by the participants who were supplemented by incentives for honest reporting (i.e., Bayesian truth serum group)

Thus there is a concern not only in clinical psychology but also in other disciplines that QRPs may be responsible for too many "false-positive" results. For example, Fanelli (2012) conducted a meta-analysis of over 4600 publications in a variety of scientific disciplines that were published between 1990 and 2007. His results found a general trend in which positive support for tested hypotheses grew over 22% during this time period. Of the most concern, Fanelli (2012) examined possible differences between disciplines in the growth of positive results and discovered that the mean frequency of positive results was "significantly higher when moving from the physical, to the biological to the social sciences, and in applied versus pure disciplines" (Fanelli, 2012, p. 893). In a more direct study of practices in psychology, Francis (2014) demonstrated through the use of a test of excess success (TES) that empirical studies published in two key psychology journals succeed at a much higher rate than should be expected given estimated effects and sample sizes.

Questionable Research Practices in Clinical Psychology

The question arises: To what extent is there such a positive bias in the publication of research in clinical psychology? Perhaps a more radical question is: With sufficient use of QRPs, can any research manufacture a positive result? In the case of psychotherapy outcome research, O'Donohue, Snipes, and Soto (2016a, b) have argued that researchers have sufficient degrees of freedom in designing and conducting research that this may be the case. For example, by choosing a better therapist for the favored experimental condition versus the control condition, choosing a weaker control condition (no attention vs treatment as usual), deciding to not use blinds, deciding to use only statistical significance versus clinical significance in reporting outcomes, deciding not to include post-treatment assessment periods to examine possible relapse, analyzing only therapy completers instead of using an intent to treat analysis, using multiple dependent variables but only reporting positive ones, using outcome measures of problematic validity, failing to include process measures to assess if changes are due to hypothesized variables, and so on, it becomes relatively easy to manufacture a positive result. Perhaps there needs to be another sort of replication—a kind of replication that eliminates as many QRPs as possible.

In fact, these authors (O'Donohue et al., 2016a) analyzed a publication of the efficacy of acceptance and commitment therapy (ACT; Hayes et al., 2012) and diabetes self-management (Gregg et al., 2007; see Gregg & Hayes, 2016, for a response). The published report was based on a dissertation and it was possible to analyze discrepancies between the dissertation and the published research report. They found the following (O'Donohue et al., 2016a p. 22):

- A failure to report several key negative results from the dissertation in a subsequent peer-reviewed journal publication.
- A series of overstatements and misstatements by the researchers in subsequent publications about the positive findings in the dissertation.

- The development of a bibliotherapeutic intervention marketed to people with diabetes (claiming to be "a proven program") in which the reader is led to believe that the bibliotherapy intervention they are buying and using has been shown to be effective and safe in past research, when the bibliotherapy intervention has not even been studied at all. In addition, the reader is not informed of the data that the intervention failed to work due to the hypothesized ACT processes.
- In addition, the reader of the self-help book is not informed of the negative results of the data relating to an intensive workshop led by an ACT therapist, numerous serious limitations of the design of this study, the fact that these results have not been replicated, or about the possible efficacy of alternative treatments.
- The failure to accurately describe in subsequent publications, particularly in a peer-reviewed journal publication, what are at best equivocal findings regarding the role of putative ACT processes as mediating these results. Instead, the opposite is found: clear, but inaccurate, statements about ACT processes producing clinically significant changes in diabetes self-management when the original data simply do not warrant this.
- A lack of appropriate caution and qualification in interpreting the data relating to the effectiveness of ACT for diabetes self-management despite numerous methodological shortcomings, including, but not limited to: therapist allegiance effects, dependent measures with unknown psychometrics, no blinds, minimal follow-up, no safety measures, significant attrition, problems with alpha rate inflation, no comparison to key treatments as usual, and no replications.
- The existence of these problems sometimes occurred in a context in which the authors were explicitly reassuring readers that they would refrain from excessive claims and would point out unresolved empirical issues, thus providing readers with a false reassurance that good scientific practices were being followed.

The concern is that such distortions in the scientific clinical literature can produce distortions and misinformation that can ultimately harm vulnerable clients. There have been too few studies of self-admission rates of clinical scientists regarding their usage of QRPs as well as too few analyses of such behavior. The methodology used by O'Donohue et al. (2016a) could be used to examine the discrepancies/ consiliences between master theses and dissertations with published articles based on these. However, although identifying some QRPs, this method admittedly can still underestimate the use of QRPs because these cannot identify those used in the original dissertation research that are not explicated in this document.

Clinical scientists consult the research literature but if this research literature has key false positives—and perhaps it does not need to have that many—incorrect treatment decisions can be made and thus the problems of clients we ought to be serving well can be prolonged. The evidence-base of clinical science becomes more a reflection of hype, marketing tactics, and rhetoric rather than science.

A related question becomes: "To what extent are peer-reviewed scientific studies in the clinical psychology literature biased by personal motivations of the scientist?" Interestingly, it is long recognized by many cognitive behavior therapists that Big Pharma has biased results in psychopharmacological research in depression as well as other problem areas (see for example Antonuccio et al., 2002). In another problem domain, Etter, Burri, and Stapleton (Etter et al., 2007) found that in an examination of all randomized controlled trials of nicotine replacement therapy for smoking cessation, industry-supported trials found more statistically significant results than non-industry trials. In addition, these studies reported larger effect sizes as well. But to a very large extent similar motivations have been largely ignored for researchers in cognitive behavioral therapy (CBT). Do proponents of some psychotherapy have personal incentives to use QRPs to manufacture positive results? Are such practices prudent and defensible? Admittedly, for cognitive behavior therapists the magnitude of the possible financial gain is less by order of magnitudes but significant personal financial consequences can still be present.

Ioannidis (2005) has made a similar but broader point about the medical literature. In his article "Why most published research findings are false," he addressed:

> **"The greater the financial and other interests and prejudices in a scientific field, the less likely the research findings are to be true.** Conflicts of interest and prejudice may increase bias, *u*. Conflicts of interest are very common in biomedical research…, and typically they are inadequately and sparsely reported…. Prejudice may not necessarily have financial roots. Scientists in a given field may be prejudiced purely because of their belief in a scientific theory or commitment to their own findings. Many otherwise seemingly independent, university-based studies may be conducted for no other reason than to give physicians and researchers qualifications for promotion or tenure. Such nonfinancial conflicts may also lead to distorted reported results and interpretations. Prestigious investigators may suppress via the peer review process the appearance and dissemination of findings that refute their findings, thus condemning their field to perpetuate false dogma. Empirical evidence on expert opinion shows that it is extremely unreliable…." (p. 0698)

Making Sense of Questionable Research Practices from the Perspective of Meta-Science

Meta-scientists such as philosophers of science and sociologists of science have attempted to study and understand science from both an "internal" perspective—that is, a focus on matters such as the logic of research—and an "external" perspective—that is, a focus on human and psychological factors influencing the behavior of the scientist. For example, the prominent philosopher of science, Sir Karl Popper (1972), focused on commonly observed cognitive biases such as "the craving to be right" (i.e., confirmation bias) as a distorting psychological influence in science.

This important distinction between logical and psychological analyses is carried further recently in a blog (corelab.blog; March 5, 2020) in what might be called a "scientist as logician" perspective versus a "scientist as human" perspective. In the scientist as logician perspective, the following are emphasized: (1) Scientists are viewed as primarily truth-seekers; (2) Scientists are seen as relying on logic to develop the most efficient ways of discovering truth and growing a knowledge base through their research; and (3) If some critic uses logic to identify flaws or errors in a scientist's current knowledge-seeking process, then, to the extent that the critic's logic is sound, that scientist will modify his or her scientific practices.

In contrast, in the "scientist as human" perspective, the following claims are instead emphasized: (1) Humans, including scientists, have a wide number and variety of goals (that include discovering truth and the growth of knowledge, being accurate and precise as possible; but in addition, scientists have other goals such as demonstrating that he or she is correct, career advancement, fame, financial gain, and so on). Scientific behavior may be associated with both sets of factors; the relative influence of each may vary across scientists; (2) In addition, all humans, including psychological researchers, are embedded in influential social systems including political, economic, and professional ones; (3) Humans, including clinical scientists, are sensitive and are influenced by the imperatives and incentives of these systems; and (4) Reformers (including those attempting to improve the quality of science) as well as other critics must attend to these human goals as well as social, political, economic, and professional imperatives that influence the clinical scientist if they want to successfully create lasting changes in scientific practice as well as the social, economic, and personal imperatives that might function to decrease the likelihood of clinical scientists from engaging in certain behaviors, especially behavior inconsistent with the scientist as logician perspective. Any reform to increase the quality of science then must work to align those imperatives with the desired behaviors.

Possible Solutions

The chapters that follow will discuss solutions that have been proposed and even tried out to better deal with QRPs in science. There has been too little attention given to this issue at present and thus too little engagement with these procedures that could improve the quality of clinical science. Briefly, some of the key improvements suggested include:

1. Pre-registration of studies (to decrease researcher's degrees of freedom that may result in the use of QRPs because the researcher has made pre-commitments about key decisions).
2. Open data (this can allow others to run or rerun analyses).
3. Adversarial research projects (in order for the research to include fewer unidirectional biases as well as allowing someone to crucially overview the process of research).
4. Decrease use of QRPs (generally, this suggestion is oriented toward increased education concerning QRPs, so perhaps individuals are less likely to employ these).
5. The development of methods to detect QRPs (some are more difficult to detect than others, but the ability to detect the use of these is invaluable).
6. Increase publication and other support of replications (attempt to persuade key individuals such as journal editors, grant administrators, and even personnel committees to increase their valuing of replications).

7. Increase activities such as the Many Lab project (see Chap. 4 in this volume) to include non-*weird* participants either to initially uncover more generalizable effects or to more quickly see demographic/cultural boundary conditions.
8. Perhaps researchers ought to be more explicit in stating the logic of their research, and examine options such as the use of *modus tollens* in Popperian inspired research that emphasizes severe testing. Severe testing may be a good way to reduce false positives (also see Chap. 2 in this volume).
9. Increase concern about the psychometric weaknesses or unknowns (initial research can be less generalizable when noisy assessment instruments are responsible for false-positive results; see Chap. 10 in this volume).
10. More clear financial disclosures especially from workshops, books sales, etc. The field has been pretty lax on this—it may be the case that millions of dollars may engender more financial influence, but individuals can also be influenced by thousands or even hundreds of dollars.
11. Perhaps more research in psychology needs to be reported using the Cochrane Collaboration's tool for assessing risk of bias in intervention outcome studies, which is a simple checklist for attempting to identify biases that includes six different types of possible bias (selection, performance, detection, attrition, reporting, and others; Higgins et al., 2011). It could improve the quality of clinical psychological science if these and other improvements receive more attention.

Once again, as discussed extensively elsewhere (e.g., Melchert et al., 2019) and throughout this volume, research practices lie at the core of psychological science and clinical psychology. However, the standard view of research practices and methods, which unfortunately allow for extensive flexibility in data analysis on the part of investigators, has recently come into question as a series of QRPs (John et al., 2012; Swift et al., 2020). In this context, we believe that promoting our awareness of and self-reflection on QRPs as well as becoming cognizant of these possible solutions will be a major first step to minimize harm to the public due to our QRPs and to advance our field as a life science.

Overview of Chapters

At a broader level, the aim of this volume is to present useful, practical information—for both consuming current research and improving one's own research—for researchers, instructors, and trainees (e.g., doctoral students) in clinical psychology. Simultaneously, at a more specific level, it is also our thesis that an improved understanding of QRPs derived from this book offers students and researchers to more accurately and deeply understand psychological science and clinical psychology and to learn to avoid errors in their own research. While taking these aims into consideration, we have organized this book into three major sections.

The first section of this volume (i.e., Chaps. 1, 2, 3, and 4) offers a general introduction to the issues of QRPs, setting them into a historical and current landscape in psychological science and clinical psychology. Following the present chapter (i.e., Chap. 1), which offers a general overview of QRPs in our field, Dylan Wong of Oregon Social Learning Center and William O'Donohue of the University of Nevada, Reno, present several models regarding the logic of research together with a philosophy of science as a meta-level, providing a guiding framework for research practice (Chap. 2). More specifically, they present an introductory overview of the philosophy of science, problems with induction, and Popperian falsificationism and its limitations, then arguing how researchers ought to think about the logic of research when designing studies and avoiding QRPs.

With the logic of research in mind, in the subsequent chapter (i.e., Chap. 3), Cory Clark of the University of Pennsylvania and her colleagues explicate major human variables influencing researchers (e.g., motivational factors and cognitive biases) that can affect the way research is conducted. They argue that scientists and researchers, like many of us humans, are susceptible to well-documented cognitive biases as well as their idiographic motivational factors. To counter these, Clark et al. also propose how *intellectual humility* can serve as a crucial disposition when engaging in research and avoiding the perils of QRPs.

The final chapter of the first section (Chap. 4) presents the historical significance of what is now called "Replication Crisis in Psychology" (Pashler & Wagenmakers, 2012). The replication crisis has made painfully evident that many of our most cherished findings may be considerably less robust than most scholars had assumed (e.g., Maxwell et al., 2015). Perhaps, it also has served as the major force in psychology and allied fields that has brought our collective focus to minimize QRPs. In Chap. 4, Alexa Tullett of the University of Alabama and her colleagues propose several possible solutions (e.g., pre-registration) to offset replication crisis by minimizing some forms of QRPs (e.g., *p*-hacking).

Some QRPs, especially data fabrication and data falsification, are blatant and intentional (Crocker, 2011). However, over the past decade, it has become apparent that many omnipresent QRPs are rather subtle and can be unintentionally practiced (e.g., John et al., 2012). The second section of this volume (Chaps. 5, 6, 7, 8, 9, 10, 11, 12, and 13) introduces some of the notable exemplars of QRPs, and QRPs in various research contexts.

One of the most well-known QRP is *p*-hacking (e.g., Wicherts et al., 2016). In Chap. 5, Dorota Reis and Malte Friese of Saarland University introduce the myriad forms of *p*-hacking, that is, non-principled decisions during data analysis that are aimed at reducing the *p*-value of a significance test to make the data look more robust than they actually are. Fortunately, QRPs, including *p*-hacking, often leave a trail of evidence that indicates they were involved in producing the reported outcomes. In Chap. 6, Gregory Francis of Purdue University and Evelina Thunell of Karolinska Institute present several data-detecting methods for identifying QRPs, such as the test for excess success (Francis, 2014) and *p*-curve analysis (Simonsohn et al., 2014).

To date, *p*-hacking has been considered as a major exemplar of QRPs. This is in part because null hypothesis significance testing (NHST) continues to remain the dominant statistical analysis method in psychological science and clinical psychology. In Chap. 7, Brian O'Connor and Nataasha Khattar of the University of British Columbia, Okanagan, provide an overview of NHST and of controversies and limitations surrounding NHST. As noted by O'Connor and Khattar, conducting NHST without knowing these controversies could easily result in the engagement in QRPs without awareness of doing so. To minimize the risk of QRP, they also offer Bayesian methods (Kruschke, 2015; O'Connor, 2017) as one of the most useful alternatives to NHST. Following the chapter on controversies surrounding NHST, Ana Bridges of the University of Arkansas presents Hypothesizing After Results Are Known (HARKing; Kerr, 1998) as another major QRP. Bridges argues that HARKing threatens a set of core values on which the scientific process rests, including objectivity, honesty, openness, accountability, fairness, and stewardship, and therefore dampens the very spirit of science.

Once again, given psychology's strong quantitative orientation, there are frequent statistical controversies pertaining to the ways in which conclusions should be drawn from data. In Chap. 9, Andrew Hales and Natasha Wood of the University of Mississippi offer a summary overview of major statistical controversies, including those that are discussed in previous chapters (e.g., NHST vs Bayesian methods). More specifically, they focus on the controversies that are most fundamental to the decisions that researchers make when planning and conducting their analyses as well as the conclusions that consumers should draw when reviewing these.

Chapter 10 then covers publication bias (Franco et al., 2014; Rosenthal, 1979), a longstanding problem in the field of psychological science and clinical psychology. Publication bias refers to the tendency for statistically significant findings to be published over non-significant findings, and it is also known to unintentionally promote many forms of QRPs, including *p*-hacking and HARKing (e.g., Ferguson & Heene, 2012). In Chap. 10, Robbie van Aert of Tilburg University and Helen Niemeyer of the Free University of Berlin (Freie Universität Berlin) present publication bias as a major threat to the validity of meta-analyses, which is often viewed as the best available quantitative summary of studies on the same topic. van Aert and Neimeyer argue that, in clinical psychology (e.g., clinical trials), a major threat of publication bias to a meta-analysis is the overestimation of the effect size, which gives a false impression with respect to the efficacy of a treatment. To minimize this risk, they also offer methods to assess publication bias in a meta-analysis (e.g., van Aert et al., 2019).

In Chap. 11, Jolynn Pek of the Ohio State University and her colleagues provide statistical justifications and illustrations of whether and when statistical power can be used to improve the conduct of psychological science, reduce QRPs, and perhaps even detect QRPs. This is a very important chapter as statistical power analysis is regarded as one of several means to reduce QRPs (e.g., Appelbaum et al., 2018), and yet it continues to be misunderstood and misapplied in research (McShane et al., 2020).

The last two chapters of this section discuss QRPs in the areas of research where they are relatively understudied. More specifically, in Chap. 12, Matt Tincani and Jason Travers of Temple University present how QRPs could manifest in single-case experimental design (SCED) research (Hayes et al., 1999; Kazdin, 2011). More specifically, they highlight several QRPs in the areas of (a) participant selection, (b) independent variable selection, (c) procedural fidelity documentation, (d) graphical depictions of behavior, and (e) effect size measures and statistics. In Chap. 13, William O'Donohue of the University of Nevada, Reno, and Akihiko Masuda and Stephen Haynes of the University of Hawai'i at Mānoa explicate QRPs in presenting the psychometric evidence for psychological measure in peer-reviewed manuscripts, advocating for a more standardized and transparent approach to reporting the psychometric evidence.

Finally, in the third and last section of this volume, newly emerging models for minimizing the impact of QRPs in research are presented, including clearer understanding of replication study and meta-analyses (Chap. 14), pre-registration of hypotheses and analyses (Chap. 15), and adversarial collaborations (Chap. 16), in which investigators holding opposing positions on a scientific issue agree to work together on a study in an effort to counteract their respective biases.

In Chap. 14, Jacob Schauer of Northwestern University presents statistical considerations for studying replication from a framework based on meta-analysis. In the chapter, Schauer focuses on direct replications, where studies are designed to be as similar as possible, as opposed to conceptual replications that (systematically or haphazardly) vary in at least one aspect of an experiment (Collins, 1992; Schmidt, 2009). In Chap. 15, Angelos-Miltiadis Krypotos of Utrecht University, Gaetan Mertens of Tilburg University, Irene Klugkist of Utrecht University, and Iris Engelhard of Utrecht University present pre-registration as one key solution to the problems of QRPs. The pre-registration offers a time-stamped documentation that describes the methodology and statistical analyses of a study before the data are collected or inspected to minimize some of the notable QRPs (e.g., *p*-hacking, HARKing, selective reporting, file drawer problems). Furthermore, readers of the study's report can evaluate whether the described research is in line with the planned methods and analyses or there are deviations from these.

In Chap. 16, Tim Rakov of King's College London presents an adversarial collaboration (Bateman et al., 2005; Matzke et al., 2015), an approach to resolving scientific disputes, wherein researchers who have different positions on the issue at hand collaborate with the aim of making progress on their disputed research question. Rokov argues that as an approach to research, adversarial collaboration sits squarely within the open science framework (Open Science Collaboration, 2015) because it puts a premium on transparency in hypothesis specification, study design, data analysis, study interpretation, and reporting—and supplies a framework that can encourage rigor in these components of the research process.

Given the recent replication crisis in psychological science and clinical psychology, conducting more replication studies seems to be an important first step. This is because a series of replication studies can identify and weed out false positives over time and increase robustness of psychological science. Only problem is that replications take considerable time and money. To respond to this dilemma, in Chap. 17, Michèle Nuijten of Tilburg University offers her "four-step robustness check" for assessing and improving the robustness of psychological research findings as an alternative to directly diving into replication studies. Her "four-step robustness check" includes checking for internal inconsistencies in reported statistics (Step 1), reanalysis of original data (Step 2), sensitivity checks (Step 3), and replication in a new sample (Step 4). Subsequently, in Chap. 18, Felix Cheung of the University of Toronto and his colleagues discuss Reproducibility Project: Psychology (RP: P; Open Science Collaboration, 2015), the resulting credibility movement, and its implications to scientific practices in clinical psychology and beyond. The RP: P is a crowdsourced collaboration of over 250 contributing authors to repeat 100 different published experimental and correlational studies. As described in detail in Chap. 18, it has led to other replication projects and the development of improved research practices.

Finally, this volume ends with the chapter by Hannah Moshontz of the University of Madison, Wisconsin, and her colleagues on Psychological Science Accelerator (PSA; Chap. 19). The Psychological Science Accelerator (PSA) is an international collaborative network of psychological scientists that facilitates rigorous and generalizable research (Moshontz et al., 2018). In this chapter, Moshontz et al. describe how the PSA can help clinical psychologists and clinical psychological science more broadly.

A Final Note

We also want to acknowledge the contributions to this volume of our beloved colleague Professor Scott Lilienfeld, PhD. Scott played a key role in designing this book and in other tasks in its inception. Tragically, he passed away on September 30, 2020, after a struggle with pancreatic cancer. Scott was an astute critique of problematic science and his illuminating writings and kind guidance are missed tremendously. His example of astute criticisms mixed with kindness serves as an inspiration to many. This volume is dedicated to the loving memory of Scott.

References

Antonuccio, D. O., Burns, D. D., & Danton, W. G. (2002). Antidepressants: A triumph of marketing over science? *Prevention & Treatment, 5*(1), Article 25. https://doi.org/10.1037/1522-3736.5.1.525c

Appelbaum, M., Cooper, H., Kline, R. B., Mayo-Wilson, E., Nezu, A. M., & Rao, S. M. (2018). Journal article reporting standards for quantitative research in psychology: The APA Publications and Communications Board task force report. *American Psychologist, 73*(1), 3–25. https://doi.org/10.1037/amp0000191

Baron, R. M., Albright, L., & Malloy, T. E. (1995). Effects of behavioral and social class information on social judgment. *Personality and Social Psychology Bulletin, 21*(4), 308–315.

Bateman, I., Kahneman, D., Munro, A., Starmer, C., & Sugden, R. (2005). Testing competing models of loss aversion: An adversarial collaboration. *Journal of Public Economics, 89*, 1561–1580.

Collins, H. M. (1992). *Changing order: Replication and induction in scientific practice*. University of Chicago Press.

Crocker, J. (2011). The road to fraud starts with a single step. *Nature, 479*, 151–151.

Darley, J. M., & Gross, P. H. (1983). A hypothesis-confirming bias in labeling effects. *Journal of Personality and Social Psychology, 44*(1), 20–33. https://doi.org/10.1037/0022-3514.44.1.20

Etter, J. F., Burri, M., & Stapleton, J. (2007). The impact of pharmaceutical company funding on results of randomized trials of nicotine replacement therapy for smoking cessation: A meta-analysis. *Addiction, 102*(5), 815–822.

Fanelli, D. (2012). Negative results are disappearing from most disciplines and countries. *Scientometrics, 90*(3), 891–904.

Ferguson, C. J., & Heene, M. (2012). A vast graveyard of undead theories: Publication bias and psychological science's aversion to the null. *Perspectives on Psychological Science, 7*(6), 555–561. https://doi-org.eres.library.manoa.hawaii.edu/10.1177/1745691612459059

Fiedler, K., & Schwarz, N. (2016). Questionable research practices revisited. *Social Psychological and Personality Science, 7*(1), 45–52.

Francis, G. (2014). The frequency of excess success for articles in psychological science. *Psychonomic Bulletin & Review, 21*. https://doi.org/10.3758/s13423-014-0601-x

Franco, A., Malhotra, N., & Simonovits, G. (2014). Publication bias in the social sciences: Unlocking the file drawer. *Science, 345*(6203), 1502–1505. https://doi-org.eres.library.manoa.hawaii.edu/10.1126/science.1255484

Gregg, J. A., & Hayes, S. C. (2016). The progression of programmatic research in contextual behavioral science: Response to O'Donohue, Snipes, and Soto. *Journal of Contemporary Psychotherapy, 46*, 27–35. https://doi.org/10.1007/s10879-015-9312-5

Gregg, J. A., Callaghan, G. M., Hayes, S. C., & Glenn-Lawson, J. L. (2007). Improving diabetes self-management through acceptance, mindfulness, and values: A randomized controlled trial. *Journal of Consulting and Clinical Psychology, 75*(2), 336.

Hayes, S. C., Barlow, D. H., & Nelson-Gray, R. O. (1999). *The scientist practitioner: Research and accountability in the age of managed care* (2nd ed.). Allyn & Bacon.

Hayes, S. C., Strosahl, K. D., & Wilson, K. G. (2012). *Acceptance and commitment therapy: The process and practice of mindful change* (2nd ed.). Guilford Press.

Higgins, J. P. T., Altman, D. G., Gøtzsche, P. C., Jüni, P., Moher, D., Oxman, A. D., Savović, J., Schulz, K. F., Weeks, L., & Sterne, J. A. C. (2011). The Cochrane Collaboration's tool for assessing risk of bias in randomisedtrials. BMJ, 343, d5928. https://doi.org/10.1136/bmj.d5928

Ioannidis, J. P. A. (2005). Why most published research findings are false. *PLoS Medicine, 2*, e124. https://doi.org/10.1371/journal.pmed.0020124

John, L. K., Loewenstein, G., & Prelec, D. (2012). Measuring the prevalence of questionable research practices with incentives for truth telling. *Psychological Science, 23*, 524–532. https://doi.org/10.1177/0956797611430953

Jussim, L., Crawford, J. T., Anglin, S. M., Stevens, S. T., & Duarte, J. L. (2016). Interpretations and methods: Towards a more effectively self-correcting social psychology. *Journal of Experimental and Social Psychology, 66*, 116–133.

Kazdin, A. E. (2011). *Single-case research designs: Methods for clinical and applied settings*. Oxford University Press.

Kerr, N. L. (1998). HARKing: Hypothesizing after the results are known. *Personality and Social Psychology Review, 2*(3), 196–217. https://doi-org.eres.library.manoa.hawaii.edu/10.1207/s15327957pspr0203_4

Kruschke, J. K. (2015). *Doing Bayesian data analysis: A tutorial with R and BUGS*. Academic Press/Elsevier.

Matzke, D., Nieuwenhuis, S., van Rijn, H., Slagter, H. A., van der Molen, M. W., & Wagenmakers, E.-J. (2015). The effect of horizontal eye movements on free recall: A preregistered adversarial collaboration. *Journal of Experimental Psychology: General., 144*(1), e1–e15.

Maxwell, S. E., Lau, M. Y., & Howard, G. S. (2015). Is psychology suffering from a replication crisis? What does "failure to replicate" really mean? *American Psychologist, 70*(6), 487–498. https://doi-org.eres.library.manoa.hawaii.edu/10.1037/a0039400

McFall, R. M. (1991). Manifesto for a science of clinical psychology. *The Clinical Psychologist, 44*(6), 75–88.

McShane, B. B., Böckenholt, U., & Hansen, K. T. (2020). Average power: A cautionary note. *Advances in Methods and Practices in Psychological Science*. https://doi.org/10.1177/2515245920902370

Melchert, T. P., Berry, S., Grus, C., Arora, P., De Los Reyes, A., Hughes, T. L., Moye, J., Oswald, F. L., & Rozensky, R. H. (2019). Applying task force recommendations on integrating science and practice in health service psychology education. *Training and Education in Professional Psychology, 13*(4), 270–278. https://doi-org.eres.library.manoa.hawaii.edu/10.1037/tep0000222

Moshontz, H., Campbell, L., Ebersole, C. R., IJzerman, H., Urry, H. L., Forscher, P. S., Grahe, J. E., McCarthy, R. J., Musser, E. D., Antfolk, J., Castille, C. M., Evans, T. R., Fiedler, S., Flake, J. K., Forero, D. A., Janssen, S. M. J., Keene, J. R., Protzko, J., Aczel, B., … Chartier, C. R. (2018). The psychological science accelerator: Advancing psychology through a distributed collaborative network Advances in Methods and Practices in Psychological Science, 1(4), 501–515. https://doi.org/10.1177/2515245918797607.

O'Connor, B. P. (2017). A first steps guide to the transition from null hypothesis significance testing to more accurate and informative Bayesian analyses. *Canadian Journal of Behavioral Science, 49*(3), 166–182.

O'Donohue, W. T., Snipes, C., & Soto, C. (2016a). A case study of the overselling of psychotherapy: ACT interventions for diabetes management. *Journal of Contemporary Psychology, 46,* 15-25. https://doi.org/10.1007/s10879-015-9308-1

O'Donohue, W., Snipes, C., & Soto, C. (2016b). The design, manufacture, and reporting of weak and pseudo-tests: The case of ACT. *Journal of Contemporary Psychotherapy, 46*, 37–40. https://doi.org/10.1007/s10879-015-9316-1

Open Science Collaboration. (2015). Estimating the reproducibility of psychological science. *Science, 349*(6251), aac4716.

Pashler, H., & Wagenmakers, E.-J. (2012). Editors' introduction to the special section on replicability in psychological science: A crisis of confidence? *Perspectives on Psychological Science, 7*(6), 528–530.

Popper, K. R. (1972). *Objective knowledge: An evolutionary approach*. Clarendon Press.

Rosenthal, R. (1979). The file drawer problem and tolerance for null results. *Psychological Bulletin, 86*, 638–641. https://doi.org/10.1037/0033-2909.86.3.638

Rotton, J., Foos, P. W., Van Meek, L., & Levitt, M. (1995). Publication practices and the file drawer problem: A survey of published authors. *Journal of Social Behavior & Personality, 10*(1), 1–13.

Schmidt, S. (2009). Shall we really do it again? The powerful concept of replication is neglected in the social sciences. *Review of General Psychology, 13*, 90–100.

Simonsohn, U., Nelson, L. D., & Simmons, J. P. (2014). P-curve: A key to the file-drawer. *Journal of Experimental Psychology: General, 143*(2), 534–547. https://doi-org.eres.library.manoa.hawaii.edu/10.1037/a0033242

Swift, J. K., Christopherson, C. D., Bird, M. O., Zöld, A., & Goode, J. (2020). Questionable research practices among faculty and students in APA-accredited clinical and counseling psychology doctoral programs. *Training and Education in Professional Psychology*. Advance online publication. https://doi-org.eres.library.manoa.hawaii.edu/10.1037/tep0000322

Van Aert, R. C. M., Wicherts, J. M., & Van Assen, M. A. L. M. (2019). Publication bias examined in meta-analyses from psychology and medicine: A meta-meta-analysis. *PLoS One, 14*(4). https://doi.org/10.1371/journal.pone.0215052

Wicherts, J. M. (2011). Psychology must learn a lesson from fraud case. *Nature, 480*, 7.

Wicherts, J. M., Veldkamp, C. L. S., Augusteijn, H. E. M., Bakker, M., van Aert, R. C. M., & van Assen, M. A. L. M. (2016). Degrees of freedom in planning, running, analyzing, and reporting psychological studies: A checklist to avoid p-hacking. *Frontiers in Psychology, 7*, Article 1832. https://doi-org.eres.library.manoa.hawaii.edu/10.3389/fpsyg.2016.01832

Chapter 2
The Logic of Research and Questionable Research Practices: The Role of Enthymemes

Dylan R. Wong and William O'Donohue

Abstract In this chapter, we argue that poor scientific reasoning in which logical errors are made is another questionable research practice. We recommend that research psychologists and consumers of psychological research pay more attention to the logic of research by identifying the relevant inferential approaches, detecting logical errors, and constructing sound reasoning. We describe some prominent types of research logic: from alogical approaches such as that of Kuhn, to deductive logical approaches of Popper, to inductive approaches and abductive/Inference to the Best Explanation (IBE) approaches. The strength and weaknesses of each approach are discussed, along with the applications of these approaches in statistical methods and Abductive Theory of Method (ATOM).

Keywords Logic · Logical error · Questionable research practice · Clinical psychology

The Logic of Research and Questionable Research Practices: The Role of Enthymemes

Questionable research practices (QRPs) have been implicated in both creating scientific conclusions that are seen as true but are actually false (Ioannidis, 2005) and in findings that fail to replicate (see Chap. 4, this volume). One construal of QRPs is that researchers can exploit what has been called "researcher's degrees of freedom" (Simmons, Nelson, Simonsohn, 2011) that reflect choices that can shape conclusions in some desired direction. There have been a variety of QRPs identified such as selective reporting of dependent variables, *p*-hacking, hypothesizing after

D. R. Wong
Oregon Social Learning Center, Eugene, OR, USA

W. O'Donohue (✉)
Department of Psychology, University of Nevada, Reno, NV, USA
e-mail: wto@unr.edu

W. O'Donohue et al. (eds.), *Avoiding Questionable Research Practices in Applied Psychology*, https://doi.org/10.1007/978-3-031-04968-2_2

the results are known, as well as the use of the file drawer for unwanted results, and many of these are covered in this book.

Logic can be broadly defined as the study of the principles of correct reasoning and how propositions relate to one another, particularly in examining the quality of inferences from one set of propositions to another. Logic answers the question, "if Propositions P, Q, and R are true, what other propositions are also true?" Valid deductive inference has been viewed as truth preserving. Say that Propositions P, Q, and R form the premise of the argument, and Propositions Y and Z are the conclusions; if the argument is logically valid, then P, Q, and R all being true entails that Y and Z must also be true. As such logic both permits (a valid inference) and constrains (e.g., disallows fallacious inferences). Without being constrained by the limitations imposed by valid inference, researchers can infer any propositions from any other set of propositions (unlimited inferential degrees of freedom)—including making the (perhaps unwarranted) conclusion that favored views are supported. Failing to adhere to the constraints of logic may be the most fundamental QRPs, and certainly facilitates many of the other QRPs.

Scientific reasoning refers to the logical inferences made in scientific work; in the empirical sciences, for instance, researchers use some sort of reasoning to make inferences about empirical states of affairs that ought or ought not be observed given a certain theory; or to make inferences about what implications the data collected have about the truth or falsity of tested theories and hypotheses. The design of one's research can be seen as a logical exercise, that is, research design involves the construction of arguments that can entail the observational consequences of some theory, and these can then be tested to see if the propositions were valid. In addition, propositions capturing observations in their data can then be used in arguments to reason regarding whether they support or falsify other propositions. Or scholars conducting a literature review can be free to conclude what they wish. However, the actual logic of research may be obscure for psychologists: either as a normative matter (what is the best or at least a sound logic of research?) or a descriptive matter (what is the logic of this particular study?). Given that to date psychologists have paid little attention to the validity of inferences in their research, it seems fair to call any incomplete argument as *enthymemes*, a technical term meaning that the argument contains missing premises or conclusions.

Valid reasoning in sound arguments sets constraints to the researcher's degrees of freedom; it allows some conclusions to be implied and disallows many others. However, psychologists rarely, if ever, explicate the logic of their published research. In this chapter we will examine the possible choices researchers have for the logic of their research and conclude that researchers ought to be more attentive to the logic of their research and explicate their arguments better, and a failure to do so is a QRP as valid constraints imposed by logic are abrogated. We review proposals for the logic of research emanating mainly from key philosophers of science and suggest that there are several possibilities. Psychologists may have their choice on the logic of their research but should explicate these choices and be aware of their respective strengths and weaknesses.

We shall argue that standard psychological research methodologies following, say, Cook and Campbell (1979) often involve a pragmatic kind of logic. On the other hand, Popper (1959) proposed a deductive logic of research. Some other accounts of the logic of research explicitly involve inductive or abductive inferences (and its related concept of "inference to the best explanation" [IBE]), such as Haig's (2005) Abductive Theory of Method (ATOM). We shall then examine the logical inferences and errors that can also be seen in the reasoning involved in statistical methods psychologists often employ and in psychologists' pragmatic application of these, such as null hypothesis significance testing (NHST) and Bayesian inference. Finally, some accounts seem to dispense with logic altogether, such as Kuhn's (1962) account of scientific revolutions. However, each of these has limitations that must be recognized, and researchers need to be strategic in their choices for the logic of their research.

The Logic of Conventional Psychological Research

The logic of conventional psychological research might be called consistent with a weak version of pancritical rationalism (Cook & Campbell, 1979; Bartley, 1990) in that it attempts to anticipate criticisms to valid inference and promotes the design and implementation of a corresponding methodological move to potentially address that criticism. For example, let us examine the logic that is employed in the conventional double-blind randomly controlled clinical trial. Each methodological move addresses and hopefully falsifies a potential criticism/plausible rival hypothesis. For example:

1. Why the methodological move of random assignment? This at least potentially addresses the criticism that the groups differed in some systematic way *before* the experimental treatment.
2. Why the methodological move of including a no-treatment control group? This potentially addresses the criticism that due to spontaneous remission the individuals would have improved even without treatment.
3. Why double-blind? This potentially addresses the criticism that either participant expectations or experimenter expectations may have altered the values on the dependent variable(s), such that expectation effects (and not treatment effects) were responsible for such values.
4. Why a statement on the psychometric properties of the measures? This potentially addresses the criticism that the measures do not validly measure the constructs under consideration.
5. And so on, for each methodological move.

These methodological moves are supposed to be made for all "plausible rival hypotheses." But note that plausibility is a pragmatic, not a logical matter. Additionally, the question of whether the methodological move is sufficient to negate the plausible rival hypothesis also involves pragmatic judgment. Finally, it is rare that research papers in psychology explicitly formalize the pragmatic logic that undergirds the design.

Another problem is that there are many potential criticisms/plausible rival hypotheses, and each often requires an expensive methodological move (in terms of time, subjects, and other resources). For individual studies, there may need to be a "meta-argument" regarding which the relevant priority of such criticisms—which are elements of a very large potential set of criticisms—is the most important and ought to be addressed methodologically. This may be one reason why programs of research are so important; across a series of studies more potential criticisms can eventually be addressed by including the requisite design move in at least some of the studies across the research program. For example, the randomly controlled trial described above did not address the criticism of treatment effects being caused by placebo effects; therefore, some of the subsequent studies could include an attention control condition that potentially addresses this criticism. In addition, due to the lack of follow-ups to assess for recidivism, some subsequent studies could include measurement periods of 6 or 12 months and so on. Some have commented that research sophistication grows over time as we "learn how to learn" (Munz, 2014) and an example of this may be the relatively recent concerns with clinical significance (versus statistical significance) or QRPs. This also creates a somewhat difficult problem of assessing the status of these potential criticisms across studies (e.g., studies out of Lab X did not carry out follow-up assessments but one study out of Lab Y did, however this study showed higher than desirable recidivism).

However, probably the most serious problem is that the "logic" of such research is not made clear, is not formal, and often is just inchoately pragmatic. It often plays out as an intellectual game: I would like to make a valid inference from my data to say something like "My treatment has caused improvement," and if you can present a criticism like "You can't say that because it is plausible that Z (e.g., the effects are due to placebo)," the desired causal inference is not valid. However, there are many potential criticisms and these can be leveled in an ad hoc and unsystematic way.

Deductive Reasoning

Deductive reasoning is characterized by its *demonstrability*—the use of a valid deductive inference rule establishes the truth of the conclusion if the premises are true. Thus, the conclusions of sound deductive arguments (true premises and valid deductive inference rule) are necessarily true. Another way of saying this is that valid deductive arguments are always *truth preserving*, that is, if all the premises of the argument are true and a valid deductive inference rule used, then this reasoning preserves the truth of the premises because the valid deductive inference rule always generates only true conclusions. However, a well-recognized and significant downside of deductive reasoning is that it is also *nonampliative*—the conclusion is not content increasing—deductive arguments simply "unpack" content that is already contained (perhaps implicitly) in the premises of the argument. For example, consider the following deductive argument:

1. All humans are mortal.
2. Barbara is a human.
3. Therefore: Barbara is mortal.

This argument is considered nonampliative because its conclusion is implicitly contained in the first premise (because to establish that all humans are mortal one must have established that a member of this set, Barbara, is also mortal). Many early philosophers of science (e.g., Carnap, 1945) have taken deduction's nonampliative character as a sure sign that science cannot rely on deduction because science seeks new information and as such it must rely on some sort of *ampliative* reasoning—the conclusion must add or increase the information in the premises. We turn first to the view that the logic of research is deductive; this view is best exemplified by the work of Sir Karl Popper (1959).

Popperian Science

Sir Karl Popper (1959) rejected the notion coming from the logical positivists that the logic of research was inductive. Popper argued that there is no such thing as a truth-preserving ampliative inductive logic. Popper claimed that the logic of research was deductive—a *hypothetico-deductive model*—and utilized the valid logical inference rule of *modus tollens*.

In general, the logical inference rule of *modus tollens* has the following (valid/truth preserving) form:

If A, then B.
Not B.
Therefore, not A.

In science, the argument may look like the following:

1. If something is a piece of copper (A) then it conducts electricity (B).
2. This piece of copper does not conduct electricity (not B).
3. Therefore, it is not the case that all copper conducts electricity (not A).

This argument is valid because it relies on the valid logical inference rule known as *modus tollens*. To determine its soundness (i.e., the truth of its premises), the question simply becomes: are Premises 1 (the hypothesis) and 2 (the evidence) true?

Popper also suggested that formulating Premise 1 and Premise 2 ought to be guided by a few considerations: it is desirable if the conjecture being tested in Premise 1 has as great as possible *empirical content*. The empirical content of a statement is basically what it rules out. The more empirical states of affairs it rules out, the greater is a statement's empirical content. In general, scientific laws have large empirical content, ruling out many states of affairs. For example, Newton's gravitational law rules out all states of affairs except gravitational attraction occurring in direct proportion to mass and inverse proportion to distance. As another example, "All folks in Reno, Nevada eat sugar" has less empirical content than "All Nevadans eat sugar" (i.e., empirical content increases as the number of cases it covers increases). Secondly, empirical content is increased by the precision of the statement: "All Nevadans eat at least 14 grams of sugar daily" has more precision and empirical content than "All Nevadans eat sugar."

There are also several key considerations for Premise 2, that is, the empirical test. Popper suggested it ought to be *severe*. The *severity of a test* is essentially an efficient search for the existence of falsificatory instances—cases that demonstrate the falsity of a proposition. For example, if a researcher is testing the proposition "Protestant leaders never swear," it is a more severe test to examine instances where people are most likely to swear (e.g., when they hit their thumbs with hammers, break something valuable, when someone cuts them off in traffic, etc.). It is a less severe test to examine the word use of religious leaders during sermons, or when they are teaching Sunday school, and so on, as people are generally much less likely to be disposed to swear in these situations. Thus, for Popper the research project itself should offer an argument that the test is severe. This might be such an example:

1. The most likely situations for people to swear are x, y, z.
2. People are also most likely to swear when they do not know they are being observed.
3. If my research consists of nice size samples of surreptitious sampling of x, y, z, then it is a severe test.
4. Therefore, my research project is a severe test.

Popper's overall conception of science implies that scientific knowledge can never be a matter of confirmation. Since Popperian falsification can never demonstrate the truth of theories, but can only falsify them, science progresses by eliminating its theories that are in error. Theories that survive severe testing are thought to be *corroborated* (not confirmed), for they can eventually be falsified by some future severe testing. Once these theories become eliminated, we are confronted with new problems and must build new tentative solutions subject to further falsification.

Popper's conception of science has been criticized on several grounds (O'Donohue, 2013). First, historians of science have argued that it does not reflect the historical record of how science has been practiced (Lakatos, 1970; Laudan, 1978). If Popper's goal had been to provide a description of how scientific research actually proceeds, then he has failed to do so. Second, Popper's account does not appear to address the Quine-Duhem thesis. The Quine-Duhem thesis suggests that when the falsifying event is observed (i.e., Premise 2: Not B), the initial hypothesis (Premise 1) need not be falsified; instead, the failure may be attributed to any *auxiliary hypotheses employed in the test*. Auxiliary hypotheses refer to the additional hypotheses that are required for the initial hypothesis to entail the observation. For instance, the premise "if something is a piece of copper, it conducts electricity" includes the additional premises that "the source of electricity is properly connected to the piece of copper," that "the copper is pure," and so on. According to the Quine-Duhem thesis, Popper's falsification should really take the following form:

1. If Theory and Aux_1 and Aux_2 and Aux_3... and Aux_n, then Observation.
2. Not Observation.
3. Therefore, Not (Theory and Aux_1 and Aux_2 and Aux_3... and Aux_n).
4. Therefore, Not Theory or not Aux_1 or not Aux_2 or not Aux_3 or not Aux_n.

This valid deductive argument now has an unsatisfying conclusion; along with the hypothesis being false, one could also conclude that any of the auxiliary hypotheses are also false. At least one of the hypotheses is false, but you cannot know which to blame.

Kuhn's Alogical Account of Science

Contrary to Popper, not all accounts of science claim that there is a logic of research. Kuhn's account (1962, 1994) is an example of this alogical approach, and it is noteworthy that psychologists have been particularly attentive and admiring of Kuhn's account (O'Donohue, 2013). It may even be the case that the alogical nature of Kuhn's account is partly what has drawn psychologists to his views, seeing as psychologists usually receive very little formal training in logic.

Kuhn (1962, 1994) suggested that sciences pass through several stages. In the first stage, which he called "pre-paradigmatic science," there is little progress in puzzle solving and those working in the field have deep disagreements about basic issues, for example, what constructs are important, how ought these be defined, what proper research methodology looks like, and so on. In Kuhn's second stage, someone solves a puzzle and in this puzzle solution a paradigm is born. In Kuhn's account, others in the field are impressed and influenced by this problem-solving exemplar and then begin to copy it to try to solve other problems. Scientists adopt many elements from the puzzle-solving exemplar such as its definitions, principles, methodological approaches, and so on. This becomes, for Kuhn, a "paradigm."

The field then enters a stage that Kuhn called "normal science" in which scientists attempt to apply this paradigm to solve other puzzles. According to Kuhn, sometimes these scientists are successful in puzzle solving and sometimes they are not. The problem-solving failures can accumulate and are generally frustrating to scientists. The final stage of science for Kuhn is when a scientific revolution occurs. According to Kuhn, a revolution occurs when someone applies a new approach to one or more of these failures of the old paradigm and achieves some problem-solving success. A new period of normal science then occurs where scientists ape the new paradigm until it starts the cycle all over again, that is, it produces success and anomalies and then a new revolution occurs and so on.

One can see that for Kuhn logic is not essentially involved in science. Certainly, a paradigm could in principle have a kind of reasoning; however, he is not explicit about this, and nowhere does he say that paradigms are defined by logical rules or preferences. Furthermore, new paradigms are thought to have different definitions, principles, and methodological approaches that are frequently inconsistent with the old paradigm. Since logical rules cannot contradict one another, the kinds of reasoning in different paradigms thus cannot all be based on logical rules. Kuhn's model of research is thus alogical.

Inductive Reasoning

Given the limitations of deductive reasoning, some have looked to inductive reasoning as a good candidate for the logic of research. Induction has been taken to be an *ampliative* but *nondemonstrative* form of reasoning, that is, the conclusions of inductive arguments contain more information than their premises. However, because inductive arguments are nondemonstrative (the truth of the premises and the use of an inductive inference rule do *not* guarantee the truth of the conclusion), at best these are only *probably* true, that is, they may still be false. In the history of philosophers studying induction, their key philosophical problem has been how to quantify how likely the conclusion of an inductive argument is, given the evidence contained in the premises. Unfortunately, this problem has resisted a clear solution.

For example, notice the following about the conclusion of the following inductive argument: (1) the scope of the conclusion (helpfully) contains more information than the scope of the premises (i.e., the conclusion refers to a previously unexamined individual), and (2) even if the premises are true, the conclusion of the inductive argument might still be false—the argument is not truth preserving because no one has been able to discover a truth-preserving inductive inference rule. For example:

1. Eighty percent of the anxious subjects were successfully treated by exposure therapy.
2. Sam is anxious and will be treated with exposure therapy (Sam was not part of the anxious subjects in Premise 1, and hence was not examined to form Premise 1).
3. Therefore, Sam's anxiety will be successfully treated.

This is not a valid argument—the truth of the conclusion is not guaranteed by the truth of the premises and the inference rule used. The "problem of induction" began to concern philosophers in the nineteenth century, starting with the Scottish philosopher David Hume (1779). Hume raises the following questions: Are inferences from what is observed in the research sample to the unobserved logically justifiable? Do observed facts give us sound evidence for conclusions about similar situations that are not observed? Or, more precisely: how much evidence, if any, does the existence of an observed regularity provide toward the claim that future observations of similar phenomena will be like these past observations? The rough idea is expressed in the folk narrative that although every morning thus far the farmer has always fed the chicken, it would be false to conclude that this invariant pattern will necessarily persist—as one day the farmer will slaughter, not feed, the chicken. The future may not always be like the past.

Hume argued that *there are no nondemonstrative inferences that are also truth preserving*. Hume noted an interesting meta-paradox to his problem of induction: One cannot justify the inference deductively, because then the inference would be nonampliative. However, if one tries to justify it inductively, then it is nondemonstrative (for example, because in the past it has worked or because the probability of it working is high) and therefore one is begging the question— in other words, one is making an appeal to the very inductive principle one wishes to justify! Hume attempted to save induction by extra-logical considerations, that is, by suggesting that although induction has no logical justification, it can be based on the "natural instinct" embedded in human psychology: namely, that humans tend to expect that observed regularities will continue to occur in the future. However, this argumentative move is called "psychologism," as it is not an appeal to the logical quality of an argument, but rather it is an appeal to an alleged contingent empirical state of affairs—a hypothesized human tendency.

Hume also argues that any number of singular observations does not entail a universal statement. That is, the observation of a thousand, or even several million black crows does not entail the truth of the statement "all crows are black" because it is still at least logically possible that some yet-to-be-observed crow will turn out not to be black.

A common response to this problem has been that although no number of observations logically entails a universal statement, observations can allow a rational assignment of *some degree of (increased) probability* to the relevant conclusion. According to this view—known as *enumerative induction*—the degree of probability of the conclusion is raised upon each consistent observation. Moreover, according to this view, with many confirming instances, the inductive conclusion becomes probable to a degree that is indistinguishable or nearly indistinguishable from the certainty of a deductive conclusion.

Problems with Inductive Reasoning

Several philosophers—particularly Sir Karl Popper (1959), a notable critic of inductive reasoning and proponent of deductive reasoning in the sciences—have raised further problems with inductive reasoning.

Popper (1959) argued that the kind of observed repetition envisaged by Hume can never be perfect: the cases he has in mind cannot in principle be cases of perfect sameness; at best they can only be cases of (perhaps very high) similarity. For example, the farmer does not display the exact feeding motions each time, and there can be numerous variations in the chicken's eating. Popper argues that these are at best "repetitions" only from a certain somewhat inexact point of view. For Popper this signifies that there must always be a point of view—embodied perhaps in a system of expectations or assumptions—before there can be any perceived repetition. Popper argued:

> We must replace, for the purposes of a psychological theory of the origin of our beliefs, the naive idea of events that are similar by the idea of events to which we react by interpreting them as being similar … For even the first repetition-for-us must be based upon similarity-for-us, and therefore upon expectations--precisely the kind of thing we wished to explain. (pp. 444–445)

Popper also disagreed with the justification of induction by enumeration. If "many" consistent observations increase the probability of the universal statement, how many do we need to raise the probability to 1.0? Popper argued that universal laws (such as "All P are Q") have a large or even an infinite number of cases. Therefore, assessing the probability of a universal statement by comparing the number of tested and confirmed instances to the number of possible tests will always result in a probability of zero or near zero. Consider the proposition "All copper conducts electricity." If one estimates the number of observations of copper conducting electricity versus the number of possible observations of copper (all copper everywhere in the universe), as well as observations of observed copper but at other points in time, just because some copper once conducted electricity does not mean it always will, one can see that this fraction would essentially equal zero. Therefore, according to Popper, false theories and well-confirmed theories will have equal probabilities, that is, zero.

Induction also involves two well-known paradoxes. The first, identified by Kyburg (1961), concerns the "*lottery paradox*." Consider the following thought experiment: Suppose that there are 1000 lottery tickets numbered consecutively from one to a thousand, and that in a fair drawing one ticket has been chosen. Now let us consider the likelihood that the winning ticket is the one numbered "1." The probability that this particular ticket is the winner is only 1/1000. Therefore, the probability that some other ticket was actually drawn is 999/1000. Assuming that 0.999 is a sufficiently high probability to justify the conclusion that "some other ticket was drawn," one infers in this inductive argument that indeed some other ticket was in fact drawn. Next let us consider the ticket numbered "2." By the same reasoning we would conclude that, again, some other ticket was drawn. But notice that we can use this same reasoning for tickets numbered 3, 4, 5 … 1000. In each case, the conclusion that some other ticket was drawn seems to be confirmed by its high probability, 0.999. However—and this is where the paradox emerges—this set of conclusions is inconsistent with our knowledge that one winning ticket was actually drawn. We are thus facing a classic dilemma. Kyburg has argued that what this dilemma shows is that we cannot validly argue that something is the case simply because it has a (very) high probability of being so. Thus, there is no logic of induction.

Carl Hempel's (1965) *paradox of the ravens* points to another problem with induction. Hempel points out that the proposition "All ravens are black" is logically equivalent to the proposition, "All non-black things are nonravens." The second proposition can be logically deduced from the first using the logical law known as the law of contraposition. The law of contraposition states that "All A's are B's" is logically equivalent to "All non-B's are non-A's." Since these two propositions are logically equivalent, evidence that confirms one proposition must also confirm the other proposition. Therefore, the observation of a white ribbon—a non-black thing that is a nonraven—would confirm the proposition that "All ravens are black." But this result is regarded as an absurdity. No one expects that a research project by an ornithologist would involve solely examining the color of, for example, ribbons. Critics of induction have taken these examples to show that certain logically proper "confirmations" seem to be substantively irrelevant.

Inference to the Best Explanation and Abduction

Inference to the Best Explanation (IBE), and its related concept of *abduction*, has been proposed as a noteworthy kind of inductive inference (Lipton, 2004). While abduction is situated in the context of *discovery* (the stage of *generating* theories and hypotheses) and IBE is situated in the context of *appraisal* (the stage of *evaluating* theories and hypotheses), both reference the same idea: namely, that one should make argumentative moves with reference to what would best explain the available evidence. With abduction, one should generate the hypotheses that have the potential to best explain the evidence, and with IBE one should evaluate the hypotheses on the basis that they best explain the evidence. We will focus primarily on IBE because the literature on IBE is far more extensive.

IBE gets its name from Gilbert Harman (1965), who defined it in the following way (p. 89):

> In making [an inference to the best explanation] one infers, from the fact that a certain hypothesis would explain the evidence, to the truth of that hypothesis. In general, there will be several hypotheses which might explain the evidence, so one must be able to reject all such alternative hypotheses before one is warranted in making the inference. Thus one infers, from the premise that a given hypothesis would provide a "better" explanation for the evidence than would any other hypothesis, to the conclusion that the given hypothesis is true.

The idea is intuitively appealing for two reasons. First, we typically believe that a fundamental aim of science is to provide explanations for phenomena. Psychologists frequently ask questions that demand explanations. Why is alcohol addictive? Why do children who experience abuse grow up to abuse their own children? Why do we think of people in social outgroups as homogenous, but see people in social ingroups as diverse? Scientific progress seems to be driven by the pursuit of explanations to such questions (Lipton, 2004), and the formation of explanations is thought to be an essential guide to the logic of research.

Second, if we consider common examples of the thoughts we have each day, we can find many examples that appear to involve IBE. Imagine hearing a voice coming from inside your house as you approach the front door. You see your housemate's car parked on the road alongside the house and notice that the door is unlocked. You infer, therefore, that the voice belongs to your housemate—perhaps they are talking on the phone. Why should you make this inference, however, and not that someone stole your housemate's car and keys, unlocked the door, and placed a speaker inside the house playing back your housemate's voice? IBE would be the answer. This is not a deductive inference, since none of these facts *necessarily logically entail* that the voice belongs to your housemate—instead, you infer that the voice belongs to your housemate because it would be the best explanation of all the evidence available to you. It is *nondemonstrative* and *ampliative* like other inductive inferences. Lipton (2004) suggests that despite what Sherlock Holmes says he is doing in his detective work (i.e., "the art of deduction"), Holmes is actually using IBE to make his claims. He observes facts and infers to the explanation that best explains them.

Since Harman, theorists of IBE have aimed to render it in a precise analytical structure and come to a consensus regarding its validity. These efforts have been extremely challenging for two reasons. First, because IBE is an inductive kind of inference, it does not follow demonstrative rules like *modus tollens*, and as such it is unclear whether there can ever be clear rules for how IBE is to be applied. Second, it is not at all clear what is meant by *best*, and what is meant by *explanation*. While the literature on IBE has helped to develop some sense of the former, defining the latter has been extremely challenging. The history of defining explanation is quite convoluted and defining explanation remains an active topic of discussion among philosophers of science today—see Salmon's (2006) *Four Decades of Scientific Explanation* or Woodward and Ross (2021) for a comprehensive review of this history. While we often have an intuitive grasp of what it means to explain something, different explanations often have different characteristics, and developing rules for IBE that encompass all these possibilities is a tremendously ambitious project.

Despite these difficulties, several explications of IBE have been constructed: Vogel (1998), Psillos (1999), and Lipton (2004) are some prominent recent examples. The next section will briefly review some of the more prominent models of explanation, primarily drawn from Salmon (2006). After that, we describe a conception of IBE that attempts to describe what is meant by a *best* explanation through well-defined criteria and principles: Thagard's (1993) Theory of Explanatory Coherence (TEC).

Scientific Explanation

The first prominent philosopher to discuss explanation was Aristotle; he made the key distinction between *knowing-that* and *knowing-why*. The former simply involves a description of some phenomenon—"my shadow is longer in the evening"—and the latter elucidates the phenomenon—"because the sun's angle to the ground gets narrower as it sets, light travels in a straight line and my shadow is produced when my body obstructs light." This was an important first step toward understanding explanation.

However, it fell to Hempel and Oppenheim (1948; henceforth referred to as H-O)—which was further developed by Hempel (1965)—to produce first comprehensive and precise notion of explanation. According to Hempel (1965), there are four categories of scientific explanations, shown in the table (taken from Salmon, 2006) below:

	Particular facts	General regularities
Universal scientific laws	D-N (deductive-nomological)	D-N (deductive-nomological)
Statistical scientific laws	I-S (inductive-statistical)	D-S (deductive-statistical)

The term "nomological" refers to a basic, universal scientific law. While the D-N model applies to both particular facts and general regularities, H-O only discusses the former. A D-N explanation is comprised of an *explanans* (the sentences that are to account for the phenomenon, or "the explaining sentences") and the *explanandum* (the sentence describing the phenomenon to be explained, or "the fact"). They state that an adequate D-N explanation—or a "correct answer to an explanation-seeking question" (p. 42)—must fulfill three logical conditions and one empirical condition:

1. The explanans must be a valid deductive argument (logical).
2. The explanans must contain essentially at least one general law (logical).
3. The explanans must have empirical content (logical).
4. The sentences constituting the explanation must be true (empirical).

Thus, so far we see that scientific explanation for H-O is a deductive enterprise. We have seen that Conditions 1 and 4 combined make for a valid deductive argument that leads to a true conclusion. As for Condition 2, the details of how H-O constitutes a general law are quite complicated and have been a notable weakness of the D-N model. You might notice the symmetry between the hypothetico-deductive model of Popper and the D-N model of explanation—the former uses (contradictory) evidence to falsify the law, while the latter uses the law to account for the (compatible) evidence.

One major criticism of D-N explanations as a universal model for explanations is that many satisfactory explanations do not contain scientific laws. For instance, the explanandum "I slipped on the floor" has the satisfactory explanans "the floor was wet," where both are particular facts. Though a defender of D-N explanations might suggest that the explanation is incomplete without referencing the universal scientific laws of friction, it seems like telling someone that "the floor was wet" serves as a fine explanation, and the law just serves to justify the explanation. Second, D-N explanations cannot be damaged by any number of additional premises—yet typical explanations do seem to become less useful when irrelevant premises are added to it. Third, D-N explanations are *bidirectional*, which leads to the absurdity of the explanandum explaining the explanans (e.g., my shadow being longer in the evening explains why the sun's angle is narrow). Finally, that not every fact can be explained as a scientific necessity; rather, some facts are merely probable, or statistical. Hempel (1962) attempts to address this last problem.

Statistical explanations, according to Hempel, are split into D-S and I-S. The D-S explanation is a statistical law that is deductively derived from other laws, at least one of which is statistical. For example, explaining the outcome of a set of dice throws involves arithmetically deriving the probability of the outcome using generalizations about the dice (e.g., the probability of getting any particular die face is 1/6). The explanans of D-S explanations need not contain empirical data, and hence they are not D-N explanations. The I-S explanation follows the structure of the D-N explanation, but the law being used is statistical, and hence the explanandum is probabilistic. In Hempel's example, we might explain that someone recovered from a strep infection because they were administered penicillin, and treatment with penicillin leads to a high (e.g., 90% but not 100%) chance of recovery.

One major issue with I-S explanations is that two I-S explanations could have compatible premises but contradictory conclusions. If an individual's strep infection is resistant to penicillin, then the probability of the person recovering would be low. This would make the reference to penicillin in the explanation untenable, but the definition of the I-S explanation does not prevent this reference. Hempel attempts to correct this by, among other things, adding a further condition that the explanans must make the explanandum highly probable. But Salmon argues that the real problem lies with whether explanans changes the probability of the explanandum (e.g., whether the penicillin made the recovery from strep infection more probable). His later model of statistical relevance takes this as the fundamental condition of explanation; however, because it was later shown that causal relationships cannot be reduced to statistical relevance relationships, causal theories of explanation were later developed. Other models of explanation developed since include unificationist and pragmatic theories of explanation; the pursuit of a general model of explanation continues today (Woodward & Ross, 2021).

In response to these difficulties, some have suggested that explanation is a *primitive* concept; this means that the concept cannot be defined in terms of other concepts, and instead appeals to intuition for its characterization. There is good reason to believe this is so; after all, we can explain things ordinarily without appealing to what we mean by an explanation (see Poston, 2014, for a fuller justification for defining explanation as primitive). The Theory of Explanatory Coherence considers explanation a primitive; hence, in applying this model practically to research, we have some justification in relying on our judgment to decide whether the hypotheses we have formulated constitute an explanation.

Thagard's Theory of Explanatory Coherence

The *Theory of Explanatory Coherence* (TEC; Thagard, 1993) suggests that the "best-ness" of an explanation depends on its explanatory coherence. A theory has explanatory coherence if the propositions in the theory have explanatory relations. For instance, Propositions P and Q have explanatory coherence if one or more of the following propositions are true (p. 65):

1. P is part of the explanation of Q.
2. Q is part of the explanation of P.

3. P and Q are together part of the explanation of some R.
4. P and Q are analogous in the explanations they respectively give of some R and S.

As mentioned earlier, TEC considers explanation to be a primitive. However, Thagard also argues that TEC may be compatible with the future integration of various strands of explanation—deductive, statistical, schematic, analogical, causal, and linguistic.

TEC relies on the seven principles and three criteria. The criteria of *consilience*, *simplicity*, and *analogy* are contained within the seven principles. A theory is the most consilient if it explains the largest range of facts; he distinguishes between static consilience (the theory explains all the different types of facts) and dynamic consilience (the theory explains more types of facts than it did when it was first generated). A theory is simpler if it makes fewer "special or ad hoc assumptions" (Haig, 2005, p. 381) than other theories; this is a "check" on the consilience criterion because simpler theories tend to have lower consilience. Finally, a theory that is better supported by an analogy to previous theories is more coherent.

TEC's seven principles (Thagard, 2000, p. 43) are offered below:

1. *Symmetry.* Explanatory coherence is a symmetric relation, unlike, say, conditional probability. That is, two propositions *p* and *q* cohere with each other equally. For example:
2. *Explanation.* (a) A hypothesis coheres with what it explains, which can either be evidence or another hypothesis. (b) Hypotheses that together explain some other proposition cohere with each other. (c) The more hypotheses it takes to explain something, the lower the degree of coherence.
3. *Analogy.* Similar hypotheses that explain similar pieces of evidence cohere.
4. *Data priority.* Propositions that describe the results of observations have a degree of acceptability on their own.
5. *Contradiction.* Contradictory propositions are incoherent with each other.
6. *Competition.* If *p* and *q* both explain a proposition, and if *p* and *q* are not explanatorily connected, then *p* and *q* are incoherent with each other (*p* and *q* are explanatorily connected if one explains the other or if together they explain something).
7. *Acceptance.* The acceptability of a proposition in a system of propositions depends on its coherence with them.

As mentioned, TEC de-emphasizes prediction over explanatory coherence; instead of concerning itself with whether the theory has good predictive power *for the future* (i.e., it anticipates a set of data that is yet to be observed), it concerns itself with whether the theory has explanatory coherence *now* (based on past data and theoretical propositions). Although explanations clearly lead to predictions of certain empirical outcomes, TEC considers the latter secondary and would not abandon a theory if it led to failed predictions; this represents a contrast with Popperian science, which values predictions because they allow for possible subsequent falsification. Thagard argues that if falsification is not a good description of how the sciences actually operate (as Popperian science has been accused of; for example, see Kuhn, 1962, 1994), predictions lose value in the scientific enterprise.

A related value of explanatory coherence is that it allows one to evaluate how to modify a theory once it is falsified: if doing so would reduce the explanatory coherence of the theory (e.g., as per principle 2 of TEC, the hypothesis goes together with another hypothesis to explain another proposition), then there is stronger reason not to modify the hypothesis. Likewise, if a hypothesis contradicts a more explanatory hypothesis within the theory, then there is stronger reason to modify or remove the former hypothesis.

TEC is directly applicable to examples in psychology; for instance, Durrant and Haig (2001) apply TEC to the comparative evaluation of two theories of language. This paper compares the adaptationist hypothesis regarding language development—accordingly, humans developed language because of natural selection—and the non-adaptationist hypothesis—that it was not because of natural selection. Accordingly, they find that adaptationist accounts of language development have strong consilience (it explains many of the features of language) and simplicity (it can account for all the features of language with that hypothesis) and are supported by analogy (language resembles the development of other well-understood biological adaptations, such as the eye). Conversely, non-adaptationist hypotheses have poor explanatory coherence: they have poor consilience because they cannot account for as many features, and they have poor simplicity because they require many hypotheses to explain how each feature of language arose separately.

In short, TEC is a conception of IBE that features as a method of appraising theories. Theories with greater explanatory coherence (evaluated on the criteria and principles described above) garner greater support, and vice versa.

Applications of Scientific Reasoning

Scientific reasoning is ubiquitous in the methods we use to conduct and analyze our research. Here, we discuss one ubiquitous feature of conventional psychological research methods—Null Hypothesis Significance Testing (NHST)—and its logical flaws. We then discuss the logic of Bayesian statistics, which addresses some of the major flaws of NHST and its status as a possible alternative to NHST. Finally, we describe a recent and promising theory of science that is grounded in abductive and explanatory reasoning: Haig's Abductive Theory of Method (2005).

The Logical Flaws of Null Hypothesis Significance Testing

NHST is another ubiquitous feature of conventional psychological research, and it features centrally as a QRP. As noted in a chapter in this book by O'Connor and Khattar (Chap. 7, this volume, 2022), NHST continues to be employed by many psychologists despite its numerous and well-documented problems. In their chapter, the authors thoroughly explore the problems associated with using NHST. In this section, we lay out NHST in the context of scientific reasoning and provide some reasons for why the use of NHST is logically flawed.

A null hypothesis is the hypothesis that the difference between two means of some variable in some population being compared is zero. NHST is thus defined in O'Connor and Khattar as follows:

> Conventionally, researchers make such decisions by assuming the null hypothesis to be true and, given this assumption, attempting to make inferences based on the probability of obtaining the actual pattern of results observed. Specifically, a statistical test yields the probability of a given results (or one more extreme) being produced by chance if the null hypothesis is true. … If this (probability) is less than a threshold probability or alpha level (typically .05), then chance is concluded to be a sufficiently unlikely explanation of the outcome, and the existence of an effect is held to be supported by the data. (Pollard & Richardson, 1987, p. 159)

NHST also relies on the following assumptions:

> 1) the null hypothesis is exactly true; 2) the sampling method is random sampling; 3) all distributional requirements, such as normality and homoscedasticity, are met; 4) the scores are independent; 5) the scores are also perfectly reliable; and 6) there is no source of error besides sampling or measurement error. (Kline, 2013, p. 74)

Homoscedasticity means that the variance in the relation between the dependent and independent variables across the different values of the independent variable is the same. For example, the relation between age and weight is often not homoscedastic, because at younger ages, the variance in weight is generally much lower than the variance in weight at adulthood.

NHST is founded on a frequentist view of probability that takes probability to be "the likelihood of an outcome over repeatable events under constant conditions except for random error" (Kline, 2013, p. 40). In other words, the probability of an event is the proportion of events occurring if the same circumstances were repeated many times (this is called the law of large numbers). This contrasts with the subjectivist view, which takes probability to be the researcher's subjective state of belief regarding the likelihood of an event—this does not rely on the event being repeatable. Frequentists consider the probability of the data given a set parameter (in NHST, this parameter is a "difference of zero"); subjectivists consider the probability of a parameter being true given that data that is set.

Conventional NHST involves both deductive and inductive reasoning. In the deductive portion, NHST assumes that the null hypothesis is true, and then deductively infers what the expected value of the test statistic should be under that assumption; the p value then represents the probability of obtaining the test statistic with reference to a distribution of results from simulated hypothesis studies. This is a matter of deductive logic because if the null hypothesis is true, the distribution of simulated results necessarily follows, and the p value follows accordingly. In the inductive portion, the researcher generalizes the comparison of the test statistic, drawing an analogy from the distribution of results of the simulated hypothesis studies to the distribution of our sample data, and drawing conclusions accordingly.

This is an inductive move because there is no guarantee that our sample data are similar to the simulated distribution; we assume this based on the fact that our sample data exhibit the six properties mentioned earlier.

As O'Connor and Khattar suggest, most problems with NHST arise from its misuse. Despite its ubiquity in psychological science, NHST is frequently misinterpreted and misapplied, and the conclusions drawn from its use are often invalid. What psychologists often hope to obtain from NHST are simply not produced by it. Some of the logical flaws in the application of NHST are listed below.

First, NHST simply tells us how likely the test statistic is likely to be obtained with reference to a distribution of imaginary test statistic values. The data from which the *p* value arises are drawn from an imaginary distribution, developed via simulation: "mathematical formulas that mimic the results from a long series of identical hypothetical studies in which the null hypothesis is true" (O'Connor & Khattar, this volume, 2022). This is problematic because the null hypothesis may not be true for our sample dataset. This disconnect leads us to make all sorts of misinterpretations regarding the *p* value: that is, that it represents "the probability of making a Type I error," or that it tells us that "5% of all published findings are Type I errors," where a Type 1 error refers to the rejection of a hypothesis when it is actually true. As O'Connor and Khattar state, if the null hypothesis is true for our sample data, then the probability of a Type I error must be zero. NHST relies on our assumption that the null hypothesis is true. Thus, the argument assumes that the null hypothesis is true, yet a conclusion is drawn about the truthfulness of the null hypothesis—this common misinterpretation is thus a logical error.

Second, psychological researchers can consistently violate the assumptions of NHST in their research: researchers may not randomly sample, sample sizes may be too small to achieve the distributional requirements of NHST (normality and homoscedasticity), and the scores obtained are never perfectly reliable (O'Connor & Khattar, this volume, 2022). For instance, Szucs and Ioannidis (2017), conducting an empirical assessment of published effect sizes and estimated power among psychology and cognitive neuroscience journals, found that the power of these studies was "unacceptably low" (p. 13). Power is defined as the probability of finding statistical significance when there is a real effect. Significance tests are constructed to produce valid results only when all the assumptions are met; as such, it is likely that the results obtained from many of the studies using NHST are biased. Viewing NHST as a form of argument, it is an invalid argument to reach the conclusion implied by the argument when the premises are untrue.

Third, the results of NHST would be insufficient for researchers to definitively attribute the difference between means to the effect that is hypothesized. The American Statistical Association statement on NHST notes that "by itself, a *p value* does not provide a good measure of evidence regarding a model or hypothesis" because it "provides limited information" (Wasserstein & Lazar, 2016, pp. 131–132). The results from NHST would still need to be combined with other background information and assumptions regarding our experiments. These assumptions may be related to our experimental design (e.g., in a between-group experiment, the variables being controlled for did not systematically differ across groups) or our background knowledge (e.g., the background literature provides evidence that the effect being tested for in NHST is plausible). For example, consider a clinical trial investigating a novel psychotherapy for depression. If it finds that there was a significant difference between the treatment and control groups, the conclusions drawn regarding the trial still rely on other aspects of the trial: the way in which depression outcomes are operationalized, the timeframe for measuring outcomes, and so on. This idea can also be seen in the distinction between "statistically significant" and "clinically significant"—a result may be statistically significant but have little to no clinical implications. Yet the conclusions we draw frequently use the *p* value to adjudicate between the falsity and truth of the hypothesized effect in question without considering these other premises (O'Connor & Khattar, 2022). In the language of logic, it is an invalid argument to conclude that the effect is present when the truth of the premises is not clearly established.

A related key flaw of NHST is its failure to take background information regarding the effect into account. The process of NHST begins with a "blank slate" assumption—no prior information regarding the effect under consideration is considered. Each experiment and its results are considered in isolation, and the conclusion is taken as a definitive answer to the question (albeit technically subject to replication). The logical flaw underlying this problem is that prior information—such as results from previous experiments for a related hypothesis—is part of the pool of evidence from which one should infer. Using only a subset of the available evidence would likely lead researchers to a conclusion that contradicts other aspects of the pool of evidence. For example, consider that high-powered experiments have been conducted to test Hypothesis H_1, of which 1 had a positive result and 9 had a negative result (under NHST). This suggests that the prior probability of H_1 being true is 1/10. If you were to ignore this prior evidence and conduct an experiment that yields a positive result, you might wrongly conclude that H_1 is true, when it is far more likely to be a false positive (Szucs & Ioannidis, 2017). Techniques such as meta-analysis (Glass, 1976), which aggregate previous results about the effect size of a particular hypothesis to determine its robustness and value, have been developed to overcome this problem.

Finally, NHST results in an "all or nothing" outcome: the null hypothesis is either significant or not significant, and researchers often (erroneously) draw the conclusion that the effect in question is "true" or "false." By collapsing the outcome into a "clean" binary (Gelman & Carlin, 2017, p. 901), the researcher risks obscuring the uncertainty of the statistical conclusions drawn. While there is truth or falsity to a hypothesized effect, there is uncertainty inherent to every psychological experiment—some sources of uncertainty arise in experimental error, imperfect reliability and validity of measurements, and uncertainty regarding the validity of previous experiments. As such, to come to a binary conclusion regarding an experimental outcome ignores the truth that scientific methods are inherently uncertain.

Bayesian Inference

The Bayesian statistical approach, sometimes called Bayesian inference, addresses some of the flaws of NHST and has been proposed as an alternative to NHST for data analysis in psychological science. Bayesian statistics takes a subjectivist view of statistics and is founded on the mathematically precise Bayes' theorem:

$$P(H \mid E) = \frac{P(H)P(E \mid H)}{P(E)}$$

where *P*(*x*) refers to the probability of *x*, *H* is the hypothesis, and *E* is the evidence. The notation " | " means "given"; hence, *P*(*H*|*E*) is the probability of *H* given *E*, which is also known as the "posterior" probability of *H*. *P*(*H*) is the prior probability of the hypothesis, *P*(*E*|*H*) is the probability that the data are generated given that the hypothesis is true—this is sometimes called the likelihood function—and *P*(*E*) is the probability of the data according to the model. *P*(*E*) is also known as the "normalizing constant," which simply divides the probabilities obtained across the distribution to ensure that the distribution sums to 1. Because *P*(*E*) does not figure into determining the relative probabilities of different hypotheses, the equation is sometimes depicted as:

$$P(H \mid E) \propto P(H)P(E \mid H)$$

where $\propto$ means "proportional to." By this formula, the probability of *H*, given the evidence, is proportional to the probability of the hypothesis multiplied by the probability of the evidence given the hypothesis. This tells us exactly how to change our degrees of belief in a hypothesis.

Once *P*(*H*|*E*) is obtained, it "updates" the model, becoming the new prior probability for the hypothesis. Upon receiving a new set of evidence, the new prior probability replaces *P*(*H*), and Bayesian updating occurs again. If $P(H|E_1)$ refers to the posterior probability of the hypothesis after receiving evidence E_1, then upon receiving new evidence E_2:

$$P(H \mid E_2) = \frac{P(H \mid E_1)P(E_2 \mid H)}{P(E_2)}$$

The subsequent equation allows the researcher to update their posterior probability of *H* once again. *P*(*H*|*E*) is typically represented as a probability distribution, wherein each potential value of a given parameter implied by the hypothesis (i.e., the effect size) has a discrete probability attached to it, representing the subjective belief of the researcher in each potential value of the parameter.

The logic of Bayesian statistics is founded on the rigor of Bayes' theorem. Because Bayes' theorem holds true across any potential set of probability distributions, it allows the researcher to determine the probability of any hypothesis so long as the probabilities corresponding to the researcher's beliefs are input into the equation. As a form of argument, it is valid because the conclusion (the posterior) will be true if the premises (the priors and the likelihood) are true.

Bayesian statistics thus holds several advantages over NHST. First, as a subjectivist view of statistics, it is a more direct way of determining the probability of a hypothesis being true given the data. Second, Bayesian estimation requires the researcher to specify their priors, which makes the researcher's preconceptions regarding the hypothesis transparent. NHST does not account for the researcher's biases, which can and do influence the results obtained (see Chap. 5 on *p*-hacking in this volume). Third, the product of Bayesian estimation is a probability distribution of parameter values representing the degrees of belief in the effect being tested; it does not commit to a binary yes/no outcome. This thus allows for uncertainty to be represented. Fourth, Bayes' theorem explicitly incorporates prior information in the estimation by requiring that prior probabilities are introduced as inputs. The information from previous experiments regarding a hypothesized effect can thus be accounted for. Fifth, it provides a way of precisely updating the probabilities upon receiving new evidence, allowing the researcher to determine exactly how their degrees of belief should change considering the new evidence. Meta-analyses can estimate the effect size of a hypothesis, but they often rely on hundreds of studies to do so. Finally, the fundamental logic of Bayesian statistics implies that the posterior distribution found will probably change as new evidence is introduced through further research; the researcher cannot help but be reminded that the results they obtain are "pending." Results from NHST have often been presented and understood as the definitive answer regarding a hypothesis, despite the rhetorical emphasis on replication.

Here is a simple example of Bayesian inference, borrowed from Kruschke (2015). Suppose that you are trying to find the bias of a coin. Based on prior information, for example, this coin came from a magician's shop, and the shopkeeper tells you that the coin mostly lands on heads, you suspect that the coin is strongly biased toward heads. As such, you might hypothesize that the coin's bias is 0.9, where 0 represents tails and 1 represents head. However, because you are not completely confident about this, you construct a prior distribution wherein the prior probability distribution is densest in the region around 0.9—in this example, a distribution known as the beta distribution is the most appropriate. This probability distribution represents your prior knowledge and your confidence in your hypothesis, $P(H)$. Subsequently, you conduct some tests—you flip the coin ten times, finding that it lands on heads eight times. What should you believe about the bias of the coin?

Bayesian inference, through Bayes' theorem, allows you to determine this precisely. To find $P(E|H)$, you ask: what is the probability of obtaining eight heads in ten coin flips if we suspect the true bias of the coin is 0.9? For example, calculating the discrete probability for the point estimate of 0.9:

$$P(E \mid H) = 0.9^8 \times 0.1^2$$
$$= 4.3 \times 10^{-3}$$

The equation above represents the fact that the event with a probability of 0.9 (heads) occurs eight times and the event with the probability of 0.1 (tails) occurs twice. You would then ask the same question for every discrete point estimate in the prior probability distribution, obtaining *P(E|H)* for each value. Finally, using Bayes' theorem, you can input the values of *P(H)* and *P(E|H)* into the proportional relationship, allowing you to find the precise probabilities for each point estimate in the distribution. Doing these calculations by hand is computationally intensive, but many statistical programs now have implementations of Bayesian statistics that are quite efficient at applying it.

One major criticism of Bayesian statistics is that the choice of the prior distribution is arbitrary. Since the prior distribution strongly influences the effect of evidence on the posterior distribution, the posterior distribution obtained may be heavily (and incorrectly) biased. For instance, assume that a coin is strongly biased toward heads, that is, the coin has a true parameter value of 0.9, where 0 represents tails and 1 represents head. If I have a strong belief that the coin has no bias, for example, centers the prior probability distribution narrowly around the parameter value of 0.5, then the appearance of nine heads among ten coin flips will still lead to a posterior distribution that clusters near the initial prior (e.g., 60% biased toward heads). This strong belief is referred to as a "strongly informative prior."

However, in practice, researchers using Bayesian estimation would choose a prior based on the background information available to them. For instance, if they had no information regarding the coin, they might choose a flat prior, making all possible parameter values have equal probability. Nine heads in ten coin flips would then lead to a posterior distribution centered around the true parameter value. Alternatively, being aware that the coin belonged to a magician whose trick relied on the coin turning up heads, I might set a prior probability distribution that clusters around a parameter value of 1. This is also close to the true parameter value of the coin. Additionally, the prior distribution becomes more likely to converge on the true value of the parameter over time, suggesting that there exists a sufficient number of observations for the likelihood function to overwhelm even a strongly informative prior.

There is some controversy as to the status of Bayesian inference as a deductive or inductive method (see Gelman, 2011; Talbott, 2008). Bayes' theorem is clearly deductive since it relies on the deductive rules of mathematics. It follows the logic of *modus ponens*: namely, that so long as the premises are true, the conclusion is true. The truth of the conclusion is contained in the truth of the premises. So long as the premises are true, the deductions obtained from Bayes' theorem are valid.

Hawthorne (1993) argues that Bayesian inference is a form of *eliminative induction*, or "induction by deduction." As evidence that is deductively entailed by the hypothesis builds up, some hypotheses get eliminated (probability reduced to zero) and one (the true hypothesis) rises to the top as all other alternatives are eliminated. In cases where evidence is only probabilistically related to the hypotheses, some hypotheses get "highly refuted" and one (the true hypothesis) becomes "highly confirmed" as its plausibility increases. In both cases, Bayesian inference is thought to result in convergence to agreement regarding the posterior probabilities of hypotheses. The former may appear to be Popperian falsification, which implies that a potentially infinite number of hypotheses to be falsified prevents us from ever knowing the true hypothesis. In response, Hawthorne suggests that if hypotheses are "ordered" in plausibility, so long as the hypotheses above the true hypothesis in the order are "evidentially distinguishable" (evidence exists that can deductively show that one hypothesis is true and another false), the true hypothesis will eventually rise to the top of the order and remain there.

IBE in Haig's Abductive Theory of Method

Brian D. Haig is a cognitive psychologist and research methodologist who advocates the use of both abduction and IBE in behavioral sciences. These feature in his *Abductive Theory of Method* (ATOM; Haig, 2005), whereby abduction is the primary tool for generating theories and IBE is the primary tool for appraising these. Haig's theory is comprehensive and detailed, and space limitations prevent a complete discussion of it here—as such, a quick sketch of ATOM will be laid out, focusing on the place of IBE within it. ATOM uses TEC for its grounding, committing to its notions of explanation as a primitive and the distinctions between explanation and prediction.

ATOM centers on the principle that explanatory considerations play a role across the three stages of theory construction: theory generation, theory development, and theory appraisal. Theory development occurs after theory generation, and theory appraisal occurs throughout the generated theory's lifespan. Unlike hypothetico-deductive models of science such as Popper's (1959)—that begins with a problem and a theory aimed at solving this problem—ATSM begins with the phenomena to be explained and suggests that theories are constructed based on the phenomena. Accordingly, "phenomena exist to be explained rather than serve as the objects of prediction in theory testing" (Haig, 2005, p. 371). Importantly, Haig distinguishes phenomena from data—data are the raw observations, while phenomena are the "robust empirical regularities" (p. 372) that are abstracted from the data—in ATSM, they are also called *phenomenal laws*. Haig provides some examples of phenomenal laws from psychology: "the matching law, the Flynn effect in intergenerational gains in IQ, and the recency effect in human memory" (p. 374).

In the theory generation stage, abductive inference—reasoning to underlying causal mechanisms to explain phenomena—is used to judge the plausibility of potential causal mechanisms, and the best of these are then selected as a "plausible enough" theory. This judgment of plausibility is based on its explanatory value, which is evaluated using the criteria of explanatory coherence. Here, *existential abduction* is applied, wherein the existence of previously unknown objects and constructs is hypothesized. This is contrasted with *analogical abduction*, whereby models of the mechanisms are developed based on analogy to other known mechanisms; this is used in the theory development stage. For instance, if it is theorized that anatomical and physiological patterns in different generations of animals can be explained by the theory of natural selection, an analogical model would be that of artificial selection.

Finally, theory appraisal involves the use of IBE. In contrast to the Popperian model of the logic of research, that evaluates the theory based on its survival from falsification attempts, the ATOM model judges the theory based on its "explanatory breadth" (p. 380), which is synonymous with Thagard's (1993) criterion of consilience. Furthermore, unlike the Bayesian model of confirmation that relies on assigning probability to various hypotheses in light of evidence, the ATOM model judges on Thagard's (1993) qualitative explanatory criteria, not quantitative statistical criteria—note that this contrasts with the justification of IBE based on simulations previously explored.

ATOM is thought to be a particularly useful philosophical contribution for clinical psychologists for three reasons. First, it was developed for application to the behavioral sciences; Popper, Kuhn, and other prominent theorists of science based their models on the physical sciences, especially physics (O'Donohue, 2013), and as such their models may not be applicable to the behavioral sciences. Second, it is a theory founded in the practice of science; it pays attention to all of the steps involved in scientific activity (theory generation, theory development, theory appraisal)—again, Popper and others have been accused of not basing their models of science on the actual practice of scientists. Finally, and quite intriguingly, Ward et al. (2016) have elaborated that ATOM can be integrated into the practices of clinical psychologists as a conceptual framework for psychological assessment.

One criticism of the theory appraisal stage of ATOM (Romeijn, 2008) is that it is subject to two common objections to IBE, labeled by Lipton (2004) as "Hungerford's objection" and "Voltaire's objection." Hungerford's objection suggests that the notion of "best-ness" of explanations is too subjective and varied. However, given the grounding of ATOM in Thagard's (1993) IBE, which has been naturalistically justified (subject to empirical testing), Romejin is willing to concede this point. Voltaire's objection suggests that there is no reason to believe that the theories chosen by IBE are true or approximately true, for we have no reason to believe that the world accords with our explanatory criteria. To that point, Haig (2008) responds that ATOM does not claim to be a method for revealing truths; instead, the explanatory criteria in TEC are guides to truth, or at least would bring us toward the goal of "maximizing true propositions and minimizing false ones" (p. 1042).

Conclusions

We argue that poor scientific reasoning in which logical errors are made is another questionable research practice. We recommend that research psychologists and consumers of psychological research pay more attention to the logic of research by identifying the relevant inferential approaches, detecting logical errors, and constructing sound reasoning. We describe some prominent types of research logic: from alogical approaches such as that of Kuhn, to deductive logical approaches of Popper, to inductive approaches and abductive/IBE approaches. The strength and weaknesses of each approach are discussed, along with the applications of these approaches in statistical methods and ATOM.

References

Bartley, W. W. (1990). The retreat to commitment. Open Court.

Carnap, R. (1945). On inductive logic. *Philosophy of Science, 12*(2), 72–97. https://doi.org/10.1086/286851

Cook, T. D., & Campbell, D. T. (1979). *Quasi-experimentation: Design & analysis issues for field settings*. Houghton Mifflin.

Durrant, R., & Haig, B. D. (2001). How to pursue the adaptationist program in psychology. *Philosophical Psychology, 14*(4), 357–380. https://doi.org/10.1080/09515080120088067

Gelman, A. (2011). Induction and deduction in Bayesian data analysis. *Rationality, Markets and Morals, 2*, 67–78, 1999.

Gelman, A., & Carlin, J. (2017). Some natural solutions to the *p* -value communication problem – And why they won't work. *Journal of the American Statistical Association, 112*(519), 899–901. https://doi.org/10.1080/01621459.2017.1311263

Glass, G. V. (1976). Primary, secondary, and meta-analysis of research. *Educational Researcher, 5*(10), 3–8. https://doi.org/10.3102/0013189X005010003

Haig, B. D. (2005). An abductive theory of scientific method. *Psychological Methods, 10*(4), 371–388. https://doi.org/10.1037/1082-989X.10.4.371

Haig, B. D. (2008). On the permissiveness of the abductive theory of method. *Journal of Clinical Psychology, 64*(9), 1037–1045. https://doi.org/10.1002/jclp.20507

Harman, G. (1965). The inference to the best explanation. *The Philosophical Review, 74*, 88–95.

Hawthorne, J. (1993). Bayesian induction is eliminative induction. *Philosophical Topics, 21*(1), 99–138. https://doi.org/10.5840/philtopics19932117

Hempel, C. G. (1965). *Aspects of scientific explanation*. Free Press.

Hume, D. (1779). An enquiry concerning human understanding. In D. Hume (Ed.), *Essays and treatises on several subjects, Vol. 2. Containing an enquiry concerning human understanding, a dissertation on the passions, an enquiry concerning the principles of morals, and the natural history of religion* (pp. 3–212). Unknown Publisher. https://doi.org/10.1037/11713-001

Hempel, C. G. (1962). Deductive-nomological vs. statistical explanation. University of Minnesota Press, Minneapolis. Retrieved from the University of Minnesota Digital Conservancy, https://hdl.handle.net/11299/184632

Ioannidis, J. P. A. (2005). Why most published research findings are false. *PLoS Medicine, 2*(8), e124. https://doi.org/10.1371/journal.pmed.0020124

Kline, R. B. (2013). *Beyond significance testing: Statistics reform in the behavioral sciences* (2nd ed.). American Psychological Association. https://doi.org/10.1037/14136-000

Kruschke, J. K. (2015). *Doing Bayesian data analysis: A tutorial with R, JAGS, and Stan (Edition 2)*. Academic Press.
Kuhn, T. S. (1962). *The structure of scientific revolutions*. University of Chicago Press.
Kuhn, T. S. (1994). *The structure of scientific revolutions* (2nd ed.). University of Chicago Press.
Kyburg, H. E. (1961). *Probability and the logic of rational belief*. Wesleyan University Press.
Lakatos, I. (1970). Falsification and the methodology of scientific research programmes. In I. Lakatos & A. Musgrave (Eds.), *Criticism and the growth of knowledge*. Cambridge University Press.
Laudan, L. (1978). *Progress and its problems: Towards a theory of scientific growth* (1st ed.). University of California Press.
Lipton, P. (2004). *Inference to the best explanation* (2nd ed.). Routledge/Taylor and Francis Group.
Munz, P. (2014). *Our knowledge of the growth of knowledge: Popper or Wittgenstein?* Routledge.
O'Connor, B. P. O., & Khattar, N. (2022). Controversies regarding null hypothesis significance testing. In M. Lillienfeld & O'Donohue (Eds.), *Questionable research practice: Designing, conducting, and reporting sound research in clinical psychology*. Springer.
O'Donohue, W. (2013). Clinical psychology and the philosophy of science. *Springer International Publishing*. https://doi.org/10.1007/978-3-319-00185-2
Pollard, P., & Richardson, J. T. E. (1987). On the probability of making Type I errors. *Psychological Bulletin, 102*, 159–163. https://doi.org/10.1037/0033-2909.102.1.159
Popper, K. R. (1959). *The logic of scientific discovery*. Hutchinson.
Poston, T. (2014). *Reason and explanation*. Palgrave Macmillan. https://doi.org/10.1057/9781137012265
Psillos, S. (1999). *Scientific realism: How science tracks truth*. Routledge.
Romeijn, J.-W. (2008). The all-too-flexible abductive method: ATOM's normative status. *Journal of Clinical Psychology, 64*(9), 1023–1036. https://doi.org/10.1002/jclp.20516
Salmon, W. C. (2006). *Four decades of scientific explanation* (1st ed.). University of Pittsburgh Press.
Szucs, D., & Ioannidis, J. P. A. (2017). Empirical assessment of published effect sizes and power in the recent cognitive neuroscience and psychology literature. *PLOS Biology, 15*(3), e2000797. https://doi.org/10.1371/journal.pbio.2000797
Simmons, J. P., Nelson, L. D., & Simonsohn, U. (2011). False-positive psychology: Undisclosed flexibility in data collection and analysis allows presenting anything as significant. *Psychological Science, 22*(11), 1359–1366. https://doi.org/10.1177/0956797611417632
Talbott, W. (2008). Bayesian epistemology. In Zalta, E. N. (Ed.), *The Stanford encyclopedia of philosophy* (Winter 2016 Edition). https://plato.stanford.edu/archives/win2016/entries/epistemology-bayesian
Thagard, P. (1993). *Conceptual revolutions* (1st ed.). Princeton Univ. Press.
Thagard, P. (2000). *Coherence in thought and action*. MIT Press.
Vogel, J. (1998). Inference to the best explanation. In E. Craig (Ed.), *Routledge encyclopedia of philosophy*. Routledge. https://www.rep.routledge.com/articles/inferenceto-the-best-explanation
Ward, T., Clack, S., & Haig, B. D. (2016). The abductive theory of method: Scientific inquiry and clinical practice. *Behaviour Change, 33*(4), 212–231. https://doi.org/10.1017/bec.2017.1
Wasserstein, R. L., & Lazar, N. A. (2016). The ASA statement on *p* -values: Context, process, and purpose. *The American Statistician, 70*(2), 129–133. https://doi.org/10.1080/00031305.2016.1154108
Woodward, J., & Ross, L. (2021). Scientific explanation. In Stanford Encyclopedia of Philosophy. https://plato.stanford.edu/archives/sum2021/entries/scientific-explanation

Chapter 3
Replicability and the Psychology of Science

Cory J. Clark, Nathan Honeycutt, and Lee Jussim

Abstract Scholars have much to gain by forwarding flashy, socially important, self-promotional, group-promotional, timely results. However, in the new era of Open Science, such gains could be short-lived if findings are not also accurate—replicable, with correct interpretations and conclusions. Accepting that we ourselves are humans who are vulnerable to unconscious motivations that influence the ways we conduct science and the conclusions we come to should motivate us to place regulations on ourselves (e.g., refusing to file drawer our own studies, searching for information that challenges our beliefs and hypotheses, working with scholars with whom we disagree). Unfortunately, even when people are presented with reasonably compelling evidence that they might have biases that steer their judgments away from accuracy, they seem unable to recognize these tendencies in themselves. If you wish to be the exception to the rule, start not by denying that you are human and prone to biases and motivations, but instead by having a conscience that bravely admits this to yourself.

Keywords Clinical science · Questionable research practice · Replicability · Clinical psychology

> They all pose as though their real opinions had been discovered and attained through the self-evolving of a cold, pure, divinely indifferent dialectic… whereas, in fact, a prejudiced proposition, idea, or 'suggestion,' which is generally their heart's desire abstracted and refined, is defended by them with arguments sought out after the event. They are all advocates who do not wish to be regarded as such, generally astute defenders, also, of their prejudices, which they dub 'truths,'—and VERY far from having the conscience which bravely admits this to itself… (Friedrich Nietzsche, *Beyond Good and Evil*, p. 14, 1886/2017).

C. J. Clark (✉)
University of Pennsylvania, Philadelphia, PA, USA
e-mail: cjclark@sas.upenn.edu

N. Honeycutt · L. Jussim
Rutgers University, New Brunswick, NJ, USA

W. O'Donohue et al. (eds.), *Avoiding Questionable Research Practices in Applied Psychology*, https://doi.org/10.1007/978-3-031-04968-2_3

Scientists are humans. They are smart, ambitious humans, with a peculiar desire to explain and understand the world and a set of principles and procedures that help steer them toward truth. They are humans nonetheless. Their psychology is therefore human psychology.

Psychological discoveries in the social sciences—human errors, heuristics, biases, motivations, psychological needs—all apply to scientists in similar if not equal (or possibly even greater) measure. For example, people with greater education and science literacy are *more* polarized in their views of scientific controversies (such as climate change), raising the possibility that education increases the extent to which reasoning is influenced by preferred worldviews (Drummond & Fischhoff, 2017).

Although such biases and errors of reasoning are frequently investigated *by* scientists, they are rarely applied *to* scientists to understand how the reasoning patterns discovered by scientists likely influence scientists' own reasoning and discoveries. The present chapter will apply psychological science to explain why, when, and how scholars engage in questionable research practices (QRPs), advance dodgy or erroneous conclusions (sometimes for decades on end), and suppress accurate or useful information. Although certainly some scholars consciously and purposefully engage in fraud or data suppression, we suspect the vast majority of these non-optimal truth-seeking strategies occur outside of researchers' awareness in the sense that they genuinely believe their research practices are more optimal than they are in reality. We first review bases for concluding that scientists are vulnerable to *motivated research.* Next, we argue that it is in the best interest of truth-seeking for scientists to acknowledge these tendencies in themselves and vigilantly and proactively defend against them. We also suggest some concrete correctives.

A Primer on Motivated Reasoning

Reasoning—the ways in which we approach and avoid information, evaluate information, and construct our attitudes and beliefs about information—is motivated (Kunda, 1990). Sometimes it is motivated by desires to reach the most accurate conclusion. This is the scientific ideal. Unfortunately, however, reasoning can also be motivated by desires to reach particular conclusions rather than truth. This can undercut scientific validity.

Imagine a trial in which a defendant was accused of robbing a locally owned mini mart and there were numerous pieces of evidence to evaluate, including a slightly blurry surveillance video, an eyewitness who claims the robber was of similar height and physique as the defendant, and a suspiciously timed bank deposit from the defendant shortly after the robbery took place. The prosecution attorney would be motivated to view this as clear and conclusive evidence of the defendant's guilt, the defense attorney would be motivated to view this as ambiguous and circumstantial evidence, and the judge and the jury would be motivated to make the most accurate evaluation of the defendant's likely guilt. Although humans prefer to see themselves as the judge—carefully weighing evidence and coming to conclusions most consistent with the data—humans often reason more like the lawyers, evaluating evidence in ways that allow them to reach conclusions most beneficial to themselves (Ditto, Clark, et al., 2019; Ditto, Liu, et al., 2019; Haidt, 2001).

Humans likely evolved to reason this way because accuracy is not always the most important goal for reproductive success (Clark et al., 2019). Sometimes it is more beneficial to persuade others of one's own greatness, to demonstrate commitment and value to one's social group, to avoid a possibly correct but risky or costly conclusion, to protect one's own reputation or the reputation of one's kin, to secure a mate, or to deceive an enemy than to be correct. In the social sciences, the consequences that flow from many research findings are so difficult to evaluate that inaccuracies can go undetected for decades. Popularity (of a scientific finding) can produce citations, grants, awards, and, therefore, career success. By the time invalidities of highly popular findings are discovered, the scientists producing them will have had wonderful careers. Thus, the current academic system is plausibly described as incentivizing popularity more than accuracy.

In science, motivated reasoning, or rather, *motivated research*, happens when extraneous concerns beyond accuracy influence how scientists familiarize themselves with extant data, reach hypotheses, collect and analyze observations, come to conclusions, and report those conclusions to other scientists and the public. Researchers do not merely forward their own conclusions however; they are also the gatekeepers (the editors, the peer reviewers, the hiring committee members, the peer commentators, etc.) for their peers' research, and thus motivated research can also happen when concerns beyond accuracy influence how scientists accept, elevate, reject, and suppress the work of their peers or the very peers themselves. The replication crisis has focused largely on how scholars advance erroneous conclusions by producing unreplicable results, but scholars may also obstruct accurate conclusions or useful information, which is problematic for advancing knowledge in the social sciences.

Social Sciences Supply Especially Fertile Ground for Motivated Research

Ambiguous, noisy, and difficult information environments increase the likelihood of motivated reasoning (Kopko et al., 2011; Munro, Lasane, & Leary, 2010; Munro, Weih, & Tsai, 2010). Accuracy motivations decrease because one cannot know with much certainty which conclusions are accurate, and thus other motivations take their place (Clark & Winegard, 2020). Science, and perhaps especially social science, generally deals with these ambiguous, noisy, and difficult information environments. Most if not all social phenomena cannot be studied in a vacuum. There is rarely if ever one clear best methodological strategy for testing a social science question, and even when scientists discover seemingly robust and replicable data patterns, there are often numerous ways of interpreting those patterns. Meehl (1990, p. 196) captured this state of affairs beautifully: "...theories in 'soft areas' of psychology have a tendency to go through periods of initial enthusiasm leading to large amounts of empirical investigation with ambiguous over-all results."

For example, any time some negative parenting behavior correlates with some negative outcome for children, did the parenting behavior have any causal influence or is there simply a genetic confound (e.g., Maranges et al., 2021)? Any time scholars discover an association between negative stereotypes or implicit attitudes and negative outcomes for the groups those stereotypes or implicit attitudes are about, did the stereotypes or implicit attitudes have any causal force on those negative outcomes, or did the negative stereotypes and implicit attitudes exist because people are reasonably skilled at detecting existing patterns in the world (e.g., Hehman et al., 2018; Payne et al., 2017; Reber, 1989)? Even best practices in social science require scholars to make numerous at least somewhat arbitrary decisions at each step of the research process, from generating hypotheses to drawing conclusions. These characteristics of the social sciences make it very difficult for an accuracy-motivated social scientist to reach correct conclusions and simultaneously make it very easy for a social scientist motivated by extraneous concerns to reach the conclusion they desire (Duarte et al., 2015; Simmons et al., 2011). Consequently, social sciences as a discipline are vulnerable to motivated research practices.

Beyond the ambiguous information environment problem, there is even more reason to believe that motivated reasoning creates unique obstacles for the social sciences. The investigators and the objects of investigation are one and the same thing—humans—and humans *care* about human things. It likely makes little difference to the average human whether flying squirrels are fluorescent or whether there is a maximum speed of light, but average humans *might* care if middle-aged men are sexually attracted to 15-year-old females, if altruism is "selfish," and if grandparents evolved to love their daughters' kids more than their sons' kids. It is likely impossible to eliminate human desires from an understanding of humankind; thus, social scientists likely have more extraneous motivations influencing their work than scientists who deal with amoebae, polymers, quarks, or any nonhuman objects.

Moreover, *morality* is frequently tangled up in the social sciences, and *moral concerns* are powerful motivators of reasoning (Clark et al., 2019; Tetlock et al., 2000). Sometimes accurate conclusions in the social sciences might cause concerns about morally undesirable implications, and people and scholars may then wish to avoid, ignore, disparage, or censor this kind of information, even when it could plausibly be correct (Campbell & Kay, 2014; Clark et al., 2020; Stewart-Williams et al., 2021; von Hippel & Buss, 2017; Winegard, Clark, et al., 2018). For just one recent example, a paper by AlShebli, Makovi, and Rahwan (AlShebli et al., 2020) collected a very large sample of mentor and protégé pairs in scientific collaborations and found evidence that female protégés with higher proportions of female mentors were less impactful later in their careers. After widespread outrage among the scientific community, on November 19, 2020, the editors of the journal released a statement, "Readers are alerted that this paper is subject to criticisms that are being considered by the editors. Those criticisms were targeted to the authors' interpretation of their data that gender plays a role in the success of mentoring relationships between junior and senior researchers, *in a way that undermines the role of female mentors and mentees…*" (emphasis added). Although there are plenty of legitimate criticisms of this paper (as there are of probably every published article in the social sciences), the investigation by editors of the journal occurred because of concerns about *undermining* the role of female mentors and mentees. Thus, this investigation is explicitly *morally* motivated. And these moral concerns might cause suppression of a real pattern and exploration into the causes of this pattern.

We are not saying that moral concerns are *never* a legitimate reason to suppress research findings (that is a difficult debate). But, in many cases, outrage mobs of academics bear a striking resemblance to a mob stirring up a moral panic, as they cause the suppression of data in the absence of evidence of harm or thoughtful consideration of alternatives (Stevens, Jussim & Honeycutt, in press). Moreover, scholars often assert themselves and their comrades as the authorities on such matters. Thus, while their intentions may be noble, such suppression is often ochlocratic and advances the interests of a subset of outraged scholars to the detriment of knowledge accruement. Occasionally, empirical reality will lead scholars to arrive at conclusions that trigger our moral alarms, and because scientists are humans and evolved to minimize certain harms, occasionally they will wish to suppress accurate information by suppressing their own findings (Zigerell, 2018) or creating obstacles for their peers' findings (Stevens et al., in press).

Similarly, occasionally, scientists will discover false patterns that are morally desirable, or real patterns but then explain these with false but morally desirable explanations. Such erroneous patterns or erroneous explanations may persist in the psychological canon for years or decades because they are morally desirable to scholars and thus few scholars will wish to challenge them (Jussim et al., 2019). To give a couple of examples, the ideas that stereotype threat could explain certain group disparities (e.g., Jussim et al., 2016) or that implicit bias could explain subtle but impactful prejudices against certain groups (Forscher et al., 2019) are arguably some of the most prominent social-psychological findings of all time, yet the effects are weak to non-existent and there is little if any evidence of their importance in the real world (e.g., Clark & Winegard, 2020). It seems likely that these effects were overblown and proper scrutiny was decades delayed because the findings were morally and thus socially desirable by scientists. Many scholars would want to forward such results themselves and few would want to challenge them.

Because the social sciences deal most directly with problems and questions with significance to humans, social scientific conclusions are vulnerable to morally motivated data suppression and morally motivated data elevation. Being a purely accuracy-driven social scientist will occasionally require an unnatural detachment from normal human concerns and motivations.

Human Motivations

We discuss four human motivations that likely influence the ways in which scholars conduct their research. We also discuss how those motivations can produce severely biased scientific research literatures.

Status Desires

Humans desire status and behave in ways that increase their chances of attaining status in social groups (e.g., Anderson & Kilduff, 2009). Scientists desire to attain status within their discipline—to be respected and admired by their peers—but also, given the relative status of scientists in society (Pew, 2020), they likely desire to use their roles as scientists to gain high status among society at large. Becoming a social scientist requires relatively high investment in education and a relatively high workload to attain a tenure-track position at a research institution, and the job actually pays relatively little compared to other degrees that require similar amounts of education and time investment (e.g., medical doctors). Therefore, it is plausible that social scientists are *more* motivated by desires for status than even the average highly educated person. Thus, it seems plausible that the social sciences likely attract the kinds of people who are *especially* incentivized by status attainment and especially likely to engage in research behaviors that would allow them to attain status.

Perhaps the chief way people attain status is by creating the appearance of providing benefits for others (e.g., Durkee et al., 2020). Although actually providing such benefits is one route to creating this appearance, it is not necessary. One can engage in virtue signaling or moral grandstanding without actually doing much else and this can sometimes be very effective at persuading others that one is a force for moral good (Grubbs et al., 2019). Providing benefits to others is also not sufficient to increase status (e.g., if it is done in a matter where few notice).

Therefore, regardless of whether anyone actually benefits, creating the appearance of providing benefits is highly incentivized. This would create a motivation to produce information that can be perceived as *new* or *novel* (Baumeister et al., 2018) or to produce "Wow Effects" (Jussim et al., 2016) with seemingly broad implications. It may take years or even decades to do the hard work to evaluate whether the findings are replicable and generalizable, and then to test them in the real world; and, at the end of that process, the entire enterprise may be found to be worthless (findings unreplicable) or trivial (replicable but only with effect sizes so small no one cares). There are few incentives to wait 15 years for such a payoff: people have jobs, tenure, promotions, and grants to obtain; bestselling books to write; workshop fees to collect; and consulting fees to garner. Put differently, the incentives all line up to create the impression that one has benefited society *now*, not 15 years from now.

Many of the most overblown findings in the social sciences fit this analysis exquisitely well (e.g., stereotype threat, implicit bias, growth mindset, various kinds of priming). We now know these findings were overblown and, in some cases, seem to be entirely invalid. Stereotype threat, priming, and growth mindset have all been subject to a series of preregistered failures to replicate and/or findings that the effects are plausibly viewed as trivially small (Doyen, Klein, Pichon, & Cleeremans, 2012; Finnigan & Corker, 2016; Flore, Mulder & Wicherts, 2019; Pashler, Rohrer & Harris, 2013; Sisk, Burgoyne, Sun, Butler & Macnamara, 2018). After almost 20 years of "implicit bias" and the Implicit Association Test (IAT) in particular being wildly overstated and oversold, in recent years, critical reviews have described the construct as "delusive," identified a slew of psychometric problems with the IAT, and shown that its predictive validity and ability to explain racial gaps is limited at best and possibly non-existent (Blanton et al., 2009; Corneille & Hütter, 2020; Forscher et al., 2019; Jussim et al., in press; Jaccard, Oswald, Mitchell, Tetlock, & Blanton, 2013; Schimmack, 2021).

Although this is not the place to review all the debunking of the last five to ten years, one example should suffice. Blanton et al. (2009) characterized a slew of studies making strong claims about racial discrimination produced by implicit bias as measured by the IAT as actually providing weak evidence. In a response, Jost et al. (2009) published a paper titled "The existence of implicit bias is beyond reasonable doubt: A refutation of ideological and methodological objections and executive summary of ten studies that no manager should ignore." In a recent review, we did a deep dive into those ten studies and found something quite startling: those ten studies supposedly refuting the "weak evidence" charge provided almost no evidence of racial discrimination (Jussim et al., in press). Put differently, there was little or no racial bias to be explained (whether by IAT scores or anything else). Indeed, most of the studies did not even address racial discrimination *at all.*

Despite the extraordinary enthusiasm for these "discoveries" (as evidenced by the massive number of papers that use the terms and measures and by the eminence and awards given to their promoters and acolytes), the fullness of time (combined with the eventual emergence of vigorous scientific skepticism) has shown them to be far less than they were cracked up to be. This may help explain why diversity and implicit bias trainings based on these (nearly) non-existent or poorly understood measures and phenomena are rarely demonstrably effective (Paluck, Porat, Clark & Green, 2021). Thus, these phenomena are all exquisite examples of how scholarship can create the impression that *AMAZING! WORLD-CHANGING!* phenomena have been discovered that will benefit humanity, without actually providing any noticeable benefit to humanity, and at great human cost in wasted effort, grant dollars, and time spent in useless trainings.

Nonetheless, selling *AMAZING! WORLD-CHANGING!* findings to an unsuspecting public and insufficiently critical scientists has been highly rewarded with status, promotions, grants, and consulting fees. And, to be clear, although scientists love to point to others (such as the media) as the culprits in overselling their findings, it is usually the scientists themselves who bear primary responsibility (Mitchell, 2018; Sumner et al., 2016). Regardless, scholars are heavily incentivized to create the appearance that their findings lead to simple and easy-to-implement interventions that will change the world. Unfortunately, many social problems persist in affluent societies precisely *because* they are extremely difficult or perhaps even impossible to fix, and so the demand for such interventions inevitably creates a low-quality supply. Unlike behavioral genetics or personality psychology, social psychology delivers simple environmental manipulations that ostensibly can create desirable changes in human behavior. The desire for effects that create potential for interventions and behavior change may even explain why social psychology is such an attractive discipline to normal people (McPhetres, 2019), despite its many flaws and embarrassments over the past decade (e.g., Nosek et al., 2015).

Scholars can also achieve media and public attention by generating findings with significance to current events and hot topics and so are likely motivated to study such topics, and to forward results quickly when they do. In a society where science often does and should move quite slowly and hot topics often change rather rapidly, this could cause scholars to draw hasty conclusions in order to be timely in their research. Moreover, quick movement to publicize *AMAZING! WORLD-CHANGING!* findings (see Mitchell, 2018, for a review of the wild overselling of implicit bias after the publication of the first IAT paper, in 1998) makes it difficult for other scholars to check such findings before they reach the broader public.

Of course, status motives will also lead scholars to pursue *accuracy* in their work, for two reasons. First, more accurate information is more useful to other people, and thus accuracy is a direct route to status attainment, and second, being *inaccurate* (if detected) can be costly. Having one's own research fail to replicate, or worse, being caught for outright research fraud are huge blows to status, and so scientists should be motivated to both appear accurate and be accurate. Given new developments in Open Science, it has become easier for other scholars to detect errors and other suspicious research practices in their peers' work, and so the current cohort of scholars should be *more* motivated to be accurate (or at least avoid certain types of errors) than the cohort existing a decade or more ago.

Open Science practices have made certain types of QRPs more difficult to get away with. For example, preregistration makes it more difficult to HARK (Kerr, 1998) and cherrypick variables, conditions, and even entire studies. However, many papers still report studies that are not preregistered leaving the door wide open to such practices. Furthermore, if studies provide narratively or theoretically "inconvenient" findings, they can still be file drawered. When acting as a reviewer, it is easy enough to suppress others' inconvenient findings or arguments—simply come up with scientifically plausible justifications for declaring the work to be sub-par.

Ostracization Avoidance

Just as people wish to gain status within their social groups, they wish to avoid being ostracized (Ouwerkerk et al., 2005). People tend to punish those who violate group norms or generate costs to the social group (Bowles & Gintis, 2004). Scholars are likely motivated to avoid these punishments, and, therefore, avoid violating group norms.

The extreme politically liberal homogeneity among social scientists (Duarte et al., 2015; Langbert, 2018) renders the entrenchment of liberal norms—such as support for parties, policies, candidates, and causes on the left; hostility to those on the right; and equalitarianism (the assumption that, but for discrimination, all demographic groups would have identical outcomes)—virtually inevitable (Clark & Winegard, 2020; Honeycutt & Jussim, 2020; Prentice, 2012). Thus, for either or both of two reasons, scientists should be motivated to avoid advancing scientific findings that challenge a liberal political agenda: (1) They share that agenda and do not wish to oppose it or (2) They correctly discern these norms and believe (probably correctly) that work challenging those norms will be more difficult to publish and fund than work that advances those norms. For example, some research has found that liberals are described more positively than conservatives in social scientific research (Eitan et al., 2018), that conservative social scientists fear ostracization and that other social scientists openly report that they would discriminate against conservatives (Honeycutt & Freberg, 2017; Inbar & Lammers, 2012), and that more liberal ideology predicts working at more prestigious universities, even after controlling for academic productivity, suggesting that ideological conformity helps one advance in their career (Rothman et al., 2005).

Another recent paper that sought to explore the relationship between ideological slant of research and replicability identified almost no papers at all in their analysis that violated liberal values, suggesting that such papers rarely come into existence (Reinero et al., 2020). Similarly, Zigerell (Zigerell, 2018) discovered 17 unpublished experiments with nationally representative samples finding either no anti-Black bias among White respondents and/or anti-White bias among Black respondents. Although we may never know exactly why those studies were never published, one possibility is that they would risk violating liberal equalitarian norms and would, therefore, either be seen as not worth publishing, or not worth the (expected extraordinary) effort, and concomitant risk of being ostracized, to do so.

Arguably, these dynamics—political skew, bias and intolerance toward individuals or ideas that conflict with mainstream liberal views—have a direct connection to censorious behaviors (Honeycutt & Jussim, 2022). This connection is not inevitable—bias does not automatically produce direct or indirect censorship. But when academic fields such as the social sciences become so heavily skewed, excluding ideas or data that conflict with the norms and worldview of the majority becomes an increasing threat to the validity of the scientific literature. This is not to say that scholarship can never be rejected—papers are routinely rejected for flaws and weaknesses that have nothing to do with political content or motivations. But ideologically motivated rejection can often be dressed up as legitimate critique, often manifested in selective calls for rigor, illusions of bad science, or claims of harm and danger (Honeycutt & Jussim, 2022). One obvious casualty is the suppression of otherwise legitimate scholarship (Stevens et al., 2020).

Scholars are likely motivated to reject information that could be perceived as opposing a politically liberal agenda both in their own research and in their peers' research. And they are likely motivated to frame their findings in ways that misleadingly portray liberals in a favorable light when their findings could just as easily or more easily be framed in ways that portray conservatives in a favorable light. For example, Lilienfeld (2015) critiqued the framing and description of conservatives having a "negativity bias" or "motivated closed-mindedness" when the findings on sensitivity to threat could have just as easily been framed as a liberal "motivated blindness to danger." More recently, a paper by Baltiansky, Jost, and Craig (Baltiansky et al., 2021) chose to highlight that high system-justifiers (correlated with more conservative ideology) found jokes targeting low-status groups to be funnier than low system-justifiers in their abstract, portraying conservatives as being insensitive toward low-status groups. However, high system-justifiers found jokes targeting low- and high-status groups similarly funny, whereas low system-justifiers found jokes targeting low-status groups to be less funny than those targeting high-status groups (Pursur & Harper, 2020). One could interpret such findings as showing that conservatives treated low- and high-status groups with equal consideration, whereas liberals were particularly condescending toward low-status groups by suggesting they need protection from jokes. Similarly, a recent paper by Brady, Wills, Burkart, Jost, and Van Bavel (Brady et al., 2019, p. 1802) highlighted that "conservative elites (on Twitter) gained greater diffusion when using moral-emotional language compared to liberal elites" portraying conservatives as vulnerable to emotional appeals. However, this effect was mainly driven by *joy-related* content, which was misleadingly labeled "moral emotion expression related to religion and patriotism" in the abstract.

Scholars likely know that to frame results in ways that portray conservatives more favorably than liberals would make the results more difficult to publish. So, the easier route to attaining publications (and status), and avoiding ostracization, is to create misleading characterizations of findings. Thus, scholars who wish to avoid ostracization among overwhelmingly liberal social scientists will engage in motivated research to generate findings and frames palatable to their liberal peers. Academia operates as a social-reputational system, whereby one's success is highly contingent upon the favorable evaluations and references of others at all career stages: to obtain admission to graduate school, publish in peer-reviewed academic journals, obtain grants, get a job, or obtain tenure/promotions. As such, there are strong incentives for doing work and staking out positions that will garner social approval from peers, and often strong disincentives surrounding the expression of ideas that colleagues reject or vehemently disagree with (Honeycutt & Jussim, 2022).

Social scientists are even more homogenous in a domain other than politics—every last one of them is a social scientist. Thus, social scientists should be motivated to avoid harming social scientists and the social scientific enterprise. The types of scholars who critique the field, for example, by suggesting the field is politically biased, or by accusing the field of shoddy methods and unreliable findings, are likely to be revered by some and loathed by others. In an effort to protect the field and their own reputations, some scholars (likely, especially older and more established scholars with more to lose) might seek to create obstacles for scholars who forward data and arguments that challenge the field. Many scholars would avoid criticizing the field, the field's theories, and the field's prominent scholars, even if they believe such criticisms are warranted, because it can be costly to them by virtue of incurring the hostility of colleagues on whom their success depends (via peer review). By writing this chapter in which we suggest that social scientists engage in *motivated research*, we risk making enemies who will dismiss us, have a lower opinion of us, or subtly punish us with ostracization.

Self-Enhancement

People are motivated to self-enhance—or to perceive and portray themselves more positively than reality would suggest (Sedikides & Gregg, 2008). Of course, social science is rarely directly about the self, but it is often indirectly about the self by being about "people like me" (sometimes referred to, only half-jokingly, as "mesearch").

Social scientists likely have some tendencies to avoid advancing data and theories that portray their own social groups unfavorably or to create obstacles for others who do. This will not always be the case because there may be competing motives for why people might want to perceive different groups in different ways (e.g., men might be more strongly motivated to portray women in a positive light than to portray men in a positive light for ideological reasons, protective reasons, or desires to earn female approval), but absent competing motives, scholars are likely motivated to reject findings that portray their own groups in a negative light. This is one reason to support numerous kinds of diversity among scientists, because these preferences cancel out in the broader literature when numerous scientists have competing motives. These self-enhancement tendencies are more likely to create systematic biases in the field if most social scientists fall within one category (i.e., men, heterosexual, liberal, etc.).

Error Management

When faced with complicated information and a noisy environment in which truth cannot be confidently discerned, people have a tendency to favor less costly errors over costlier ones. A classic example found that men have a tendency to overestimate a woman's sexual interest in them because it is costlier to miss out on a mating opportunity than to make an unwanted sexual advance, whereas women have a tendency to underestimate a man's commitment to her because it is costlier for her to risk pregnancy from a man who will abandon her after sex than to miss out on a sexual opportunity from a man who might commit to and support her (Haselton & Buss, 2000).

This is not a motive separate from the others (desires to gain status, avoid ostracization, and self-enhance), but rather one constantly interacting with the other motives. Imagine, for example, that you have run two studies that found interesting and novel pattern X. You decide to run one more study to really solidify your set of studies before submitting for publication, and you fail to replicate your first two studies even though this third study was very similar to the first two. This is an ambiguous piece of new information—you do not know *why* the third study failed to replicate. Maybe the effect is not real. Or maybe, it was because you ran this third study late at night or toward the end of the semester or because the first two studies used up all the conscientious subjects in the subject pool. In the first case, you miss out on a publication and have wasted time and perhaps money running studies that will never be published. In the latter cases, you can—with justification to yourself—file drawer your third "flawed" study and move forward with just the two. (In such a situation, the right thing to do would be to run a fourth study to test which of your hypotheses about your own findings is correct, but some scholars would not want to risk confirming that the first two studies were flukes.)

Motivated Research in Practice

Thus far we have explained why the social sciences create an environment ripe for motivated research and why scholars will often have preferences for certain kinds of conclusions over others—occasionally, though not always—to the detriment of accuracy. But how might motivated research work in practice?

Selective Exposure and Selective Avoidance

At the information recruitment stage, people have a tendency to seek out information that confirms their desired conclusions and avoid information that challenges their desired conclusions (DeMarree et al., 2017; Frimer et al., 2017; Stroud, 2010). These are referred to as selective exposure and selective avoidance, respectively. Although such patterns are often explored in media consumption among everyday people (Stroud, 2010), scholars likely engage in selective exposure in selecting which articles to read. But people also engage in selective exposure by creating social information environments that are likely to deliver information that confirms their desired beliefs, by surrounding themselves with other people who share their cherished beliefs (McPherson et al., 2001) both in person (Motyl et al., 2014) and on social media (Bakshy et al., 2015). In academia, scholars likely "follow" the scholarly and social media outputs of particular scholars whose research and research interests support their own research agenda. Further, one novel source of selective exposure among academics lies within their ability to *create* information that supports particular conclusions. By selecting certain materials and methods that they believe are most likely to confirm their hypotheses and avoiding the use of materials and methods they have less confidence in, they can often generate their own confirmatory information.

Consequently, scholars will be more aware of information that supports their preferred hypotheses than information that challenges it, making their hypotheses appear more plausible and correct than perhaps a more balanced understanding of the literature would predict. Such tendencies would be particularly problematic for review papers, as scholars likely overrepresent information consistent with their theory and underrepresent contradictory information. These same tendencies can happen with editors and reviewers, who may have imbalanced information about the phenomenon under investigation. If the reviewers have the same blind spots as the authors, they will be unable to point these out. Given the aforementioned and discussed ideological lopsidedness of social science disciplines, blind spots are likely more common than many in the field are willing to concede.

Selective exposure and avoidance can therefore create biased citation patterns, which can continue to perpetuate biased understandings of different domains of research (for recent examples, see Honeycutt & Jussim, 2020). If scholars have preferences for certain conclusions, scholars will be more aware of those findings and thus more likely to cite those findings, and then those highly cited articles become accepted as the authority on the particular issue. Discordant findings, in contrast, are ignored and eventually forgotten. Ideally, the findings in these highly cited articles are valid, and the relevant knowledge improves theory and applications. But, if biased citation patterns result in the canonization of invalid findings, this can produce a reign of error (Honeycutt & Jussim, 2020) whereby socially desirable, but nonetheless flawed, work is propped up to reflect the field's general knowledge. This, in turn, creates dynamics and crises of confidence such as those that have stemmed from psychology's replication crisis. Under these dynamics, biased citation patterns can also contribute to ignoring valid findings, which produces a loss in understanding and deprivation of relevant knowledge. Science strives to be self-correcting, but if invalid findings are canonized and continue to be highly cited, and valid findings (e.g., failed replications) go ignored, self-correction does not occur.

If scholars can acknowledge these tendencies in themselves, they should be motivated to overcome them. A biased awareness of extant data will make it harder to generate hypotheses that are likely to be confirmed by data collection. Exposing oneself to unpalatable information will help scholars identify dead-end hypotheses before they sink time and money into testing them.

Motivated Skepticism and Credulity

Once people are exposed to information (whether they sought it out or could not avoid it), they engage in motivated skepticism and credulity, or the tendencies to be highly skeptical and critical of discordant information and relatively credulous and uncritical of concordant information (e.g., Ditto & Lopez, 1992; Taber & Lodge,

2006). For example, people are more critical of the methods of a scientific study when the results come to an inconvenient conclusion than the *same methods* when the results come to a preferred conclusion (Lord et al., 1979). This can also be conceptualized as a selective call for rigor, whereby one rejects work they do not like on supposedly scientific grounds, but then fail to apply those same standards to work they do like or agree with (Honeycutt & Jussim, 2022). People also make more mistakes with both numeric and logical reasoning when conclusions are discordant (Gampa et al., 2019; Kahan et al., 2017). Among scholars, this likely happens both in evaluations of one's own findings and in evaluations of others' findings in peer review, acceptance into conferences, awards, decisions to cite, and decisions to hire.

Running experiments on the peer review process can be difficult with tightly controlled methods, but there have been a couple of examinations that have found that reviewers tended to judge research as higher quality when the findings supported their prior beliefs and theoretical orientations than when the findings challenged their prior beliefs and theoretical orientations (Koehler, 1993; Mahoney, 1977). This suggests scholars may evaluate research more leniently when findings support their own research agendas. Some research has identified how personal values interfere with the human subjects review process (Ceci et al., 1985). Similarly, research suggests that ideological and moral concerns influence scholars' evaluations of research (Abramowitz et al., 1975) and perhaps even their understanding of empirical reality. For example, von Hippel and Buss (2017) found that social psychology professors were more likely to believe that women could have evolved to be more verbally talented than men than that men could have evolved to be more mathematically talented than women. To our knowledge, there is no legitimate scientific reason to believe that one evolved gender difference is more plausible than the other, which suggests their beliefs may be partially motivated by ideological or moral concerns. Moreover, some scholars even openly admit that they would discriminate against conservative research and conservative scholars (Honeycutt & Freberg, 2017; Inbar & Lammers, 2012; Peters et al., 2020).

Other extraneous concerns influence the reviewing process as well. For example, conference submissions from more prestigious scholars and institutions are evaluated more favorably in single-blind than double-blind reviews, which suggests that either scholars are using a heuristic about prestige and quality or that perhaps scholars are hesitant to give negative evaluations to people and institutions with high status (Tomkins et al., 2017). Such biases, sometimes also referred to as an eminence obsession (Vazire, 2017), are likely quite common in reviews of peers and research because, as noted above, there is a great deal of noise and ambiguity in evaluating the quality of work. Some scholars have pointed out that the interrater reliability of peer review is barely above chance (Lee et al., 2013). On the one hand, this suggests the possibility that editors are selecting diverse reviewers with different strengths and perspectives, which in many cases could help cancel out systematic biases. On the other hand, it is a reminder that scientific evaluations—even among experts—are not perfectly objective, and that features of the reviewers influence the perceived quality of science, not merely the science itself.

One report found that reviewer agreement on funding applications was higher for low-scoring applications than for top-scoring applications (Gallo et al., 2016). Differentiating between a handful of top candidates is likely more subjective—all the top candidates are high quality, so there is no clear "accurate" or "best" decision, and thus extraneous concerns of the individual scholars have greater influence on their evaluations. Given how frequently scholars are differentiating between high-quality content for limited outcomes (journal and conference acceptances, awards, hiring), many of these important decisions that determine scholars' success depend on the idiosyncratic motivations of the reviewers and committee members (so long as applicants reach a certain quality threshold to be considered in the first place). Of course, scholars understand this, and that is why such decisions are usually made among multiple people. This strategy will be more useful when the panel of decision-makers have diverse motivations and preferences, for example, different theoretical and ideological orientations.

Some scholars have contended that these biased information processes are more likely to occur among "experts" or the cognitively sophisticated (e.g., Kahan, 2012; Kahan et al., 2012). People who are more cognitively sophisticated or more knowledgeable would be better able to justify their own biased reasoning processes to themselves and to other people, and thus could get away with more bias than less sophisticated people. Other scholars have challenged this hypothesis, finding that greater cognitive sophistication is instead associated with converging toward accuracy (McPhetres & Pennycook, 2019). Future research will shed more light on these patterns. It may be that expertise and cognitive sophistication simultaneously increase motivated reasoning and ability to detect accurate patterns (and perhaps motivation to detect accurate patterns), and so in some cases scholars will be more biased than the average person and in other cases, less. There also could be individual differences in whether people tend to "use" their cognitive sophistication more to approach accuracy or to advance their own interests. At minimum, there is no strong evidence that experts and those high in cognitive sophistication are immune to biases.

The Protective Powers of Science

Although scientists themselves are but mere mortals, the *institution* of science can mitigate against scientists' human fallibilities. Peer review requires scholars to convince two to five other scholars who do not (necessarily) share the same motives of the scientist and thus who are not particularly motivated to enhance the importance or quality of whatever manuscript they are reviewing. Sometimes these peers are actually competitors (there is only so much journal space), and so, in some cases, reviewers might be strongly motivated to find flaws, which requires authors to be particularly impressive (although, this could also incentivize *p*-hacking to generate impressive results).

The (mostly) shared mission of seeking true and accurate information incentivizes truth and accuracy-seeking in scientists. All else equal, scholars would prefer to forward *true* impressive results rather than *false* impressive results, because both contribute to status, but the latter creates reputational risk of being discovered as a phony. Scholars likely feel some shame and embarrassment when their own theories fail to hold up and their findings fail to replicate, and much more shame and humiliation when they are caught in outright fraud. Science has created a culture in which the social response to indicators of dishonest research practices likely disincentivizes the most obvious transparent forms of data manipulation and fraud. However, it remains unclear whether that culture has disincentivized more subtle influences and tactics (e.g., using positions of power, such as organization leadership roles and journal editorships, to benefit one's own and one's friends' research and careers).

The Open Science movement has also done a lot to mitigate motivated research, primarily through incentivizing transparency. It is now much more difficult to get away with certain dishonest or questionable research practices. Preregistering studies binds scholars to distinguishing transparently their initial hypotheses from post hoc fishing expeditions and to their methods and analysis plan, and requires them to indicate when they deviate. Making data publicly available is a big step toward transparency, and likely increases accountability for data tampering. The new "replication movement" has created an atmosphere where all scholars must consider the possibility or probability that some other scholar will try to replicate their findings. This might render scholars more hesitant to publish papers they have little confidence in, because the status and esteem reward could be short-lived and the long-term consequences of work failing to replicate or being labeled a fraud could be far costlier. However, it may be years before other scholars attempt to replicate one's work, let alone publish it, so that the short-term rewards of publishing may still overwhelm the costs of others failing to replicate, which might not be felt for a very long time. By that time, the original researcher may be a full professor with a large grant portfolio, lots of graduate assistants, and a *New York Times* bestselling book.

But We Can Do Better

Science has an impressive history of generating accurate information over long stretches of time (i.e., converging upon truth), but most of this progress was made by scientists being completely wrong for long periods of time (young Earth, bleeding to cure illnesses, spontaneous generation of life, all of which were believed for centuries). Some norms of scientific practice in psychology are improving and we hope replication rates in the future will confirm that these new procedures are effective at minimizing researcher degrees of freedom to pursue preferred results and effective in generating a more reliable body of information. But, there are many problems these norms either do not help at all or help only very little.

File Drawering

Open Science practices do very little to minimize file drawering. Depending on preregistration platform, even preregistered studies can be file drawered without notice. One solution would be to ask scholars to declare in their papers that they have no file drawer studies or any other studies that tested the same hypothesis tested in the paper. Of course, scholars could still lie, but requiring an explicit, public, and published declaration of the lie turns the act of omission into an act of commission, which could create additional psychological barriers. If it is discovered that there were other studies, this act can be considered active fraud rather than a more ambiguous questionable practice. This also increases the likelihood that at least one co-author on a multi-authored paper would object to the outright lie.

There are also selfish reasons not to file drawer your own studies. When scholars file drawer, they inflate their own effect sizes. If and when there are replication attempts, and the findings either fail to replicate or have smaller effect sizes, this will raise suspicion. The more surprising findings are, the more likely it is that other scholars will attempt to replicate the findings, and so by exaggerating the size of one's own effect, scholars likely increase the odds that they will be caught and viewed with suspicion by peers.

Updated Replication Tracking

A more laborious, but perhaps beneficial, strategy would be for journals to include replication sections on their journal pages for each article where scholars can link their successful or failed replications of the published study and code their own replication study as failed, successful, or ambiguous/semi-successful. Published studies could then have a live "replication score" attached to them that is easily visible to other scholars who read those published articles. This would help scholars know whether they should take a particular study seriously when theorizing, developing hypotheses, and deciding whether to conduct further replication attempts. '

A replication tracking system within the journals might, over the long run, influence the reputation of a journal, and, therefore, incentivize editors to publish robust science rather than flashy science. Such a system might also disincentivize authors from publishing science they have little confidence in because their publication could end up being flagged with a low replication score, which would be embarrassing. This would also provide a greater incentive to those scholars who do fail to replicate a particular study to write up their failed replications. Their replications would be more visible to other scholars interested in the particular effect (rather than buried on some other website) and thus increase the chances that they will at least receive citations for their work (if not publications). This, in turn, would also make it much easier for scholars who wish to conduct meta-analyses to detect successful and failed replications.

Review Papers

Review papers are often highly cited and help solidify many broader theories and ideas in the social sciences that are then used by other scholars to generate new hypotheses. Therefore, more than sets of experiments, it is important that scholars get review papers right so they do not waste the time and resources of their peers. Yet Open Science practices do little to help review papers be more accurate and portray the full range of relevant data (rather than a biased subset).

One corrective for review papers would be for editors and reviewers to require explicit and clearly labeled sections containing a mini review of findings that are inconsistent with the present theory or hypothesis. If the scholars know of no research that contradicts their hypothesis, they could be required to say this explicitly in their paper. This should incentivize them to do a dedicated search of inconsistent findings so that other scholars do not accuse them of being unfamiliar with the topic even after writing a review.

Such papers could also be required to include statements of falsification. If they present a new theory, it will be important to know not only what it explains or what it predicts or the conditions under which such predictions apply, it will be crucial to know what would falsify the theory's predictions. If stereotypes are declared to be the default basis of person perception (Fiske & Neuberg, 1990), how would we know if this was wrong? Would it be falsified by evidence showing powerful individuating information effects? Weak biasing effects? Easily eliminated biasing effects? Even better, scholars could be required to identify the most severe test of the hypothesis—that is, the test that would be *most likely* to detect the falsity of the theory or finding under investigation (O'Donohue, 2021). Theories that generate non-falsifiable predictions are plausibly considered non-scientific, so that one means of elevating the scientific credibility and validity of psychological science would be to articulate explicit statements of what it would take to falsify a theory.

Evaluations of the *quality* of the evidence and not just the presence/absence or even size of some effect of phenomenon would also be valuable, as is common in Cochrane reviews (the gold standard in medical research). Do studies have large or small N's? Are they experimental or non-experimental? Are they preregistered or not? If so, did they follow the preregistration closely or not? The reviews could also use the *new forensics* (*p*-curves, *R*-Index, etc.) to evaluate the quality of the evidence they reviewed (Bartoš & Schimmack, 2020; Simonsohn, 2015; Simonsohn et al., 2014a; Simonsohn et al., 2014b). Evaluations of the quality of the evidence might reduce the wild overclaiming that has characterized so many conclusions in social psychology for decades (Jussim et al., 2016).

Academic Reviewing, Gatekeeping, and Data Suppression

Scholars have little ability to criticize the gatekeepers in academia. Calling attention to the flaws of reviewers or editors risks alienation, making it more, not less, difficult to get one's work published and funded. Similarly, accusing a hiring committee, conference committee, or award committee of bias would violate norms in the field; and generally, it is difficult or impossible to know or prove when another scholar is supporting or opposing a particular finding or scholar for non-accuracy reasons. Consequently, there are almost no ways to hold the gatekeepers of academia accountable. Although accountability to reviewers constitutes a check on author biases, there is no comparable check on reviewers or editors' biases.

Open peer review is, however, one way to mitigate some of those biases. Reviews, with or without reviewer identifying information can be publicly posted. Therefore, the entire scientific community at least has the opportunity to evaluate for itself whether a set of reviews are themselves valid and whether their evaluations of a paper have been fair. As public information, it might even become possible for authors to criticize reviews without fear of retaliation.

One growing trend in academia are mob demands to retract papers that have already passed peer review. Unless fraud or statistical errors that change the conclusions are detected, these are data suppression attempts, usually in the service of some moral or political goal (Stevens et al., in press). Attempts to suppress research by mob can be plausibly interpretable as inability to refute the work—because if the work could be refuted, the solution would be to publish the supposed refutation and allow readers to judge for themselves which is stronger. Our view is that *building* the discussion, rather than erasing it, is far more likely to advance science. Nonetheless, many journals and editors may feel extreme pressure to give in to such demands because they fear their own reputation or the reputation of the journal. And numerous recent examples exist attesting to this trend (described at length in Honeycutt & Jussim, 2022, and Jussim, 2020).

To guard against mob retraction demands, journals should have explicit guidelines for when they will consider retractions. We recommend the Committee on Publication Ethics' guidelines (https://publicationethics.org/retraction-guidelines), which include unreliable findings resulting from major errors or fabrication of data, plagiarism, redundant publications, unauthorized material or data, copyright infringement, research that violated ethics, compromised peer review, or failure to disclose competing interests. Journals should adhere to their guidelines without exception, thus disincentivizing calls for retraction based on other concerns such as alternate explanations, concerns about possible moral implications of the data, or methodological weaknesses. With the rise of retraction-by-outrage-mob procedures, it would be especially useful for journals whose policy is to retract only in cases of fabrication or massive data errors to explicitly state in their instructions to authors that "under no conditions will we retract a paper that has passed peer review and been accepted for publication, or published, on grounds other than those articulated here, no matter how many people sign petitions or open letters or send us outraged emails to do so."

Journals, of course, could create their own guidelines, including, for example, "public concerns about the moral implications." This would then signal that they are willing to retract papers that are objected to by outrage mobs, even when they have passed peer review. This would permit scholars to make their own decisions about whether to publish in journals with retraction policies that violate their own standards for science. Just as transparency will improve the work of authors, it will also improve the work of journals and editors.

Use of Strong Theories

The idea that human minds and human behavior are the product of evolution and thus what humans think and do should generally promote reproductive success is one of the few broad theories within the social sciences that has withstood substantial criticism and has been very useful for generating countless other more specific hypotheses (Lewis et al., 2017). If, for example, a particular finding seems inconsistent with natural and sexual selection in human cognition and behavior, skepticism would be warranted. Some may contend that any hypothesis or finding could be made consistent with an evolutionary account, but we doubt this is so. For example, Freud's Oedipus complex, or the idea that young boys would have sexual desire for their mothers and jealousy and hostility toward their fathers, makes almost no sense at all from an evolutionary perspective (e.g., Daly & Wilson, 1990). Using strong theories, those which we can have high confidence in, can help scholars generate better hypotheses, which is advantageous both for scientists and scientific progress.

Adversarial Collaborations

Working with others with whom you disagree might be temporarily unpleasant, but it will make you a better scientist (see, e.g., Bateman et al., 2005; Mellers et al., 2001). Those who disagree with you have a different perspective, and possibly different motives and biases, that can help cancel out systematic error in your own work. Adversarial collaborations require scholars to articulate their hypothesis in a clear and testable way, understand their adversary's hypothesis as their adversary understands it (and not as a caricature), identify actual points of disagreement (rather than imagined ones based on caricatures), and generate methods that could differentiate between the two hypotheses and feasibly falsify either hypothesis. These kinds of collaborations constrain researcher degrees of freedom because adversaries will not approve of methodological approaches that provide (even if unintentionally) rigged tests of hypotheses or that appear designed to confirm a scholar's hypothesis.

They also have greater potential to advance debates and change minds. Because scholars commit to a methodological approach before testing their competing hypotheses, this minimizes scholars' ability to post-hoc criticize methods, explain away unexpected results, and file drawer undesired results. Third parties should be more persuaded by results of adversarial collaborations knowing that a scholar who made opposite predictions stands behind the methods and findings.

Although adversarial collaborations might feel like an unnecessary constraint in the short term, it will likely improve research in the long run (Ellemers et al., 2020). If your hypothesis is correct, it likely will win out in an adversarial collaboration. If it is incorrect, likely it will eventually be falsified regardless of whether you discover this on your own in an adversarial collaboration or whether other scholars discover this in failed replications or failed conceptual replications. Delaying the inevitable by refusing to participate in adversarial collaborations only risks wasting more time and money and lowering the ratio of science that *will* withstand the test of time.

Nonetheless, adversarial collaborations can also be quite difficult. Especially if adversaries have been openly hostile with one another, forging the cooperative bonds needed to work together on a project may be a bridge too far. Even without personal antipathy, however, bridging differences in assumptions, perspectives, and motives can be a formidable and effort-intensive task. We all have only limited time and resources, and, sometimes, such a project may not be viewed as worth the effort.

On the other hand, we can also imagine a scientific world in which adversarial collaborations were incentivized, thereby rewarding researchers who succeed at bridging these divides. Given the higher confidence we can have in the findings resulting from adversarial collaborations, editors and reviewers should consider these a methodological strength, similar to preregistrations, large sample sizes, and meta-analyses. Given the self-discipline and commitment to rigor required to participate in adversarial collaborations, such efforts should be rewarded by other scholars when making hiring, funding, and award decisions. Participation in these collaborations indicates that a scientist is committed to truth-seeking rather than in advancing flashy results that may not hold up to higher scrutiny.

Reward Rigor

Grants and awards in the social sciences should prioritize scholars who produce robust effects—those that are reliable and replicable and stand up to severe scrutiny. Truth-telling and rigor should be prioritized over flash, drama, novelty, counter-intuitiveness, and supposedly easy solutions to complex problems. By providing resources to researchers who produce findings that are true, powerful, and robust, psychology will wander down far more scientifically productive paths than if it follows every bright shiny object that shows up flashing $p < 0.05$ and a compelling narrative. Of course, *sometimes*, even those findings *will* be flashy or dramatic. But flash and drama should not detract from the value of work, and might be valuable add-ons, *if* that work was conducted in such a manner as to lead to high confidence that it is true, powerful, and robust.

The Case for Accuracy Motivations

Scholars have much to gain by forwarding flashy, socially important, self-promotional, group-promotional, timely results. However, in the new era of Open Science, such gains could be short-lived if findings are not also accurate—replicable, with correct interpretations and conclusions. Accepting that we ourselves are humans who are vulnerable to unconscious motivations that influence the ways we conduct science and the conclusions we come to should motivate us to place regulations on ourselves (e.g., refusing to file drawer our own studies, searching for information that challenges our beliefs and hypotheses, working with scholars with whom we disagree). Unfortunately, even when people are presented with reasonably compelling evidence that they might have biases that steer their judgments away from accuracy, they seem unable to recognize these tendencies in themselves (Pronin et al., 2002). If you wish to be the exception to the rule, start not by denying that you are human and prone to biases and motivations, but instead by having a conscience that bravely admits this to yourself.

References

Abramowitz, S. I., Gomes, B., & Abramowitz, C. V. (1975). Publish or politic: Referee bias in manuscript review. *Journal of Applied Social Psychology, 5*, 187–200.

AlShebli, B., Makovi, K., & Rahwan, T. (2020). The association between early career informal mentorship in academic collaborations and junior author performance. *Nature Communications, 11*, 1–8.

Anderson, C., & Kilduff, G. J. (2009). The pursuit of status in social groups. *Current Directions in Psychological Science, 18*, 295–298.

Bakshy, E., Messing, S., & Adamic, L. A. (2015). Exposure to ideologically diverse news and opinion on Facebook. *Science, 348*, 1130–1132.

Baltiansky, D., Craig, M. A., & Jost, J. T. (2021). At whose expense? System justification and the appreciation of stereotypical humor targeting high-versus low-status groups. *Humor, 34*(3), 375–391.

Bartoš, F., & Schimmack, U. (2020). Z-Curve.2.0: Estimating replication rates and discovery rates. *PsyArXiv*. https://psyarxiv.com/urgtn/Gampa

Bateman, I., Kahneman, D., Munro, A., Starmer, C., & Sugden, R. (2005). Testing competing models of loss aversion: An adversarial collaboration. *Journal of Public Economics, 89*(8), 1561–1580.

Baumeister, R. F., Maranges, H. M., & Vohs, K. D. (2018). Human self as information agent: Functioning in a social environment based on shared meanings. *Review of General Psychology, 22*, 36–47.

Blanton, H., Jaccard, J., Klick, J., Mellers, B., Mitchell, G., & Tetlock, P. E. (2009). Strong claims and weak evidence reassessing the predictive validity of the IAT. *Journal of Applied Psychology, 94*, 567–582.

Bowles, S., & Gintis, H. (2004). The evolution of strong reciprocity: Cooperation in heterogeneous populations. *Theoretical Population Biology, 65*, 17–28.

Brady, W. J., Wills, J. A., Burkart, D., Jost, J. T., & Van Bavel, J. J. (2019). An ideological asymmetry in the diffusion of moralized content on social media among political leaders. *Journal of Experimental Psychology: General, 148*, 1802–1813.

Campbell, T. H., & Kay, A. C. (2014). Solution aversion: On the relation between ideology and motivated disbelief. *Journal of Personality and Social Psychology, 107*, 809–824.

Ceci, S. J., Peters, D., & Plotkin, J. (1985). Human subjects review, personal values, and the regulation of social science research. *American Psychologist, 40*, 994–1002.

Clark, C. J., Liu, B. S., Winegard, B. M., & Ditto, P. H. (2019). Tribalism is human nature. *Current Directions in Psychological Science, 28*, 587–592.

Clark, C. J., & Winegard, B. M. (2020). Tribalism in war and peace: The nature and evolution of ideological epistemology and its significance for modern social science. *Psychological Inquiry, 31*, 1–22.

Clark, C. J., Winegard, B. M., & Farkas, D. (2020). *A cross-cultural analysis of censorship on campuses*. Unpublished manuscript.

Corneille, O., & Hütter, M. (2020). Implicit? What do you mean? A comprehensive review of the delusive implicitness construct in attitude research. *Personality and Social Psychology Review, 24*(3), 212–232.

Daly, M., & Wilson, M. (1990). Is parent-offspring conflict sex-linked? Freudian and Darwinian models. *Journal of Personality, 58*, 163–189.

DeMarree, K. G., Clark, C. J., Wheeler, S. C., Briñol, P., & Petty, R. E. (2017). On the pursuit of desired attitudes: Wanting a different attitude affects information processing and behavior. *Journal of Experimental Social Psychology, 70*, 129–142.

Ditto, P. H., Clark, C. J., Liu, B. S., Wojcik, S. P., Chen, E. E., Grady, R. H., … Zinger, J. F. (2019). Partisan bias and its discontents. *Perspectives on Psychological Science, 14*, 304–316.

Ditto, P. H., Liu, B. S., Clark, C. J., Wojcik, S. P., Chen, E. E., Grady, R. H., … Zinger, J. F. (2019). At least bias is bipartisan: A meta-analytic comparison of partisan bias in liberals and conservatives. *Perspectives on Psychological Science, 14*, 273–291.

Ditto, P. H., & Lopez, D. F. (1992). Motivated skepticism: Use of differential decision criteria for preferred and nonpreferred conclusions. *Journal of Personality and Social Psychology, 63*, 568–584.

Doyen, S., Klein, O., Pichon, C. L., & Cleeremans, A. (2012). Behavioral priming: it's all in the mind, but whose mind?. *PloS one, 7*(1), e29081.

Drummond, C., & Fischhoff, B. (2017). Individuals with greater science literacy and education have more polarized beliefs on controversial science topics. *Proceedings of the National Academy of Sciences, 114*, 9587–9592.

Duarte, J. L., Crawford, J. T., Stern, C., Haidt, J., Jussim, L., & Tetlock, P. E. (2015). Political diversity will improve social psychological science 1. *Behavioral and Brain Sciences, 38*, 1–13.

Durkee, P. K., Lukaszewski, A. W., & Buss, D. M. (2020). Psychological foundations of human status allocation. *Proceedings of the National Academy of Sciences, 117*, 21235–21241.

Eitan, O., Viganola, D., Inbar, Y., Dreber, A., Johannesson, M., Pfeiffer, T., … Uhlmann, E. L. (2018). Is research in social psychology politically biased? Systematic empirical tests and a forecasting survey to address the controversy. *Journal of Experimental Social Psychology, 79*, 188–199.

Ellemers, N., Fiske, S. T., Abele, A. E., Koch, A., & Yzerbyt, V. (2020). Adversarial alignment enables competing models to engage in cooperative theory building toward cumulative science. *Proceedings of the National Academy of Sciences, 117*, 7561–7567.

Finnigan, K. M., & Corker, K. S. (2016). Do performance avoidance goals moderate the effect of different types of stereotype threat on women's math performance? *Journal of Research in Personality, 63*, 36–43.

Fiske, S. T., & Neuberg, S. L. (1990). A continuum of impression formation, from category-based to individuating processes: Influences of information and motivation on attention and interpretation. In Advances in experimental social psychology (Vol. 23, pp. 1–74). Academic Press.

Flore, P. C., Mulder, J., & Wicherts, J. M. (2019). The influence of gender stereotype threat on mathematics test scores of Dutch high school students: a registered report. *Comprehensive Results in Social Psychology*, 1–35.

Forscher, P. S., Lai, C. K., Axt, J. R., Ebersole, C. R., Herman, M., Devine, P. G., & Nosek, B. A. (2019). A meta-analysis of procedures to change implicit measures. *Journal of Personality and Social Psychology, 117*, 522–559.

Frimer, J. A., Skitka, L. J., & Motyl, M. (2017). Liberals and conservatives are similarly motivated to avoid exposure to one another's opinions. *Journal of Experimental Social Psychology, 72*, 1–12.

Gallo, S. A., Sullivan, J. H., & Glisson, S. R. (2016). The influence of peer reviewer expertise on the evaluation of research funding applications. *PLoS One, 11*, e0165147.

Gampa, A., Wojcik, S. P., Motyl, M., Nosek, B. A., & Ditto, P. H. (2019). (Ideo) logical reasoning: Ideology impairs sound reasoning. *Social Psychological and Personality Science. 10*(8), 1075–1083.

Grubbs, J. B., Warmke, B., Tosi, J., James, A. S., & Campbell, W. K. (2019). Moral grandstanding in public discourse: Status-seeking motives as a potential explanatory mechanism in predicting conflict. *PLoS One, 14*, e0223749.

Haidt, J. (2001). The emotional dog and its rational tail: A social intuitionist approach to moral judgment. *Psychological Review, 108*, 814–834.

Haselton, M. G., & Buss, D. M. (2000). Error management theory: A new perspective on biases in cross-sex mind reading. *Journal of Personality and Social Psychology, 78*, 81–91.

Hehman, E., Flake, J. K., & Calanchini, J. (2018). Disproportionate use of lethal force in policing is associated with regional racial biases of residents. *Social Psychological and Personality Science, 9*, 393–401.

Honeycutt, N., & Freberg, L. (2017). The liberal and conservative experience across academic disciplines: An extension of Inbar and Lammers. *Social Psychological and Personality Science, 8*, 115–123.

Honeycutt, N., & Jussim, L. (2022). On the connection between bias and censorship in academia. Pre-print. https://doi.org/10.31234/osf.io/4f9va

Honeycutt, N., & Jussim, L. (2020). A model of political bias in social science research. *Psychological Inquiry, 31*, 73–85.

Inbar, Y., & Lammers, J. (2012). Political diversity in social and personality psychology. *Perspectives on Psychological Science, 7*, 496–503.

Jaccard, J. J., Oswald, F. L., Mitchell, G., Tetlock, P. E., & Blanton, H. (2013). Reassessing the predictive power of the race IAT: A new meta-analysis of criterion studies. *Journal of Personality and Social Psychology, 105*, 171–192.

Jost, J. T., Rudman, L. A., Blair, I. V., Carney, D. R., Dasgupta, N., Glaser, J., & Hardin, C. D. (2009). The existence of implicit bias is beyond reasonable doubt: A refutation of ideological and methodological objections and executive summary of ten studies that no manager should ignore. *Research in organizational behavior, 29*, 39–69.

Jussim, L. (2020). *The threat to academic freedom ... from academics*. Retrieved from: https://medium.com/@leej12255/the-threat-to-academic-freedom-from-academics-4685b1705794

Jussim, L., Careem, A., Goldberg, Z., Honeycutt, N., & Stevens, S. (in press). IAT scores, racial gaps, and scientific gaps. In J. A. Krosnick, T. H. Stark, & A. L. Scott (Eds.), *The future of research on implicit bias*. Cambridge University Press.

Jussim, L., Crawford, J. T., Anglin, S. M., Stevens, S. T., & Duarte, J. L. (2016). Interpretations and methods: Towards a more effectively self-correcting social psychology. *Journal of Experimental Social Psychology, 66*, 116–133.

Jussim, L., Krosnick, J. A., Stevens, S. T., & Anglin, S. M. (2019). A social psychological model of scientific practices: Explaining research practices and outlining the potential for successful reforms. *Psychologica Belgica, 59*, 353–372.

Kahan, D. M. (2012). Ideology, motivated reasoning, and cognitive reflection: An experimental study. *Judgment and Decision making, 8*, 407–424.

Kahan, D. M., Peters, E., Dawson, E. C., & Slovic, P. (2017). Motivated numeracy and enlightened self-government. *Behavioural Public Policy, 1*, 54–86.

Kahan, D. M., Peters, E., Wittlin, M., Slovic, P., Ouellette, L. L., Braman, D., & Mandel, G. (2012). The polarizing impact of science literacy and numeracy on perceived climate change risks. *Nature Climate Change, 2*, 732–735.

Kerr, N. L. (1998). HARKing: Hypothesizing after the results are known. *Personality and social psychology review, 2*(3), 196–217.

Koehler, J. J. (1993). The influence of prior beliefs on scientific judgments of evidence quality. *Organizational Behavior and Human Decision Processes, 56*, 28–55.

Kopko, K. C., Bryner, S. M., Budziak, J., Devine, C. J., & Nawara, S. P. (2011). In the eye of the beholder? Motivated reasoning in disputed elections. *Political Behavior, 33*, 271–290.

Kunda, Z. (1990). The case for motivated reasoning. *Psychological Bulletin, 108*, 480–498.

Langbert, M. (2018). Homogenous: The political affiliations of elite Liberal arts college faculty. *Academic Questions, 31*, 186–197.

Lee, C. J., Sugimoto, C. R., Zhang, G., & Cronin, B. (2013). Bias in peer review. *Journal of the American Society for Information Science and Technology, 64*, 2–17.

Lewis, D. M., Al-Shawaf, L., Conroy-Beam, D., Asao, K., & Buss, D. M. (2017). Evolutionary psychology: A how-to guide. *American Psychologist, 72*, 353–373.

Lilienfeld, S. O. (2015). Lack of political diversity and the framing of findings in personality and clinical psychology. *Behavioral and Brain Sciences; New York, 38*, n/a.

Lord, C. G., Ross, L., & Lepper, M. R. (1979). Biased assimilation and attitude polarization: The effects of prior theories on subsequently considered evidence. *Journal of Personality and Social Psychology, 37*, 2098–2109.

Mahoney, M. J. (1977). Publication prejudices: An experimental study of confirmatory bias in the peer review system. *Cognitive Therapy and Research, 1*, 161–175.

Maranges, H. M., Hasty, C. R., Maner, J. K., & Conway, P. (2021). The behavioral ecology of moral dilemmas: Childhood unpredictability, but not harshness, predicts less deontological and utilitarian responding. *Journal of Personality and Social Psychology, 120*(6), 1696–1719.

McPherson, M., Smith-Lovin, L., & Cook, J. M. (2001). Birds of a feather: Homophily in social networks. *Annual Review of Sociology, 27*, 415–444.

McPhetres, J. (2019). A perspective on the relevance and public reception of psychological science. *Collabra: Psychology, 5*, 34.

McPhetres, J., & Pennycook, G. (2019). *Science beliefs, political ideology, and cognitive sophistication*. Unpublished manuscript.

Meehl, P. (1990). Why summaries of research on psychological theories are often uninterpretable. *Psychological Reports, 66*, 195–244.

Mellers, B., Hertwig, R., & Kahneman, D. (2001). Do frequency representations eliminate conjunction effects? An exercise in adversarial collaboration. *Psychological Science, 12*, 269–275.

Mitchell, G. (2018). Jumping to conclusions: Advocacy and application of psychological research In Crawford, J. T., & Jussim, L. (Eds.), The politics of social psychology. New York: Psychology Press.

Motyl, M., Iyer, R., Oishi, S., Trawalter, S., & Nosek, B. A. (2014). How ideological migration geographically segregates groups. *Journal of Experimental Social Psychology, 51*, 1–14.

Munro, G. D., Lasane, T. P., & Leary, S. P. (2010). Political partisan prejudice: Selective distortion and weighting of evaluative categories in college admissions applications. *Journal of Applied Social Psychology, 40*, 2434–2462.

Munro, G. D., Weih, C., & Tsai, J. (2010). Motivated suspicion: Asymmetrical attributions of the behavior of political ingroup and outgroup members. *Basic and Applied Social Psychology, 32*, 173–184.

Nosek, B. A., Aarts, A. A., Anderson, J. E., Kappes, H. B., & Open Science Collaboration. (2015). Estimating the reproducibility of psychological science. *Science, 349*, aac4716–aac4716.

O'Donohue, W. (2021). Are psychologists appraising research properly?: Some Popperian notes regarding replication failures in psychology. *Journal of Theoretical and Philosophical Psychology, 41*, 233.

Ouwerkerk, J. W., Kerr, N. L., Gallucci, M., & Van Lange, P. A. M. (2005). Avoiding the social death penalty: Ostracism and cooperation in social dilemmas. In K. D. Williams, J. P. Forgas, & W. von Hippel (Eds.), *Sydney symposium of social psychology series. The social outcast: Ostracism, social exclusion, rejection, and bullying* (pp. 321–332). Psychology Press.

Paluck, E. L., Porat, R., Clark, C. S., & Green, D. P. (2021). Prejudice reduction: Progress and challenges. *Annual Review of Psychology, 72*, 533–560.

Pashler, H., Rohrer, D., & Harris, C. R. (2013). Can the goal of honesty be primed?. *Journal of Experimental Social Psychology, 49*(6), 959–964.

Payne, B. K., Vuletich, H. A., & Lundberg, K. B. (2017). The bias of crowds: How implicit bias bridges personal and systemic prejudice. *Psychological Inquiry, 28*, 233–248.

Peters, U., Honeycutt, N., Block, A. D., & Jussim, L. (2020). Ideological diversity, hostility, and discrimination in philosophy. *Philosophical Psychology, 33*(4), 511–548.

Pew. (2020). Science and scientists held in high esteem across global publics. Retrieved on 25 Nov 2020 from https://www.pewresearch.org/science/2020/09/29/scientists-are-among-the-most-trusted-groups-in-society-though-many-value-practical-experience-over-expertise/

Prentice, D. A. (2012). Liberal norms and their discontents. *Perspectives on Psychological Science, 7*, 516–518.

Pronin, E., Lin, D. Y., & Ross, L. (2002). The bias blind spot: Perceptions of bias in self versus others. *Personality and Social Psychology Bulletin, 28*, 369–381.

Pursur, H., & Harper, C. (2020). *Low system justification drives ideological differences in joke perception: A critical commentary and re-analysis of Baltiansky et al. (2020)*. Unpublished manuscript.

Reber, A. S. (1989). Implicit learning and tacit knowledge. *Journal of Experimental Psychology: General, 118*, 219–235.

Reinero, D. A., Wills, J. A., Brady, W. J., Mende-Siedlecki, P., Crawford, J. T., & Van Bavel, J. J. (2020). Is the political slant of psychology research related to scientific replicability? *Perspectives on Psychological Science, 15*, 1310–1328.

Rothman, S., Lichter, S. R., & Nevitte, N. (2005). Politics and professional advancement among college faculty. *The Forum, 3*, 2.

Schimmack, U. (2021). The implicit association test: A method in search of a construct. *Perspectives on Psychological Science, 16*(2), 396–414.

Sedikides, C., & Gregg, A. P. (2008). Self-enhancement: Food for thought. *Perspectives on Psychological Science, 3*, 102–116.

Simmons, J. P., Nelson, L. D., & Simonsohn, U. (2011). False-positive psychology: Undisclosed flexibility in data collection and analysis allows presenting anything as significant. *Psychological Science, 22*, 1359–1366.

Simonsohn, U. (2015). Small telescopes: Detectability and the evaluation of replication results. *Psychological Science, 26*, 559–569.

Simonsohn, U., Nelson, L. D., & Simmons, J. P. (2014a). P-curve: A key to the file-drawer. *Journal of Experimental Psychology: General, 143*, 534–547.

Simonsohn, U., Nelson, L. D., & Simmons, J. P. (2014b). P-curve and effect size: Correcting for publication bias using only significant results. *Perspectives on Psychological Science, 9*, 666–681.

Sisk, V. F., Burgoyne, A. P., Sun, J., Butler, J. L., & Macnamara, B. N. (2018). To what extent and under which circumstances are growth mind-sets important to academic achievement? Two meta-analyses. *Psychological Science, 29*, 549–571.

Stevens, S. T., Jussim, L., & Honeycutt, N. (2020). Scholarship suppression: Theoretical perspectives and emerging trends. *Societies, 10*, 82.

Stewart-Williams, S., Thomas, A., Blackburn, J. D., & Chan, C. Y. M. (2021). Reactions to male-favoring vs. female-favoring sex differences: A preregistered experiment. *British Journal of Psychology, 112*(2), 389–411.

Stroud, N. J. (2010). Polarization and partisan selective exposure. *Journal of Communication, 60*, 556–576.

Sumner, P., Vivian-Griffiths, S., Bolvin, J., Williams, A., Bott, L., Adams, R., Chambers, C. D., et al. (2016). Exaggerations and caveats in press releases and health-related science news. *PloS One, 11*, e0168217.

Taber, C. S., & Lodge, M. (2006). Motivated skepticism in the evaluation of political beliefs. *American Journal of Political Science, 50*, 755–769.

Tetlock, P. E., Kristel, O. V., Elson, S. B., Green, M. C., & Lerner, J. S. (2000). The psychology of the unthinkable: Taboo trade-offs, forbidden base rates, and heretical counterfactuals. *Journal of Personality and Social Psychology, 78*, 853–870.

Tomkins, A., Zhang, M., & Heavlin, W. D. (2017). Reviewer bias in single-versus double-blind peer review. *Proceedings of the National Academy of Sciences, 114*, 12708–12713.

Vazire, S. (2017). Our obsession with eminence warps research. *Nature News, 574*(7661), 7.

von Hippel, W., & Buss, D. M. (2017). Do ideologically driven scientific agendas impede the understanding and acceptance of evolutionary principles in social psychology. In J. T. Crawford & L. Jussim (Eds.), *Frontiers of social psychology series: The politics of social psychology* (pp. 7–25). Routledge.

Winegard, B. M., Clark, C. J., Hasty, C., & Baumeister, R. F. (2018). *Equalitarianism: A source of liberal bias*. Unpublished manuscript.

Zigerell, L. J. (2018). Black and white discrimination in the United States: Evidence from an archive of survey experiment studies. *Research and Politics, 5*, 2053168017753862.

Chapter 4
History of Replication Failures in Psychology

Cassie M. Whitt, Jacob F. Miranda, and Alexa M. Tullett

Abstract In this chapter, we document notable failed replications in psychology. According to a pioneering project conducted by the Open Science Collaboration, less than half of a sample of 100 studies successfully replicated. Since that time, other large-scale replication attempts have echoed the worrisome state of psychological science. Dubbed "the replication crisis," this dilemma in the sciences is two-fold; not only are replication studies rarely conducted but the results of original studies are often difficult to replicate. The crisis has been challenging for psychological science in many ways, but one particular quagmire it has revealed is a body of knowledge potentially fraught with Type I errors (i.e., rejection of a true null hypothesis)—a sentiment some researchers suggested before the crisis even began. The crisis has also functioned to highlight systemic biases and problematic incentive structures in our scientific enterprise, which we will discuss in greater detail later in the chapter. We conclude this chapter with the hopes that learning more about the historical context of the replication crisis helps readers participate in discourse on the subject and motivates them to be active participants in improving psychological science.

Keywords Psychological science · Questionable research practices · Replication failures

This Chapter Starts with a Story about a Cat

In 2007, geriatrician Dr. David Dosa published an article in *The New England Journal of Medicine* titled "A Day in the Life of Oscar the Cat." In this perspective piece, Dosa describes the astounding capabilities of a therapy cat named Oscar who lived among the residents of a nursing and rehabilitation center in Rhode Island. The center primarily cared for patients with end-stage illnesses like Alzheimer's

C. M. Whitt · J. F. Miranda · A. M. Tullett (✉)
University of Alabama, Birmingham, AL, USA
e-mail: atullett@ua.edu

W. O'Donohue et al. (eds.), *Avoiding Questionable Research Practices in Applied Psychology*, https://doi.org/10.1007/978-3-031-04968-2_4

disease, and according to Dosa's report, Oscar possessed the uncanny ability to predict the deaths of residents living there. Stories of the cat's ability to prophesy the passing of a patient began when a staff member noticed that Oscar, who was reportedly not an overly friendly cat, began to persistently hang around patients who passed soon after the cat's appearance. The staff noticed that this happened repeatedly over a period of several months.

We can imagine that after the first patient passed in Oscar's presence, the staff members might have thought nothing of it, or perhaps considered it a coincidence, but then, after it occurred a second time, a third time, and eventually, a twenty-fifth time, they became more confident that what they saw was not a fluke (or a false positive) but rather a phenomenon with some logical explanation—the cat knows when death is imminent. Importantly, however, we may also speculate that the staff members unwittingly failed to register instances in which Oscar did *not* accurately predict someone's death; surely, not every patient with whom he came into repeated contact passed in his presence. In our scientific endeavors, our dealings with feline psychopomps are disappointingly rare, but the consequences of ignoring failed replications are serious; nevertheless, we risk maintaining beliefs that are (possibly wildly) false.

In this chapter, we document notable failed replications in psychology. According to a pioneering project conducted by the Open Science Collaboration (2015), less than half of a sample of 100 studies successfully replicated. Since that time, other large-scale replication attempts have echoed the worrisome state of psychological science (e.g., Klein et al., 2018; Ebersole et al., 2020). Dubbed "the replication crisis," this dilemma in the sciences is twofold; not only are replication studies rarely conducted (Makel et al., 2012) but the results of original studies are often difficult to replicate (Open Science Collaboration, 2015). The crisis has been challenging for psychological science in many ways, but one particular quagmire it has revealed is a body of knowledge potentially fraught with Type I errors (i.e., rejection of a true null hypothesis; Funder et al., 2014)—a sentiment some researchers suggested before the crisis even began (Ioannidis, 2005). The crisis has also functioned to highlight systemic biases and problematic incentive structures in our scientific enterprise (Nosek et al., 2012), which we will discuss in greater detail later in the chapter.

Before delving into the history of replication attempts, it behooves us to clarify what we mean by "replication." Often, people label replications as either direct or conceptual. Direct (also called exact) replications are defined as an attempt to implement a research protocol that is as similar to an original study as possible in terms of materials, procedures, analyses, and sample demographics. Conceptual (also called indirect) replications identify the fundamental hypothesis in an original study, and then test the research question with a novel protocol (e.g., new operationalizations of variables, different study design, demographically distinct sample; Crandall & Sherman, 2016). Although some have contested this distinction and noted that truly "direct" replications are strictly impossible (Brandt et al., 2014; Crandall & Sherman, 2016; Nosek & Errington, 2020), evaluating the implications of a replication often depends on an assessment of the differences between the replication and the original study.

Replication failures are the clear focus of this chapter; however, what scholars mean by "failure" is less clear. Failure to replicate has been defined in a variety of ways, with some scholars noting that the "success/failure" dichotomy obscures the fact that replications may provide inconclusive results (Nelson et al., 2018). Later in this chapter, we will elaborate on different ways one can evaluate a replication.

The replication crisis is a critical topic in the field of psychology across varied sub-disciplines and occupations. Because science is cumulative, and inherently requires psychologists to possess some level of dependence on the expertise of their peers (Oreskes, 2014), addressing the state of the field is imperative for the work that we all do. This is perhaps most apparent for those who work as career researchers but is still at play for those psychologists in applied or clinically based positions. After all, the implementation of evidence-based practices requires trust in the research that goes into developing clinical procedures, and while the replication crisis is often linked most closely to the field of social psychology (Earp & Trafimow, 2015; Open Science Collaboration, 2015), it is important to keep in mind that social psychology's issues are not isolated. Difficulty with replicability, and many of the underlying causes of the problem, are shared with clinical psychology (Tackett et al., 2017, 2019). We assert that there are important lessons to be learned from examining these problems more closely. In particular, one problem that has significantly contributed to the replication crisis is the propensity of researchers to engage in Questionable Research Practices, also referred to as QRPs (John et al., 2012). Broadly, the present chapter will focus on the history of replication failures in psychology and how questionable research practices have contributed to issues with replicability. A relative lack of replications (and possible failures to replicate) may lead to a false belief that clinical psychology has less of a problem compared to other sub-fields.

Psi, Fraud, and the Many Labs to Follow

There are good reasons to be highly skeptical that humans—or felines—can predict the future (Wagenmakers et al., 2011). However, in a paper published in the *Journal of Personality and Social Psychology (JPSP)*, Bem (2011) reported a series of nine experiments providing empirical evidence for *psi*—the ability to predict a future event. In response, many psychologists and journalists asked, "What went wrong?" How could a researcher use the tried-and-true research and statistical methods of the times, yet draw such implausible conclusions (Engber, 2017)?

It is not hard to imagine the ripples Bem's claims caused across the psychological discipline. Other papers in related fields, such as one documenting "voodoo" correlations within neuroscience, questioned if most associations found were spurious (Vul et al., 2009). Meanwhile, Diederek Stapel's famous fraud case—in which he fabricated data for several widely disseminated studies—prompted further scrutiny of the kinds of counter-intuitive findings that were commonplace on the pages of social psychology journals (Borsboom & Wagenmakers, 2012; Verfaellie & McGwin, 2011). For example, Stapel was exposed for faking data corresponding to a study published in *Science* that concluded that trash-filled environments lead to racist tendencies (Bhattacharjee, 2013). Questions started to arise: Does a problem exist in our field? If so, what's the extent of the problem? Should *I* be worried about *my* research? What does it even mean for an effect to successfully replicate?

Ways of Evaluating Replications

When evaluating the success of a replication effort, one could simply assess whether the replication yields a significant effect in the same direction as the original. According to this approach, if an original study showed that a meditation intervention reduced self-reported stress, a successful replication would be one that yields any significant reduction in stress, regardless of effect size. There are some shortcomings of this approach: If the original study reported that the effect size of their intervention was extremely large (e.g., Cohen's $d = 0.9$), and the replication found a statistically significant effect that is very small (e.g., $d = 0.1$), would we say that we have successfully replicated the effect? The answer may depend on the kinds of conclusions one is interested in drawing. If seeing *any* reduction in stress would be considered a success, then this approach is appropriate. More probable, there is some point at which the reduction in stress is so small that it does not justify the time and money invested in the intervention. In this case, significance would be an overly simplistic metric of replication success.

Alternatively, rather than a focus on p-values, one could look at the uncertainty surrounding the point estimates of the effect size to see if the replication's estimated magnitude significantly differs from the original. One could do this by calculating a 95% confidence interval around the replication study's point estimate and seeing if the original effect size estimate was "captured" in that range. If so, this could be interpreted as a successful replication in that the original effect size was not significantly different from that observed in the replication. There are limitations with this approach as well. Replication studies (i.e., the subsequent studies, not the original) are usually extremely well powered, collecting sample sizes in the 1000s (e.g., Klein et al., 2018). With greater power, the more precise and narrower the confidence intervals will be. Thus, even slight deviations between the original effect size and the replication's estimate would be considered significantly different, even though they could be conceptually comparable.

A third way researchers have commonly evaluated and used replication studies is in the form of a meta-analysis. Specifically, one could combine both the original and replication studies estimated magnitudes together to have an aggregated predicted effect size. This combination of multiple studies together may at first glance seem strong, especially with the recent advocation of meta-analytic thinking (Cumming, 2014). Unfortunately, meta-analyses are not immune to bias; the old adage of "garbage in-garbage out" captures concerns that meta-analyses cannot correct for systematic biases in a sample of studies (Nelson et al., 2018). If the original studies have inflated effect sizes, then meta-analysis will still provide a bloated aggregate not reflecting reality. Nelson et al. (2018) described this with the metaphor of combining glasses of juice into a pitcher: if even one of the contributing glasses has been poisoned, the resulting mixture will also be tainted.

These three common markers are not the only approaches to evaluate a replication. In more recent years, a variety of different methods of evaluation have been proposed focusing on detectability capacity of the original study as well as Bayesian inferential analyses (Verhagen & Wagenmakers, 2014; Simonsohn, 2015). There is not currently, and likely will never be, a one-size-fits-all for what counts as a dichotomous success or failure to replicate. Rather, we encourage readers to assess each large-scale replication attempt holistically, using multiple indicators and critically thinking about how (or if) the original and replication studies inform the underlying theory of what is being proposed (Stroebe, 2019; Nosek & Errington, 2020). Ultimately, scientists are not interested in any single study, or protocol, but rather, in developing frameworks for understanding the world. Uri Simonsohn (2016) perhaps stated it best, "Each reader decides if a replication counts."

A History of Notable Replication Projects

Many Labs 1

Many Labs based projects consist of crowdsourcing a large number of researchers, who have access to different samples, and orchestrating them to conduct replications using similar procedures. By running identical studies in numerous contexts, a greater amount of information can be gleaned regarding the factors that influence the effect of interest, and the meta-factors that influence effect sizes generally. Many Labs 1 (ML1) was one of the first large-scale replications conducted concurrently with the Open Science Collaboration's Reproducibility Project: Psychology, or RP: P (Klein et al., 2014; OSC, 2015).

The main contribution of ML1 has little to do with the 13 specific effects that were selected as a target for replication. Instead ML1 accomplished the extraordinarily difficult task of getting hundreds of researchers to work together. ML1 set a precedent for subsequent large-scale initiatives (e.g., Many Labs 2 through 5) and collaborations such as the Psychological Science Accelerator (Moshontz et al., 2018). ML1 serves as a proof of concept that experts could come together, isolate a research question of interest, set parameters for selecting a study to replicate, and coordinate a global research endeavor. Notably, ML1 also served as a catalyst for the growing interest in the field of meta-science. Researchers began asking empirical questions regarding the influence of various methodological characteristics of research studies on replicability. For example, researchers began to investigate the roles of factors, such as lab setting, mode of instruction (e.g., lab vs. online), and the language stimuli were presented in. Arguably, these meta-scientific questions, and the unique insights they can provide, are two of the great strengths of this project.

Ten out of the 13 (77%) effects investigated in ML1 were labeled by the project's authors as "successful" replications of the original work. However, we must consider that the Many Labs projects all explicitly stated that the original studies used were chosen selectively—not randomly. As an example, ML2 studies were chosen based on criteria such as feasibility of implementation through web browser, length (shorter procedures were preferred), number of citations the original study had received (more citations were desired), and general interest in the effect. In order for any study to generalize beyond its sample, random selection that is representative of a population is a key feature that is necessary. Thus, even if all 13 effects had been replicated, we could not conclude that the field of psychology is in an ideal state because the studies replicated were not randomly chosen nor are they representative of the field as a whole. Likewise, even if not a single study replicated in a Many Labs project, we could not generalize that the field is in shambles. In other words, it is important that psychologists do not overgeneralize beyond the scope of the project.

Reproducibility Project: Psychology (RP: P)

Many Labs projects were not designed to provide an estimate of replicability for the field of psychological research. Some researchers posit that this estimate is impossible to calculate, as one is unlikely to obtain a truly random representative sample of all studies across the disciplines in the field (Stroebe, 2019). However, the 2015 RP: P effort aspired to provide at least a rough estimate (Open Science Collaboration, 2015).

Beginning in November 2011, the project began recruitment of 270 authors to attempt to replicate 100 studies. These 100 studies were chosen from three influential journals, with two-thirds having either a social psychology or cognitive science scope and all from the year 2008: *Journal of Experimental Psychology: Learning, Memory, and Cognition* (*JEP: LMC*), *Journal of Personality and Social Psychology* (*JPSP*), and *Psychological Science* (*PSCI*). Although these journals are not representative of the field of psychology as a whole, nor is the work done in 2008 necessarily indicative of later work, this selection of studies was intended to provide at least some insight into the health of the field as a whole.

RP: P was a leading initiative that contributed to our understanding of how to evaluate replication studies. As such, they provided several operationalizations of what counts as successful replication described in the previous section. Across the three journals they sampled from, 35 out of 97 showed a significant effect in the original direction (36%), 47% of the original effect sizes were within the 95% CIs of the replication, and the meta-analytic mean of effect sizes (all transformed to a standardized correlation coefficient) was an $r = 0.31$. The average effect size reported in the original studies was equal $r = 0.40$, whereas the average effect size reported in the replications was an r of 0.20. This led the authors to conclude: "The present results suggest that there is room to improve reproducibility in psychology" (Open Science Collaboration, 2015, p. aac4716-7). Regardless of the replicability estimate used (36% and 47%), these initial rates were seen as cause for concern.

Many Labs 2 and 5

Although some of the criticisms of the RP: P have been described as "completely invalid" (Lakens, 2016), one common point of discussion was the possibility that failed replications might actually be reflecting "hidden moderators" rather than challenging the strength of original effects (e.g., Van Bavel et al., 2016). It is impossible to do a complete replication of an original study; if the replications were implemented using the same staff, in the same lab, using the exact same procedure, there would still be differences between the original study and the replication attempt. For example, college students in the mid-2000s might possess different cultural beliefs and values than college students in the early 2020s. Some other moderators that have been attributed to failed replications include temperature, time of day, and weather. There are also considerations such as demographic composition of the sample or individual differences among research assistants. In short, a prominent critique of the RP: P, and indeed, a recurring argument of almost every unsuccessful replication attempt is: did you consider X moderator?

Fortunately, these concerns can be addressed empirically by putting the scientists under the microscope—one example of the broader practice of scientifically studying scientific practices, or "meta-science." Many Labs 2 and 5 provide some initial answers to these questions (Klein et al., 2018; Ebersole et al., 2020).

Many Labs 2 attempted replications on 28 findings across 125 collected samples with a total of 15,305 research participants covering over 30 countries. Meta-scientific variables were collected from the 125 samples: both traditional demographics (e.g., gender, political ideology, education level, and socioeconomic status), commonly referenced individual differences as potential hidden moderators (e.g., cognitive reflection capacity, Big 5 personality traits, self-esteem, and mood), as well as characteristics of the labs conducting the replication (e.g., some labs conducted the study online only while others had participants come in-person).

What they found was surprising; initial evidence suggested that most of the potential "hidden moderators" that were investigated explained little variation in replicability. In other words, studies had similar replication outcomes regardless of lab characteristics, demographics, and other individual differences. For example, if one lab found weak evidence of an effect, it was likely that other labs, did too, regardless of the sample. Likewise, if one lab found little to no evidence of an effect, then other labs, with different types of demographics, also found little to no effect. Moreover, 27 of the 28 studies had statistically equivalent results regardless of whether surveys were conducted completely remotely (self-administered) or had participants come in person. As mentioned earlier, it is fortunate that the Many Labs projects are well-equipped to address such meta-science questions; less fortunately is that it seems failures to replicate cannot simply be dismissed as replicators not successfully achieving the impossible ideal of a truly "direct" replication.

Many Labs 5 goes one step further and attempts to answer: Does including the original author's input when creating the design of a replication protocol improve replicability? The goal was to address the critique that replicators don't consult the original authors or lack expertise in the content of the original work (Gilbert et al., 2016). This project does not have as clear-cut an answer. Within the RP: P, eleven studies were flagged by the original authors as having potential design flaws in the replication protocol. Many Labs 5 worked with ten of those protocols and randomly assigned labs to one of two conditions: in-house protocols vs. original-author-approved protocols. The goal was to examine if the protocol that was intensively reviewed by the original authors significantly improved the replicability rate compared to protocols whose designs were left up to the replicating labs. These ten were also thought to be a good selection because in the first replication attempt (RP: P), the chosen studies showed supportive evidence of the original study.

What makes the results of Many Labs 5 difficult to interpret is this: Out of the 20 protocols (10 studies × 2 versions)), only 2 found evidence of a significant effect in the same direction as the original. Both of the studies that found supportive evidence of the original effect were in the original-author-approved condition. One could argue that including original authors does improve replicability rates, if we simply compare scores: 2 vs. 0. On the other hand, 2 out of 10 successful replications is not particularly encouraging support for the effects more broadly. This investigation provides ambiguous evidence about the extent to which replication studies benefit from the original authors' input. It also demonstrates that an initial single large-scale replication attempt, such as the RP: P, cannot provide the final word on a topic.

Social Sciences Replication Project

Meta-scientific introspection is not limited to psychology. Many large-scale replications have been conducted in the fields of economics (Camerer et al. 2016), cancer biology (Nosek & Errington, 2020), and experimental philosophy (Cova et al. 2018). As such, we can now consider a more recent evaluation on the overall replicability of social science experiments in the journals of *Science* and *Nature* from 2010 to 2015. Between these two journals 13 out of 21 showed a significant effect in the original direction (62%), 14 out of 21 (66.7%) of the original effect sizes were within the 95% CIs of the replication, and the mean effect size in a replication study was on average half the size of the original study (identical to the effect size reduction reported in RP: P).

Camerer et al.'s (2016) prevalence estimate faces the same limitations of not being generalizable to the field as a whole due to its small sample of replicated studies and the selective, non-random criteria used. Nevertheless, their replication initiative as a whole is contributing information on some of the highest impact journals that should be representative of the most rigorous social science published.

Registered Replication Reports (RRRs)

A registered report (RR) is a relatively new publishing format that runs counter to traditional publishing (Simons et al., 2014; Nosek et al., 2018). Historically, publishing involves researchers conducting a study, writing everything up into a manuscript, and sending that finished manuscript to journals for peer review. At that point, peer reviewers decide whether to accept it for publication or not. RRs represent an alternative in that they require researchers to first develop an idea and study design which they then submit to a journal for peer review. These submitted protocols, which include a literature review, hypotheses, methods, and an analysis plan are known as stage 1 manuscripts. If a stage 1 manuscript is approved, the authors will receive what is called "in-principle acceptance." This means they are guaranteed a publication in the journal, as long as they run the study and write up a final report (known as the stage 2 manuscript).

This general format served as the template for Registered Replication Reports (RRRs), formally defined by the Association for Psychological Science (APS) as, "... a collection of independently conducted, direct replications of an original study, all of which follow a shared, predetermined protocol." These RRRs target important, cornerstone effects (e.g., ego-depletion effects) and explicitly address the concerns that replication attempts are "destructive." The journal *Advances in Methods and Practices in Psychological Science* (AMPPS) publishes RRRs and highlights that RRRs aid in understanding the true size of a selected effect, as well as if it is replicable, robust, and generalizable. AMPPS' acceptance of RRRs creates an incentive for researchers to engage in more replications with the guarantee of publication regardless of the outcome.

All published RRRs involved consulting subject matter experts (typically the original authors) to vet the replication protocol. This engagement creates a culture where scientists are coming together to build knowledge and a shared goal to get at the truth, without accusations of malice or "destructive" intent being attributed to the replicators. Another strength of conducting large-scale replications in an RRR format is that it allows researchers to investigate an effect without the influence of publication bias from the picture. Recall that one way to evaluate replications is through a meta-analytic approach, in which one could aggregate multiple estimates of an effect size to get a more precise idea of how large an effect is. Also recall the main limitation of the meta-analytic approach was "garbage-in, garbage out." The beauty of an RRR is that it meta-analyses all point estimates provided across the participating labs, not including the original study. With the guarantee of publication, the poison of bias has not contaminated the water; thus, RRRs report the meta-analytic criterion of their replication attempts.

As of June 2021, a list of 9 published RRRs (10 including Many Labs 2) are available to the public (Association for Psychological Science, 2014). We provide a table, starting in the year 2014, that lists these RRRs, along with the general conclusions drawn by the replicators (Table 4.1). Readers will note that only the first published report (Alogna et al., 2014) provided a meta-analyzed effect size for the verbal overshadowing effect in the same direction as the original but at only a modest size. The other RRRs have the modal outcome of not being significant (i.e., having zero included in the meta-analyzed confidence interval), or if significant, in the *opposite* direction the original study found. For example, in 2018, an RRR was conducted of Mazar, Amir, & Ariely's (Mazar et al., 2008) study that found that people who were primed with the 10 biblical commandments were less likely to cheat on a task (Verschuere et al., 2018). The RRR which includes 25 labs and a total of 5786 participants found a meta-analytic effect that priming with the 10 commandments lead to a (modest) *increase* in cheating.

Recognizing that most of the published RRRs suggest null to opposite findings for well-known effects, what should we take away? One positive consideration is that this is a clear demonstration of psychology engaging in self-correction. Psychologists are taking the time to pause and evaluate our body of knowledge and consequently losing confidence in effects that do not seem to replicate. From that perspective, we are also cleaning up our literature so future research can be built upon a more solid foundation. Furthermore, RRRs are helping to change how researchers approach replication by incentivizing replications via publication opportunities and dispelling misperceptions that replications are destructive or done with malicious intent. Original authors have the opportunity to respond to any conclusions the replicators report, showcasing the constructive, collaborative nature of RRRs (e.g., Strack, 2016).

That being said, we also cannot ignore the significant number of failures to replicate cornerstone effects. When extremely influential effects such as ego-depletion fail to replicate (Hagger et al., 2016), the results compel us to adopt a more skeptical stance on "classic," popular findings, especially when educating students and communicating to the public. Perhaps the greatest takeaway from the RRRs is that they motivate the community to continue checking in on the health of our field. We will soon take a closer look at the root causes, and possible solutions, to the replication problems that ail us. Before we do though, we will briefly explore the case of one of the most popular effects shared with the public and its failure to replicate: power posing.

Table 4.1 Summary of registered replication reports (RRRs)

Original study	RRR authors	Theory/effect of interest	Conclusions drawn from RRR
Schooler and Engstler-Schooler (1990), Study 4 (RRR1) and Study 1 (RRR2)[a]	Editors of the Association for Psychological Science (2014)	Verbal overshadowing effect on eye-witness testimony	Clear, but weaker evidence effect
Hart & Albarracin (Hart & Albarracín, 2011), Study 3	Eerland et al. (2016)	Influence of grammar on perceived intentionality	Little evidence of original effect, with significant results in the opposite direction
Sripada et al. (2014)	Hagger et al. (2016)	Ego-depletion effect	Little evidence of original effect (as operationalized in the study)
Finkel et al. (2002), Study 1	Cheung et al. (2016)	Priming "commitment" influences forgiveness responses	Little evidence of original effect, with significant results in the opposite direction
Strack et al. (1988), Study 1	Wagenmakers et al. (2016)	Facial feedback hypothesis: Forcing a smile with pen in mouth increase perceived funniness of cartoons	Little evidence of original effect (as operationalized in the study)
Rand et al. (2012), Study 7	Bouwmeester et al. (2017)	Social heuristics hypothesis: People under time constraints are more generous	Little evidence of original effect, with significant results in the opposite direction
Dijksterhuis and van Knippenberg (1998), Study 4	Nelson & O'Donnell et al. (2018)	Priming of professor (compared to "soccer hooligan") improves trivia quiz perform	Little evidence of original effect (as operationalized in the study)
Srull and Wyer (1979), Study 1	Verschuere et al. (2018)	Priming of "hostility" increases perceived hostility of vignette protagonist	Inconclusive evidence, with modest significant effects in similar/opposite directions of the original
Mazar et al. (2008), Study 1	Verschuere et al. (2018)	People provided the opportunity to cheat are less likely to do so if reminded of the 10 biblical commandments	Little evidence of original effect, with significant results in the opposite direction

[a]The first published RRR unintentionally implemented the original study similar to Study 4, and thus has two parts RRR1/RRR2, with the second attempt in the intended presentation of study materials to the order described in Study 1

Power Posing

TED (Technology, Entertainment, and Design) is a nonprofit organization that provides a wide-reaching platform for speakers, including researchers, to share ideas to the lay public. Among the thousands of talks and events, the second most viewed TED talk in the history of the organization, with nearly 62 million views as of 2021, is on power posing (Cuddy, 2012). In this talk, social psychologist Amy Cuddy shares her research on the power posing effect, based in part on a paper she co-authored in 2010 (Carney et al., 2010). In the original power posing study, the idea was that, if one changes their body language to an expansive (open) posture, the physical change leads to increased feelings of power. In turn, these feelings have a downstream effect, which leads to psychological, physiological, and behavioral changes. The crux of the effect is that even if we do not feel confident, there are simple actions capable of providing agency in our lives. This notion captivated the public.

Ranehill et al. (2015) conducted an independent conceptual replication of the power posing effect with a larger sample than the original. Ranehill's group found that the open, expansive poses did increase self-reported feelings of power. However, they failed to replicate the effect on hormonal levels or behavioral tasks. In 2016, Dana Carney, the first author of the original 2010 study, wrote an open letter in which she explicitly stated, "I do not believe that 'power pose' effects are real." She also stated that she does not teach the concept in her classes and encourages fellow researchers to stop pursuing embodied effects (Carney, 2016).

Carney likely did not change her confidence in power posing because of the single attempt by Ranehill's group. Instead, it was likely a revision of her beliefs in the face of mounting evidence. For example, Simmons and Simonsohn (2015) conducted a p-curve analysis, a method of detecting bias in a group of studies and found red flags of a compromised literature. Garrison, Tang, and Schmeichel (Garrison et al., 2016) also attempted an independent replication and failed to find an effect on risk-taking behaviors. *Comprehensive Results in Social Psychology* (CRSP) released a special issue on power posing in an attempt to directly replicate and test possible moderators through seven transparent, pre-registered studies (Cesario et al., 2017). Gronau et al. (2017) published an evaluation in the CRSP issue utilizing a novel Bayesian model-averaged meta-analysis method. Specifically, Gronau's analysis investigated the effect of posing on felt power (not the downstream effects those feelings have on hormones) tested in six of the seven studies and found evidence supporting the original effect. Exploring this further, one variable also measured was the participant's familiarity of power posing, and when excluded, the evidence becomes modest at best. This could suggest that the replicated effect may be a demand characteristic of what participants *think* they should report based on what they have previously heard.

Interestingly, in Carney's (2016) reflection sharing her updated position, she noted some details about the original study: data from 25 subjects were initially analyzed, then re-analyzed as more participants were collected. She also self-disclosed that they selectively reported only some p-values but not others and that participants were excluded as outliers in only specific analyses. The practices Carney described were not unique to her. Rather, they are the same commonplace practices that gave rise to Bem's (2011) previously cited claim of psychic abilities: QRPs.

How Do QRPs Influence Reported Findings?

Considering the reported instances of replication failures discussed above, one question inevitably arises: What causes a failed replication? One potential origin of the problem harkens back to maladaptive incentive structures in the scientific community. For example, scientists are often rewarded for pursuing and publishing "sexy" findings. That is, researching novel, counter-intuitive effects (e.g., washing your hands will wash away guilt; Zhong & Liljenquist, 2006) are often very beneficial for individual researchers. These kinds of attractive studies are more likely to get funded (e.g., the National Health Institute assigns grant applications an "innovation score"), published (Giner-Sorolla, 2012), and disseminated to lay circles via media outlets (e.g., many of the studies that failed to replicate in the Open Science Collaboration (2015) project were widely reported in the media).

This deeply ingrained preference for what is counter-intuitive and trendy has two primary consequences. One consequence is that replication studies, as non-novel forms of research, are rarely incentivized; if the system favors novel findings, why would a career researcher waste time and resources running replication studies that probably won't result in publications? A second consequence is that we have ended up with a body of literature filled with fun, novel effects may not be robust. It is impossible to measure the exact type I error rate in psychology (as previously discussed), but it is reasonable to assume that many of the effects we consider "real" or empirically supported could be statistical flukes. Thus, this problematic incentive, combined with psychology's preoccupation with obtaining statistically significant results, means that researchers face a persistent and ubiquitous pressure to produce novel, significant work. It also means that studies that do not meet these qualifications are deemed less important for the field (e.g., replication studies and studies with null results). We suggest that attempting to achieve research results that possess these qualities is also capable of pushing researchers to engage in QRPs.

Discussion about the issue of QRPs has been ongoing for decades (Crocker, 2011; Marshall, 2000; Sovacool, 2008; Wicherts, 2011), but for psychology, the discussion largely gained attention when John et al. (2012) published a paper outlining some common, but problematic, practices that researchers engage in. Specifically, they investigated the rates of self-reported QRPs among psychologists. Some examples of QRPs the authors asked participants to admit to blatant falsification of data, deciding whether to exclude data after looking at the impact of doing so on the results, failing to report all of a study's dependent measures, and deciding to collect more data after looking to see whether the results were significant. Notably, for most of these QRPs, rates of self-admission were high, and some statistically inferred estimates were close to 100%. For obvious reasons, these results were concerning to the collective field.

It is important to keep in mind when considering QRP rates that most researchers are likely motivated to do good science and want replicability rates to be higher than current estimates indicate. However, QRPs often happen unintentionally. Certain practices, such as continuing to collect data until a significant result was obtained, were arguably the norm as well. However, in light of our new knowledge about the impact of, and varied forms of QRPs, this kind of ignorance has become difficult to defend. We assume that most researchers are probably now aware of QRPs like p-hacking (flexibly analyzing data until the results are significant), data peeking, and HARKing (Hypothesizing After Results are Known). Still, in many cases, QRPs may also be rare (or even singular) occurrences rather than regular practices among researchers (Fiedler & Schwarz, 2015).

Addressing QRPs is important for the pragmatic reason that these practices can cause researchers to waste time and money exploring effects that are not real. The bigger concern is that QRPs forestall replicability by inflating the type I error rate in our literature. That is, QRPs make it much more likely that false positives are pervading the literature, and this has serious consequences for the science of psychology. As alluded to in the introduction of this chapter, a conventional perspective in the philosophy of science is that science is a cumulative enterprise that requires researchers to work together. Consequently, this places psychologists in a position where we must depend on other researchers and experts to provide us with information necessary to help us build theory and generate meaningful hypotheses. Simply put, researchers are required to place trust in their intellectual peers—trust that those peers are producing replicable work.

Examples of QRPs in a Failed Replication

Beyond simply stating that QRPs can lead to failed replication studies, we offer a brief analysis of a psychological effect that is thought to have been influenced by QRPs. Specifically, we will use the ego depletion effect (the idea that self-control is a limited mental resource) to showcase how QRPs work to diminish replicability. Ego depletion is an effect that comes from the social psychology literature and has been the subject of various large-scale replication attempts (see Dang et al., 2021; Hagger et al., 2016). The consensus reached from these studies was that ego depletion effects fail to replicate across multiple operationalizations of the construct (Dang et al., 2021 and Hagger et al., 2016). Metascience experts looking at these replication results have offered the explanation that part of the reason ego depletion effects are declining is because of the way that researchers have attempted to depict the construct in the literature. Researching ego depletion has become an endeavor mostly likely fraught with QRPs (Vadillo, 2019). As a result, some researchers have attempted to use metascience to better understand the extent to which QRPs contributed to an inability to reproduce such a widely studied and well-funded line of research.

One such metascience study conducted by Wolff et al. (2018) investigated rates of both QRPs and replication in ego depletion work. The authors surveyed 1721 ego depletion researchers (i.e., individuals who had previously published work on ego depletion) and found 39.2% were aware that their peers engaged in QRPs (i.e., selectively reporting subgroups, dropping data points based on a gut feeling, rejecting "outliers" without statistical support, excluding data after looking at their impact on results, and selectively reporting outcomes). Furthermore, 37.7% of participants self-admitted to having engaged in those same QRPs themselves. These data provide evidence that failures to replicate ego-depletion effects may reflect the fact that the ego-depletion literature has a high prevalence of false-positives as a consequence of QRPs.

Can Pre-registration Combat QRPs?

What Is Pre-registration?

Looking at the information we have presented on the replication crisis can easily prompt one to become disheartened with psychological science. There are many examples of replication failures and evidence of QRPs is not difficult to spot in many papers. However, we maintain the optimistic perspective that the influence of QRPs can potentially be overcome. Indeed, one positive outcome of intensely highlighting replicability problems in psychological science has been the resulting "credibility revolution" it spurred. This movement, which focuses on improving the methods of psychological science and pushing for more open science practices (Vazire, 2018), has the central aim of addressing—and correcting—issues that contributed to the replication crisis—like QRPs. Some suggestions for accomplishing this include establishing norms for, and promoting, activities like making data and materials publicly available on data repository sites, publishing replication studies, and publishing studies with statistically non-significant results.

Another suggestion that has gained increasing popularity is the idea of public pre-registration. This practice refers to creating an open, time-stamped document that outlines a researcher's a priori predictions and hypotheses prior to data collection, or in the case of secondary data analysis, pre-registration statements are prepared prior to viewing or analyzing data. Essentially, pre-registration represents a scientific record that allows for an easy comparison between a study's original plan and subsequent reports of that same study (e.g., a published manuscript). For the interested reader who is motivated to create a pre-registration document themselves, templates and recommendations are available for general social science research (e.g., Christensen et al., 2019; Simmons et al., 2021), social psychology research, (e.g., van't & Ginger-Sorolla, 2016), and psychopathology research (e.g., Krypotos et al., 2019).

Broadly, pre-registration also helps to establish clear boundaries for what is confirmatory research and what is exploratory research. In other words, one purpose of pre-registration is to "distinguish prediction from postdiction" (Nosek et al., 2018). This distinction is important to make because prediction represents a situation in which data are collected to test a particular idea; specifically, data are used to test if a prediction can be falsified. Postdiction, however, is "characterized by the use of data to generate hypotheses about why something occurred" (Nosek et al., 2018). Both are critical to scientific progress. Predictions obviously provide scientists with information about the validity of explanatory claims, and postdictions allow for the detection of previously unexplored, and often unexpected, effects. The distinction between these two forms of research is important to make at the very beginning of the research process, however, because without it, researchers may exhibit overconfidence in postdictions, consequently inflating the likelihood of a false positive. Essentially, presenting postdictions as predictions can cause us to falsely reduce uncertainty, and this decreases reproducibility (Nosek et al., 2018).

Pre-registration and QRPs

The type of QRP most likely to be combated by pre-registration is p-hacking. One example of how p-hacking takes place was demonstrated in a paper by Simmons, Nelson, and Simonsohn (Simmons et al., 2011), in which the authors describe an experiment in which the central hypothesis was that listening to a song by the Beatles would make the listener younger. By changing the way data were analyzed, they were able to show how it was possible to obtain statistical evidence for this ludicrous effect. Some specific examples of p-hacking include collecting participants until statistically significant effects are obtained (i.e., having no set stop criteria for data collection). Another common example of p-hacking takes the form of cherry picking, in which researchers highlight evidence that supports their hypotheses and conceal results that are inconsistent. Notably, as with most QRPs, these things can happen with explicit intent, but more than likely they occur without the researcher meaning for them to. Pre-registration offers an easy opportunity for researchers to reduce the likelihood that they will engage in such practices (Mellor & Nosek, 2018). By recording a specific hypothesis and the specific statistical analysis that will be used to test it, a researcher is willingly tying their own hands, so to speak, in order to prevent themselves from manipulating data until statistically significant results are ascertained.

HARKing also represents a common QRP that can be mitigated through pre-registration. This practice occurs when a researcher generates hypotheses after data collection has already ceased and the results of the study are already known. In other words, it means that a researcher has already analyzed their data and is now generating hypotheses consistent with the results. By coming up with post-hoc predictions for certain findings researchers are leaving themselves quite vulnerable to a type I error. By stating predictions a priori, however, researchers leave little room to concoct hypotheses consistent with their results.

Some Recommendations for Pre-registering a Scientific Study

For someone new to pre-registration, we understand that the practice can seem intimidating. Thus, we aim to provide some explicit recommendations for how to pre-register a scientific study, in hopes that it will make the practice seem more accessible and less daunting.

One of the first considerations when pre-registering a study is the basic issue of where to do it. Since pre-registrations are meant to be public, accessible scientific records, it is important to find a space where these features are available. Luckily, several online platforms host pre-registration, which makes it incredibly easy to make a registration reachable to others in the scientific community (and lay audiences, too). One of the most popular pre-registration websites is Open Science Framework (OSF; osf.io/). It is a flagship product of the non-profit organization Center for Open Science—a technology startup focused on increasing the replicability of science. Some of the advantages of using OSF are that it is free to use and is structured so that researchers can easily collaborate, archive, and pre-register projects. The website also lists a variety of different pre-registration templates for researchers to use, in order to take some of the guesswork out of what to include. Some templates are simple, while others are more involved, but there are plenty of accessible options depending on the scope of the project. Finally, OSF also has the advantage of being well-known. As of 2018, OSF pre-registration rates have been doubling every year, and from 2012 to 2018, the site registered 18,000 unique research projects. This means that if you are looking to pre-register in a commonly used and established web location, OSF is a good option.

As an alternative to OSF, some researchers use AsPredicted (aspredicted.org). This platform has the primary advantages of being succinct and standardized. All pre-registrations on the platform require a researcher to answer the same nine questions, and this generates a time-stamped PDF document with the responses that comes attached with its own URL for sharing purposes. The website also includes a pre-registration practice feature that allows one to create a pre-registration document that self-destructs 24 hours after its creation.

Registered Reports: A Unique Form of Pre-registration

Earlier in the chapter we introduced the concept of RRs and specifically discussed examples of them. Here, we want to note that registered reports help to prevent QRPs in two major ways. One way is through pre-registration; the stage 1 manuscript represents a record of hypotheses and planned analyses, which has all the benefits of pre-registration outlined above. The second way is that by providing in-principle acceptance, journals are removing the pressure to reach statistically significant results. Researchers are given the freedom to conduct their research without the pressure to obtain certain results, and we argue that removing this problematic incentive also removes some of the motivation to engage in QRPs. Promisingly, over 250 journals (Center for Open Science, n.d.) have begun to use the registered report format and this hypothetico-deductive approach is becoming increasingly encouraged for the field (Mellor, 2021).

Limitations to Pre-registration

We argue that pre-registration is a beneficial practice that should be normalized and incentivized, however, we also recognize the importance of calibrating our expectations about what pre-registration can reasonably accomplish. While it is beneficial for decreasing QRPs, the practice is not a panacea for QRPs—or the totality of consequences that stem from engaging in them. Below we outline some concluding considerations for utilizing pre-registration as a means to mitigate the prevalence of QRPs in our science.

One important consideration for pre-registration is that it is not required. This means that the efficacy of this practice for reducing rates of QRPs is significantly limited. While some journals are beginning to explicitly recommend and support pre-registration (e.g., *Nature Human Behaviour* and *Psychological Science*), the majority do not. This likely means that pre-registrations represent an added step to the already lengthy research process with little incentive to write and post them.

Another important consideration is that pre-registration (and other open science practices) functions under the assumption that others in the scientific community will hold individual researchers accountable. In the case of pre-registration, it specifically means that other researchers will help to ensure that pre-registration documents provide the necessary information and match what is reported in published manuscripts. In practice, however, we do not know how common this system of peer-checks is. We speculate that for most papers, it is rare for other scientists to do this extra work unless they have a self-motivated reason for viewing the open materials posted by an independent research team (e.g., they plan to conduct a replication study). Even when studies link pre-registrations in their manuscripts during the peer-review process, it is unclear how often reviewers do quality checks on pre-registration documents. Thus, in considering making pre-registration a normative practice, we must also consider where the onus of checking for their quality, and utility in holding other researchers accountable, lies.

A third consideration is that not everyone—even those who support the open science movement—supports the practice of pre-registration. In these circles, the primary argument for pushing back against the practice is that it inherently discourages people from conducting exploratory research (Coffman & Niederle, 2015). Hypothesis generating research, conducted without strong a priori predictions, is an important component of scientific inquiry, and some psychologists have expressed concern that by requiring pre-registration, we may inadvertently cause researchers to be afraid to explore their data and look for potential, unknown relationships between investigated constructs. There are several arguments used to dispel the misconception that pre-registration cripples exploratory research (e.g., Nosek et al., 2018); notably, pre-registration does not prohibit exploratory research, but instead prevent researchers from presenting exploratory research as confirmatory.

A fourth consideration is that pre-registration does not make a study methodologically sound. Even studies with registered hypotheses can have weak theoretical accounts and methodological flaws such as using unvalidated measures or failing to randomly assign. Pre-registration can be used to combat QRPs, but it is important to remember that it is not an index of a study's overall quality. It is also important to remember that pre-registration is limited in terms of the specific QRPs it can address. For example, it can be quite useful in preventing HARKing, but it cannot do much for faking data, and in its simplest form (i.e., a list of hypotheses), it also cannot prevent actions like selectively reporting outcomes. Bigger picture QRPs, such as sweeping generalizations to different populations and settings also cannot be prevented with pre-registration.

In sum, adding pre-registration to our current research process cannot, by itself, bring us out of the replication crisis and usher us into a world free of QRPs. However, we do believe it is a step in the right direction, and pre-registration offers the clear benefit of reducing the occurrence of certain QRPs. Significantly reducing QRPs and reducing the number of replication failures will require additional changes to the way science is currently conducted, especially changes that tackle larger, institutionalized incentive structures.

Including Clinical Psychology in the Conversation

We have briefly reviewed the history of major replication attempts in psychological science, and connected failed replications to QRPs. Astute readers may have noticed that many of these large-scale replications have been conducted within the subfields of social and cognitive psychology. But what about clinical psychology? Why have they not been more prominent in discussions about the replication crisis when, arguably, the consequences of replication failures in their research have the most direct, potentially disastrous effects of any other subfield (e.g., advocating for an ineffective therapeutic approach)?

In recent years, clinical practitioners and researchers have called for a broadening of the replicability conversation to include clinical psychology and the unique challenges clinicians face (Leichsenring et al., 2016; Tackett et al., 2017, 2019; Hengartner, 2018). Tackett and colleagues (2017) delve into the different reasons for clinical psychologists' apparent exclusion from major discussions about the replication crisis and open science movement, pointing out that core features advocated to remedy low replicability rates are often at odds with the research culture within the discipline of clinical psychology. Open science initiatives call for the use of large samples, specific forms of data analysis, and being transparent with data and materials, and for the many clinical psychologists, these recommendations are not easy to adopt for several reasons.

One reason current open science initiatives are challenging is that clinical scientists often rely on data that are difficult to collect. The populations they work with are rarer than most accessible to social psychology researchers, who are historically reliant on undergraduate samples. Additionally, access to clinical patients is often limited by geographic and community resources. As such, tiny samples that provide "noisy, messy" data are the norm. Another related concern is that access to patients may change rapidly when working within a community, and committing to any specific sample size without the flexibility to modify such a commitment is not desirable. The nature of the samples is also unique for clinicians; they work with people who are dealing with private health issues. To share all data with the public would violate HIPAA guidelines and their clients' right to privacy. It is not hard to imagine that if you are working in a small, rural clinic with a hyper-specific population, any demographic information can be used to identify people seeking help. Thus, a lot of data collected in clinical research are not easily anonymized or ethically appropriate to post to public data repositories.

Do these challenges mean that clinical psychologists should be left out of the conversation? We do not think so. The same incentive structures that gave rise to questionable research practices in other fields of psychology also exist within clinical psychology, and there is no reason to think that the field is immune to replication issues. So, what can be done? It has been advocated that clinical psychology researchers should stay educated on the replicability crisis going on in other fields; again, there are many parallels and important lessons to be shared. We also suggest that clinical psychologists start thinking creatively about potential versions of preregistrations that more appropriately fit the needs of the discipline (e.g., more adaptability to modify data collection). It has also been suggested that clinical researchers should consider working collaboratively with other labs to collect larger samples together, and start addressing the concern of power and small samples (Tackett et al., 2017; Hengartner, 2018).

We conclude this chapter with the hopes that learning more about the historical context of the replication crisis helps readers participate in discourse on the subject and motivates them to be active participants in improving psychological science. We also hope that the history we have discussed prompts readers to think about the abstract enterprise of science, recognizing that, in the purest sense, science is a process to discover truths about our world, but unfortunately, that goal can become perverted with other incentives. We do not believe the current state of the field of psychology is healthy, but we also don't believe Oscar the cat has come to take us away. We believe there is room for improvement and a growing community of researchers from a range of subfields committed to progressing our science. Ultimately, a healthy psychological science is not one devoid of replication failures, but one that acknowledges those failures and incorporates them into our understanding of human behavior.

References

Alogna, V. K., Ataya, M. K., Aucoin, P., Bahnik, S., Birch, S., Birt, A. R., ... Zwaan, R. A. (2014). Registered replication report: Schooler and Engstler-Schooler (1990). *Perspectives on Psychological Science, 9*(5), 556–578. https://doi.org/10.1177/1745691614545653

Association for Psychological Science – APS. (2014). *Ongoing Replication Projects*. Retrieved from https://www.psychologicalscience.org/publications/replication/ongoing-projects

Bem, D. J. (2011). Feeling the future: Experimental evidence for anomalous retroactive influences on cognition and affect. *Journal of Personality and Social Psychology, 100*(3), 407–425. https://doi.org/10.1037/a0021524

Bhattacharjee, Y. (2013). The mind of a con man. *The New York Times Magazine*. https://www.nytimes.com/2013/04/28/magazine/diederik-stapels-audacious-academic-fraud.html.

Borsboom, D., & Wagenmakers, E.-J. (2012). *Derailed: The rise and fall of Diederek Stapel*. Retrieved from https://www.psychologicalscience.org/observer/derailed-the-rise-and-fall-of-diederik-stapel.

Bouwmeester, S., Verkoeijen, P. P. J. L., Aczel, B., Barbosa, F., Begue, L., Branas-Garza, P., ... Wollbrant, C. E. (2017). Registered replication report: Rand, Greene, and Nowak (2012). *Perspectives on Psychological Science, 12*(3), 527–542. https://doi.org/10.1177/1745691617693624

Brandt, M. J., IJzerman, H., Dijksterhuis, A., Farach, F. J., Geller, J., Ginger-Sorolla, R., ... van't Veer, A. (2014). The replication recipe: What makes for a convincing replication? *Journal of Experimental Social Psychology, 50*, 217–224. https://doi.org/10.1016/j.jesp.2013.10.005

Camerer, C. F., Dreber, A., Forsell, E., Ho, T.-H., Huber, J., Johannesson, M., ...Wu, H. (2016).Evaluating replicability of laboratory experiments in economics. *Science, 351*(6280), 1433–1436. https://doi.org/10.1126/science.aaf0918

Carney, D. R., Cuddy, A. J. C., & Yap, A. J. (2010). Power posing: Brief nonverbal displays affect neuroendocrine levels and risk tolerance. *Psychological Science, 21*(10), 1363–1368. https://doi.org/10.1177/0956797610383437

Carney, D. R. (2016). *My position on "Power Poses"*. Retrieved from http://faculty.haas.berkeley.edu/dana_carney/pdf_My%20position%20on%20power%20poses.pdf

Center for Open Science (n.d.). *Registered reports: Peer review before results are known to align scientific values and practices*. Retrieved 24 June 2021, from https://www.cos.io/initiatives/registered-reports

Cesario, J., Jonas, K. J., & Carney, D. R. (2017). CRSP special issue on power poses: What was the point and what did we learn? *Comprehensive Results in Social Psychology, 2*(1), 1–5. https://doi.org/10.1080/23743603.2017.1309876

Cheung, I., Campbell, L., LeBel, E., ... Yong, J. C. (2016). Registered replication report: Study 1 from Finkel, Rusbult, Kumashiro, & Hannon (2002). *Perspectives on Psychological Science, 11*, 750–764. https://doi.org/10.1177/1745691616664694

Christensen, G. S., Freese, J., & Miguel, E. (2019). *Transparent and reproducible social science research: How to do open science*. University of California Press.

Coffman, L. C., & Niederle, M. (2015). Pre-analysis plans have limited upside, especially where replications are feasible. *Journal of Economic Perspectives, 29*(3), 81–98. https://doi.org/10.1257/jep.29.3.81

Cova, F., Strickland, B., Abatista, A. G. F., Allard, A., Andow, J., Attie, M., ... Xiang, Z. (2018). Estimating the reproducibility of experimental philosophy. *Review of Philosophy and Psychology, 12*(1), 9–44. https://doi.org/10.1007/s13164-018-0400-9

Crandall, C. S., & Sherman, J. W. (2016). On the scientific superiority of conceptual replications for scientific progress. *Journal of Experimental Social Psychology, 66*, 93–99. https://doi.org/10.1016/j.jesp.2015.10.002

Crocker, J. (2011). The road to fraud starts with a single step. *Nature, 479*(7372), 151. https://doi.org/10.1038/479151a

Cuddy, A. (2012). *Your body language may shape who you are*. [Video]. TED Conferences. https://www.ted.com/talks/amy_cuddy_your_body_language_may_shape_who_you_are?referrer=playlist-the_most_popular_talks_of_all#t-7174

Cumming, G. (2014). The new statistics: Why and how. *Psychological Science, 25*(1), 7–29. https://doi.org/10.1177/0956797613504966

Dang, J., Barker, P., Baumert, A., Bentvelzen, M., Berkman, E., Buchholz, N., … Zinkernagel, A. (2021). A multilab replication of the ego depletion effect. *Social Psychological and Personality Science, 12*(1), 14–24. https://doi.org/10.1177/1948550619887702

Dijksterhuis, A., & van Knippenberg, A. (1998). The relation between perception and behavior, or how to win a game of trivial pursuit. *Journal of Personality and Social Psychology, 74*, 865–877.

Dosa, D. (2007). A day in the life of Oscar the cat. *The New England Journal of Medicine, 357*, 328–329. https://doi.org/10.1056/NEJMp078108

Earp, B., & Trafimow, D. (2015). Replication, falsification, and the crisis of confidence in social psychology. *Frontiers in Psychology, 6*. https://doi.org/10.3389/fpsyg.2015.00621

Ebersole, C. R., Mathur, M. B., Baranski, E., Bart-Plange, D.-J., Buttrick, N. R., Chartier, C., … Nosek, B. A. (2020). Many labs 5: Testing pre-data-collection peer review as an intervention to increase replicability. *Advances in Methods and Practices in Psychological Science*, 309–331. https://doi.org/10.1177/2515245920958687

Eerland, A., Sherrill, A. M., Magliano, J. P., Zwaan, R. A., Arnal, J. D., Aucoin, P., … Prenoveau, J. M. (2016). Registered replication report: Hart & Albarracín (2011). *Perspectives on Psychological Science, 11*, 158–171.

Engber, D. (2017). Daryl Bem proved ESP is real. Which means science is Broken. *Slate*. https://slate.com/health-and-science/2017/06/daryl-bem-proved-esp-is-real-showed-science-is-broken.html

Fiedler, K., & Schwarz, N. (2015). Questionable research practices revisited. *Social Psychological and Personality Science, 7*(1), 45–52. https://doi.org/10.1177/1948550615612150

Finkel, E. J., Rusbult, C. E., Kumashiro, M., & Hannon, P. A. (2002). Dealing with betrayal in close relationships: Does commitment promote forgiveness? *Journal of Personality and Social Psychology, 82*, 956–974.

Funder, D., Levine, J. M., Mackie, D., Morf, C. C., Vazire, S., & West. S. G. (2014). Improving the dependability of research in personality and social psychology. *Personality and Social Psychology Review, 18*(1), 3–12. https://doi.org/10.1177/1088868313507536

Garrison, K. E., Tang, D., & Schmeichel, B. J. (2016). Embodying power: A preregistered replication and extension of the power pose effect. *Social Psychological and Personality Science, 7*, 623–630. https://doi.org/10.1177/1948550616652209

Gilbert, D. T., King, G., Pettigrew, S., & Wilson, T. D. (2016). Comment on "Estimating the reproducibility of psychological science". *Science*. Retrieved from https://science.sciencemag.org/content/351/6277/1037.2

Gronau, Q. F., Van Erp, S., Heck, D. W., Cesario, J., Jonas, K. J., & Wagenmakers, E.-J. (2017). A Bayesian model-averaged meta-analysis of the powerpose effect with informed and default priors: The case of felt power. *Comprehensive Results in Social Psychology, 2*(1), 123–138. https://doi.org/10.1080/23743603.2017.1326760

Giner-Sorolla, R. (2012). Science or art? How aesthetic standards grease the way through the-publication bottleneck but undermine science. *Perspectives on Psychological Science, 7*(6), 562–571. https://doi.org/10.1177/1745691612457576

Hagger, M. S., Chatzisarantis, N. L. D., Alberts, H., Anggono, C. O., Batailler, C., Birt, A. R., … Zwienenberg, M. (2016). A multilab preregistered replication of the ego-depletion effect. *Perspectives on Psychological Science, 11*, 546–573.

Hart, W., & Albarracín, D. (2011). Learning about what others were doing: Verb aspect and attributions of mundane and criminal intent for past actions. *Psychological Science, 22*, 261–266.

Hengartner, M. P. (2018). Raising awareness for the replication crisis in clinical psychology by focusing on inconsistencies in psychotherapy research: How much can we rely on published findings from efficacy trials? *Frontiers in Psychology, 9*, 256. https://doi.org/10.3389/fpsyg.2018.00256

Ioannidis, J. P. A. (2005). Why most published research findings are false. *PLoS Medicine, 2*(8), e124. https://doi.org/10.1371/journal.pmed.0020124

John, L. K., Loewenstein, G., & Prelec, D. (2012). Measuring the prevalence of questionable research practices with incentives for truth telling. *Psychological Science, 23*(5), 524–532. https://doi.org/10.1177/0956797611430953

Klein, R. A., Ratliff, K. A., Vianello, M., Adams, R. B., Jr., Bahník, Š., Bernstein, M. J., ... Nosek, B. A. (2014). Investigating variation in replicability: A "many labs" replication project. *Social Psychology, 45*(3), 142–152. https://doi.org/10.1027/1864-9335/a000178

Klein, R. A., Vianello, M., Hasselman, F., Adams, B. G., Adams, R. B., Alper, S., ... Nosek, B. A. (2018). Many labs 2: Investigating variation in replicability across samples and settings. *Advances in Methods and Practices in Psychological Science, 443–490.* https://doi.org/10.1177/2515245918810225

Krypotos, A.-M., Klugkist, I., Mertens, G., & Engelhard, I. M. (2019). A step-by-step guide on preregistration and effective data sharing for psychopathology research. *Journal of Abnormal Psychology, 128*(6), 517–527. https://doi.org/10.1037/abn0000424

Lakens, D. (2016). The statistical conclusions in Gilbert et al (2016 are completely invalid. *The 20% Statistician.* http://daniellakens.blogspot.com/2016/03/the-statistical-conclusions-in-gilbert.html

Leichsenring, F., Abbass, A., Hilsenroth, M. J., Leweke, F., Luyten, P., Keefe, J. R., Midgley, N., Rabung, S., Salzer, S., & Steinert, C. (2016). Biases in research: Risk factors for non-replicability in psychotherapy and pharmacotherapy research. *Psychological Medicine, 47*(6), 1000–1011. https://doi.org/10.1017/S003329171600324X

Makel, M. C., Plucker, J. A., & Hegarty, B. (2012). Replications in psychology research: How often do they really occur? *Perspectives on Psychological Science, 7*(6), 537–542. https://doi.org/10.1177/1745691612460688

Marshall, E. (2000). How prevalent is fraud? That's a million-dollar question. *Science, 290*(5497), 1662–1663. https://doi.org/10.1126/science.290.5497.1662

Mazar, N., Amir, O., & Ariely, D. (2008). The dishonesty of honest people: A theory of self-concept maintenance. *Journal of Marketing Research, 45*, 633–644. https://doi.org/10.1509/jmkr.45.6.633

Mellor, D. (2021). Improving norms in research culture to incentivize transparency and rigor. *Educational Psychologist, 56*(2), 122–131. https://doi-org.libdata.lib.ua.edu/10.1080/00461520.2021.1902329

Mellor, D. T., & Nosek, B. A. (2018). Easy preregistration will benefit any research. *Nature Human Behavior, 2*, 98. https://doi.org/10.1038/s41562-018-0294-7

Moshontz, H., Campbell, L., Ebersole, C. R., IJzerrman, H., Urry, H. L., Forscher, P. S., ... Chartier, C. R. (2018). The psychological science accelerator: Advancing psychology through a distributed collaborative network. *Advances in Methods and Practices in Psychological Science*, 501–515. https://doi.org/10.1177/2515245918797607

Nelson, L. D., Simmons, J., & Simonsohn, U. (2018). Psychology's renaissance. *Annual Review of Psychology, 69*(1), 511–534.

Nosek, B. A., Spies, J. R., & Motyl, M. (2012). Scientific utopia: II. Restructuring incentives and practices to promote truth over publishability. *Perspectives on Psychological Science, 7*(6), 615–631. https://doi.org/10.1177/1745691612459058

Nosek, B. A., Ebersole, C. R., DeHaven, A. C., & Mellor, D. T. (2018). The preregistration revolution. *Proceedings of the National Academy of Sciences, 115*(11), 2600–2606. https://doi.org/10.1073/pnas.1708274114

Nosek, B. A., & Errington, T. M. (2020). What is replication? *PLoS Biology, 18*(3), E3000691. https://doi.org/10.1371/journal.pbio.3000691

O'Donnell, M., Nelson, L. D., Ackermann, E., Aczel, B., Akhtar, A., Aldrovandi, S., ... Babincak, P. (2018). Registered replication report: Dijksterhuis and van Knippenberg (1998). *Perspectives on Psychological Science, 13*(2), 268–294. https://doi.org/10.1177/1745691618755704

Open Science Collaboration. (2015). Estimating the reproducibility of psychological science. *Science, 349*(6251), Aac4716-7. https://doi.org/10.1126/science.aac4716

Oreskes, Naomi. (2014). *Why we should trust scientists*. [Video]. TED Conferences. https://www.ted.com/talks/naomi_oreskes_why_we_should_trust_scientists

Rand, D. G., Greene, J. D., & Nowak, M. A. (2012). Spontaneous giving and calculated greed. *Nature, 489*, 427–430.

Ranehill, E., Dreber, A., Johannesson, M., Leiberg, S., Sul, S., & Weber, R. A. (2015). Assessing the robustness of power posing: No effect on hormones and risk tolerance in a large sample of men and women. *Psychological Science, 26*(5), 653–656. https://doi.org/10.1177/0956797614553946

Schooler, J. W., & Engstler-Schooler, T. Y. (1990). Verbal overshadowing of visual memories: Some things are better left unsaid. *Cognitive Psychology, 22*, 36–71.

Simmons, J. P., Nelson, L. D., & Simonsohn, U. (2011). False-positive psychology: Undisclosed flexibility in data collection and analysis allows presenting anything as significant. *Psychological Science, 22*(11), 1359–1366. https://doi.org/10.1177/0956797611417632

Simmons, J., & Simonsohn, U. (2015). *Power posing: Reassessing the evidence behind the most popular TED talk.* Retrieved from http://datacolada.org/2015/05/08/37-power-posing-reassessing-the-evidence-behind-the-most-popular-ted-talk/

Simmons, J. P., Nelson, L. D., & Simonsohn, U. (2021). Pre-registration: Why and how. *Journal of Consumer Psychology, 31*(1), 151–162. https://doi.org/10.1002/jcpy.1208

Simons, D. J., Holcombe, A. O., & Spellman, B. A. (2014). An introduction to registered replication reports at perspectives on psychological science. *Perspectives on Psychological Science, 9*(5), 552–555. https://doi.org/10.1177/1745691614543974

Simonsohn, U. (2015). Small telescopes: Detectability and the evaluation of replication results. *Psychological Science, 26*(5), 559–569. https://doi.org/10.1177/0956797614567341

Simonsohn, U. (2016). Each reader decides if a replication counts: Reply to Schwarz and Clore (2016). *Psychological Science, 27*(20), 1410–1412. https://doi.org/10.1177/0956797616665220

Sovacool, B. (2008). Exploring scientific misconduct: Isolated individuals, impure institutions, or an inevitable idiom of modern science? *Journal of BioEthical Inquiry, 5*(4), 271–282. https://doi.org/10.1007/s11673-008-9113-6

Sripada, C., Kessler, D., & Jonides, J. (2014). Methylphenidate blocks effort-induced depletion of regulatory control in healthy volunteers. *Psychological Science, 25*(6), 1227–1234. https://doi.org/10.1177/0956797614526415

Srull, T. K., & Wyer, R. S. (1979). The role of category accessibility in the interpretation of information about persons: Some determinants and implications. *Journal of Personality and Social Psychology, 37*, 1660–1672. https://doi.org/10.1037/0022-3514.37.10.1660

Strack, F., Martin, L. L., & Stepper, S. (1988). Inhibiting and facilitating conditions of the human smile: A nonobtrusive test of the facial feedback hypothesis. *Journal of Personality and Social Psychology, 54*, 768–777.

Strack, F. (2016). Reflection on the smiling registered replication report. *Perspectives on Psychological Science, 11*(6), 929–930. https://doi.org/10.1177/1745691616674460

Stroebe, W. (2019). What can we learn from many labs replications? *Basic and Applied Social Psychology, 41*(2), 91–103. https://doi.org/10.1080/01973533.2019.1577736

Tackett, J. L., Lilienfeld, S. O., Patrick, C. J., Johnson, S. L., Krueger, R. F., Miller, J. D., Oltmanns, T. F., & Shrout, P. E. (2017). It's time to broaden the replicability conversation: Thoughts for and from clinical psychological science. *Perspectives on Psychological Science, 12*(5), 742–756. https://doi.org/10.1177/1745691617690042

Tackett, J. L., Brandes, C. M., King, K. M., & Markon, K. E. (2019). Psychology's replication crisis and clinical psychological science. *Annual Review of Clinical Psychology, 15*, 579–604. https://doi.org/10.1146/annurev-clinpsy-050718-0957104

Vadillo, M. A., (2019). Ego depletion may disappear by 2020. Social Psychology, 50(5–6), 282–291. https://doi.org/10.1027/1864-9335/a000375

van't Veer, A. E., & Ginger-Sorolla, R. (2016). Pre-registration in social psychology – A discussion and suggested template. *Journal of Experimental Social Psychology, 67*, 2–12. https://doi.org/10.1016/j.jesp.2016.03.004

Van Bavel, J., Mende-Siedlecki, P., Brady, W., & Reinero, D. A. (2016). Contextual sensitivity in scientific reproducibility. *Proceedings of the National Academy of Sciences, 113*(23), 113.201521897. https://doi.org/10.1073/pnas.1521897113

Vazire, S. (2018). Implications of the credibility revolution for productivity, creativity, and progress. *Perspectives on Psychological Science, 13*(4), 411–417. https://doi.org/10.1177/1745691617751884

Verfaellie, M., & McGwin, J. (2011). The case of Diederik Stapel. *Psychological Science Agenda.*. http://www.apa.org/science/about/psa/2011/12/diederik-stapel.

Verhagen, J., & Wagenmakers, E.-J. (2014). Bayesian tests to quantify the result of a replicationattempt. *Journal of Experimental Psychology: General, 143*(4), 1457–1475. https://doi.org/10.1037/a0036731

Verschuere, B., Meijer, E. H., Jim, A., Hoogesteyn, K., Orthey, R., McCarthy, R. J., … Bakos, B. E. (2018). Registered replication report on Mazar, Amir, and Ariely (2008). *Advances in Methods and Practices in Psychological Science*, 1–19. https://doi.org/10.1177/2515245918781032

Vul, E., Harris, C., Winkielman, P., & Pashler, H. (2009). Puzzlingly high correlations in fMRI studies of emotion, personality, and social cognition. *Perspectives on Psychological Science, 4*(3), 274–290. https://doi.org/10.1111/j.1745-6924.2009.01125.x

Wagenmakers, E.-J., Beek, T., Dijkhoff, L., Gronau, Q. F., Acosta, A., Adams, R. B., Jr., … Zwaan, R. A. (2016). Registered replication report: Strack, Martin, & Stepper (1988). *Perspectives on Psychological Science, 11*, 917–928. https://doi.org/10.1177/1745691616674458

Wagenmakers, E.-J., Wetzels, R., Borsboom, D., & van der Maas, H. L. J. (2011). Why psychologists must change the way they analyze their data: The case of psi: Comment on Bem (2011). *Journal of Personality and Social Psychology, 100*(3), 426–432.

Wicherts, J. (2011). Psychology must learn a lesson from fraud case. *Nature, 480*(7375), 7. https://doi.org/10.1038/480007a

Wolff, W., Baumann, L., & Englert, C. (2018). Self-reports from behind the scenes: Questionable research practices and rates of replication in ego depletion research. *PLoS One, 13*(6), e0199554. https://doi.org/10.1371/journal.pone.0199554

Zhong, C.-B., & Liljenquist, K. (2006). Washing away your sins: Threatened morality and physical cleansing. *Science, 313*(5792), 1451–1452. https://doi.org/10.1126/science.1130726

Part II
Questionable Research Practices

Chapter 5
The Myriad Forms of *p*-Hacking

Dorota Reis and Malte Friese

Abstract In the present chapter, we are going to discuss several *p*-hacking practices as part of the broader category of questionable research practices. It has become clear that *p*-hacking can have detrimental consequences—particularly an increase in false-positive rates—that ultimately damage the trustworthiness and robustness of psychological science. What can any researchers do to confirm that they did not engage in questionable research practices? The solution is surprisingly simple. It lies in the transparent distinction between a priori planned, confirmatory steps of data analysis and exploratory, additional steps. The line between the two can be drawn easily by adhering to the open science practices outlined in this chapter, particularly the detailed preregistration of all measures, manipulations, hypotheses, and planned analysis steps. Open science practices are surely not the solution to all challenges psychological science currently faces, but they are a pretty good and easy-to-implement solution to prevent *p*-hacking. Let's do it.

Keywords Meta science · *p*-hacking · Questionable research practices

Credibility Concerns about (Clinical) Psychological Science

Alice is an experienced psychotherapist. For many years, she has worked in an outpatient facility specializing in the treatment of chronic pain. Being a passionate practitioner, Alice is continuously educating herself on how to use state-of-the-art treatment methods to best benefit her patients. Therefore, Alice is thrilled when she reads about a new therapy and its impressive treatment response in a prestigious scientific clinical journal. "With this new approach," Alice feels, "I will be able to

---

Dorota Reis and Malte Friese are contributed equally to this work.

D. Reis (✉) · M. Friese (✉)
Saarland University, Saarbrücken, Germany
e-mail: dorota.reis@uni-saarland.de; malte.friese@uni-saarland.de

W. O'Donohue et al. (eds.), *Avoiding Questionable Research Practices in Applied Psychology*, https://doi.org/10.1007/978-3-031-04968-2_5

have a substantial additional impact on the well-being of my patients!" She invests time and money to be certified as a specialist in this new approach and starts implementing the new intervention strategy in her clinical work.

A few months later, Alice receives the treatment evaluations. They are sobering. Although she closely adhered to the therapy manual, the desired reduction of symptoms remains far behind her expectations. Even more, the new therapy appears to be less effective than the conventional "gold standard" treatment previously applied in the facility. Although the evaluations confirm the subjective impressions she obtained during the therapy sessions and match those reported by colleagues who have also implemented the new technique, Alice is frustrated. Instead of improving the treatment for her patients, the changes to the protocol appear to have backfired. The success rate even falls below that of previous treatments. Some patients begin dropping out early, whereas others begin to take even longer than before to attain noticeable treatment results. What happened here?

In the last decade, scientific psychology has seen a multitude of scenarios similar to the one described in the opening paragraphs. Large-scale replication projects (Klein et al., 2014; Open Science Collaboration, 2015, see also Chap. 18, this volume) and countless primary studies have shown disturbingly low replication rates (see also Chap. 4, this volume). Psychology is not alone. Other disciplines have reported similar problems, including the neurosciences (Button et al., 2013), economics (Camerer et al., 2016), cancer research (Begley & Ellis, 2012), and drug research (Prinz et al., 2011), to just name a few. Although some disciplines are more affected than others, low replicability appears to be a problem in many fields.

In psychology, what began as a "replication crisis" has quickly become a more general "credibility crisis." As a result, psychological science is under scrutiny (Lilienfeld & Waldman, 2017). This is not just academic ivory tower talk. The credibility of psychological research has profound real-world consequences. Particularly in clinical research and practice, unreliable findings can affect the (mental) health of people who rely on the trustworthiness of the science that informed their treatments. Interventions that were believed to be effective but actually are not imply that patients and clients will experience less symptom reduction and need more time to reduce distress than necessary.

Various issues have been discussed as undermining the credibility of psychological science. These include low statistical power (Bertamini & Munafò, 2012), an over-reliance on *p*-values (Wasserstein & Lazar, 2016, see also Chap. 7, this volume), maladaptive incentives (Lilienfeld, 2017; Nosek et al., 2012), hypothesizing after the results are known (HARKing, Kerr, 1998, see also Chap. 8, this volume), publication bias (Bakker et al., 2012, see also Chap. 10, this volume), and *p*-hacking (John et al., 2012; Simmons et al., 2011), among others.

We cannot know why the new treatment that Alice was so enthusiastic about in our fictitious introductory example was less effective in Alice's facility than what seemed to be realistic on the basis of the associated scientific publication. In the present chapter, we will focus on one specific issue that jeopardizes the credibility and robustness of psychological science: *p*-hacking. We will discuss what *p*-hacking is, which scientific practices are subsumed under this umbrella term, its prevalence and detection (see also Chap. 6, this volume), its consequences, and how it can be prevented. We will close by making a case for open science practices that we argue are an effective remedy for several of the challenges that scientific psychology currently faces.

If you are a clinical psychologist, you may wonder why you should go on reading. Isn't the replication crisis and the associated use of problematic research practices something for other psychological subdisciplines to worry about? It is true that the extent of replicability problems and the use of problematic research practices vary across subdisciplines (John et al., 2012), but this does not mean that clinical psychology is free of concerns (Leichsenring et al., 2017; Tackett et al., 2019). In fact, there are a few weak spots that endanger the replicability and robustness of clinical research as well. (For an overview of research biases in psychotherapy research, in particular, see Leichsenring et al., 2017.)

For example, statistical power is often low, particularly for treatment/intervention research and clinical neuroscience (Button et al., 2013; Cuijpers, 2016; Reardon et al., 2019; Sakaluk et al., 2019). One reason for low power is small sample sizes. As later sections of this chapter will clarify, some *p*-hacking practices are particularly "effective" in small samples, making such studies vulnerable to considerable bias. In addition, in a scientific culture that values novel, statistically significant findings so much more than less novel and/or statistically nonsignificant findings, incentives to "find" something in the data of a given study are high, and this is particularly true when the study cannot be easily repeated or extended because it is very resource-intensive or relies on a difficult-to-reach sample. This applies to a lot of clinical psychology intervention studies, which is one reason why a lot of important studies cannot be easily replicated in single studies or in large-scale replication projects (Tackett & Miller, 2019). Based on novel evidential value metrics, such as rates of misreported statistics, power, and Bayes Factors, the replicability of empirically supported treatments seems to be remarkably low (Sakaluk et al., 2019). Thus, it cannot be assumed that actual therapy results will achieve the effectiveness that could be assumed on the basis of the scientific articles the interventions were published in. On the basis of their analysis, Sakaluk et al. (2019) concluded that whereas there is strong evidence behind a few therapies, "the evidence is mixed or consistently weak for many, including some classified by Division 12 of the APA as 'Strong'" (p. 500). Finally, there is evidence for considerable publication bias (also) in clinical psychology, another factor that undermines the robustness of published findings (Cuijpers et al., 2010; Rapport et al., 2013). Thus, yes, we are afraid the credibility crisis in general, and *p*-hacking in particular, are topics that should also be of interest to clinical psychologists.[1]

---

[1] We were initially invited to contribute a chapter that focuses on the implications of *p*-hacking for clinical psychology specifically. This is why most of our examples in this chapter come from this area of research. However, the general issues covered in this chapter also apply to other areas of (applied) psychology.

What Is *p*-Hacking?

When researchers collect and analyze data, they have many decisions to make. Unless they commit a priori to a specified and exhaustive set of decision outcomes, they have many so-called researchers' degrees of freedom (Wicherts et al., 2016). These degrees of freedom invite *p*-hacking. The term *p*-hacking (also called data dredging or inflation bias) refers to a family of practices that can be responsible for substantial biases. It is an umbrella term that "refers to nonprincipled decisions during data analysis that are aimed at reducing the *p*-value of a significance test and thus make the data look more robust than they actually are" (Friese & Frankenbach, 2020, p. 457). Researchers have known about such nonprincipled decisions for a long time: As early as 1956, De Groot described this approach as an "attempt to let the material speak [that] leads to ad hoc decisions in terms of processing" (de Groot, 1956/2014, p. 191). *p*-Hacking can take various forms (also known as "*p*-hacks") and can occur at various points during the data analysis process, even before formal data analysis has even begun (see Fig. 5.1 for a schematic overview). The main characteristic of these practices is that they are selectively employed to bring an originally nonsignificant *p*-value below the alpha level of 5%.

p-Hacking can occur intentionally and with full awareness that one is not following best scientific practices. However, importantly, *p*-hacking can also occur largely without awareness of the potentially detrimental consequences or even with genuinely honest intentions. To illustrate, we believe it is quite likely that before the seminal paper by Simmons et al. (2011) demonstrated the tremendous effect of *p*-hacking on false-positive rates, many researchers were aware that playing with the data too much would likely increase the chances of false positives, but they probably had little idea about the extent of this problem. Even more, researchers who are convinced of the validity of a particular hypothesis may want to uncover what they think is hidden like a precious, but hard-to-detect signal in the data (Heene & Ferguson, 2017; Nelson et al., 2018). Because of confirmation and hindsight bias,

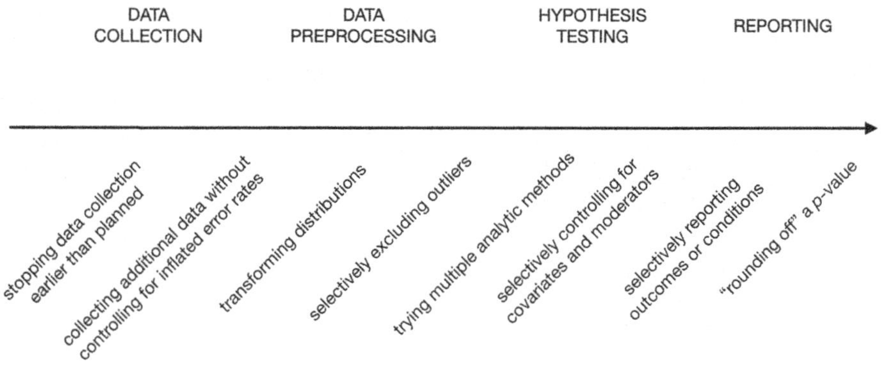

Fig. 5.1 Schematic depiction of exemplary *p*-hacking practices according to when they typically occur during the research process

researchers may believe that undesired outcomes result from suboptimal analyses (Munafò et al., 2017), so they try to optimize their analyses without any bad intent. Thus, the general notion of the present chapter is not to blame or denounce researchers for their presumably ill-intentioned behavior. Instead, we intend to provide information about the nature, consequences, and prevention of *p*-hacking, whether it occurs intentionally or not.

p-Hacking Practices

Various different practices are considered *p*-hacking. Our overview is not exhaustive. In Fig. 5.1, we arranged some exemplary strategies according to when they typically occur in the research process.

p-Hacking during Data Collection

During the data collection process, two types of *p*-hacking can occur: first, stopping data collection earlier than planned because preliminary analyses appear to reveal the result that one is looking for, and, second, collecting additional data without controlling for inflated error rates.

Each of these strategies can be problematic. When stopping early (i.e., with lower power than planned), it is more difficult to distinguish random variation from a true effect. Underpowered samples not only have a reduced chance of detecting a true effect, but the likelihood that a statistically significant effect reflects a true effect is also reduced (Button et al., 2013; Ioannidis, 2005). Moreover, if the true effect is zero, *p*-values are uniformly distributed: Every possible *p*-value is equally likely (Simonsohn et al., 2014). Hence, stopping earlier than intended can lead to uninformative results if statistical power remains low.

Collecting more data is generally a good thing as it increases power and the chances of revealing a true effect. However, looking at the data multiple times (and deciding to continue data collection) also increases the danger of false positives if researchers do not statistically control for the increased Type I error rate. If the true effect is zero, *p*-values are uniformly distributed. Hence, they will zigzag endlessly, and as a consequence, a result that is "approaching significance" may turn significant when a few more data points are added without actually revealing a true effect. Fortunately, because no *p*-value is more likely than another if the true effect is zero, in these cases, larger samples will also often reveal larger *p*-values. If there is a true effect, a larger sample increases the chance of obtaining a particularly small *p*-value, not one that barely crosses the 0.05 mark.

For trial researchers, both the practices of stopping early and collecting more data after looking at the results are known as sequential investigation (Armitage et al., 2002) or the sequential stopping rule (Dienes, 2008; Lakens, 2014). Sequential (group) investigations are particularly important for clinical trials because it may be ethical to terminate the trial early when there is strong evidence in favor of, or against, the treatment under investigation. The determination of when data collection will end is defined as the stopping rule for a study. Problems arise if the decision of when to terminate—or collect more data—is not specified a priori. Checking data more often (than once) increases the actual α level because each test that is conducted offers a new chance to reject the null hypothesis. In the extreme, a stopping rule that implies "I will continue running the experiment until the test is significant" guarantees a significant finding even when the null hypothesis is true (Dienes, 2008).

p-Hacks During Data Preprocessing

Several *p*-hacks can occur during data analysis. We arranged these *p*-hacks according to whether they most typically happen during data preprocessing or during hypothesis testing. In reality, the separation between these stages is not strict. All strategies can be employed at any point when trying to make the data reveal the most about the proposed hypotheses.

We discuss two types of *p*-hacking that refer to data preprocessing and exploratory data analysis: transforming distributions and (selectively) excluding outliers. Data transformations can be useful for normalizing the data. For example, log-transforming the data may give a parametric test more power and—as a result—lower *p*-values. However, such transformations must be specified in advance. *p*-Hacking occurs when an analyst runs the analyses on raw data first and, after trying one or even several transformations, reports the results with the smallest *p*-value (Lew, 2020).

Similar problems can occur when outliers are excluded. Excluding a few data points that are not representative of the rest of the distribution can be useful when these data points exert an extraordinarily strong influence on the inferences researchers draw from the data. Problems arise when the decisions about whether to exclude data points and which ones to exclude are based on how much the various decisions change the *p*-value toward significance. Admittedly, decisions about the exclusion of outliers can be inherently ambiguous. There are several approaches that explain how to exclude outliers (e.g., standard deviations from the mean, median absolute deviation, Boxplot analysis), and within each approach, there are several choices (e.g., 2.5 or 3 standard deviations/median absolute deviations), opening up an extensive array of options. It can be challenging to decide which of these paths is the best choice in a particular study. However, what is clear is that picking the specific path on the basis of the resulting *p*-value will increase the danger of believing that the effect that was found is more robust than it actually is. We will discuss the detrimental consequences of this and other *p*-hacking practices in a later section as well as how to prevent them from occurring.

p-Hacks During Hypothesis Testing

When researchers test confirmatory hypotheses, they may try out multiple analytical approaches (e.g., a *t* test for dependent measures, a robust *t* test such as the Yuen-Test, and an analysis based on change scores). Again, applying diverse methods may happen in good faith in an attempt to identify the most adequate method for analyzing the particular data set. However, running a bunch of analyses and reporting only the method yielding the lowest *p*-value capitalizes on chance. Indeed, recent endeavors have shown that even different well-intentioned analysts can analyze the same data in widely different ways and arrive at conclusions that differ greatly (Silberzahn et al., 2018).

A similar reasoning applies when researchers' decisions to selectively control for covariates or moderators (e.g., gender, age) are based on whether or not this reduces their focal *p*-value instead of a priori theoretical reasoning that it will be advisable to do so. This practice highlights the similarities between *p*-hacking and overfitting. Overfitting refers to situations in which sample-specific noise is misinterpreted as a true signal that can be generalized to the population. Yarkoni and Westfall (2017) consider *p*-hacking to be a form of procedural overfitting because it takes place either prior to or in parallel with hypothesis testing or model estimation.

p-Hacks During the Reporting of Studies

Another subset of *p*-hacking practices refers to decisions about whether to exclude data after looking at the impact of doing so on the results. These decisions may pertain to single data points (e.g., outliers, see above), ghost variables (i.e., dependent variables assessed during data collection but not reported in the publication itself; Bishop & Thompson, 2016), or experimental conditions (e.g., dropping, combining, splitting groups). Conceptually, the dropping of conditions is a borderline case that falls between *p*-hacking and publication bias (i.e., dropping whole studies). Similarly, failing to disclose experimental conditions (e.g., when the results are inconsistent with theoretical predictions) is considered *p*-hacking because this may impact the *p*-value of some analyses (e.g., dropping a condition in a one-way ANOVA).

The term *ghost variables* describes the situation when researchers do not specify in advance their hypotheses about which specific measure will differ between groups or will show a substantial association with a chosen predictor. If researchers report only the significant ones and assign a ghost status to the remaining variables, this is considered *p*-hacking (Bishop & Thompson, 2016). This type of *p*-hacking is problematic because, due to the problems that arise from multiple testing, the inferential statistics reported in such cases will be misleading. Conceptually, dropping dependent variables is also linked to publication bias on the outcome level instead of the study level.

Multiple testing refers to situations in which researchers perform a "family" of tests. When performing one *t* test, the null hypothesis can be rejected at the 5% α level. But when performing two *t* tests, the probability of making a Type I error increases to 9.75%. Therefore, when running more than one test on the same hypothesis, researchers need to control the overall Type I error rate (i.e., the family-wise error). This can be done by correcting (i.e., reducing) the α level for every single test in the family. Researchers sometimes try to avoid having to make such a correction by not reporting some of the tests they ran or by dropping some of their (dependent) variables. Consequently, they report results as significant even when these tests would have missed the corrected threshold if the proper procedures had been followed.

A final *p*-hacking practice that may occur during the reporting of studies is the rounding off of *p*-values. In practice, this means reporting values slightly above 0.05 as equal to or even less than 0.05, and hence, reporting the results as significant when in fact they are not. Thus, other than the previously discussed techniques, this *p*-hack does not even lead to a formally significant result. It only pretends to do so.

The Consequences of *p*-Hacking

p-Hacking has a whole range of implications. In this section, we will discuss two: an increase in false-positive findings and an overestimation of effect sizes.

Increase in False-Positive Findings

The most tangible consequence of *p*-hacking is a sharp increase in false-positive findings—hence, the reference to *p*-hacking as "inflation bias." Put differently, *p*-hacking practices "wreak havoc with a method's error probabilities. It becomes easy to arrive at findings that have not been severely tested" (Mayo, 2018, p. 439). *p*-Hacking leads researchers to believe they found a real effect when in reality there was none, at least not one strong enough to reach statistical significance.

In an impressive demonstration of this consequence, Simmons et al. (2011) convincingly showed how strongly *p*-hacking inflates actual false-positives rates. These authors simulated scenarios for four *p*-hacking practices: choosing from among outcome variables, optional stopping, including covariates, and excluding experimental conditions. Also, they evaluated various combinations of these four practices by taking into account the possibility that several practices may occur jointly. Their simulations showed that applying a single *p*-hack can easily double the factual false-positive rate (under the specific conditions employed by Simmons et al. in their simulation). At the same time, researchers still assume that their results are quite unlikely (≤ 0.05) under the null hypothesis. The most disturbing finding from the analyses demonstrated that false-positive rates increased to 61% when the four *p*-hacking practices were combined. Consequently, in this situation, the probability that researchers would erroneously conclude and report a significant finding was higher than the probability that they would correctly reject the null hypothesis. They would even have been better off by flipping a coin.[2]

On a larger scale, one may wonder what it means if a literature is built substantially on studies that, in reality, did not reveal significant findings but were false positives. False-positive findings may be particularly detrimental in small, emerging literatures with a few landmark studies that may give the impression of a robust effect when it is much weaker in reality. One may hope that in the long run, as a literature grows and matures, the weight of individual studies will decrease. Still, if a literature (for whatever reason) remains small, a few false-positive findings can bias perceptions of this literature for a long time. Obviously, the more prevalent and severe *p*-hacking is in a given literature, the more damaging the consequences.

Overestimation of Effect Sizes

A second detrimental consequence of *p*-hacking involves the overestimation of effect sizes. *p*-Hacking means using any of the aforementioned practices to bring an originally nonsignificant *p*-value down to significance. For many of these practices, this essentially means obtaining a larger effect size estimate that will cross the significance threshold (e.g., by including a covariate or excluding some outliers). This effect will be particularly pronounced in small studies with low power because only large effect sizes become significant in such studies.

Imagine a research group that ran a relatively small study with a striking result: evidence for a hitherto unknown effect Y. They published the study in a high-impact journal. When analyzing the data, they tried many different things (i.e., they used researchers' degrees of freedom) and settled on a solution that they believed was most appropriate (in fact, they overfitted the analysis to the data, resulting in an effect size that overestimated the true effect). In addition to (in this case unintended) *p*-hacking, the study is haunted by another problem: Effect sizes are additionally exaggerated in small, underpowered studies such as the one our fictitious research group ran, a statistical phenomenon called the winner's curse (Button et al., 2013). It means that the research group is "cursed" by overestimating the magnitude of the effect in the population due to random error that is more pronounced in underpowered studies (see also Chap. 11, this volume). Other researchers trying to replicate the initial finding Y will then suffer from a "decline effect" (Protzko & Schooler, 2017), indicating that attempts to replicate the effect will likely find considerably

[2]You can experience the power of *p*-hacking yourself by using the *p*-hacker Shiny app (Schönbrodt, 2016).

smaller effects or even end up with a null finding. In this situation, further *p*-hacking when analyzing the replication attempt becomes more likely, particularly in a culture that incentivizes significant results. The researchers' assumption that there must be a true effect and that it's only hidden due to a suboptimal analysis will motivate them to dig deeper and "dredge" the data. This illustrates how nonrobust findings can initiate a vicious cycle that—in combination with maladaptive incentive structures—motivates the continued use of *p*-hacking and other questionable research practices.

A recent large-scale simulation study examined the extent to which *p*-hacking can bias effect size estimates of whole literatures, either on its own or in combination with publication bias (Friese & Frankenbach, 2020). Publication bias is another questionable research practice that arises when studies that did not produce the desired outcomes are less likely to be published than studies that "worked" (Franco et al., 2014; Ioannidis et al., 2014). Hence, publication bias occurs at the level of studies (is a study published or not?), whereas *p*-hacking refers to the data collection practices and analyses used in a study.

The results of Friese and Frankenbach's (2020) study revealed that *p*-hacking and publication bias result in different threats to the robustness of findings. Whereas *p*-hacking can dramatically increase the rate of false positives in a given literature, high levels of publication bias can lead to a considerable distortion of (meta-analytic) effect size estimates. Perhaps surprisingly, in the absence of publication bias, *p*-hacking does little to distort meta-analytic effect size estimates. However, the two phenomena interact: *p*-hacking adds considerable bias to effect size estimates at medium levels of publication bias—particularly in literatures where the true effects in question are small. At low and high levels of publication bias, *p*-hacking hardly contributes any bias to meta-analytic effect size estimates. With increasing true effect sizes, literatures are more and more shielded against the effect size bias introduced by publication bias and *p*-hacking (Fig. 5.2).

Increased rates of false-positive findings and bias in effect size estimates have palpable implications for the literature's meta-analytical record. Large numbers of seemingly positive (but factually false-positive) results create the impression of a robust and accurate literature. As the inflated effect sizes from single studies will be included and summarized in meta-analyses, they bias the meta-analytical effect size (provided that there is some publication bias). As a consequence, researchers conducting new work who base their expectations on these biased estimates will inadvertently run underpowered studies because they believe that the effects of interest are more robust than they actually are. This combination of increased rates of false-positive findings and biased meta-analytical effect size estimates impedes the accumulation of knowledge. For one, it leads researchers who are trying to build upon previous work astray. In addition, practitioners relying on biased literature might not be able to provide the best solutions to those they are working with. The result is a lamentable waste of resources.

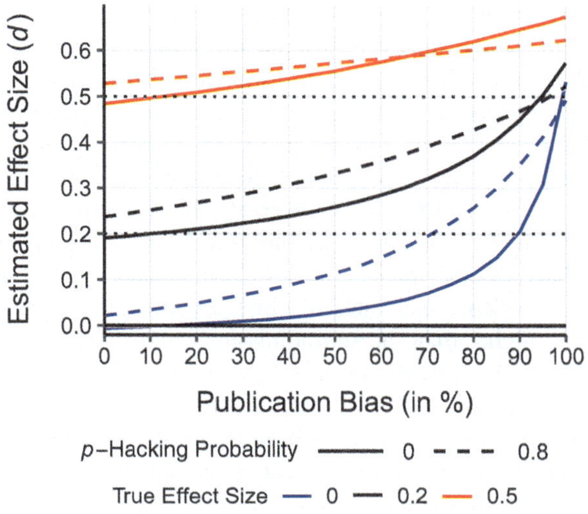

Fig. 5.2 Meta-analytic effect size estimates as a function of degrees of *p*-hacking, publication bias, and true effect size

Note. In the absence of publication bias, *p*-hacking does little to distort meta-analytic effect size estimates. By contrast, high degrees of publication bias do distort these estimates even in the absence of *p*-hacking. Together, *p*-hacking and publication bias interact such that *p*-hacking adds considerable bias when publication bias is moderate. Bias is greatest when the true effect size is very small. Larger true effects act as a shield against the deleterious effects of *p*-hacking and publication bias on meta-analytic effect size estimates. Figure reprinted from Friese and Frankenbach (2020).

The Prevalence and Detection of *p*-Hacking

The prevalence of *p*-hacking in different disciplines has been discussed repeatedly (Fiedler & Schwarz, 2016; John et al., 2012, see also Chap. 6, this volume). There are essentially three approaches that seek to determine how frequently researchers *p*-hack. One directly surveys researchers about their practices, whereas the others attempt to obtain statistical indicators of *p*-hacking based on published literature or by comparing planned with reported analyses.

Prevalence Estimates Based on Self-Reports

John et al. (2012) aimed to estimate the prevalence of *p*-hacking (and other Questionable Research Practices, QRPs) in a few ways (see also Chaps. 1 and 2, this volume). One way was to acquire self-admission rates for 10 QRPs (e.g., "rounding off" a *p*-value, deciding whether to exclude data after determining how such an exclusion would impact the results, or deciding whether to collect more data after looking to see whether the results were significant; John et al., 2012, p. 525). The second way involved asking participants to estimate the percentage of other psychologists who had engaged in the behavior. There was large variability across the 10 QRPs for both indicators. For example, up to 58% of participants indicated that they had at least once decided "whether to collect more data after looking to see whether the results were significant." Conversely, for claiming in a paper "that results are unaffected by demographic variables (e.g., gender) when one is actually unsure (or knows that they do)" (p. 525), only 4.5% indicated that they had done this at least once. The prevalence estimates of other psychologists engaging in these practices were often higher than the self-admissions. With respect to participants working in clinical psychology, the mean self-admission rate across all 10 QRPs was 27%.

The findings by John et al. (2012) have been frequently cited but also criticized for exaggerating actual prevalence rates because the authors used lifetime prevalence rates to conclude that some research practices "constitute the prevailing research norm" (p. 524). However, lifetime prevalence rates are unable to distinguish between researchers who engaged in a QRP once and only once in their lifetime and researchers who engage in the same QRP regularly. A conceptual replication among German psychologists decomposed the prevalence of QRPs into their two multiplicative components: the proportion of researchers who ever committed a given practice and, if so, how frequently (Fiedler & Schwarz, 2016). This survey found prevalence estimates that were a lot lower than those reported by John et al. (2012).

John et al. (2012) also suggested that prevalence rates of QRPs are not uniformly distributed across the subdisciplines of psychology. Even within subdisciplines, there are different subfields with potentially different research cultures that may be more or less susceptible to certain QRPs. Of particular relevance for the present purposes is a recent survey among faculty and students in clinical and counseling psychology doctoral programs (Swift et al., 2020). In this survey, over 64% of faculty and 48% of students indicated engaging in at least one of 12 QRPs at least once during their career. The *p*-hacking practices that participants admitted to engaging in included rounding off a *p*-value (12.8% of faculty and 8.2% of doctoral students) and excluding data after looking at the impact of doing so (11.8% of faculty and 9.1% of doctoral students). These admission rates were considerably lower than the rates reported in other surveys (e.g., 22% for at least once rounding-off a *p*-value and 40% for at least once excluding data; Fiedler & Schwarz, 2016).

Prevalence Estimates Based on Analyses of the Published Literature

Assuming that self-reported data underestimate socially undesirable behavior, other approaches attempt to obtain statistical indicators of *p*-hacking on the basis of published literature. For example, some researchers have suggested that when looking at empirical p-value distributions, clusters of p-values just below 0.05 may indicate that researchers engaged in p-hacking strategies until their results were (barely) significant. Indeed, this pattern was found by some large-scale analyses of p-value distributions across multiple sciences, suggesting that p-hacking is widespread (e.g., Head et al., 2015).

These findings have been disputed for two reasons: First, other researchers have argued that a bump in the number of p-values just below 0.05 is a sufficient but not necessary condition for the presence of specific forms of p-hacking (Hartgerink, 2017) and that p-value distributions that reveal evidence of p-hacking likely look different (Lakens, 2015a). p-Value distributions depend on additional factors, such as power and publication bias. With some types of p-hacking (e.g., multiple testing and reporting the analysis that yielded the smallest p-value), the p-value distributions are not likely to reveal clusters just below 0.05 (Hartgerink, 2017; Lakens, 2015a, b). Second, re-analyses of the data by Head et al. (2015) and other studies did not find convincing evidence of a bump in the number of p-values just below 0.05 (Hartgerink, 2017; Lakens, 2015a, b).

Prevalence Estimates Based on Planned Versus Reported Analyses

A third approach is somewhat broader and does not apply to all of the p-hacking practices we discussed, but only to a subset. It compares records of studies that were openly available before publication with the final published paper. Thereby, this approach can reveal so-called selective reporting practices because it can detect the omission of variables or conditions that yielded undesired results, the underreporting of null results (publication bias), and HARKing (Cairo et al., 2020).

In one study, Franco et al. (2014) looked at a database of empirical studies that had been submitted for review at a National-Science-Foundation-sponsored program. They found that the publication probability of null findings was remarkably lower than for studies that yielded the desired results (a difference of approximately 40%). Hence, Franco et al.'s results indicate the presence of publication bias.

This approach has been further refined within organizational and management research (O'Boyle et al., 2017) and social psychology (Cairo et al., 2020). O'Boyle et al. (2017) vividly labeled the process of initial results (the ugly caterpillar) turning into a journal article (the beautiful butterfly) the "Chrysalis Effect." The authors compared 1978 hypotheses proposed in dissertations with hypotheses published in journal articles that were based on these dissertations. They found that 1000 hypotheses (!) were dropped in the process. The proportion of significant findings (the ratio of supported hypotheses to all contained hypotheses) increased by 21.0% (from 44.9% in dissertations to 65.9% in published articles). This inflation happened not only because hypotheses that did not yield the desired (i.e., significant) result had been dropped but also because new hypotheses had been added, the direction of predicted effects had been reversed, data had been altered, or variables had been selectively deleted or added (O'Boyle et al., 2017).

In social psychology, Cairo et al. (2020) looked at 100 dissertations, 373 published studies, and 1136 hypotheses and found that selective reporting practices were widespread. Supported hypotheses were four times more likely to end up in published journal articles than unsupported hypotheses and three times more likely to be reported unchanged. Again, the dropping of unsupported hypotheses alone resulted in a 20% inflation of significant findings in the published literature. In conclusion, the prevalence of *p*-hacking has been tackled via different approaches. All approaches have some merits but also some unresolved issues. Therefore, the actual frequency of *p*-hacking is unknown to date.

The Prevention of *p*-Hacking

Researchers have proposed, developed, and refined several solutions to prevent *p*-hacking practices. Some of them have been around for many years, for example, randomized controlled trials (RCTs). Others, such as preregistration, Registered Reports, and multiverse analyses, are more recent. In the following, we will describe some of the proposed solutions and discuss their good points and challenges.

RCTs

Medicine was one of the first disciplines to use registries. Initially, registries for clinical trials were aimed at facilitating the recruitment of patients, speeding up the dissemination of information, and reducing bias in the reporting of trials (Dickersin & Rennie, 2003). This idea is as timely as ever. Unfortunately, evaluations of RCTs have suggested that they often fall short of their potential. Of all registered trials, only about 50–60% end up in a journal, and those that find significant results have a higher probability of being published (Easterbrook et al., 1991; Tackett et al., 2019). This evidence of publication bias and low replication rates in registered clinical trials (e.g., Begley & Ellis, 2012; Prinz et al., 2011) has led to various attempts to improve the registration processes. New legal regulations, official statements (e.g., the Declaration of Helsinki), and technical advances have promoted centralized registries. These developments seem to have been successful at reducing selective reporting practices. For example, Kaplan and Irvin (2015) looked at large RCTs in drug research published between 1970 and 2012. They showed that after making the registration of primary outcomes obligatory on ClinicalTrials.gov in 2000, the percentage of positive results in the published trials dropped from 57% to 8%. They argued that both the obligatory prospective declaration of outcomes and improvements in transparency in the reporting standards may be responsible for this decline in the proportion of positive findings. Although one cannot be entirely certain that the inflation of positive findings before 2000 is purely due to *p*-hacking, it stands to reason that a flexible determination of the primary outcome after looking at the data may have played a role here. Therefore, Kaplan and Irvin concluded that the required registration of studies accompanied by improvements in the transparency of the RCTs were the key for the sharp increase in null findings.

Preregistration

Several clinical research questions cannot be addressed with RCTs, but alternative solutions can help researchers avoid *p*-hacking. One of them is preregistration (see also Chap. 15, this volume). Preregistrations state research objectives, report the study design, describe the planned sample (size), and detail the planned analyses. Thus, they allow for a comparison between the studies that have been conducted with the studies that have been published. For the prevention of *p*-hacking, preregistration has numerous benefits. For one, it allows confirmatory research to be distinguished from exploratory research. Specifying which analyses were planned a priori and which were run post hoc helps to prevent (or at least to detect) practices such as including covariates or excluding or switching outcomes.

The idea of registering empirical studies a priori is itself not new. For example, de Groot (1956/2014) determined that "it is essential that these hypotheses have been precisely formulated and that the details of the testing procedure (which should be as objective as possible) have been registered in advance" (p. 188). Similarly, concerning decisions about whether to perform a one-sided versus a two-sided test, Bakan (1966) stated: "How should this be handled? Should there be some central registry in which one registers one's decision to run a one- or two-tailed test before collecting the data? Should one, as one eminent psychologist once suggested to me, send oneself a letter so that the postmark would prove that one had pre-decided to run a one-tailed test?" (p. 431). Hence, the awareness that a priori predictions are essential in science has been around for over 60 years. But only the recent development of online tools that allow for the time stamping and freezing of research plans, accompanied by the acknowledgment of an imperative change in culture, have substantially improved the feasibility of preregistrations. Researchers may now use platforms such as the Open Science Foundation (OSF) or AsPredicted.org to share their research plans openly. Moreover, several templates tailored for different purposes (e.g., experimental research, longitudinal and experience sampling studies, analysis of existing data) may lower the threshold for undertaking a registration (see https://osf.io/zab38/wiki/home/).

Recently, Benning et al. (2019) and Krypotos et al. (2019) introduced helpful guidelines directed at clinical psychology. Benning et al. (2019) spoke of a continuum of registration because study registrations may vary in the timing and their disclosure. Whereas preregistrations (such as clinical trial registrations, Registered Reports or grant proposals) occur before the data are collected, *coregistrations* disclose decisions made after researchers began collecting data but before any data were analyzed. *Postregistrations* occur after data analysis has begun but still offer the opportunity to disclose specific analytic choices. In all types of registrations, researchers may register anything from a single specific aspect of data collection and analysis to complex decision trees that illustrate a series of decisions. Hence, Benning et al. (2019) presented registrations as a flexible framework for helping clinical researchers to increase the credibility of their work. To do this, Krypotos et al. (2019) provide a hands-on approach. They offer a step-by-step guide on preregistration, anonymizing data, and sharing both materials and data in psychopathology studies. The authors developed an open-source application based on the (free) statistical software package R (R Core Team, 2020) and git (a toolkit for tracking and merging changes) to facilitate version control and the time stamping of each step during the study. Researchers may thus use the same files throughout the study and easily track changes throughout the project.

One final remark about preregistration strikes us as important: A preregistration is more useful and effective at preventing *p*-hacking the more clearly and precisely it lays out the plan for a study. At the same time, it always has to be clear that a preregistration is a "plan, not a prison" (DeHaven, 2017). Making a plan to the best of one's ability is great, but there can always be reasons why it became necessary to deviate from this plan. This poses no problem as long as these deviations are transparently reported and explained.

Registered Reports

A particular type of preregistration is a Registered Report (Chambers et al., 2014). Registered Reports refer to a type of preregistration that is presented in an article format and undergoes peer review before the data are collected. In a first step, the authors submit a Stage 1 part of the manuscript, including the Introduction, Method, and pilot study results if available (Chambers et al., 2014). After revisions proposed by reviewers and the editor, the authors are offered an in-principle acceptance if the Stage 1 manuscript is sound. An in-principle acceptance guarantees the final paper's publication regardless of the results as long as the authors adhere to the approved protocol. After collecting and analyzing the data, the authors submit their initial Stage 1 manuscript along with the Results and Discussion sections. This Stage 2 manuscript may contain any unplanned, additional analyses labeled as "exploratory."

This publishing model prevents *p*-hacking—in addition to HARKing, problems with low power (because Stage 1 manuscripts are only accepted if the planned study seems adequately powered), and publication bias. Registered Reports alleviate the pressure to produce novel and astounding results and emphasize rigor and reproducibility instead (Chambers et al., 2014). Moreover, due to the in-principle acceptance, Registered Reports help (early career) researchers disseminate their ideas more quickly and increase the visibility of these ideas. Given these benefits, it is not surprising that this new submission category has been introduced in over 250 journals by now (https://www.cos.io/initiatives/registered-reports).

Multiverse Analyses

Whereas unreviewed and reviewed preregistration types provide a solution for distinguishing confirmatory from exploratory research, multiverse analyses (Steegen et al., 2016) help to prevent biases in exploratory research. They involve running the same analyses across all reasonable combinations of different transformations, exclusions, and inclusions of data and variables to examine how they affect the results and conclusions. Thus, multiverse analyses address all the arbitrary decisions that have to be made during data processing. They demonstrate the sensitivity of the results to analysts' arbitrary choices. Hence, transparent multiverse analyses leave it to the scientific community to gauge the fragility of the conclusions and their credibility.

Concluding Remarks: A Case for Open Science

In the present chapter, we discussed several *p*-hacking practices as part of the broader category of questionable research practices. Throughout the sections, it became clear that *p*-hacking can have detrimental consequences—particularly an increase in false-positive rates—that ultimately damage the trustworthiness and robustness of psychological science. In these concluding remarks, we would like to add a final nuance to the previous considerations by asking: Are *p*-hacking practices—or any questionable research practices for that matter—necessarily blameworthy after all?

For some practices, the answer is clear. They are simply wrong. For example, there is no justification for generously rounding off a *p*-value to 0.05 to make the result look significant if the actual value is higher. The *p*-value should be reported precisely to the third decimal place (APA, 2020). However, other practices might not be inherently wrong. In fact, they can be quite sensible, useful, or even necessary. For example, in general, more data are better than less data. So, continuing data collection after peeking at the data may be a good idea. Including a covariate can make a lot of sense. Trying many different ways to analyze a data set can be highly informative and a sign of conscientiousness instead of a questionable research practice and so on. What can make these practices bad scientific practice is not that they are conducted at all. Rather, researchers are engaging in bad practice when their actions are not transparently reported to make clear what parts of the data analysis were planned a priori and what parts were added as exploratory analyses. In addition, bad practice occurs when the increased Type I error rates that result from massaging the data are not controlled for.

If researchers transparently disclose their a priori data analysis plan, where they deviated from this plan, why they did so, and how this affected the results, there is nothing wrong with amply exploring the data and reporting emerging insights that seem interesting. In fact, we encourage all researchers to explore their data sets, run unplanned analyses, and come up with post hoc reasoning and new theoretical ideas—as long as these steps are labeled as such: post hoc. They can then be tested with confirmatory analyses in future research.

When appraising what is and is not questionable about questionable research practices, it becomes clear that some are not questionable, they are simply indefensible. Others might be better termed "questionable reporting practices," indicating that the problem lies in a lack of transparency more than in engaging in these practices per se (Wigboldus & Dotsch, 2016).

What can any researchers do to confirm that they did not engage in questionable research practices? The solution is surprisingly simple. It lies in the transparent distinction between a priori planned, confirmatory steps of data analysis and exploratory, additional steps. The line between the two can be drawn easily by adhering to the open science practices outlined above, particularly the detailed preregistration of all measures, manipulations, hypotheses, and planned analysis steps. Preliminary evidence suggests that this practice is remarkably effective. A recent analysis found that manuscripts published in the Registered Report format outperformed comparison papers published in the traditional format on 19 different criteria, including improvements in novelty, creativity, methodological rigor, and overall paper quality, among others (Soderberg et al., 2020).

Open science practices are surely not the solution to all challenges psychological science currently faces, but they are a pretty good and easy-to-implement solution to prevent *p*-hacking. Let's do it.

References

APA. (2020). *Publication manual of the American Psychological Association* (7th ed.). APA. https://apastyle.apa.org/products/publication-manual-7th-edition

Armitage, P., Berry, G., & Matthews, J. N. S. (2002). *Statistical methods in medical research* (4th ed.).

Bakan, D. (1966). The test of significance in psychological research. *Psychological Bulletin, 66*(6), 423. https://doi.org/10.1037/h0020412

Bakker, M., van Dijk, A., & Wicherts, J. M. (2012). The rules of the game called psychological science. *Perspectives on Psychological Science, 7*(6), 543–554. https://doi.org/10.1177/1745691612459060

Begley, C. G., & Ellis, L. M. (2012). Raise standards for preclinical cancer research. *Nature, 483*(7391), 531–533. https://doi.org/10.1038/483531a

Benning, S. D., Bachrach, R. L., Smith, E. A., Freeman, A. J., & Wright, A. G. C. (2019). The registration continuum in clinical science: A guide toward transparent practices. *Journal of Abnormal Psychology, 128*(6), 528–540. https://doi.org/10.1037/abn0000451

Bertamini, M., & Munafò, M. R. (2012). Bite-size science and its undesired side effects. *Perspectives on Psychological Science, 7*(1), 67–71. https://doi.org/10.1177/1745691611429353

Bishop, D. V., & Thompson, P. A. (2016). Problems in using p-curve analysis and text-mining to detect rate of p-hacking and evidential value. *PeerJ, 4*, e1715. https://doi.org/10.7717/peerj.1715

Button, K. S., Ioannidis, J. P. A., Mokrysz, C., Nosek, B. A., Flint, J., Robinson, E. S. J., & Munafò, M. R. (2013). Power failure: Why small sample size undermines the reliability of neuroscience. *Nature Reviews Neuroscience, 14*(5), 365–376. https://doi.org/10.1038/nrn3475

Cairo, A. H., Green, J. D., Forsyth, D. R., Behler, A. M. C., & Raldiris, T. L. (2020). Gray (literature) matters: Evidence of selective hypothesis reporting in social psychological research. *Personality and Social Psychology Bulletin, 46*(9), 1344–1362. https://doi.org/10.1177/0146167220903896

Camerer, C. F., Dreber, A., Forsell, E., Ho, T.-H., Huber, J., Johannesson, M., Kirchler, M., Almenberg, J., Altmejd, A., Chan, T., Heikensten, E., Holzmeister, F., Imai, T., Isaksson, S., Nave, G., Pfeiffer, T., Razen, M., & Wu, H. (2016). Evaluating replicability of laboratory experiments in economics. *Science, 351*(6280), 1433–1436. https://doi.org/10.1126/science.aaf0918

Chambers, C. D., Feredoes, E., Muthukumaraswamy, S. D., & Etchells, P. (2014). Instead of "playing the game" it is time to change the rules: Registered reports at AIMS neuroscience and beyond. *AIMS Neuroscience, 1*(1), 4–17. https://doi.org/10.3934/Neuroscience.2014.1.4

Cuijpers, P. (2016). Are all psychotherapies equally effective in the treatment of adult depression? The lack of statistical power of comparative outcome studies. *Evidence-Based Mental Health, 19*(2), 39–42. https://doi.org/10.1136/eb-2016-102341

Cuijpers, P., Smit, F., Bohlmeijer, E., Hollon, S. D., & Andersson, G. (2010). Efficacy of cognitive-behavioural therapy and other psychological treatments for adult depression: Meta-analytic study of publication bias. *The British Journal of Psychiatry: the Journal of Mental Science, 196*(3), 173–178. https://doi.org/10.1192/bjp.bp.109.066001

de Groot, A. D. (1956). The meaning of "significance" for different types of research [translated and annotated by Eric-Jan Wagenmakers, Denny Borsboom, Josine Verhagen, Rogier Kievit, Marjan Bakker, Angelique Cramer, Dora Matzke, Don Mellenbergh, and Han L. J. van der Maas]. *Acta Psychologica, 148*, 188–194. https://doi.org/10.1016/j.actpsy.2014.02.001

DeHaven, A. (2017). *Preregistration: A plan, not a prison*. https://www.cos.io/blog/preregistration-plan-not-prison.

Dickersin, K., & Rennie, D. (2003). *Registering clinical trials. Jama, 290*(4), 516–523. https://doi.org/10.1001/jama.290.4.516

Dienes, Z. (2008). *Understanding psychology as a science: An introduction to scientific and statistical inference*. Palgrave Macmillan.

Easterbrook, P. J., Gopalan, R., Berlin, J. A., & Matthews, D. R. (1991). Publication bias in clinical research. *The Lancet, 337*(8746), 867–872. https://doi.org/10.1016/0140-6736(91)90201-y

Fiedler, K., & Schwarz, N. (2016). Questionable research practices revisited. *Social Psychological and Personality Science, 7*(1), 45–52. https://doi.org/10.1177/1948550615612150

Franco, A., Malhotra, N., & Simonovits, G. (2014). Publication bias in the social sciences: Unlocking the file drawer. *Science, 345*(6203), 1502–1505. https://doi.org/10.1126/science.1255484

Friese, M., & Frankenbach, J. (2020). P-hacking and publication bias interact to distort meta-analytic effect size estimates. *Psychological Methods, 25*(4), 456–471. https://doi.org/10.1037/met0000246

Hartgerink, C. H. J. (2017). Reanalyzing Head et al. (2015): Investigating the robustness of widespread p-hacking. *PeerJ, 5*, e3068. https://doi.org/10.7717/peerj.3068

Head, M. L., Holman, L., Lanfear, R., Kahn, A. T., & Jennions, M. D. (2015). The extent and consequences of P-hacking in science. *PLoS Biology, 13*(3). https://doi.org/10.1371/journal.pbio.1002106

Heene, M., & Ferguson, C. J. (2017). Psychological science's aversion to the null, and why many of the things you think are true, aren't. In *Psychological science under scrutiny* (pp. 34–52). Wiley. https://doi.org/10.1002/9781119095910.ch3

Ioannidis, J. P. A. (2005). Why most published research findings are false. *PLoS Medicine, 2*(8), e124. https://doi.org/10.1371/journal.pmed.0020124

Ioannidis, J. P. A., Munafò, M. R., Fusar-Poli, P., Nosek, B. A., & David, S. P. (2014). Publication and other reporting biases in cognitive sciences: Detection, prevalence, and prevention. *Trends in Cognitive Sciences, 18*(5), 235–241. https://doi.org/10.1016/j.tics.2014.02.010

John, L. K., Loewenstein, G., & Prelec, D. (2012). Measuring the prevalence of questionable research practices with incentives for truth telling. *Psychological Science, 23*(5), 524–532. https://doi.org/10.1177/0956797611430953

Kaplan, R. M., & Irvin, V. L. (2015). Likelihood of null effects of large NHLBI clinical trials has increased over time. *PLoS One, 10*(8), e0132382. https://doi.org/10.1371/journal.pone.0132382
Kerr, N. L. (1998). HARKing: Hypothesizing after the results are known. *Personality and Social Psychology Review, 2*(3), 196–217. https://doi.org/10.1207/s15327957pspr0203_4
Klein, R. A., Ratliff, K. A., Vianello, M., Adams, R. B., Bahník, Š., Bernstein, M. J., Bocian, K., Brandt, M. J., Brooks, B., Brumbaugh, C. C., Cemalcilar, Z., Chandler, J., Cheong, W., Davis, W. E., Devos, T., Eisner, M., Frankowska, N., Furrow, D., Galliani, E. M., … Nosek, B. A. (2014). Investigating variation in replicability: A "many labs" replication project. *Social Psychology, 45*(3), 142–152. https://doi.org/10.1027/1864-9335/a000178
Krypotos, A.-M., Klugkist, I., Mertens, G., & Engelhard, I. M. (2019). A step-by-step guide on preregistration and effective data sharing for psychopathology research. *Journal of Abnormal Psychology, 128*(6), 517–527. https://doi.org/10.1037/abn0000424
Lakens, D. (2014). Performing high-powered studies efficiently with sequential analyses. *European Journal of Social Psychology, 44*(7), 701–710. https://doi.org/10.1002/ejsp.2023
Lakens, D. (2015a). Comment: What p-hacking really looks like: A comment on Masicampo and Lalande (2012). *Quarterly Journal of Experimental Psychology, 68*(4), 829–832. https://doi.org/10.1080/17470218.2014.982664
Lakens, D. (2015b). On the challenges of drawing conclusions from p-values just below 0.05. *PeerJ, 3*, e1142. https://doi.org/10.7717/peerj.1142
Leichsenring, F., Abbass, A., Hilsenroth, M. J., Leweke, F., Luyten, P., Keefe, J. R., Midgley, N., Rabung, S., Salzer, S., & Steinert, C. (2017). Biases in research: Risk factors for non-replicability in psychotherapy and pharmacotherapy research. *Psychological Medicine, 47*(6), 1000–1011. https://doi.org/10.1017/S003329171600324X
Lew, M. J. (2020). A reckless guide to P-values. In A. Bespalov, M. C. Michel, & T. Steckler (Eds.), *Good research practice in non-clinical pharmacology and biomedicine* (pp. 223–256). Springer International Publishing. https://doi.org/10.1007/164_2019_286
Lilienfeld, S. O. (2017). Psychology's replication crisis and the Grant culture: Righting the ship. *Perspectives on Psychological Science, 12*(4), 660–664. https://doi.org/10.1177/1745691616687745
Lilienfeld, S. O., & Waldman, I. D. (2017). *Psychological science under scrutiny: Recent challenges and proposed solutions*. Wiley.
Mayo, D. G. (2018). *Statistical inference as severe testing*. Cambridge University Press.
Munafò, M. R., Nosek, B. A., Bishop, D. V. M., Button, K. S., Chambers, C. D., Percie du Sert, N., Simonsohn, U., Wagenmakers, E.-J., Ware, J. J., & Ioannidis, J. P. A. (2017). A manifesto for reproducible science. *Nature Human Behaviour, 1*(1), 1–9. https://doi.org/10.1038/s41562-016-0021
Nelson, L. D., Simmons, J., & Simonsohn, U. (2018). Psychology's renaissance. *Annual Review of Psychology, 69*(1), 511–534. https://doi.org/10.1146/annurev-psych-122216-011836
Nosek, B. A., Spies, J. R., & Motyl, M. (2012). Scientific utopia: II. Restructuring incentives and practices to promote truth over publishability. *Perspectives on Psychological Science, 7*(6), 615–631. https://doi.org/10.1177/1745691612459058
O'Boyle, E. H., Banks, G. C., & Gonzalez-Mulé, E. (2017). The chrysalis effect: How ugly initial results metamorphosize into beautiful articles. *Journal of Management, 43*(2), 376–399. https://doi.org/10.1177/0149206314527133
Open Science Collaboration. (2015). Estimating the reproducibility of psychological science. *Science, 349*(6251), aac4716–aac4716. https://doi.org/10.1126/science.aac4716
Prinz, F., Schlange, T., & Asadullah, K. (2011). Believe it or not: How much can we rely on published data on potential drug targets? *Nature Reviews Drug Discovery, 10*(9), 712–712. https://doi.org/10.1038/nrd3439-c1
Protzko, J., & Schooler, J. W. (2017). Decline effects: Types, mechanisms, and personal reflections. In *Psychological science under scrutiny: Recent challenges and proposed solutions* (pp. 85–107). Wiley Blackwell. https://doi.org/10.1002/9781119095910.ch6
R Core Team. (2020). *R: A language and environment for statistical computing*. R Foundation for Statistical Computing. https://www.R-project.org/.

Rapport, M. D., Orban, S. A., Kofler, M. J., & Friedman, L. M. (2013). Do programs designed to train working memory, other executive functions, and attention benefit children with ADHD? A meta-analytic review of cognitive, academic, and behavioral outcomes. *Clinical Psychology Review, 33*(8), 1237–1252. https://doi.org/10.1016/j.cpr.2013.08.005

Reardon, K. W., Smack, A. J., Herzhoff, K., & Tackett, J. L. (2019). An N-pact factor for clinical psychological research. *Journal of Abnormal Psychology, 128*(6), 493–499. https://doi.org/10.1037/abn0000435

Sakaluk, J. K., Williams, A. J., Kilshaw, R. E., & Rhyner, K. T. (2019). Evaluating the evidential value of empirically supported psychological treatments (ESTs): A meta-scientific review. *Journal of Abnormal Psychology, 128*(6), 500–509. https://doi.org/10.1037/abn0000421

Schönbrodt, F. D. (2016). *p-hacker: Train your p-hacking skills!.* http://shinyapps.org/apps/p-hacker/.

Silberzahn, R., Uhlmann, E. L., Martin, D. P., Anselmi, P., Aust, F., Awtrey, E., Bahník, Š., Bai, F., Bannard, C., Bonnier, E., Carlsson, R., Cheung, F., Christensen, G., Clay, R., Craig, M. A., Dalla Rosa, A., Dam, L., Evans, M. H., Flores Cervantes, I., … Nosek, B. A. (2018). Many analysts, one data set: Making transparent how variations in analytic choices affect results. *Advances in Methods and Practices in Psychological Science, 1*(3), 337–356. https://doi.org/10.1177/2515245917747646

Simmons, J. P., Nelson, L. D., & Simonsohn, U. (2011). False-positive psychology: Undisclosed flexibility in data collection and analysis allows presenting anything as significant. *Psychological Science, 22*(11), 1359–1366. https://doi.org/10.1177/0956797611417632

Simonsohn, U., Nelson, L. D., & Simmons, J. P. (2014). P-curve: A key to the file-drawer. *Journal of Experimental Psychology: General, 143*(2), 534–547. https://doi.org/10.1037/a0033242

Soderberg, C. K., Errington, T., Schiavone, S. R., Bottesini, J. G., Thorn, F. S., Vazire, S., Esterling, K. M., & Nosek, B. A. (2020). Initial evidence of research quality of registered reports compared to the traditional publishing model. *MetaArXiv*. https://doi.org/10.31222/osf.io/7x9vy

Steegen, S., Tuerlinckx, F., Gelman, A., & Vanpaemel, W. (2016). Increasing transparency through a multiverse analysis. *Perspectives on Psychological Science, 11*(5), 702–712. https://doi.org/10.1177/1745691616658637

Swift, J. K., Christopherson, C. D., Bird, M. O., Zöld, A., & Goode, J. (2020). Questionable research practices among faculty and students in APA-accredited clinical and counseling psychology doctoral programs. *Training and Education in Professional Psychology*. https://doi.org/10.1037/tep0000322

Tackett, J. L., Brandes, C. M., King, K. M., & Markon, K. E. (2019). Psychology's replication crisis and clinical psychological science. *Annual Review of Clinical Psychology, 15*(1), 579–604. https://doi.org/10.1146/annurev-clinpsy-050718-095710

Tackett, J. L., & Miller, J. D. (2019). Introduction to the special section on increasing replicability, transparency, and openness in clinical psychology. *Journal of Abnormal Psychology, 128*(6), 487. https://doi.org/10.1037/abn0000455

Wasserstein, R. L., & Lazar, N. A. (2016). The ASA statement on p-values: Context, process, and purpose. *The American Statistician, 70*(2), 129–133. https://doi.org/10.1080/00031305.2016.1154108

Wicherts, J. M., Veldkamp, C. L. S., Augusteijn, H. E. M., Bakker, M., van Aert, R. C. M., & van Assen, M. A. L. M. (2016). Degrees of freedom in planning, running, analyzing, and reporting psychological studies: A checklist to avoid p-hacking. *Frontiers in Psychology, 7*. https://doi.org/10.3389/fpsyg.2016.01832

Wigboldus, D. H. J., & Dotsch, R. (2016). Encourage playing with data and discourage questionable reporting practices. *Psychometrika, 81*(1), 27–32. https://doi.org/10.1007/s11336-015-9445-1

Yarkoni, T., & Westfall, J. (2017). Choosing prediction over explanation in psychology: Lessons from machine learning. *Perspectives on Psychological Science, 12*(6), 1100–1122. https://doi.org/10.1177/1745691617693393

Chapter 6
Data Detective Methods for Revealing Questionable Research Practices

Gregory Francis and Evelina Thunell

Abstract There are many types of Questionable Research Practices (QRPs) that all tend to generate statistical information that misrepresents reality. This chapter discusses some methods for detecting the presence of QRPs, mostly by looking for conflicts in different sources of information. These methods typically cannot identify precisely which QRPs were used, and sometimes the conflicts are due to typos or simple mistakes, but either way readers should be skeptical about the validity of studies with inconsistent statistical information. An appropriate mindset for identifying inconsistencies is that of a "data detective" who looks for patterns that do not make sense. We start by describing mathematical inconsistencies between sample sizes and the degrees of freedom in hypothesis tests, which are easy to detect and indicate either a QRP, unreported outlier removal, or sloppiness in reporting. A similarly easy check is the use of the STATCHECK program to identify inconsistencies between reported test statistics and *p*-values, which may indicate sloppiness in reporting or improper rounding to conclude statistical significance. Similar problems can also be discovered with the GRIM test, which identifies situations where reported means or proportions are impossible for the given measurement and sample size(s). Two additional tests explore inconsistencies across experiments. First, the Test for Excess Success compares the frequency of reported successful outcomes to the expected frequency if the tests were run properly, fully reported, and analyzed without QRPs. Too much success indicates a problem with the reported results (possibly because of QRPs). Second, the *p*-curve analysis examines the distribution of reported *p*-values for properties that indicate invalid data sets (that are perhaps the result of QRPs).

G. Francis (✉)
Department of Psychological Sciences, Purdue University, West Lafayette, IN, USA
e-mail: gfrancis@purdue.edu

E. Thunell
Department of Psychological Sciences, Purdue University, West Lafayette, IN, USA

Department of Clinical Neuroscience, Karolinska Institutet, Stockholm, Sweden

W. O'Donohue et al. (eds.), *Avoiding Questionable Research Practices in Applied Psychology*, https://doi.org/10.1007/978-3-031-04968-2_6

Keywords Questionable research practice · Clinical psychology · Excess success · Data detective methods for revealing questionable research practices · STATCHECK program · GRIM test

Introduction

As discussed in other chapters, questionable research practices (QRPs) and *p*-hacking can turn non-conclusive data sets into seemingly interesting findings. While such practices might be tempting for a researcher who is desperate to publish their work in fancy journals, they come at the expense of the credibility and reproducibility of the findings. Examples of QRPs are publication bias (reporting significant findings but not reporting relevant non-significant findings), inappropriate sampling (e.g., adding data points until achieving statistical significance), inappropriate analyses (e.g., trying various analyses and reporting only the ones that give the wanted result), and hypothesizing after the results are known (HARKing; inventing a new theory and hypothesis that matches your results). Hypothesis testing is the dominant statistical analysis method in clinical psychology, and it comes with strict requirements and rules that are violated in different ways by QRPs. The impact of using QRPs is a kind of bias that misrepresents reality.

QRPs can make studies appear to provide strong support for effects that do not exist in reality. That is, the results seem to support the alternative hypothesis, but the null hypothesis is actually true. How then can we distinguish scientific results that are valid from results that are based on QRPs? Luckily, QRPs tend to leave a pattern of statistical evidence that can be used to identify their presence. In this chapter, we show how to detect and interpret such patterns.

In many respects, revealing the patterns generated by QRPs is similar to a detective trying to crack a case. The information may not be right in front of you, but different clues can be combined to demonstrate problems with experimental results and conclusions that are based on QRPs. In this chapter, we describe a number of methods that help you act like a data detective and identify problems in reported statistics.

Mathematical Inconsistencies and Data Gleaning

A valuable skill for a data detective is recognizing how to extract relevant information from what the authors themselves report. Here, we review some approaches that have proven useful for identifying problems with reported results.

A simple approach for detecting errors in reported results is to look for numerical inconsistencies. For example, many statistical tests (e.g., *t* and *F* tests) are based on distributions with a "degrees of freedom" (df) value. For example, a one-sample *t*-test has df $= n - 1$, where n is the sample size, while a two-sample *t*-test has df = $n_1 + n_2 - 2$, where n_1 and n_2 are the sizes of the two samples. Likewise, an independent one-way ANOVA F-test has two degrees of freedom terms called $df_{numerator} = K - 1$ and $df_{denominator} = N - K$. Here, K is the number of conditions and N is the sum of sample sizes across all conditions. Scientific papers usually report the sample sizes and the number of conditions, so it is relatively easy to calculate the degrees of freedom. Thus, you can easily check the following text: "As predicted, with $n_1 = 35$ and $n_2 = 27$, we found a significant difference between the control and experimental means $t(58) = 2.1$, $p = 0.04$." The authors report 58 degrees of freedom, but using the formula above for the two-sample t-test you know that the degrees of freedom should actually be $n_1 + n_2 - 2 = 60$. An inconsistency of this type might indicate that the authors removed some participants from their data set without reporting this, but still properly reported the degrees of freedom for the remaining data. Outlier removal is not necessarily a QRP, but sometimes participants are removed because their absence allows the remaining data to show a significant ($p < 0.05$) result. At any rate, data removal should be fully reported and justified. Errors of this type are rather common. At best they indicate sloppiness, and regardless of their source should prompt you to feel less confident in the reported results and their associated conclusions. The next section describes a conceptually similar check for inconsistencies that often have more severe consequences.

STATCHECK

Most statistical analyses in psychology use hypothesis testing to determine whether there is an "effect." Typically, this is done by defining an "alternative hypothesis" that there is a true effect and a null hypothesis that indicates "no effect." For example, when testing whether a drug is effective at reducing the duration of a cold, the null hypothesis H_0 might look like:

$$H_0 : \mu_1 = \mu_2$$

where μ_1 and μ_2 denote the duration of the cold with and without the drug, respectively. Thus, the null hypothesis states that there is no difference in the population mean durations with or without the drug whereas the alternative hypothesis states that the drug does change the duration. The goal of the hypothesis test is to decide whether to reject the null hypothesis. This decision is based on "statistical significance," which is determined by a test statistic that is derived from the experimental data. A two-sample t test for independent equal means has a test statistic of:

$$t = \frac{\bar{X}_1 - \bar{X}_2}{s_{\bar{X}_1 - \bar{X}_2}} = \frac{\bar{X}_1 - \bar{X}_2}{s\sqrt{\frac{1}{n_1} + \frac{1}{n_2}}}$$

where $\bar{X}_1$ and $\bar{X}_2$ are the sample means and $s_{\bar{X}_1-\bar{X}_2}$ is the standard deviation of the sampling distribution of the difference of means, which is a function of the standard deviation of each sample, s, and the sample sizes n_1 and n_2. If the null hypothesis is true, the t-value is usually close to 0. The hypothesis test will then not reject the null hypothesis. If the alternative hypothesis is true and the sample sizes are large enough, the t-value will typically deviate substantially from 0. In this case, the researcher rejects the null hypothesis and can argue for their alternative hypothesis. However, just due to random sampling, the t-value will sometimes deviate from 0 even if the null hypothesis is true, and the researcher will then erroneously reject the null hypothesis. How often this so-called Type I error happens is controlled by the researcher through a significance criterion, α.

Oftentimes, the criterion is set to $\alpha = 0.05$, meaning that the probability of concluding that an effect exists when it truly does not is 5%. The decision about whether to reject the null hypothesis and thus conclude that an effect exists (concluding statistical significance) is based on the p-value (the area under the tail, beyond the observed t-value, of the t sampling distribution if the null hypothesis is true). If $p < \alpha$, then the observed t-value deviates more from 0 than what should be common if the null hypothesis is true. Therefore, the researchers conclude that there seems to be an effect: they reject the null hypothesis and claim that the observed difference of means is "statistically significant."

When reporting the results of a hypothesis test it is common to report the computed t-value, the corresponding degrees of freedom (which depends on the sample size(s)), and the p-value. It often looks like: $t(48) = 2.55$, $p = 0.014$. It is actually redundant to report both the t- and p-values, as there is a one-to-one relationship between them for a given degrees of freedom. This redundancy can be used to check the reported statistics.

For example, suppose you read an article that reports: "As predicted we found a significant difference between the control and experimental conditions, $t(22) = 2.00$, $p < 0.05$." For the given degrees of freedom (df = 22) and t-value, one can recompute the corresponding p-value[1] to discover that $p = 0.058$. Thus, the reported t-value is incompatible with the statement $p < 0.05$. Instead, the result is actually not statistically significant (because $p > \alpha = 0.05$). The mathematics in the original text, therefore, indicates that something is wrong with the numbers. p-value inconsistencies can come about from simple typos (e.g., typing 0.014 instead of 0.14), or honest mistakes (e.g., copying the wrong line from the output of statistical software). In some cases (as in the above example), p-value inconsistencies might be because authors "round down" a reported p-value in order to make readers believe an experiment produced statistical significance. This kind of inappropriate rounding is a QRP. Regardless of how they appear, p-value inconsistencies should raise concerns about the reported results and their associated conclusions.

---

[1] For example, with the online calculator at https://introstatsonline.com/chapters/calculators/t_dist.shtml

STATCHECK is an online program (http://statcheck.io) that automates this kind of consistency check. To use it, simply upload a copy of an article and let STATCHECK scan it for statistical information. Just as we did above, STATCHECK identifies test statistics and their accompanying degrees of freedom, recomputes the p-value based on these numbers, and then compares it to the reported p-values. STATCHECK includes some additional computations (such as checking on whether the recomputed p-value is close enough to the reported p-value for appropriate rounding to be an explanation, and identifying whether an inconsistency in p-values changes the decision on statistical significance). STATCHECK works for a variety of statistical tests.

Some limitations of STATCHECK include that it cannot process certain file formats, it typically does not distinguish between one- and two-sided tests, and it cannot parse non-standard formats for reporting statistical outcomes. These limitations cause STATCHECK to sometimes omit or misinterpret statistical test results, and it is therefore always advisable to manually check the statistics flagged by STATCHECK.

Errors of this type are shockingly common. Systematic investigations of scientific articles have found that around half of them have at least one inconsistent p-value and that around 12–14% of the articles contain an inconsistency that alters the interpretation of statistical significance.

GRIM Tests

Another way of identifying inconsistencies in statistical reporting is to notice a relationship between sample sizes and measured values. Let's consider a simple case. Suppose you receive a marketing report for a survey to evaluate how many people might be interested in a new product (a macaroni-and-cheese pizza) at your restaurant. One of your employees runs a survey on $n = 37$ people and reports that 56% of the people expressed interest in the new product. Your first reaction might be that the survey seems pretty promising for your new product. A bit of data detective work, however, suggests that you should assign the survey task to a different employee. The percentage calculation is computed from the following formula

$$\%\text{Interest} = \frac{f}{n} \times 100$$

where f is the number of survey respondents who are interested in your product and $n = 37$ is the number of people who participated in the survey. Let's deduce the value for f by plugging in the values reported by your employee

$$56 = \frac{f}{37} \times 100$$

With a bit of algebra, we find that $f = 20.72$. We know this value for f cannot be quite right because there cannot be fractions of respondents. Could the reported percentage have been rounded from the true value? We can check by looking at nearby values of f. For example, if $f = 21$ then we would get

$$\%\text{Interest} = \frac{21}{37} \times 100 = 56.76$$

Unfortunately, this value does not explain why your employee reported 56% because rounding of 56.76% would produce 57%. What if $f = 20$? Then we get

$$\%\text{Interest} = \frac{20}{37} \times 100 = 54.05,$$

which is too small to be rounded up to 56%. In fact, with $n = 37$ people in the survey it is impossible for the percentage to equal 56%, even after rounding. So, either your employee misreported the number of people in the survey or simply made up the numbers. At any rate, you should hold off on making changes to your menu until you resolve the inconsistency.

Similar logic applies to reported values of means. For example, suppose a survey asks people to rate, on an integer scale from 1 to 7, how much interest they have in a macaroni-and-cheese pizza. A rating of 1 indicates no interest at all and a rating of 7 indicates that they want it *now!* The employee responsible for the survey reports that 55 responders gave a mean value of 4.74, which indicates interest above the middle point of the scale. The computation of the mean, $\overline{X}$, is based on the following formula:

$$\overline{X} = \frac{\Sigma X_i}{n},$$

where X_i refers to the score for responder number i, and the capital sigma indicates to sum the scores of all the responders. Thus, with the reported mean and sample size, we can solve for the sum of scores:

$$\Sigma X_i = (n)\overline{X} = (55)4.74 = 260.7$$

Importantly, the scores can only take integer values (1, 2, 3, 4, 5, 6, or 7) because that is the nature of the rating scale. This means that the sum of scores must also be an integer value, which it is not in the above calculation. Did we get a decimal value for the sum because the reported mean was rounded from its true value? We can check this possibility by considering nearby values for the sum of scores and seeing if the corresponding mean value matches what was reported. For example, using $\sum X_i = 261$ (e.g., rounding up to the nearest integer) gives

$$\bar{X} = \frac{\Sigma X_i}{n} = \frac{261}{55} = 4.7455$$

which would round up to 4.75 and so cannot correspond to the reported mean of 4.74. Likewise, using a smaller value for the sum of scores such as 260 would give

$$\bar{X} = \frac{\Sigma X_i}{n} = \frac{260}{55} = 4.727$$

which would round up to 4.73, and thus is too small to match the reported mean of 4.74. Once again, a mean value of 4.74 is mathematically impossible for a sample of size $n = 55$ when measuring ratings with this kind of 1–7 scale.

Note that this kind of inconsistency is sometimes explained by rounding of reported statistics. If the sample size was $n = 46$, a mean of $\bar{X}$ =4.74 would be fine because

$$\Sigma X_i = (n)\bar{X} = (46)4.74 = 218.04$$

which rounds down to 218. A re-computation of the sample mean gives:

$$\bar{X} = \frac{\Sigma X_i}{n} = \frac{218}{46} = 4.739$$

which rounds up to match the reported value of 4.74. Thus, here the reported mean is consistent with the sample size, the nature of the scale, and a bit of rounding for reported values.

These types of calculations are referred to as exploring the Granularity-Related Inconsistency of Means (GRIM). Many of the calculations described above can be automated in a spreadsheet. We have provided such a spreadsheet, GrimTest.xls, at the Open Science Framework (https://osf.io/k8yjc/). Enter a reported mean (or proportion) and a sample size, and the spreadsheet indicates whether the numbers make sense.

With a bit of ingenuity and algebra, one can apply the GRIM analysis also to other situations. For example, sometimes an article reports the combined sample size across two samples and proportions or means for each sample but not the specific size of each sample. A variation of the GRIM test might consider all possible sample size combinations that add up to the reported combined sample size and see if any combination is consistent with the reported means or proportions. In some cases, it is possible to use both means and standard deviations to identify inconsistencies.

GRIM inconsistencies can occur because of typos or other forms of sloppiness. They can also happen through QRPs such as removing data from the sum of scores but not taking their removal into account when reporting the sample size. In some cases, a GRIM inconsistency may indicate that the reported data is simply "made up." Whether based on fraud, tinkering, or a typo, readers of data with a GRIM inconsistency should be skeptical about the reported results and their implications.

Data Extraction Techniques

Many GRIM inconsistencies could be easily resolved if scientists shared their data and analysis code. Regrettably, this is not the norm. Even though many journals formally require authors to share their data, it is uncommon for authors to do so, and the journals often do not ensure that authors follow the rules.

Data sharing also has other advantages, such as allowing a scientific field to take full advantage of a scientist's empirical work by allowing other researchers to explore additional aspects of the data or use it to guide new experiments. Until data sharing becomes common, data detectives can use a variety of techniques to glean some statistical information from reported statistics. Here, we show how some of these techniques can be combined.

Figure 6.1a schematizes the stimuli in a spatial cuing experiment. On each trial, a participant looks at a computer screen that briefly flashes a central arrow pointing to the left or to the right and then shows a target letter either to the left or to the right. The observer's task is to identify the target letter as quickly as possible by making a button-press, and the computer measures their response time. On 80% of the trials, the arrow points to where the target letter is about to appear, so observers learn to attend to the indicated location. The experiment investigates how much such

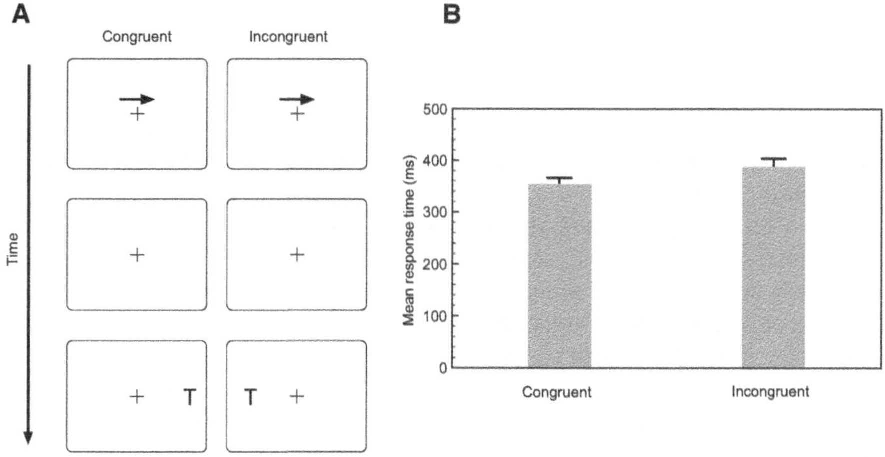

Fig. 6.1 Stimuli and results for a spatial cueing experiment. (**a**) shows stimuli for congruent and incongruent trials. In congruent trials the arrow points to the location of a subsequent target letter. In incongruent trials the arrow points to the opposite location of the target letter. (**b**) shows the results. Mean response time is shorter for congruent trials than for incongruent trials. The error bars indicate one standard error of the mean

attention affects the speed of letter identification. Each observer produces a mean response time for congruent trials (when the arrow points to the location of the target letter) and for incongruent trials (when the arrow points to the opposite side of the location of the target letter). Figure 6.1b shows typical data from $n = 31$ observers (the data are available in a spreadsheet, SpatialCueingData.xlsx, at the Open Science Framework). It indicates that the mean response time is shorter for the congruent than for the incongruent trials. The error bars indicate the standard error across observers for each condition.[2]

It is common to present findings with a data plot (like Fig. 6.1b) along with a summary of a statistical test. Here, the test is a dependent t test that compares mean response times for congruent and incongruent conditions: $t(30) = 2.13, p = 0.04$. For a data detective, there is more quantitative information than what is directly reported. For example, you might want to know the standard deviations for the conditions and the correlation across observers. The standard error for a condition, $S_{\overline{X}}$, is related to the standard deviation of the data, S, by the formula:

$$S_{\overline{X}} = \frac{S}{\sqrt{n}}$$

So if we know the standard error, we can easily solve for the standard deviation. The error bars in Fig. 6.1b indicate the standard error, so we just need to extract the information from the plot. We do this using a program called *Plot Digitizer*, which prompts the user to identify the ends of each axis and then click on points of interest in the plot. The program computes the position of each marked point in the plot. Figure 6.2 shows the two windows from *Plot Digitizer* that report the height of each bar and its associated error bar.

The values under the "Condition" column in the small window to the left indicate the x-value of each point, in the order they were clicked. In this bar plot, the x-values simply indicate the two conditions. We are more interested in the values in the "Response Time" column. The first two values refer to the mean and top of the error bar for the congruent condition, and the last two values refer to the mean and top of the error bar for the incongruent condition. The height of the error bar above the mean is the standard error, and so we can compute the standard error $S_{\overline{X}}$ with the following formula:

$$S_{\overline{X}} = \text{Error bar height} - \text{Mean}$$

We can then easily compute the standard deviation S for each condition as:

$$S = S_{\overline{X}}\sqrt{n}$$

[2] Sometimes authors compute an error bar using the standard deviation across observers or the range of a 95% confidence interval; it is typical for the figure caption to indicate the basis of each error bar.

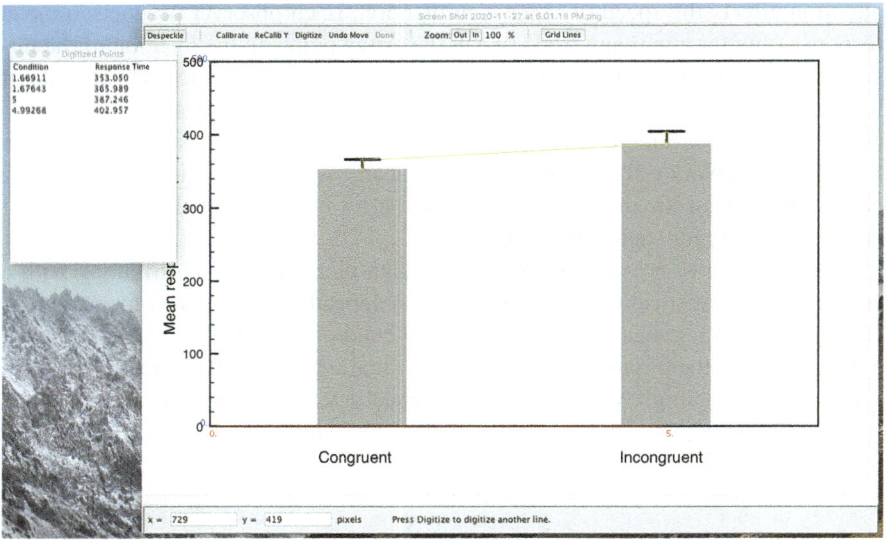

Fig. 6.2 Data gleaning of spatial cueing data using the *Plot Digitizer* program. The yellow lines on the plot connect selected points

Finally, we can compute the correlation between the congruent and incongruent conditions by using the variance sum law, which describes how the variance of difference scores S^2_{x-y} is related to the variance of each score and their correlation r:

$$S^2_{x-y} = S^2_x + S^2_y - 2rS_xS_y$$

Here, we use variables x and y to refer to the two correlated measures (e.g., congruent and incongruent response times). Some algebra shows that the correlation must be:

$$r = \frac{S^2_{x-y} - S^2_x - S^2_y}{-2S_xS_y}$$

We can compute the variance of difference scores from the means and t-value because the t-value is given by:

$$t = \frac{\overline{X} - \overline{Y}}{S_{\overline{X}-\overline{Y}}} = \frac{\overline{X} - \overline{Y}}{S_{x-y} / \sqrt{n}}$$

A bit of algebra results in a formula for the standard deviation of the difference scores S_{x-y}:

$$S_{x-y} = \frac{\overline{X} - \overline{Y}}{t}\sqrt{n}$$

Table 6.1 compares the values gleaned from Fig. 6.1b and the above computations against the values computed directly from the raw data. One can see that the gleaned values are quite close to the actual values. Small discrepancies exist because it is difficult to place the clicks directly on the top of the bars in the plot and because the reported t-value is rounded to two decimal places. Using the gleaned values to estimate the correlation between congruent and incongruent response times gives $r = 0.385$. The true correlation (computed from the raw data) is $r = 0.391$.

These extraction techniques can also be used to identify non-obvious inconsistencies in a data set. For example, suppose the text describing a dependent samples t-test reported the following, "As predicted, there was a significant difference, $t(30) = 2.8$, $p = 0.009$, between the control ($\overline{X} = 45$, $s = 7.3$) and experimental ($\overline{X} = 55$, $s = 7.6$) conditions." While this result might seem like convincing support for there being a difference in means, it actually makes no sense at all. The reported degrees of freedom for the dependent t-test indicates that $n = 31$. Combining this sample size with the reported mean and t-values, the standard deviation of the difference scores can be computed using the formula above, $S_{x-y} = 19.88$. Now, we can check whether this value is possible with the standard deviations given for each condition. Solving for the correlation between scores in the control and experimental conditions using the formula above gives $r = -2.56$, which violates the constraint that correlations must always be between plus and minus one. Thus, we can conclude that the reported numbers cannot be correct.

This section has mostly dealt with mathematical inconsistencies in statistical reports. Standard reporting formats include redundant information that sometimes allow data detectives to identify inconsistencies. With these methods, the data detective checks for inconsistencies in the reported results of a single experiment. In the next section we identify two methods for characterizing inconsistencies *across* experiments.

Table 6.1 True and gleaned values for the means, standard errors, and standard deviations of the spatial cueing data

Statistic	Congruent		Incongruent	
	True	Gleaned	True	Gleaned
Mean	353.64	353.05	387.85	387.25
SE	12.45	12.94	16.10	15.71
SD	69.30	72.04	89.63	87.48

Experimental Inconsistencies

Hypothesis testing is often presented as a way of drawing conclusions within a single experiment. However, sometimes conclusions are based on statistical outcomes *across* experiments, and the properties of hypothesis testing impose important constraints in such situations. We will describe two analysis methods that look for violations of these constraints. Conceptually, identifying inconsistencies across experiments is similar to identifying mathematical inconsistencies within an experiment. However, there are two important differences. First, mathematical inconsistencies could potentially be due to typos or calculation errors rather than QRPs. The same interpretation is usually not plausible for inconsistencies across experiments. Second, mathematical inconsistencies are definitive in the sense that there is no way for the numbers to make logical sense. Inconsistencies across experiments, on the other hand, are defined as improbable (rather than impossible) inferential outcomes. These inconsistencies suggest the involvement of QRPs because observed outcomes would be very rare if QRPs were not involved.

Test for Excess Success

In most experiments in clinical psychology, conclusions are based on hypothesis testing. Due to how samples are randomly selected for such tests, it sometimes happens that a test draws the wrong conclusion. For example, it is possible that a population with a true null hypothesis produces a significant outcome simply due to the scientist happening to get an unusual sample of data. The hypothesis testing procedure for drawing a conclusion controls the rate of making such a Type I error; and scientists typically set that rate to be 5%. Likewise, it is possible that a population with a true effect produces a non-significant outcome due to the scientist happening to get an unusual data sample. The probability of making such a Type II error is not directly controlled in hypothesis testing, unless the scientist has a good idea of the size of the true effect and gathers a large enough sample of data.

An important implication of drawing conclusions based on hypothesis tests is that mistaken conclusions are *inevitable*. Even when doing everything correctly (in terms of random sampling, analyzing the data, and reporting the results), scientists *must* sometimes make the wrong decision. Consider the *power* of an experiment. Power is the complement of Type II error, meaning that it refers to the probability that a hypothesis test based on a random sample of data will reject the null hypothesis when this is the correct conclusion (there really is an effect). Power depends on the size of the effect and on the size of the sample, in that larger effects and larger samples give higher power. Oftentimes, scientists do not try to control power, because the effect size is unknown. When power is considered, scientists often aim for sample sizes that give at least 80% power. However, this is an arbitrary target, and it is sometimes inappropriate. Consider a scientist who plans two independent experiments, and will draw a conclusion in favor of some theoretical conclusion only if both experiments show significant effects. If each experiment has 80% power, then the probability of both experiments producing significant results is $0.8^2 = 0.64$. Thus, even though the power of each experiment is acceptable when considered alone, the odds of the scientist finding support for their theoretical conclusion are only slightly better than a coin flip.

As additional successful experiments are added to the list of requirements for drawing a theoretical conclusion, the probability of consistent success decreases. Out of 20 experiments, one should expect on average $0.8 \times 20 = 16$ significant outcomes. The probability of all 20 experiments producing significant outcomes is only $0.8^{20} \approx 0.01$. Thus, if a scientist reports that 20 out of 20 experiments each with a power of 0.8 produced significant outcomes, this should not be interpreted as strong evidence for the theoretical conclusions but instead as an indication that something has gone wrong; in particular it suggests that the scientist engaged in some types of QRPs. The absence of non-significant findings in experiments with limited power is a marker for flaws in the scientific process because the reported findings seem "too good to be true."

These observations can be quantitatively formalized with the Test for Excess Success (TES). By estimating effect magnitudes from the reported experiments, this method estimates the success rate of future experiments that use the same sample sizes. The success rate is an estimate of the probability of future replication experiments to produce the same degree of success as the original experiments. If this rate is low (0.1 is a common, if arbitrary, threshold), then the reported results of the original studies are deemed problematic (too good to be true).

To demonstrate how to perform a TES analysis, consider a prominent paper that reported six experiments investigating the impact of poverty on cognitive performance. The main claim was that poverty-related concerns use mental resources that would otherwise be available for other tasks. This claim implies that poor people make bad choices partly because they are poor, rather than being poor because they make bad decisions. If true, this finding has many important policy implications. When deciding on how to best help poor people, one needs to consider their lower cognitive capabilities, which may vary with their financial situation. The paper describing these six experiments was published in the journal *Science*, which is widely regarded as the most prestigious scientific academic journal, and the findings were considered important enough to merit mention in the *New York Times* and numerous other media outlets. Below, we use a TES analysis to show that these results actually do not adequately support the theoretical claims. Arguably, some of the findings were produced with QRPs.

For each of the six studies, we can extract the statistics for the relevant hypothesis tests. For most of the studies, multiple hypothesis tests were performed. However, to keep the current analysis simple, we estimate an upper limit of the success rate for each experiment by considering only the statistically weakest relevant test. This approach is conservative, since an experiment is always less likely to produce multiple specific outcomes than only one of the outcomes.

A key result from Experiment 1 was an interaction between income (rich or poor, defined by a median split) and condition (scenarios describing hard or easy to manage financial difficulties). The measurements included performance on a Raven's matrices task (a measure of fluid intelligence) and a cognitive control task. To estimate an upper limit of the power of Experiment 1, we used the weaker of the results from these two measures. The calculation of power is done in an R program (TESAnalysis.R) that is available for download at the Open Science Framework. Without going into the specific formulas, the program converts the sample sizes and test statistics (F- or t-value) into a standardized effect size (Hedges' g). This standardized effect size is then used to estimate the probability that a new experiment with the same sample size as the original experiment would produce a significant outcome. As the first row of Table 6.2 indicates, the power is around 0.6. So, if the effect is real and similar to what was originally reported, future replication studies with the same sample size have around a 60% chance of producing a significant outcome.

Experiment 2 was similar to Experiment 1, but with nonfinancial scenarios. The prediction of the authors was that this design would *not* produce a significant difference between rich and poor participants; and that was precisely what they reported. The success probability for this experiment is computed as one minus power, which gives the probability of a random sample *not* producing statistical significance. As shown in Table 6.2, the success probability is rather high because it is easy to not produce a significant outcome with a small sample.

Experiment 3 added monetary incentives for correct responses and found similar effects as for Experiment 1. Namely, there was a significant interaction between income and scenario for measures of cognitive control. If the effect is similar to what is reported in Experiment 3, then the power of a replication experiment with the same sample sizes is just above 0.5.

Table 6.2 Estimated success probabilities for six experiments investigating poverty and cognition. The probability of all six experiments producing successful outcomes is so low (0.065) that the results seem too good to be true

Experiment	Test	Reported statistics	Success probability
1	Interaction for Raven's matrices	$F(1,97) = 5.12$, $p = 0.03$	0.602
2	Non-significant difference for rich and poor on cognitive control	$F(1,35) = 1.69$, $p = 0.20$	0.764
3	Interaction for cognitive control	$F(1,98) = 4.31$, $p = 0.04$	0.532
4	Interaction for Raven's matrices	$F(1,92) = 4.04$, $p = 0.04$	0.505
Field 1	Pre- and post-harvest differences	$p < 0.001$	~1
Field 2	Pre- and post-harvest heart rate (stress)	$t(187) = 1.715$, $p = 0.088$	0.523
P_{TES}			0.065

Experiment 4 was very similar to Experiment 1, but with a different order of some tasks. A key result is an interaction between income and condition for the Raven's matrices task. Power for a replication experiment is barely above 0.5. We should note that the reported statistics for Experiment 4 show a p-value inconsistency. A recalculation shows that $F(1,92) = 4.04$ corresponds to $p = 0.047$ rather than the reported $p = 0.04$. For the TES analysis, we assume that the reported F-value is correct.

To explore the generality of the findings beyond the controlled settings of Experiments 1–4, two field studies were run to investigate cognitive performance for farmers in India. The first field study found strong differences in cognitive performance for farmers pre-harvest (when they are relatively poor) compared to post-harvest (when they are relatively wealthy). The original text does not report sufficient statistical information to compute power of a replication study, but the reported p-values are small, so the estimated power will be close to 1.

The second field study also found cognitive effects pre- and post-harvest, and this study concluded that the effect is not because of nutritional differences (food consumption was similar pre- and post-harvest) but seemed to be due to stress (farmers had a higher heart rate pre- compared to post-harvest). The authors of the study used a non-typical significance criterion of 0.1 rather than the usual 0.05. In our analysis, we suppose that a deviation from the norms of hypothesis testing was appropriate, and we calculated power with this atypical significance criterion. Regardless of these details, the probability of a replication study showing a significant result is only a bit above 0.5.

The probability that six independent experiments like these should *all* be successful (a non-significant test outcome for Experiment 2 and significant test outcomes for the other studies) is the product of the probabilities in Table 6.2, which is 0.065. Thus, if the effects are real and similar to what is reported, studies like these are unlikely to produce six successful outcomes. Given the rarity of the observed results, scientists should be skeptical that the reported experiments are representative of reality. The studies described in the original paper do not make a strong argument for poverty having the hypothesized impact on cognition, and it remains an open question whether this effect actually exists.

A reasonable interpretation of our TES analysis result is that the authors of the original study engaged in some kind of QRPs in order to produce their reported results. The TES analysis cannot differentiate between different types of QRPs, and it is possible that the authors themselves do not know what kinds of choices they made to produce success across their experiments. Regardless of the origins of the problems, the bottom line is that the reported results are unlikely to represent reality. We advise readers to ignore the reported findings and wait for (or plan) better experiments.

P-Curve Analysis

If the null hypothesis is true, then *p*-values across experiments are approximately uniformly distributed. That is, the *p*-value is equally likely to take any value between 0 and 1. At first glance, this might seem like a very strange claim, but it is actually intuitive once you understand how *p*-values are related to Type I error control.

Remember that in hypothesis testing the scientist defines a significance criterion, α, to set the probability of picking a random sample that rejects a true null hypothesis. The scientist then computes the *p*-value for their data and compares it to α. If $p < \alpha$, then the null hypothesis is rejected. Importantly, this procedure works for any value of α. Thus, if $\alpha = 0.05$ and the null hypothesis is true, there is a 0.05 probability of picking a random data set that produces a *p*-value smaller than 0.05. If $\alpha = 0.10$, then there is a 0.1 probability of picking a random data set that produces a *p*-value lower than 0.1. Just to continue the example, if $\alpha = 0.34294$, then there is a probability of 0.34294 of picking a random data set that produces a *p*-value below 0.34294. This property indicates that the probability of observing a *p*-value smaller than any value x is precisely x. This is the definition of a uniform probability distribution.[3]

Figure 6.3a shows the distribution of *p*-values for simulated one-sample *t*-tests when the null hypothesis is true (effect size equals zero). Here, simulated data were drawn from a standard normal distribution and then analyzed with a one-sample *t*-test for H_0: $\mu = 0$. This simulated experiment was repeated 10,000 times, and the histogram shows that the resulting *p*-values are approximately uniformly distributed across the interval 0 to 1. The gray vertical line indicates the 0.05 criterion for statistical significance. As intended for a true null hypothesis, about 5% of the *p*-values fall below this criterion. R code, pValues1.R, to reproduce the plots in Fig. 6.3a is available at the Open Science Framework.

The situation is quite different when the null hypothesis is false. When there really is an effect, the distribution of *p*-values is positively skewed, with more small *p*-values than large *p*-values. Figure 6.3b shows the distribution of *p*-values when the standardized effect size is 0.2 and the sample size is $N = 50$. The skew is intuitive if we consider the fact that increasing the effect size leads to increased power (the probability of picking a sample that rejects the null hypothesis). In this case, power is 0.284, and so 28.4% of the *p*-values must fall below the 0.05 criterion. As power increases, the distribution of *p*-values becomes more skewed with more very low *p*-values. Figures 6.3c and d show this property for larger effect sizes (and thus higher power). In fact, the shape of the *p*-value distribution for a given test is entirely determined by the power of the test. Figures 6.3e and f show *p*-value distributions

[3] There are some situations where *p*-values do not follow a uniform distribution even when the null hypothesis is true. For example, a test of proportions with a small sample size is constrained by combinatorics to produce some *p*-values and not others; therefore the *p*-values will not follow a uniform distribution. Likewise, a test of means may show a small preference for some *p*-values due to rounding characteristics of mean measurements. These issues aside, the distribution of *p*-values is close to uniform for many hypothesis tests when the null is true.

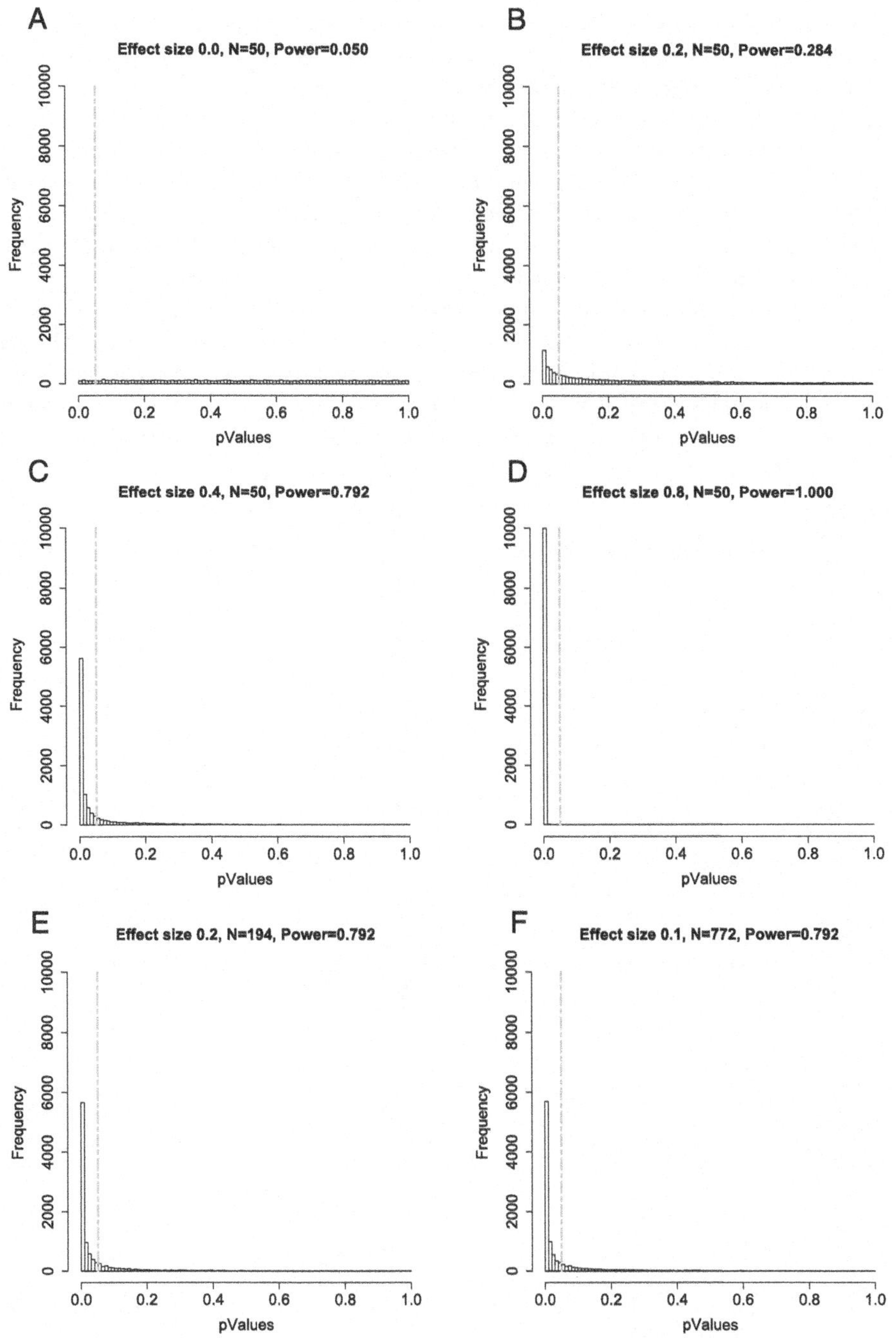

Fig. 6.3 Histograms characterizing p-value distributions for tests with different power values. The vertical gray line indicates the significance criterion (0.05). The histogram interval width is 0.01

for combinations of effect sizes and sample sizes that give the same power value as in Fig. 6.3c. The *p*-value distributions are essentially the same (small deviations are due to random sampling in the simulations).

Importantly, these properties hold even when considering only significant (e.g., $p < 0.05$) findings. Figure 6.4 plots *p*-value distributions for significant *p*-values (between 0 and 0.05). When the null hypothesis is true, the distribution is uniform (Fig. 6.4a). For non-zero (real) effects, the *p*-value distribution is skewed, with a preponderance of very small *p*-values (Fig. 6.4b–d). The code to reproduce these simulations, pValues2.R, is available at the Open Science Framework.

The *p*-values for each histogram in Figs. 6.3 and 6.4 were generated from experiments that have the same effects and sample sizes (and thus the same power). Should experiments differ in sample sizes or effect sizes (and thus in power), the curves are different, but the general shape (e.g., positive skew) continues to hold. Thus, a set of experiments with some (different) real effects should produce a distribution of

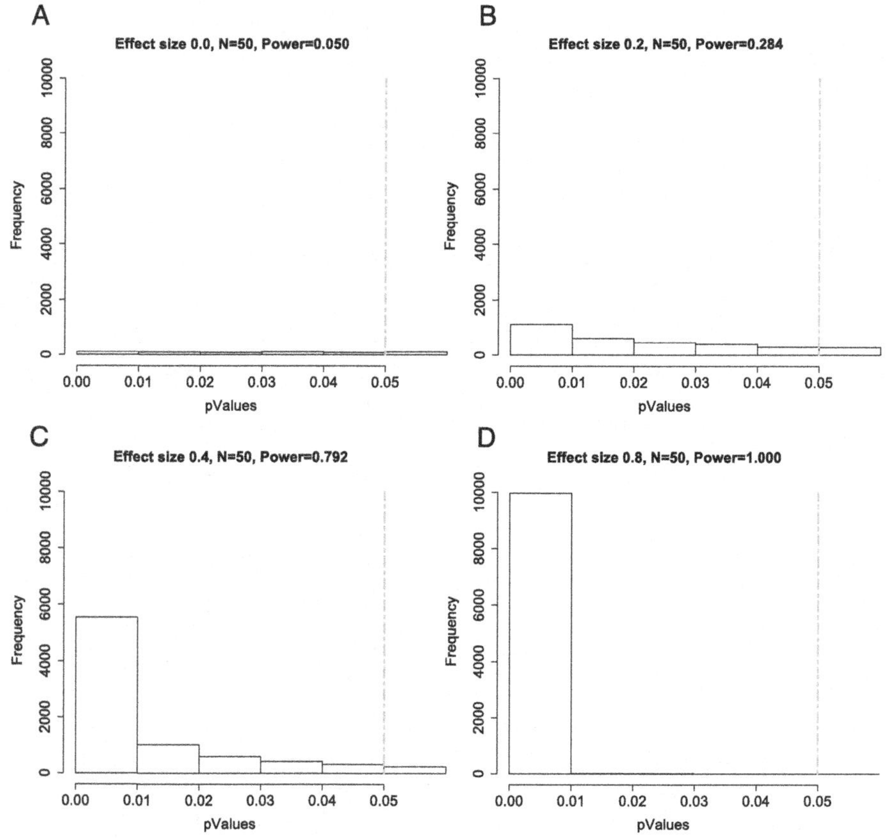

Fig. 6.4 Histograms characterizing *p*-value distributions between 0 and 0.05 for tests with different power values. The properties are the same as for the histograms in Fig. 6.3. The vertical gray line indicates the significance criterion (0.05)

p-values that has positive skew. Likewise, a set of experiments that entirely investigates (maybe different) null effects should produce a distribution of *p*-values that is flat. A set of experiments that contains some true null effects and some real effects will produce a *p*-value distribution with positive skew.

As data detectives, we can make use of the *p*-value distribution. First, we note that its shape is essentially unaffected by publication bias (a bias to only report significant outcomes): Even if only significant outcomes are published, the distribution of *p*-values below the significance criterion differs between null and real effects and true null effects will produce something close to a uniform distribution. Moreover, there are other problems that an analysis of the *p*-value distribution can efficiently identify. For example, left-skewed distributions are a sign of QRPs because such distribution shapes should be very unlikely if data collection and statistical analyses are done properly. The online app at http://p-curve.com automates analyses of the *p*-value distribution. Figure 6.5 plots the *p*-curve generated by a set of experiments that explored how the placement of calorie labels (before or after a menu item) influenced selection of foods with high calories. Across six studies (three in the main text and three in supplemental material), researchers consistently found significant effects that indicated that placing the calorie labels before a menu item led people to order lower calorie foods. The test statistics for these studies are shown in the small window in Fig. 6.5 (note that the test statistic for one study was taken from a corrigendum provided by the authors to fix a small error in their data set). The

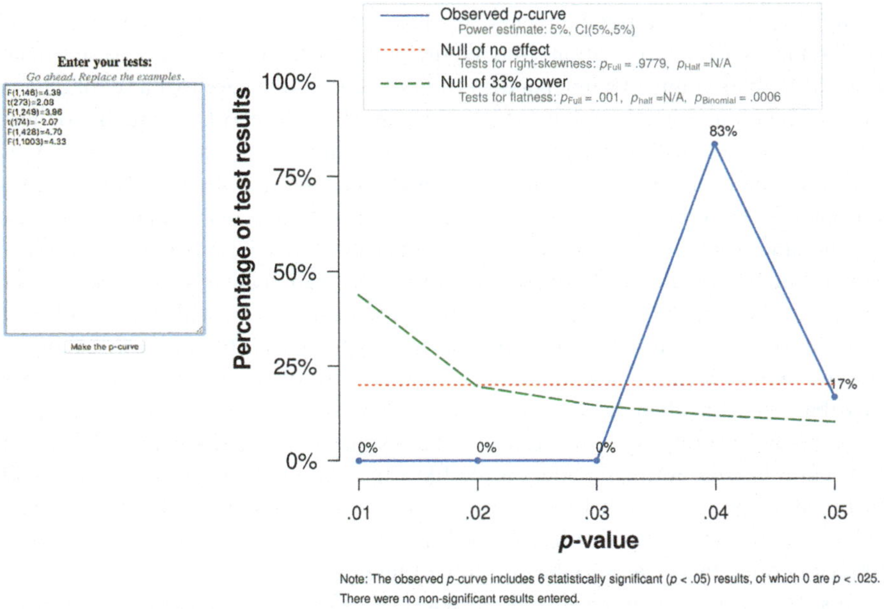

Fig. 6.5 Results of the *p*-curve app for six studies investigating the impact of calorie information on menu choices. The solid blue curve reflects the frequency of reported *p*-values. It is left skewed, which is not how *p*-values should be distributed

researchers argued that putting the calorie information in the leading position makes it more prominent in memory and therefore more influential than when it is placed after the menu item (importantly, placing the calorie information after the menu item is standard in the United States). However, the distribution of *p*-values for these six studies suggests that something is wrong with this set of results. The blue curve in Fig. 6.5 reflects the reported *p*-values, and there are none smaller than 0.04. Such a left-skewed distribution should be very unlikely if the studies are run correctly. The online app includes statistical tests for evaluating the distribution of *p*-values relative to a null (uniform) distribution and to what they refer to as an "inadequate" distribution (the *p*-value distribution for studies with power of 0.33), which is described by the green line in Fig. 6.5. The app also reports a test for whether the studies contributing to a *p*-value distribution contain "evidential value," meaning that the distribution is right skewed. For these tests, the *p*-curve analysis indicates that the evidential value is inadequate (the empirical curve is flatter than a curve with power of 0.33) and it does not indicate evidential value (the empirical curve is not right skewed).

A reasonable interpretation of the *p*-value distribution in Fig. 6.5 is that some of the experimental results were generated with QRPs. It is not possible to identify precisely what QRPs were used, but we should not trust that the reported results or corresponding conclusions reflect reality. The conclusions may yet be correct, but the reported experiments do not provide appropriate support for those conclusions. Scientists who want to investigate this topic further need to start over with better experiments.

It is fairly easy to apply the *p*-curve analysis, but it is important to understand its requirements and interpretation. One requirement is that the *p*-values that contribute to the distribution must be independent. It is sometimes the case that a set of data is analyzed with multiple hypothesis tests (e.g., an ANOVA reports an interaction and specific contrasts with the same data set). The *p*-values from these tests are (typically) not independent, and so the tests to explore the distribution shape can be misleading. To address this concern, researchers using the *p*-curve analysis use just one *p*-value from each data set or experiment. Unfortunately, it is not always clear how to select a *p*-value from the set, and the choice can make a big difference. For example, choosing the smallest *p*-value from each experiment will often result in a distribution with right skew even when the null hypothesis is true. Likewise, choosing the biggest significant *p*-value from each set will often produce a left skewed distribution, even when there is a real effect. To avoid this problem, some researchers apply an arbitrary rule, such as using the *p*-value from the first reported relevant test; but this does not really address the fundamental problem: the analysis should be based on the *p*-values that are relevant to the question of interest. It often requires subject matter expertise to identify such *p*-values, and sometimes there is not a unique *p*-value that relates to the question of interest.

For the *p*-curve graph in Fig. 6.5, the question of interest is, "does the location of caloric information influence menu choices?" and we picked the *p*-values that specifically investigated that question. The resulting left skewed *p*-curve distribution indicates that the six studies reported here were not produced by proper hypothesis tests. Importantly, this conclusion does not mean that *each* of the six tests is flawed. The identified problem is with the *set* of hypothesis tests (their distribution of *p*-values). Surely some of the individual studies are problematic as well (else the set could not be), but it is possible that some studies are flawed and some studies are fine.

This aspect of interpretation can matter quite a bit for other types of questions of interest. For example, suppose you applied a *p*-curve analysis to a specific researcher because you wonder if he engages in QRPs. You select one *p*-value from each of seven articles published by this researcher. Figure 6.6 shows the (entirely made up) *p*-curve for the selected *p*-values. It is right skewed, so the *p*-curve analysis suggests that there is "evidential value" in this set of *p*-values. Unfortunately, this conclusion does not really answer the question of whether the researcher engages in questionable research practices. It could be that the researcher does not engage in QRPs, but it could also be the case that for some investigations the researcher does use QRPS and for some investigations he does not. Publishing some studies with evidential value means that a combination of studies with evidential value and studies without evidential value (e.g., a flat distribution) might produce a right skewed distribution of *p*-values. The point is that a property of the set does not necessarily apply to each

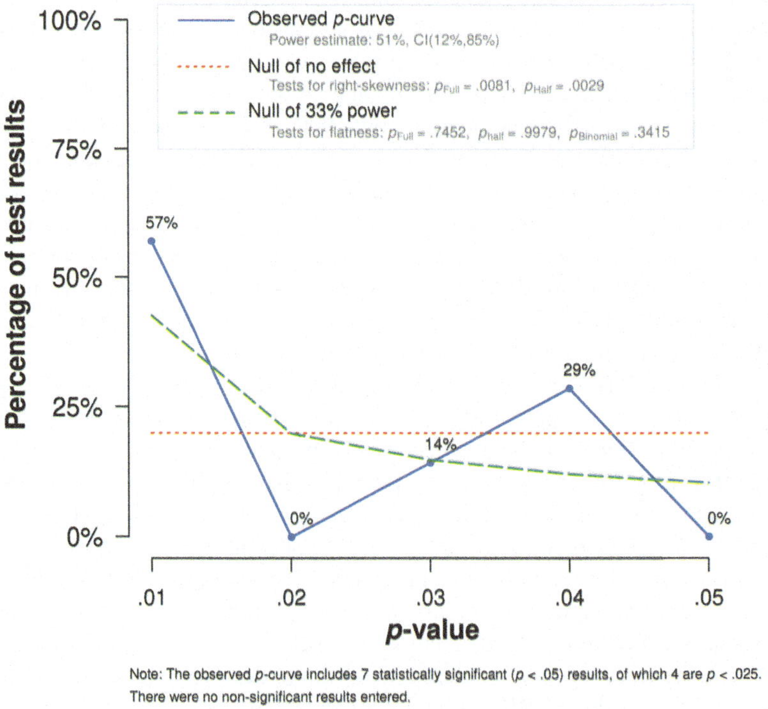

Fig. 6.6 Results of the *p*-curve app for seven (hypothetical) studies investigating a researcher who investigates two different topics. Although the distribution is right-skewed, thereby indicating some "evidential value," this finding is difficult to interpret

member of the set. A right skewed *p*-curve does not mean that every study a researcher reports is fine, and a left skewed *p*-curve does not mean that every study a researcher reports is problematic. For this reason, it usually does not make sense to apply *p*-curve analyses to an author, a specific scientific journal, or a field of study. Instead, *p*-curve analyses should be used to evaluate specific claims or conclusions, when those claims or conclusions are based on a reported set of *p*-values. For the studies producing the *p*-values in Fig. 6.6, it might make sense to look into the set of studies related to specific conclusions made by the researcher, and use the *p*-curve analysis to evaluate the evidential value of the studies relative to those claims.

Conclusions

Questionable Research Practices (QRPs) often leave a trail of evidence that indicates they were involved in producing the reported outcomes. Proper experiments (without QRPs) have fundamental properties that can be identified across experiments. One such property is how success should relate to experimental power. Excess success for a set of experiments indicates that the results were generated in a way that violates good data collection, analysis, or reporting. This discrepancy can be identified with the Test for Excess Success. A second such property is the distribution of *p*-values, which should be right-skewed for proper experiments that investigate a real effect. The distribution of *p*-values should almost never be left-skewed for experiments that were generated without QRPs. A left-skewed distribution indicates that the results were generated in a way that violates good data collection, analysis, or reporting. These problems can be identified by the *p*-curve analysis.

Within a single experiment, it is often useful to look for various discrepancies between reported statistics. Such discrepancies do not necessarily indicate the involvement of QRPs, but they do suggest that something has gone wrong in the reporting of the experiment. Thus, readers should be somewhat skeptical about the validity of the reported results and the associated conclusions.

As is the case for many types of detective work, a data detective may be able to conclude that there is something "odd" about reported results but not pinpoint exactly what has gone wrong. Inconsistencies between statistics might arise from fraud or they might be the result of simple typos. In a similar way, neither the Test for Excess Success nor the *p*-curve analysis can identify precisely *how* researchers produced results that are too-good-to-be-true or that generate a left-skewed distribution of *p*-values. Still, the burden of proof is on the scientists; they should always provide evidence to support their claims. If the reported results seem unbelievable, other scientists should dismiss the claims until sufficient evidence is produced.

While some scientists may deliberately set out to deceive others, we suspect that most scientists introduce QRPs without realizing it. Indeed, one very beneficial use of the various methods for detecting the impact of QRPs is for scientists to apply them to their own work before publishing. Hopefully, this could motivate scientists to examine their research methods in detail and root out QRPs. Such applications will greatly improve scientific work.

Further Reading

GRIM Test

Brown, N. J. L., & Heathers, J. A. J. (2016). The GRIM test: A simple technique detects numerous anomalies in the reporting of results in psychology. *Social Psychological and Personality Science, 8*(4), 363–369. https://doi.org/10.1177/1948550616673876

Heathers, J. (2017). Introducing SPRITE (and the case of the carthorse child). *Hackernoon*, https://medium.com/hackernoon/introducing-sprite-and-the-case-of-the-carthorse-child-58683c2bfeb

Heathers, J. A., Anaya, J., van der Zee, T., & Brown, N. J. (2018). Recovering data from summary statistics: Sample parameter reconstruction via Iterative techniques (SPRITE). *PeerJ Preprints, 6*, e26968v1. https://doi.org/10.7287/peerj.preprints.26968v1

Test for Excess Success

Francis, G. (2012). Too good to be true: Publication bias in two prominent studies from experimental psychology. *Psychonomic Bulletin & Review, 19*, 151–156. https://doi.org/10.3758/s13423-012-0227-9

Francis, G., & Thunell, E. (2019). Excess Success in "Ray of hope: Hopelessness increases preferences for brighter lighting". *Collabra: Psychology, 5*(1), 22. https://doi.org/10.1525/collabra.213

P-Curve

Simonsohn, U., Nelson, L. D., & Simmons, J. P. (2013). P-curve: A key to the file drawer. *Journal of Experimental Psychology: General, 143*, 534–547.

Calorie Labels

Dallas, S. K., Liu, P. J., & Ubel, P. A. (2019). Don't count calorie labeling out: Calorie counts on the left side of menu items lead to lower calorie food choices. *Journal of Consumer Psychology, 29*(1), 60–69. https://doi.org/10.1002/jcpy.1053

Francis, G., & Thunell, E. (2020). Excess success in "Don't count calorie labeling out: Calorie counts on the left side of menu items lead to lower calorie food choices". *Meta-Psychology, 4*. https://doi.org/10.15626/MP.2019.2266

Poverty and Cognition

Mani, A., Mullainathan, S., Shafir, E., & Zhao, J. (2013). Poverty impedes cognitive function. *Science, 341*, 976–980.

Chapter 7
Controversies Regarding Null Hypothesis Significance Testing

Brian P. O'Connor and Nataasha Khattar

Abstract This chapter provides an overview of null hypothesis significance testing (NHST) and of the problems involved with NHST. This is followed by a non-technical description of perhaps the most useful NHST alternative, Bayesian methods. We then provide illustrations of an unnecessary, tragic consequence of using NHST in a series of individual studies: The inability to incorporate previous findings when analyzing a new dataset. This is accompanied by illustrations, using both generated data and real data from meta-analyses in clinical psychology, of the more coherent findings that occur when NHST is ignored and when previous findings are taken into account when assessing a phenomenon across a series of studies.

Keywords Significance testing · Null hypothesis testing · Bayesian statistics · Meta-analysis

Null hypothesis significance testing (NHST) continues to be the primary data analysis method in the vast majority of research reports in psychology (data on this fact for clinical psychology will be provided below). NHST usage persists despite the fact that most researchers have surely at least some awareness of the controversies and of the serious drawbacks with NHST that have been described in numerous books and journal articles (e.g., Bakan, 1966; Carver, 1978; Cohen, 1994; Harlow et al., 1997; Hunter, 1997; Kline, 2013; Nickerson, 2000; Oakes, 1986; Schmidt, 1996; Ziliak & McCloskey, 2008). The calls for change have apparently had minimal impact on research practices and on student education. Although effect sizes and confidence intervals are more likely to appear in research reports than they were in the past, the *p* values from NHST are still the primary focus of most investigations (Cumming & Calin-Jageman, 2017; Kline, 2013; Sharpe, 2013).

This chapter provides an overview of NHST and of the problems involved with NHST. This is followed by a non-technical description of perhaps the most useful NHST alternative, Bayesian methods. We then provide illustrations of an unnecessary, tragic consequence of using NHST in a series of individual studies: The inability to incorporate previous findings when analyzing a new dataset. This is

B. P. O'Connor (✉) · N. Khattar
University of British Columbia – Okanagan, Kelowna, BC, Canada
e-mail: brian.oconnor@ubc.ca

W. O'Donohue et al. (eds.), *Avoiding Questionable Research Practices in Applied Psychology*, https://doi.org/10.1007/978-3-031-04968-2_7

accompanied by illustrations, using both generated data and real data from meta-analyses in clinical psychology, of the more coherent findings that occur when NHST is ignored and when previous findings are taken into account when assessing a phenomenon across a series of studies.

What Is NHST?

The now-entrenched version of NHST involves declaring what constitutes a null hypothesis (e.g., a zero difference between two means) and conducting a test of statistical significance. The test is used to make a binary decision about whether or not an effect exists in a population. This version of NHST is a blend of approaches developed by Fisher and by Neyman and Pearson, who had bitter quarrels over what they considered best practices (see Gigerenzer et al., 2004; Kline, 2013; Morrison & Henkel, 1970; and Salsburg, 2001 for historical descriptions and references). The conventional practice of NHST is well-captured in the following statements:

> Conventionally, researchers make such decisions by assuming the null hypothesis to be true and, given this assumption, attempting to make inferences based on the probability of obtaining the actual pattern of results observed. Specifically, a statistical test yields the probability of a given results (or one more extreme) being produced by chance if the null hypothesis is true. … If this (probability) is less than a threshold probability or alpha level (typically 0.05), then chance is concluded to be a sufficiently unlikely explanation of the outcome, and the existence of an effect is held to be supported by the data (Pollard & Richardson, 1987, p. 159).

> p actually stands for the conditional probability … which represents the likelihood of a result or outcomes even more extreme assuming (1) the null hypothesis is exactly true; (2) the sampling method is random sampling; (3) all distributional requirements, such as normality and homoscedasticity, are met; (4) the scores are independent; (5) the scores are also perfectly reliable; and (6) there is no source of error besides sampling or measurement error. In addition to the specific observed result, p values reflect outcomes never observed and require many assumptions about those unobserved data. If any of these assumptions are untenable, p values may be inaccurate (Kline, 2013, p. 74).

Statistical significance is determined by reference to the distribution of test statistic values that occur when the null hypothesis is true. The distribution is imaginary and has nothing to do with any given real dataset. To illustrate this fact, consider a study in which scores for two groups, for example, $N = 30$ for each, are randomly drawn from the same very large population. There is no treatment or intervention, and a t test value is computed for the two groups of scores. Repeat this random sampling of data for two $N = 30$ groups and the t test computation millions of times. The distribution of t values from these studies would be the sampling distribution of t values that occur when $N = 30$ per group and when the null hypothesis is true. The null hypothesis is true because the two samples are always drawn from the same population and there is no intervention for either of the groups, that is, there is no reason why they should be different. The p values from the back of statistics textbooks, or from software packages, are produced by mathematical formulas that mimic the results from a long series of identical hypothetical studies in which the null hypothesis is true. Real data (e.g., data from your study) are never involved in the production of the sampling distribution of test statistic values that are used in NHST.

Problems with NHST

Many of the problems with NHST stem from misinterpretations of the minimally satisfying information that is provided by statistical significance testing. When we obtain a test statistic value for which $p < 0.05$, all we can say is that "when the null hypothesis is true, a test statistic value of this magnitude is unlikely on the basis of sampling variability alone" (O'Connor, 2017). The *p* value that is produced by statistical software does not "know" anything about our data. It does not know if our samples were drawn from the same population or if the null hypothesis is true for our dataset. We are nevertheless prone to thinking that it does have such knowledge. One source of this mistaken belief may be that while an obtained *t* test value or regression coefficient really are values for our particular real dataset, the corresponding *p* value is not. Yet the statistics appear side by side in the software output. The misinterpretations then begin (Spence & Stanley, 2018).

A *p* value is not the probability of committing a Type I error in any given study. The statistical software does not know if the null hypothesis is true for our particular dataset. The researcher does not know if Ho is true either, which is why the study is being conducted. If the null hypothesis is false, then the probability of a Type I error is zero and not the *p* value.

The common use of $p < 0.05$ for statistical significance does not mean that 5% of all published findings are Type I errors. This could only be true if researchers were always testing true null hypotheses, which is certainly not the case (Pollard & Richardson, 1987). Surely some treatments designed by psychologists really do work. It is not reasonable to believe that researchers only ever test population associations that are truly zero.

The *p* value tells us nothing about the reliability of our findings, as in whether they will replicate. The *p* value is not an effect size. The *p* value is not the probability that our particular results are due to chance because the statistical software does not know if the null hypothesis is true for our data.

The *p* value is not a probability statement about the truthfulness of the null hypothesis (e.g., there is only a 5% chance that the null hypothesis is true). A *p* value is based on the assumption that the null hypothesis is true. It cannot be converted into a probability statement about the null hypothesis, although researchers and readers are prone to making such conversions (Cohen, 1994; Maxwell & Delaney, 2004, p. 48). For example, the probability that a population of persons with schizophrenia will generate a person who is on medication (high) is not the same as the probability that someone who is on medication was generated by a population of persons with schizophrenia (low). Similarly, the probability of obtaining a particular test statistic value given the null hypothesis (which is what *p* values do tell us) is not the same as the probability of the null hypothesis, given that a particular test statistic value was obtained (O'Connor, 2017). The *p* value is not a probability for the null hypothesis, which is often our intuitively preferred but incorrect belief (Cohen, 1994).

We tend to believe that *p* values are highly accurate, but they are accurate only in specific circumstances that never occur in real data. We never randomly sample from the populations of interest. The assumptions of statistical procedures are typically violated, at least to some degree. Scores are never perfectly reliable. Hair-splitting decisions over $p < 0.05$ vs. $p < 0.06$ are meaningless in this context, yet they determine the kinds of conclusions that are made in discussion sections and they are often the basis of getting published or not.

In NHST, all that can be concluded when a test statistic is not significant is that the null hypothesis cannot be rejected. Although a failure to reach statistical significance does not mean that the null hypothesis is true, non-significant effects are commonly and mistakenly considered to be evidence for the null hypothesis (Falk & Greenbaum, 1995; Oakes, 1986). Discussion sections commonly involve speculations about why there was "no effect" for a predicted association. Searches for moderator variables may be recommended. Failures to replicate and growing piles of apparently conflicting findings may be caused solely by natural sampling variability and by the use of NHST to evaluate raw data (O'Connor, 2017; Schmidt, 1996).

Many, if not most, studies in psychology have relatively small samples and modest statistical power (Button et al., 2013; Maxwell, 2004). Studies that happen to overestimate a true population effect size on the basis of sampling variability alone are more likely to find that $p < 0.05$ and therefore get published. Identical studies that obtain lower but accurate effect sizes are less likely to get published because they are less likely to find that $p < 0.05$. There are frustrating failures to replicate and a lack of cumulative progress. After years of research on a hypothesis, a meta-analysis (where the focus is on effect sizes and not NHST) may be conducted that finally resolves the apparently conflicting conclusions that were generated by the use of NHST in the individual studies. The inability to take previous findings into account when analyzing new datasets is a serious shortcoming with NHST that is illustrated in detail below.

After decades of preoccupation with NHST, the American Statistical Association (ASA) recently published the following declarations on statistical significance testing and *p* values:

> While the *p value* can be a useful statistical measure, it is commonly misused and misinterpreted.
>
> *P values* do not measure the probability that the studied hypothesis is true, or the probability that the data were produced by random chance alone.
>
> Scientific conclusions and business or policy decisions should not be based only on whether a *p value* passes a specific threshold.

A *p value*, or statistical significance, does not measure the size of an effect or the importance of a result.

By itself, a *p value* does not provide a good measure of evidence regarding a model or hypothesis (Wasserstein & Lazar, 2016, pp. 131–132).

Evidence for the Ongoing Use of NHST

All articles in the 2019 and 2020 volumes of the Journal of Consulting and Clinical Psychology were coded for the kinds of statistical methods that were used for the data analyses. The results are provided in Table 7.1. NHST was used in the vast majority (91.4%) of the research reports. Confidence intervals were reported in 82.8% of the articles. Bootstrapping and related methods (Good, 2010) were used in 15.6% of articles, and robust methods of computing statistical significance or standard errors (Field & Wilcox, 2017) were used in 9.7% of articles. Bayesian methods (excluding the reporting of BIC model fit coefficients) were used 7% of the time, and clinical significance statistics (Lambert & Bailey, 2012) were reported in 14.5% of the articles. NHST clearly remains the dominant statistical analysis method in the two most recent volumes of the leading journal in clinical psychology. Bootstrapping and robust methods can both be used to provide more accurate estimates of statistical significance than parametric methods while remaining within the NHST framework. But these more precise methods were not used as often as they could be. The use of non-NHST methods was related to the software that was used by the authors. The use of non-NHST methods was typically accompanied by references to software (e.g., MPlus) that has options for alternative analytic methods. Build it and they will use it.

Table 7.1 Statistical methods used in all articles in the 2019 and 2020 volumes of the *Journal of Consulting and Clinical Psychology*

	2019 $N = 97$	2020 $N = 89$	Total $N = 186$
Significance testing	87 (89.7%)	83 (93.3%)	170 (91.4%)
Confidence intervals	89 (91.8%)	65 (73.0%)	154 (82.8%)
Bootstrapping or related methods	14 (14.4%)	15 (16.9%)	29 (15.6%)
Robust methods	8 (8.3%)	10 (11.2%)	18 (9.7%)
Bayesian methods	8 (8.3%)	5 (5.6%)	13 (7.0%)
Clinical significance	15 (15.5%)	12 (13.5%)	27 (14.5%)

Alternatives to NHST

"The New Statistics"

The many problems with NHST led the American Psychological Association to strongly recommend the reporting of effect sizes and confidence intervals as the primary output from statistical analyses, rather than p values (Wilkinson & the Task Force on Statistical Inference, 1999). In 2014, the journal *Psychological Science* began recommending, for authors, the use of "the new statistics" because "the problems that pervade NHST are avoided by the new statistics—effect sizes, confidence intervals, and meta-analysis" (Eich, 2014, p. 5). The emphasis on "the new statistics" in recent introductory statistics textbooks (e.g., Cumming & Calin-Jageman, 2017) is a welcome change.

Unfortunately, the revolution has been occurring in slow motion. Progress sometimes seems barely discernible. The preoccupation with $p < 0.05$ persists. An effect is considered statistically significant when a 95% confidence interval does not include a zero-effect size. The NHST binary decision about whether there is an effect or not thus lives on in the new statistics via confidence intervals. It remains the basis for conclusions in discussion sections and in publication decisions. An effect size is often merely a supplementary finding that is considered most credible if its confidence interval does not include a zero value. Effect sizes are commonly reported only if $p < 0.05$.

Worse, confidence intervals are prone to misinterpretation. A 95% confidence interval indicates that if the study were conducted many times, 95% of the confidence intervals would contain the true population effect size (Cumming & Calin-Jageman, 2017, p. 101). This accurate, precise statement is nevertheless perplexing and unsatisfying to our brains. We are instead prone to believing that a 95% confidence interval means that there is a 95% chance that the true population effect size falls within the confidence interval. This interpretation is incorrect and unwarranted when the analyses are based on NHST (Kruschke, 2015). Fortunately, the interpretation is correct when the confidence intervals are provided by Bayesian analyses. Our intuitive, preferred statistical reasoning about confidence intervals is Bayesian and is not permitted by NHST.

Bayesian Statistics

Reverend Thomas Bayes developed his statistical methods around the year 1740. Although he published theological works, he kept his statistical side-interests to himself and did not publish his methods during his lifetime (Bellhouse, 2004). They remained largely neglected until about 40 years ago, when computational hardware and software advances made Bayesian analyses possible for all sorts of datasets, not just small ones. Introductions and tutorials have been provided by Kruschke (2015),

Kruschke et al. (2012), Kruschke and Liddell (2018), O'Connor (2017), Pruzek (2016), Quintana et al. (2017), Wagenmakers, Morey, and Lee (2016a), Wagenmakers, Verhagen, and Ly (2016b), and Zyphur and Oswald (2015). The present discussion will begin with a description of the output from Bayesian raw data analyses.

In Bayesian analyses, there are no comparisons of a statistical coefficient for the real data with any imaginary, theoretical distribution of values for a hypothesis that is not of direct interest (Ho) and whose values were not derived from the current data (O'Connor, 2017). What researchers typically most want to know is, "what is the most likely value for an effect size, and what are the other reasonable possibilities for the effect size in case the most likely value happens to not be perfectly accurate?" Bayesian raw data analyses give us the answers. For each parameter of interest (e.g., an effect size), a distribution of the most likely values is produced. It is called a "posterior distribution" because it is the distribution of the most likely parameter values after the analyses have been conducted. It represents possible beliefs about a parameter. The mean, median, or mode of the posterior distribution may be chosen as the best estimate of a parameter.

The continuous posterior distribution is a probability density function and there is usually much focus on the "highest density interval" (HDI) within this function/distribution.

> Points inside an HDI have higher probability density (credibility) than points outside the HDI, and the points inside the 95% HDI include 95% of the distribution. Thus, the 95% HDI includes the most credible values of the parameter. The 95% HDI is useful both as a summary of the distribution and as a decision tool. Specifically, the 95% HDI can be used to help decide which parameter values should be deemed not credible, that is, rejected. … One simple decision rule is that any value outside the 95% HDI is rejected. In particular, if we want to decide whether the regression coefficients are nonzero, we consider whether zero is included in the 95% HDI (Kruschke et al., 2012, p. 730).

In other words, a 95% HDI (a credibility interval) is the range of parameter estimates that captures 95% of the posterior probability distribution. Statements such as "There is a 95% chance that the parameter value falls between ___ and ___" are possible, that is, are legitimate. In contrast, NHST provides no information about the probability of a parameter. A kind of NHST conclusion can nevertheless be derived for anyone who is concerned about veering too far away from $p > 0.05$ in a research report. Ho can be rejected as improbable if the 95% HDI for a parameter does not contain a zero value.

The MCMC The Markov Chain Monte Carlo (MCMC) is the computational breakthrough that makes Bayesian raw data analyses possible for most researchers (Gill, 2015; Kruschke, 2015). This advanced, complex method will here be described using simple analogies. One analogy is that MCMC methods are like trying to produce a heat map of a previously unexplored mountain range while being both blind and drunk. A map emerges after very much stumbling around and feeling one's way.

To discover the likely parameter values for a quantitative dataset, imagine that the computer (the MCMC algorithm) is first told that the range of possible, true population values for a correlation coefficient is −1 to 1. The actual, unknown population value exists somewhere in this range, but it is not known where. On the first step, the procedure randomly selects a possible value from the range (which is the drunken stumbling part), and then randomly selects another possible value. It then determines which value is most consistent with the real raw data and the prior beliefs (described below) provided by the researcher. The winning value gets a tally. Then the procedure randomly selects another possible value and determines whether or not the new value is more consistent with the data and prior beliefs than the value from the previous step. This goes on many thousands or hundreds of thousands of times, with the winning values always being tallied along the way. The resulting frequency distribution of the winning tallies gives us the most likely values for the parameter, along with other reasonable but less likely possible values. These analogies are over-simplifications, but they help capture the essence of what is involved in building a posterior distribution via MCMC methods.

To provide an example of a posterior distribution, of an HDI, and of MCMC stumbling around steps, a 10,000-case computer-generated dataset was created for two variables in which the correlation was set at 0.22. Bayesian analyses were conducted on the raw data for the two variables. A trace plot of the MCMC correlation coefficient values for steps 1000 to 3000 is provided in the top portion of Fig. 7.1. Most of the values (in the black smudge) are close to the true population correlation value of 0.22, although more deviant values do sometimes occur. The posterior distribution from the MCMC steps, along with the 95% HDI, appears in the lower portion of Fig. 7.1. The peak of the distribution is right around 0.22. It can be concluded from the 95% HDI that, based on the MCMC analyses, there is a 95% probability that the true population correlation coefficient value is somewhere between 0.16 and 0.28, with 0.22 being the most likely value.

The Prior Distribution Perhaps the most distinctive and useful feature of Bayesian methods is that it is possible to take previous findings into account when conducting the analyses for a new dataset. Imagine that someone has already begun exploring the new mountain range and has produced a crude, initial map. In this case, the explorer (the researcher) is only partially blind. Providing the procedure with previous knowledge is called "specifying the priors" or "specifying the prior distribution." An overly simple example would be the equivalent of saying that, "based on previous findings, the correlation is likely somewhere between −0.5 and 0.5." This will cause the procedure to stumble around the more narrow region and pay less attention to other possible values. More accurately though, a range is not provided. Instead, the priors are usually the best estimate for a parameter based on previous research along with a degree of certainty estimate for this parameter, which is typically its standard error (or variance). The MCMC algorithm randomly selects values from this distribution rather than from a specified range.

Bayesian analyses require that priors be specified, but the priors can be completely uninformative. This would be equivalent to saying, "The best guess for the correlation coefficient is zero, and I am as uncertain as can be about this guess. I

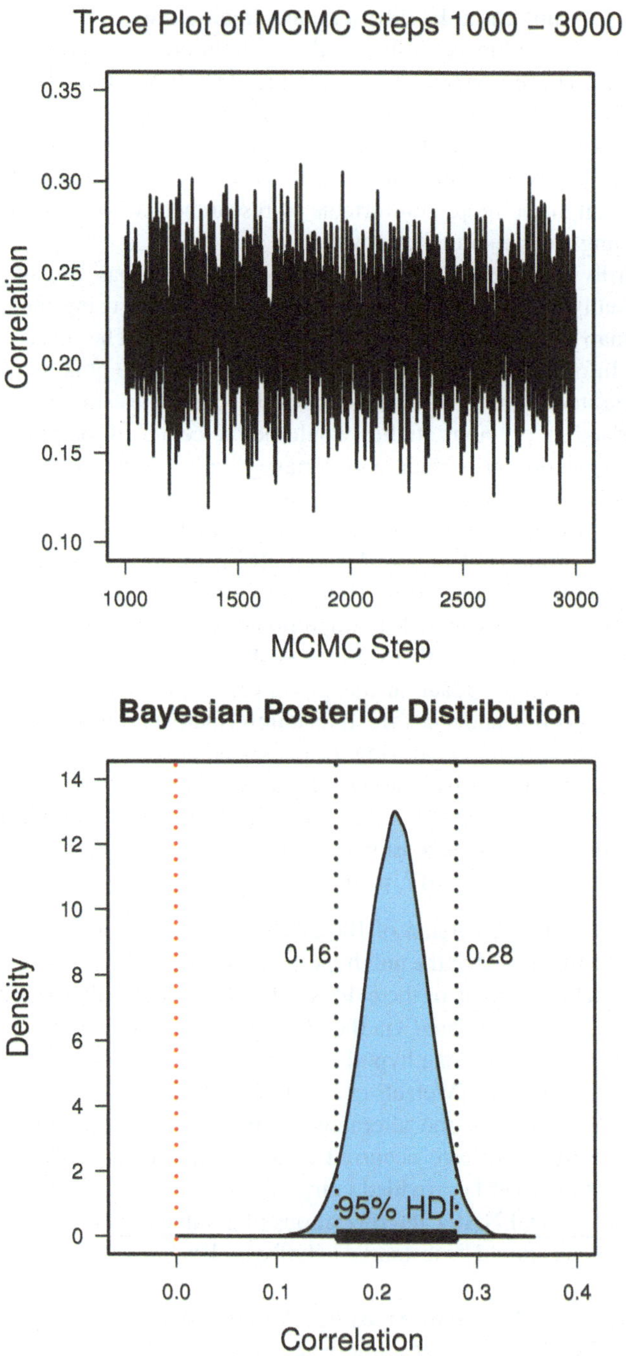

Fig. 7.1 MCMC trace plot and Bayesian posterior distribution

have no clue." In this case, the diffuse, noncommittal, non-informative priors will essentially have no influence on the analyses (uninformative priors were used for the above data analyses that are depicted in Fig. 7.1). In this case, the findings from the Bayesian analyses (e.g., the 95% HDI) will be very similar to the findings from traditional data analyses when the statistical assumptions of the conventional procedures are met. The use of uninformative priors is a way of conducting Bayesian data analyses that might be more comfortable to researchers who are concerned about previous findings influencing their results.

But it hardly makes sense to completely ignore what has already been found. Should we really try mapping the mountain range without using the existing, perhaps crude map that was provided by previous explorers? The inaccuracies can be worked out. Informative priors for raw data analyses can be obtained from a meta-analysis of the studies that have been conducted to date. The estimated meta-analysis effect size based on previous studies would be the center of the prior distribution, and the corresponding prior degree of certainty value would be its standard error. Not using available information would be analogous to police detectives refusing to jointly consider all of the available clues when evaluating any single piece of information about a crime (O'Connor, 2017, p. 169). Informed, empirical priors are especially helpful in making firmer inferences possible, and in avoiding Type II errors, in small sample research. Many additional benefits of Bayesian data analyses were described in the sources that were cited above.

Bayesian data analyses result in changes in beliefs about an effect from before to after the data collection and analyses. If an uninformative prior was used, the change in beliefs could be: "Before analyzing the data I had no idea what values the parameter might take. I conservatively assumed that the value was zero. Now that I have run the analyses on my data, I believe that the likely value for the parameter is between ___ and ___, and the most likely value is ___." Beliefs shift in the direction of the evidence (O'Connor, 2017, p. 169).

The "Bayes Factor" A version of Bayesian statistics can be used to make statements about the likelihood of the null hypothesis without using the intensive MCMC computations. The strengths of the evidence for the null and alternative hypotheses can be quantified and compared via the "Bayes factor," which is the probability of the researcher's data under one hypothesis compared to the probability of the data under the other hypothesis (Wetzels et al., 2011). The Bayes factor is an odds ratio. A Bayes factor of 2.5 for the alternative hypothesis indicates that the data are 2.5 times more likely to have occurred under the alternative than under the null hypothesis. Jeffreys (1961) provided conventions for comparing Bayes factor values to the conventional NHST interpretations of p values. Bayes factors above 3 or below 0.33 are considered "substantial" (Jeffreys, 1961; Wetzels et al., 2011).

Getting Going with Bayesian Analyses A user-friendly, free Bayesian program with a GUI is JASP, available from https://jasp-stats.org.SPSS 28 now provides an option for Bayesian analyses. There are MPlus, Matlab, and some SAS routines, and an online calculator that provides Bayes factors for entered data (e.g., http://pcl.

missouri.edu/bayesfactor). Software availability is no longer an obstacle. There are also numerous free R packages for conducting Bayesian analyses and a growing number of tutorials in books, articles, and online (Kruschke, 2015; Lee & Wagenmakers, 2013; O'Connor, 2017). NHST was developed in the pre-computer era and it served a purpose at that time. But there is no scientific justification for the ongoing use of a relatively crude and uninformative method that is so restricting and prone to misinterpretations, especially now that we have easy access to the hardware and software for better methods.

An easy way to begin would be to first run one's analyses using familiar, non-Bayesian software in order to identify patterns in the data and to generate parameter estimates. Then switch to Bayesian software and output for more informative and definitive findings. Examples of how to describe Bayesian analyses in the Methods and Results sections of journal articles have been provided by Kruschke (2015) and O'Connor (2017). If the results from the Bayesian and from the traditional ("frequentist") NHST analyses are equivalent, if the Bayesian HDI credibility intervals are essentially the same as the NHST confidence intervals, then it will be possible to state that the data were analyzed both ways and the findings were the same. Bayesian statements about the credibility of the coefficients would be permitted, while the kinds of problematic NHST misinterpretations described at the outset of this chapter could be avoided.

Illustrations of Why *Not* Taking Previous Findings into Account When Evaluating New Datasets Is a Questionable Research Practice

Evidence from previous studies is not, and cannot be, incorporated into the NHST analyses for a new dataset. Each dataset is its own separate, isolated voice. Fisher, Neyman, Pearson, and other influential figures in NHST history were vociferously opposed to letting previous findings influence one's analyses (Lehmann, 1993; Salsburg, 2001, p. 133). The consequences have been tragic for all disciplines that got hooked by NHST.

Most studies have modest sample sizes (Button et al., 2013; Maxwell, 2004). We also typically seek to find evidence for phenomena with that have low-to-moderate effect sizes (psychological phenomena are complex and have multiple predictors). This can easily result in roughly as many $p < 0.05$ findings in favor of a hypothesis as there are $p > 0.05$ findings, solely on the basis of sampling variability. Apparently conflicting findings at the individual study level are common. Followers of a sequence of studies observe a back-and-forth ping-pong of conclusions about whether there is an effect or not, and sometimes also about the direction of the effect (Meehl, 1978; O'Connor & Ermacora, 2021). The conflicting conclusions that are reached in the individual studies, based on NHST, leave observers of the literature baffled. There is not much order in the research universe on the topic. The phenomenon seems very complex. Social science research is apparently not answering important questions after all. Reviews of these issues were provided by Bakan (1966), Carver (1978), Harlow et al., 1997, Hunter (1997), Kline (2013), Oakes (1986), and Schmidt (1996).

In this section, we illustrate how incorporating previous findings when evaluating a new dataset can reduce the ping-pong-like confusion and increase the rates of correct conclusions *in individual studies* regarding whether or not an effect exists. Two methods of incorporating previous findings will be used: Updating a meta-analysis to include the new data, and Bayesian data analysis.

A new meta-analysis (MA) can be conducted every time data from an additional study become available. As a pool of effect sizes grows, the studies can be sorted into a sequence based on, for example, year of publication. A table or plot of the MA results that emerge as each study is added to the updating MA can reveal how the effect size estimate and its precision change over time (Borenstein et al., 2009). Our focus in this chapter will be on how updating an MA can be used to reach more consistent and accurate conclusions when evaluating each new, individual dataset as it becomes available. The effect size and confidence interval (or standard error) from the updated MA may often provide more solid grounds for conclusions about the existence and size of an effect in discussion sections (Tryon, 2016). The process is naturally Bayesian. The results from updating MAs and from Bayesian analyses are highly similar. Both computer-generated data and real data from the literature are used below to illustrate these points.

Datasets

Effect sizes from four previously published meta-analyses, on a diversity of topics in clinical psychology, were selected and re-analyzed for illustrative purposes. The datasets are described below. The patterns of results across multiple studies are heavily determined by the effect size for a phenomenon and by the study sample sizes. The varying effect sizes across the four selected, previously published meta-analyses help illustrate the different kinds of end results that can emerge when evidence from previous studies is taken into account when evaluating new study data.

Additional, parallel data analyses were conducted on computer-generated datasets in which the population effect sizes were set to be identical to those from the four real data meta-analyses from the literature. In each case, data for two variables with a specified correlation from a real data meta-analysis were generated for a population of 100,000 cases. The median sample size from each real data meta-analysis was identified. The data analyses were then conducted on 50 random samples, each of the median sample size, that were drawn from the same population. Each sample thus represented a possible "study" of the variables. The results from these analyses of samples that were all drawn from the same computer-generated populations serve as useful comparison points for the findings that emerged when the analyses were conducted on real, previously published meta-analysis data (in which case it can never be certain if the samples are from the same population or not).

Analytic Methods

The meta for package in R (Viechtbauer, 2010) was used to conduct the random-effects cumulative meta-analyses. The random-effects model assumes that there is possibly a variety, or mixture, of true effect sizes.

The Bayesian analyses were conducted using three different methods. First, when the raw data points from individual studies were available, the Bayesian analyses were conducted using the MCMCglmm package in R (Hadfield, 2010). Broad, noninformative priors were used for the first study in each sequence. The priors for each subsequent analysis were the effect size estimate from a random effects MA of the previous effect sizes, along with the sampling error of this effect size. The sampling error values thus served as credibility (or degree of confidence) weights. This was a conservative decision, as the error variance from a random effects MA allows for variation in true effects. Second, when only the effect sizes and sample sizes were available (as was the case for the previously published meta-analysis data), the just-described Bayesian analyses were conducted on generated data that had the exact same effect size and sample size as the real data. Third, the Bayesian analyses were also conducted using the computational methods described by Schmidt and Raju (2007). These three Bayesian methods produce essentially identical results (O'Connor & Ermacora, 2021). The method used for the Bayesian analyses does not matter, except perhaps in unusual circumstances that are not relevant to the present concerns.

Consistency and agreement rates were computed for the NHST analyses, for the updating MAs, and for the Bayesian analyses. The consistency rate was the proportion of times that the most common conclusion was reached for a pool of effect sizes. Three conclusions are possible for each effect size: a positive effect, a negative effect, and no effect. The signs of the effect sizes and the possible inclusion of a zero value in a confidence interval were used to make these categorizations (e.g., a "negative effect" conclusion was when a negative effect size had a confidence interval that did not include zero). The number of times each of the three possible conclusions occurred for a pool of effect sizes was counted, and the consistency rate was based on the most common conclusion. The agreement rate for a pool of effect sizes was the proportion of times that the conclusions for individual studies were identical to the conclusion (re: the same three categories) of the final, all-studies-combined MA. More detailed descriptions of the analytic methods were provided by O'Connor and Ermacora (2021). All of the analyses can be conducted using the NO.PING.PONG package in R (O'Connor, 2021), which also contains all of the datasets described above.

Low Self-Esteem Predicting Depression: **R** *= 0.57*

Sowislo and Orth (2013, Table 7.2) reported findings from a meta-analysis of 77 longitudinal studies that provided effect sizes for self-esteem predicting future depressive symptoms (median $N = 224$). The final effect size, in correlation coefficient metric, was 0.57, which is a large effect size. Not surprisingly, the consistency and agreement rates for NHST, cumulative MA, and Bayesian methods were all 0.99 (see Table 7.2). With regard to yes-or-no decisions about an association, NHST works well when the effect size and sample sizes are large. The confidence intervals for cumulative MA, and Bayesian methods became increasingly narrow as the study sequence progressed, while those for NHST varied due to the sample sizes of the individual studies (Fig. 7.2). Very similar findings emerged for the computer-generated data in which the population effect size was also 0.57 (Table 7.2 and Fig. 7.3).

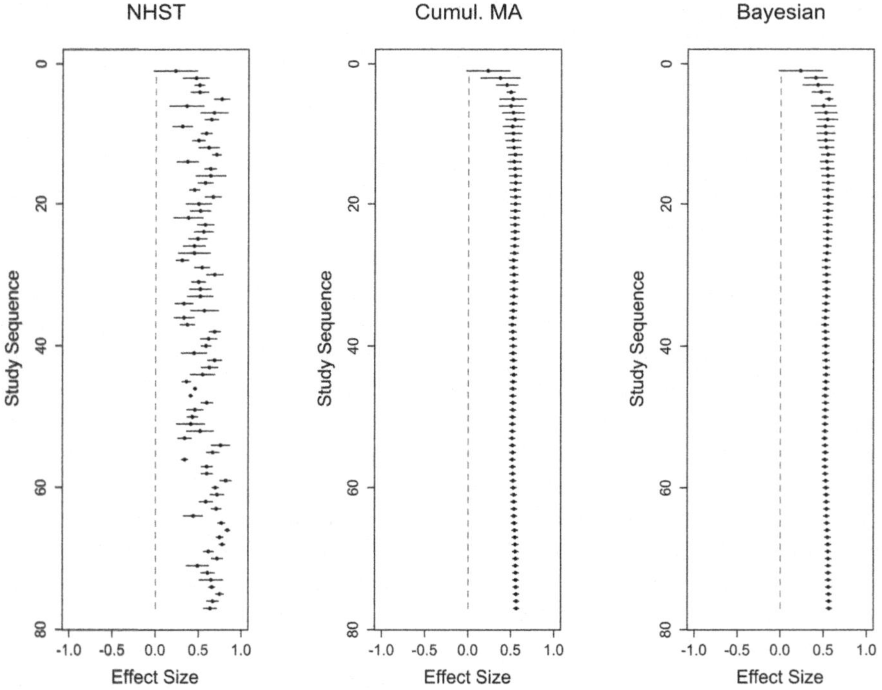

Fig. 7.2 Effect sizes and confidence intervals for low self-esteem predicting depression: $r = 0.57$

Table 7.2 Effect sizes, consistency, and agreement with the final meta-analysis findings, and heterogeneity statistics

	Final *r*			NHST		Updating Meta-Analysis		Bayesian Analysis		Q		Tau		
Dataset	*r*	*r* LB	*r* UB	Consistency	Agreement	Consistency	Agreement	Consistency	Agreement	Q	*p*	Tau	Tau LB	Tau UB
Self-Esteem & Depression	0.57	0.53	0.6	0.99	0.99	0.99	0.99	0.99	0.99	1928.4	0	0.13	0.11	0.16
Generated Data: $r = 0.57$	0.57	0.56	0.58	1	1	1	1	1	1	5.98	0.4	0.01	0	0.03
CBT for Social Anxiety	0.32	0.26	0.38	0.71	0.71	1	1	1	1	44.61	0	0.1	0.04	0.18
Generated Data: $r = 0.32$	0.32	0.29	0.35	0.86	0.86	1	1	1	1	28.34	0.99	0	0	0
Hypomanic Personality & BIS	−0.04	−0.15	0.07	0.37	0.37	0.95	0.95	0.89	0.89	429.3	0	0.23	0.17	0.34
Generated Data: $r = -0.04$	−0.04	−0.06	−0.02	0.88	0.1	0.92	0.92	0.92	0.92	54.9	0.26	0.02	0	0.06
CBT for Autism	0.11	−0.02	0.25	0.82	0.82	0.65	0.65	0.65	0.65	138.1	0	0.24	0.15	0.36
Generated Data: $r = 0.11$	0.11	0.07	0.16	0.9	0.1	0.94	0.94	0.94	0.94	41.3	0.78	0	0	0.09
Generated Data: $r = 0.077$	0.077	0.02	0.13	0.78	0.18	0.56	0.56	0.54	0.54	102.5	0	0.14	0.09	0.19

Note. *Consistency* the proportion of times that the most common conclusion was reached for an analytic method, *Agreement* the proportion of times that the conclusions for individual studies were identical to the conclusion of the final, all-studies-combined meta-analysis, *LB & UB* lower and upper bound confidence intervals, *Q* heterogeneity, *tau* the estimated standard deviations of the distributions of the true effect sizes

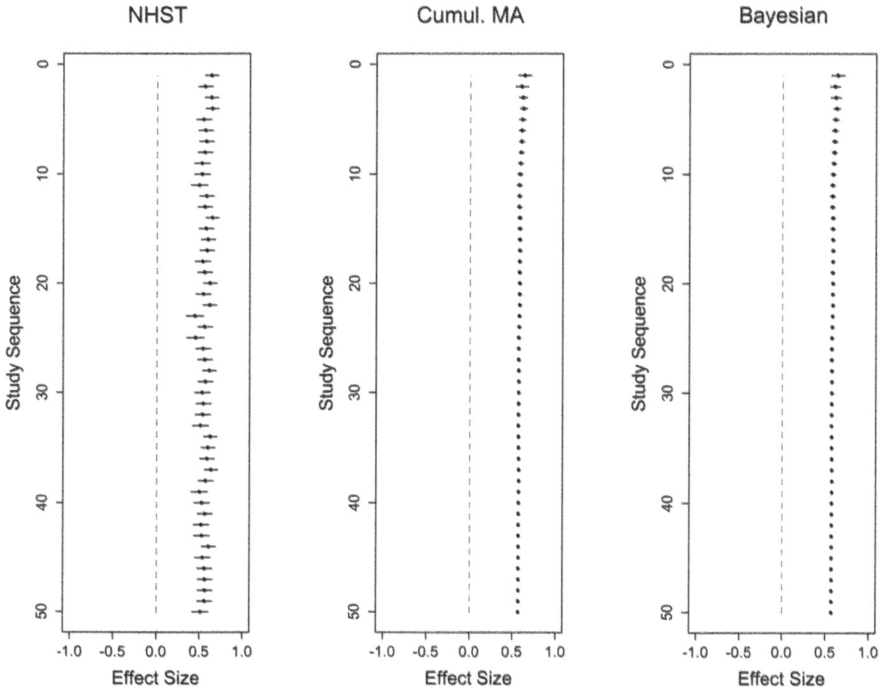

Fig. 7.3 Effect sizes and confidence intervals for generated data: Population $r = 0.57$

Internet-Delivered CBT for Social Anxiety Disorder: R = *0.32*

Kampmann, Emmelkamp, and Morina (2016, Fig. 7.4) reported findings from a meta-analysis of 24 studies on Internet-delivered cognitive behavior therapy (vs. control conditions) for social anxiety disorder (median $N = 65$). The final effect size, in correlation coefficient metric, was 0.32. The consistency and agreement levels for NHST (both 0.71) were lower than those for cumulative MA and for Bayesian analyses (for which all of the rates were 1, or 100%; see Table 7.2). The confidence intervals for the NHST analyses across the sequence of studies were relatively large and variable. In contrast, the confidence intervals for the updating meta-analyses and for the Bayesian analyses were consistent, increasingly narrow, and they did not include a zero-effect size, even early on in the study sequence (see Fig. 7.4). Similar findings emerged when the analyses were conducted on computer-generated data in which the population effect size was 0.32 (Table 7.2 and Fig. 7.5). Conclusion error rates for NHST thus begin to increase when effect sizes and sample sizes are no longer large. In contrast, the cumulative MA and Bayesian results remained stable and accurate.

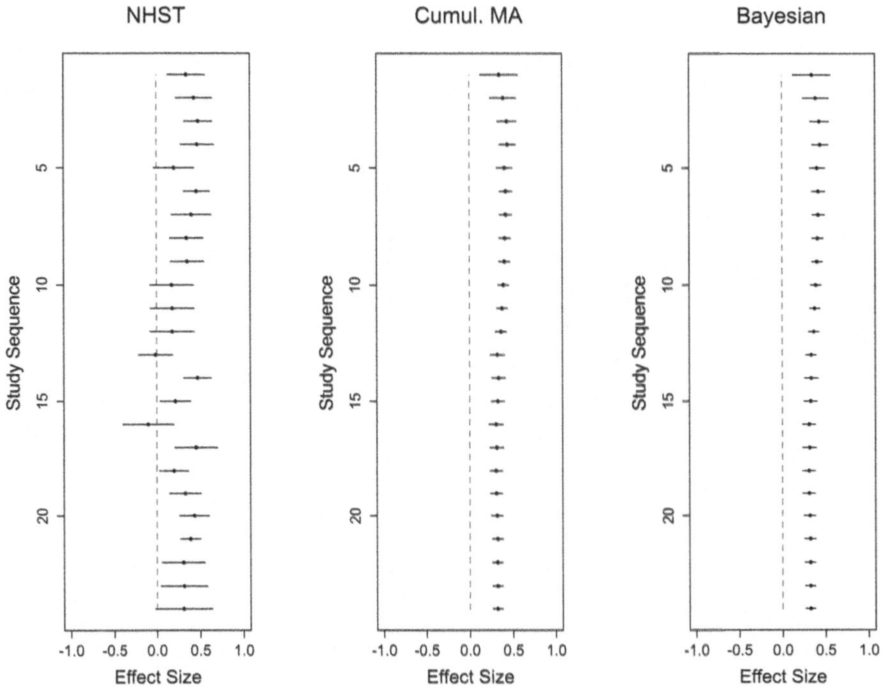

Fig. 7.4 Effect sizes and confidence intervals for Internet-delivered CBT for social anxiety disorder: $r = 0.32$

Hypomanic Personality and BIS Sensitivity: R = *−0.04*

Katz, Naftalovich, Matanky, and Yovel (2021) reported findings from a meta-analysis of 19 studies on hypomanic personality tendencies and behavioral inhibition system sensitivity (median N = 230). The final effect size, in correlation coefficient metric, was −0.04. The authors also reported a meta-analysis effect size of $r = 0.34$ for hypomanic personality tendencies and behavioral *activation* system sensitivity, but our focus will be on the near-zero effect size for BIS sensitivity.

For NHST, the consistency and agreement rates were low (both 0.37), whereas the corresponding rates for cumulative MA and Bayesian analyses were all 0.89 or higher. Viewers of this literature who see only the NHST findings (the left-most plot in Fig. 7.6) are likely to be misled and perplexed. Most studies (63%) reported a statistically significant effect, but the effect sizes bounced around considerably. A distinctly different pattern was evident when the findings from previous studies were incorporated in the analyses for each new study. The cumulative MA and Bayesian analyses indicated, early on in the study sequences and with narrow confidence intervals, that the effect size is very near zero. The study conclusions did not bounce back and forth.

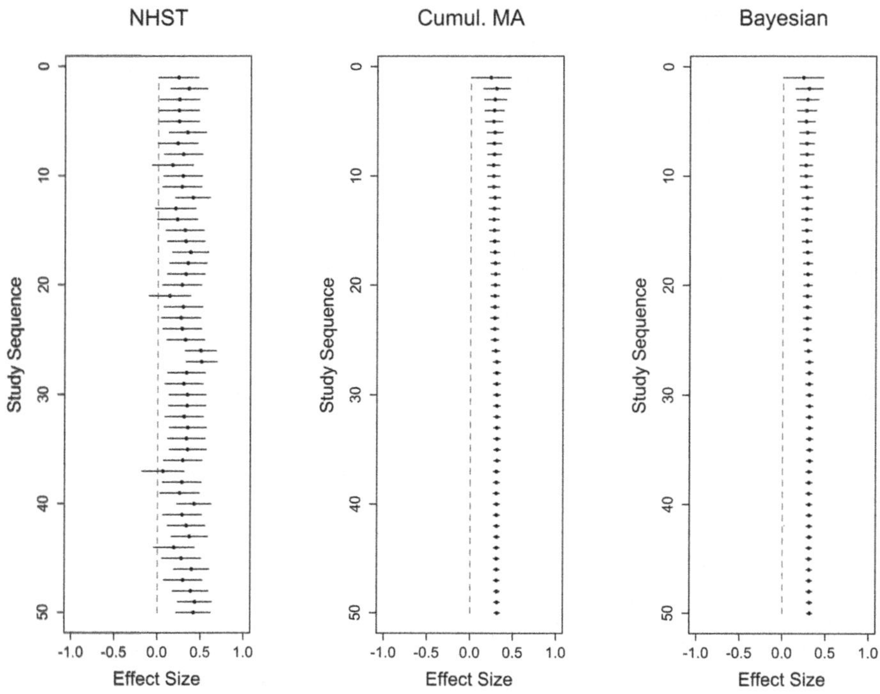

Fig. 7.5 Effect sizes and confidence intervals for generated data: Population $r = 0.32$

The findings for the computer-generated data in which the population effect size was −0.04 were somewhat different but also informative (see Table 7.2 and Fig. 7.7). There was reduced bouncing around in the NHST study conclusions, as $p > 0.05$ occurred 88% of the time for NHST. Conclusions about an effect were nevertheless thus not permitted. The null hypothesis cannot be rejected or accepted when $p > 0.05$. In contrast, the cumulative MA and Bayesian analyses revealed, correctly, that there was a very small but nonzero effect (the consistency and agreement rates were all 92%). This emerged early on in the study sequence. Viewers of this literature who see the cumulative MA and Bayesian findings would be able to reach a correct conclusion about a tiny effect size. They may well consider $r = -0.04$ to be a meaningless and practically negligible effect size, and they can be confident about the near-zero effect size that they are dismissing.

CBT for Affective Symptoms in Autism: **R** *= 0.11*

Weston, Hodgekins, and Langdon (2016) reported findings from a meta-analysis of 17 studies on the effectiveness of CBT on affective symptoms for people with autistic spectrum disorders (median $N = 36$). The final effect size, in correlation

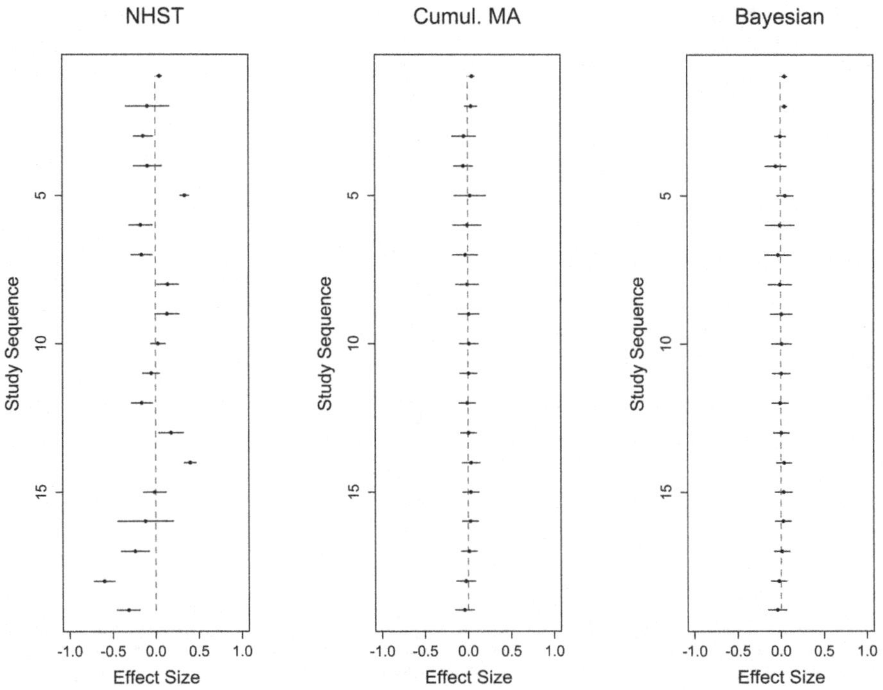

Fig. 7.6 Effect sizes and confidence intervals for hypomanic personality and BIS sensitivity: $r = -0.04$

coefficient metric, was 0.11. The effect size was thus small and the sample sizes were modest. The NHST, cumulative MA, and Bayesian analyses all generally indicated "no effect" (see Table 7.2 and Fig. 7.8). The confidence intervals nevertheless became increasingly narrow and close to excluding zero by the 17th study for the cumulative MA and Bayesian analyses. More studies were clearly needed for these procedures to identify the 0.11 effect size with greater confidence.

The findings for the computer-generated data, in which the population effect size was 0.11 and the median sample size was 36, provide context for the real-data findings (see Table 7.2 and Fig. 7.9). These findings cannot be expected to parallel those for the real autism data wherein the sample sizes were not identical for each study in the sequence (the median N was used in every case for the computer-generated data). The findings merely indicate what would happen if the true population effect size was $r = 0.11$ and the sample size for every study was 36, *and* if there were data from 50 studies instead of just 17 studies. For NHST, there would be very high consistency (90%) in the study findings, but quite low agreement with the true effect size (10%). In contrast, the consistency and agreement rates were all 0.94 for the cumulative MA and Bayesian analyses. In these cases, correct conclusions were reached most of the time when both the effect size and sample sizes were relatively modest.

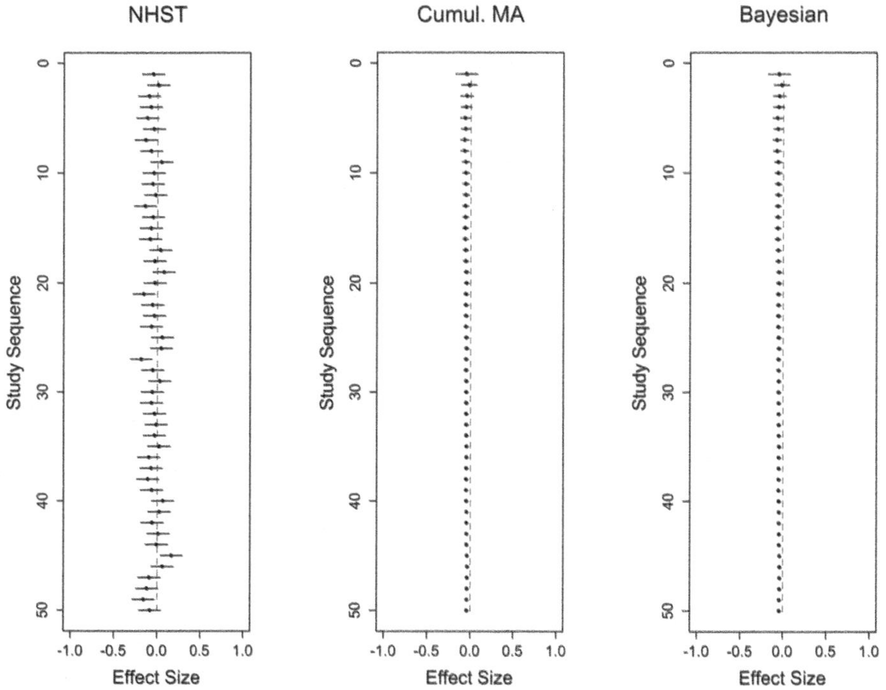

Fig. 7.7 Effect sizes and confidence intervals for generated data: Population $r = -0.04$

Generated Data: R = *0.077*

Cumulative MA and Bayesian analyses do not always result in high rates of consistency and agreement across studies. A computer-generated population wherein $r = 0.077$ was created to further illustrate this point. The sample sizes were set to $N = 50$ for every "study." The results are provided in Table 7.2 and Fig. 7.10. The consistency rates were 56% for cumulative MA and 54% for Bayesian analyses, and the corresponding agreement rates were 56% and 54%. The population effect size and the sample sizes were small. Early on in the study sequence, the confidence intervals for the cumulative MA and Bayesian analyses included zero, indicating "no effect" in NHST terminology. But about half way through the study sequences, the cumulative MA and Bayesian confidence intervals no longer included zero. Both methods began correctly detecting the small, nonzero population effect size and permitted correct conclusions. The low overall consistency and agreement rates were due to the small effect and sample sizes, in which case it took a few studies for the procedures to zoom in on the real effect. In contrast, for NHST, the findings were quite consistent (78%) and wrong (the agreement rate with the true effect was only 18%).

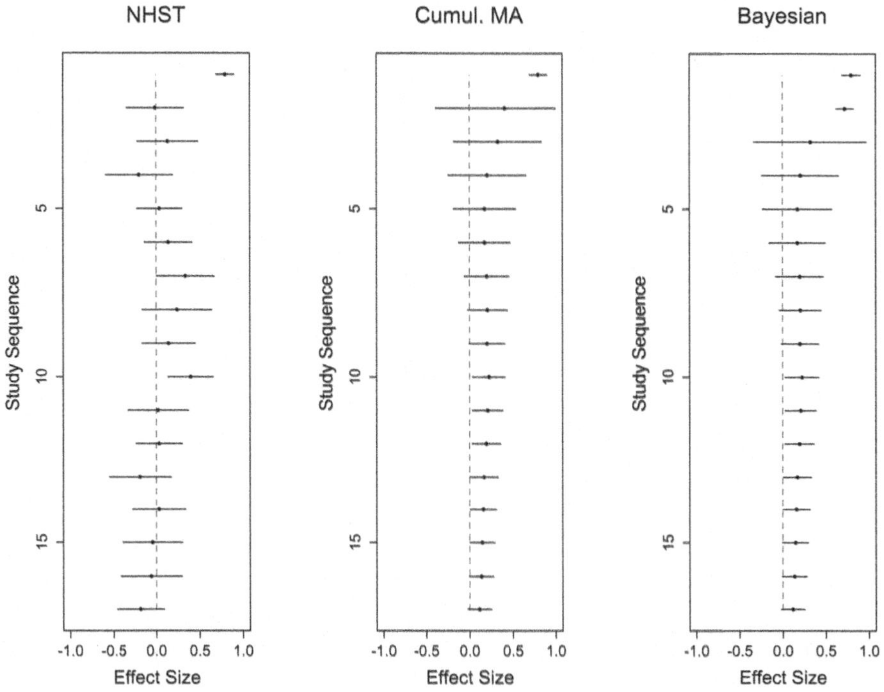

Fig. 7.8 Effect sizes and confidence intervals for CBT for affective symptoms in autism: $r = 0.11$

Heterogeneity Among Studies

The meta-analysis heterogeneity statistics are provided in Table 7.2. The heterogeneity levels (Q statistics) for the real data were significant whereas the Q values for the generated data were mostly not significant. This indicates that the variability in the pools of effect sizes for the real datasets was greater than what would be expected on the basis of sampling variability alone. The estimated standard deviations of the distributions of the true effect sizes (the tau values) were generally small, ranging from zero to 0.24, but not always negligible for the real data. The Q statistics and tau values together indicate that there is variation in the effect sizes for the real data that deserves research attention. A larger pool of studies might help reduce the apparent variability in the effect sizes, especially given the sometimes modest numbers of studies. The variability in effect sizes could also be explained by moderator variables (e.g., gender, ethnicity). Effect size variability is to be expected and is common in meta-analyses. The use of the random effects meta-analysis model, which assumes such variability, nevertheless produced confidence intervals for the estimated effect sizes that were relatively narrow. The Q statistic can thus generate alarms when there is in fact little doubt about an effect.

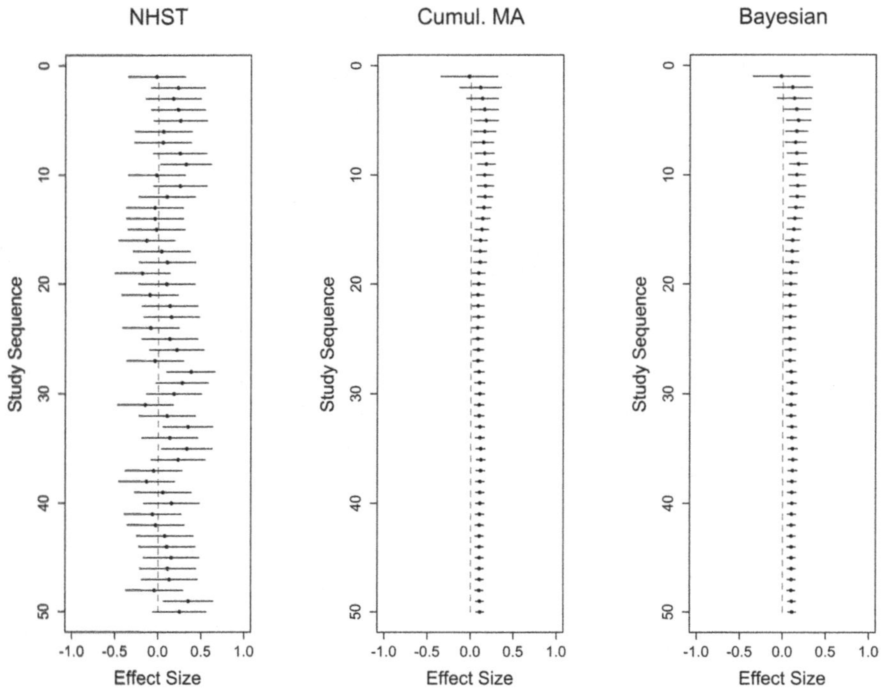

Fig. 7.9 Effect sizes and confidence intervals for generated data: Population $r = 0.11$

Discussion

It is clearly negligent to not incorporate previous findings when attempting to reach a conclusion about an effect based on a new dataset (see also Quintana et al. (2017); Wagenmakers et al., 2016a). When previous findings are incorporated, the confidence intervals around an effect size become narrow and relatively consistent as the pool of studies grows. This phenomenon occurred quite early in the study sequences (see the above Figures; and O'Connor & Ermacora, 2021). The conclusions for the updating MA and Bayesian analyses were usually in agreement with the final, all-studies-combined conclusions very early on and did not change as additional studies were added to the pools of effect sizes. The NHST confidence intervals remained wide and variable across the study sequences. They were bounced back and forth by sampling variability and by the statistical power (the Ns) of the individual studies.

NHST results in consistent, accurate *"Is there an effect or not?"*, conclusions in two scenarios: When a population effect size is large, and when a population effect size is very close to zero. However, large population effect sizes are rare in psychological research. When a population effect size is close to zero, high power levels are required to reach statistical significance. Most low and moderate powered studies will produce a "no effect" conclusion. A serious shortcoming with NHST in this case is that the failure to reach statistical significance does not permit one to

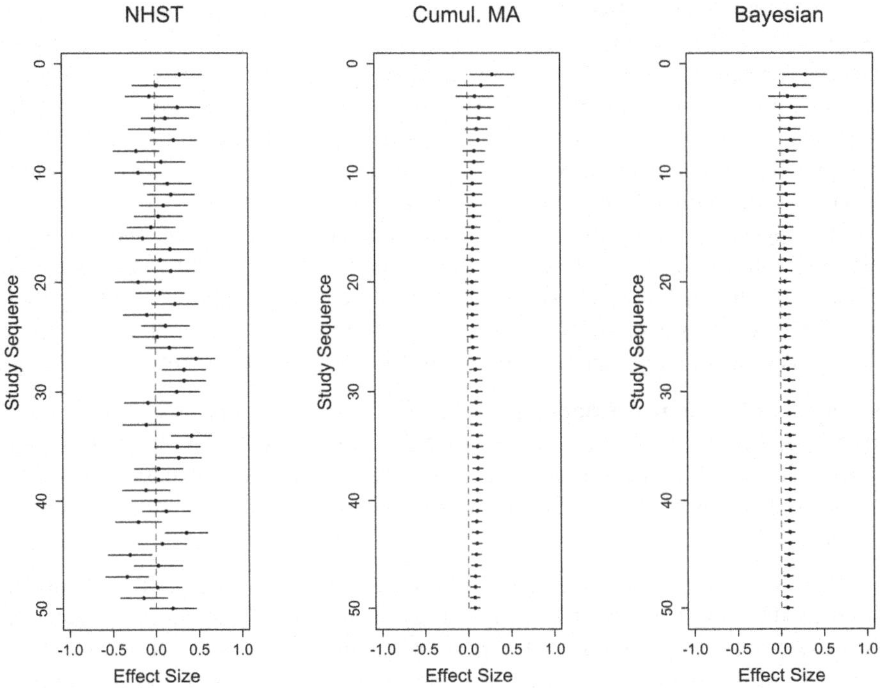

Fig. 7.10 Effect sizes and confidence intervals for generated data: Population $r = 0.077$

conclude that the null hypothesis is true. One cannot conclude that "there is no relationship," despite the fact that such statements are often made in discussion sections (Cohen, 1990, 1994; Falk & Greenbaum, 1995; Gallistel, 2009; Gigerenzer et al., 2004; Oakes, 1986). When $p > 0.05$, one can only conclude that the null hypothesis could not be rejected, which seems like an unsatisfying, uninformative conclusion after going through all of the efforts to conduct a study. A further serious problem with NHST, both when an effect is large and when an effect is near zero, is that the confidence intervals remain wide and variable across the study sequences. Firm conclusions about the magnitude of an effect size can rarely be reached when previous findings are not incorporated into the data analyses. This is in sharp contrast with the narrow confidence intervals that emerged from the updating MA and Bayesian analyses.

The benefits of incorporating previous findings are particularly evident when an all-studies-combined effect size is nonzero, but small (O'Connor & Ermacora, 2021; Rindskopf, 2016). Researchers can potentially reach accurate conclusions about small effect sizes relatively early on in a study sequence. Incorporating previous findings increases the likelihood reaching confident conclusions about weak effect sizes without having to wade through many apparently conflicting or non-significant effects.

Confusion about an effect is greatest when the most consistent finding in a literature occurs 50% of the time. Even worse than maximum confusion is reaching an incorrect conclusion more than 50% of the time. This occurred when the final effect sizes were weak, but not zero. The NHST conclusion *error* rates were 67% for the hypomanic personality tendencies and BIS dataset, and 82% for the $r = 0.077$ computer-generated dataset. NHST study conclusion consistency can occur simultaneously with a final conclusion agreement rate that is well below 50%. The agreement/accuracy rates for the analyses that incorporated previous findings were higher. Conflicts with the final, all-studies-combined conclusion rarely occurred when an individual dataset was evaluated while taking previous findings into account. There was essentially no back and forth ping-pong of conclusions in these cases. There was more order in the research universe and fewer conflicting findings.

Caveats and Practical Challenges The central findings reported in this manuscript were not meaningfully affected by publication bias. The findings from analytic methods that incorporate previous data should nevertheless always be considered alongside the evidence from corresponding tests for publication bias for the datasets.

The consistency and agreement (accuracy/error) rates for NHST and for the other methods will be affected by the study sample sizes and by the population effect sizes. As described above, larger study sample sizes will result in lower conclusion error rates for NHST when the population effect sizes are non-negligible. Larger study sample sizes will result in higher conclusion error rates for NHST when the population effect sizes are negligible because trivial effect sizes may be statistically significant in these cases. In contrast, larger study sample sizes will always result in more accurate conclusions for the updating MA and Bayesian analyses.

Effect size heterogeneity within datasets nevertheless always deserves consideration when deciding on the existence, magnitude, and the degree of certainty of an effect. Research should be conducted on the reasons for the heterogeneity when it does occur. For example, the effect sizes may well vary depending on gender or other demographic variables.

The heterogeneity statistic values must also be considered alongside the confidence intervals from the updating MA and Bayesian analyses. The confidence intervals for the updating MA and Bayesian analyses were based on the random-effects model which assumes that there is variation in the true effect sizes. This causes the confidence intervals to be wider under random effects models than under other models, such as the fixed effects model. The overall effect size, across possible moderator variables, falls within the confidence interval. Although heterogeneity statistics may indicate that moderator variables may still be found, an overall effect size is not in doubt when the confidence interval for it is narrow. More generally, use of the error variance from a random effects MA of previous effect sizes assists researchers in dealing with the fact that studies of the same research questions are rarely completely identical to one another (O'Connor & Ermacora, 2021).

The practical challenge, of course, is having the effect sizes from the previous studies. This nontrivial challenge should nevertheless be placed in context. Previous findings must always be reviewed when planning a new study. Why not make the review quantitative and precise? One might well discover that an additional study is not necessary. One can use either the raw data or the effect sizes from previous studies. Lip service attention to previous findings is asking for trouble. The process would be greatly facilitated if public repositories were maintained of all of the studies and/or datasets for a research question. Such repositories would also make it easier for reviewers to determine whether relevant studies were for some reason excluded by authors who are otherwise attempting to incorporate previous findings into their data analyses. Subjectivity (bias) in the selection of previous studies for setting the priors in Bayesian analyses was a major reason for Fisher's resistance to Bayesian methods (Lehmann, 1993; Salsburg, 2001). Comprehensive, public, online repositories for research programs would help circumvent concerns about possibly biased selections of previous studies.

Conclusions

Lack of awareness of alternative methods of data analysis and the fear of not getting published are perhaps the biggest reasons for the ongoing use of NHST. Software availability is no longer an obstacle. Substantive arguments in favor of NHST are difficult to imagine, especially now that better methods are available. Change would be greatly facilitated by explicit encouragement for non-NHST methods in journal editorial policies. Researchers will remain reluctant to learn and use alternative methods as long as they believe that it is not necessary to do so, and as long as they believe that their chances of getting published may be reduced if they do not use NHST (O'Connor, 2017). Simple statements in journal editorial policies that encourage alternative methods could be very influential.

The incorporation of previous findings when analyzing new data would clearly lead to less chaos and more coherence in studies that are conducted on the same phenomena. Having to endure years of apparently conflicting NHST findings while waiting for an eventual meta-analysis on an accumulation of studies seems like extended, pointless chaos. The ping-pong game goes on too long. The conclusions that are reached in the discussion sections of many research reports would be more accurate if previous findings were incorporated into the data analyses (O'Connor & Ermacora, 2021). The typical smaller-scale study, such as a student thesis project, is underpowered, which reduces the accuracy of NHST-based conclusions. The $p < 0.05$ level is often not obtained. Wishing to be cautious scientists, authors then feel obligated to give serious consideration to the null hypothesis or to speculate about the conditions or samples (moderator variables) for which there may be significant effects. On the other hand, if findings from previous studies had been incorporated into the data analyses, then authors would be more likely to reach correct conclusions about the existence of an effect in their discussion sections. They could focus on the width of the confidence intervals based on all available data, and on how the findings from their own individual studies affect these confidence intervals. Discussion sections would become more interesting to write and read, and more accurate. A low-risk, comfortable way forward may be to run the analyses both with, and without, the incorporation of previous findings and to give readers both sides of the picture.

References

Bakan, D. (1966). The test of significance in psychological research. *Psychological Bulletin, 66*, 423–437.

Bellhouse, D. R. (2004). The Reverend Thomas Bayes, FRS: A biography to celebrate the tercentenary of his birth. *Statistical Science, 19*, 3–43.

Borenstein, M., Hedges, L. V., Higgins, J. P. T., & Rothstein, H. R. (2009). *Introduction to meta-analysis*. Wiley.

Button, K. S., Ioannidis, J. P. A., Mokrysz, C., Nosek, B. A., Flint, J., Robinson, E. S. J., & Munafò, M. R. (2013). Power failure: Why small sample size undermines the reliability of neuroscience. *Nature Reviews Neuroscience, 14*(5), 365–376.

Carver, R. P. (1978). The case against statistical significance testing. *Harvard Educational Review, 48*, 378–399.

Cohen, J. (1990). Things I have learned (so far). *American Psychologist, 45*, 1304–1312.

Cohen, J. (1994). The Earth is round ($p < 0.05$). *American Psychologist, 49*, 997–1003.

Cumming, G., & Calin-Jageman, R. (2017). *Introduction to the new statistics: Estimation, open science, and beyond*. Routledge.

Eich, E. (2014). Business not as usual [Editorial]. *Psychological Science, 25*(1), 3–6.

Falk, R., & Greenbaum, C. W. (1995). Significance tests die hard: The amazing persistence of a probabilistic misconception. *Theory and Psychology, 5*(1), 75–98.

Field, A. P., & Wilcox, R. R. (2017). Robust statistical methods: A primer for clinical psychology and experimental psychopathology researchers. *Behaviour Research and Therapy, 98*, 19–28.

Gallistel, C. R. (2009). The importance ofproving the null. *Psychological Review, 116*(2), 439–453. https://doi.org/10.1037/a0015251

Gigerenzer, G., Krauss, S., & Vitouch, O. (2004). The null ritual: What you always wanted to know about null hypothesis testing but were afraid to ask. In D. Kaplan (Ed.), *Handbook on quantitative methods in the social sciences* (pp. 391–408). Sage.

Gill, J. (2015). *Bayesian methods: A social and behavioral sciences approach*. Chapman & Hall/CRC.

Good, P. I. (2010). *Permutation, parametric, and bootstrap tests of hypotheses*. Springer.

Hadfield, J. D. (2010). MCMC methods for multi-response generalized linear mixed models: The MCMCglmm R Package. *Journal of Statistical Software, 33*(2), 1–22. URL http://www.jstatsoft.org/v33/i02/

Harlow, L. L., Mulaik, S. A., & Steiger, J. H. (1997). *What if there were no significance tests?* Lawrence Erlbaum.

Hunter, J. E. (1997). Needed: A ban on the significance test. *Psychological Science, 8*(1), 3–7.

Jeffreys, H. (1961). *Theory of probability*. Oxford University Press.

Kampmann, I. L.,Emmelkamp, P. M., & Morina, N. (2016). Meta-analysis of technology-assisted interventions for social anxietydisorder. *Journal of anxiety disorders, 42*, 71–84. https://doi.org/10.1016/j.janxdis.2016.06.007

Katz, B. A.,Naftalovich, H., Matanky, K., & Yovel, I. (2021). The dual-system theory of bipolar spectrum disorders: A metaanalysis. *Clinical Psychology Review, 83*, Article 101945. https://doi.org/10.1016/j.cpr.2020.101945

Kline, R. B. (2013). *Beyond significance testing: Statistics reform in the behavioral sciences* (2nd ed.). American Psychological Association.

Kruschke, J. K. (2015). *Doing Bayesian data analysis: A tutorial with R and BUGS*. Academic Press/Elsevier.

Kruschke, J. K., & Liddell, T. M. (2018). The Bayesian new statistics: Hypothesis testing, estimation, meta-analysis, and power analysis from a Bayesian perspective. *Psychonomic Bulletin & Review, 25*, 178–206.

Kruschke, J. K., Aguinis, H., & Joo, H. (2012). The time has come: Bayesian methods for data analysis in the organizational sciences. *Organizational Research Methods, 15*, 722–752.

Lambert, M. J., & Bailey, R. J. (2012). Measures of clinically significant change. In H. Cooper, P. M. Camic, D. L. Long, A. T. Panter, D. Rindskopf, & K. J. Sher (Eds.), *APA handbook of research methods in psychology, Vol. 3. Data analysis and research publication* (pp. 147–160). American Psychological Association.

Lee, M. D., & Wagenmakers, E.-J. (2013). *Bayesian cognitive modeling: A practical course*. Cambridge University Press.

Lehmann, E. L. (1993). The Fisher, Neyman–Pearson theories of testing hypotheses: One theory or two? *Journal of the American Statistical Association, 88*(424), 1242–1249.

Maxwell, S. E. (2004). The persistence of underpowered studies in psychological research: Causes, consequences, and remedies. *Psychological Methods, 9*(2), 147–163.

Maxwell, S. E., & Delaney, H. D. (2004). *Designing experiments and analyzing data: A model comparison perspective* (2nd ed.). Erlbaum.

Meehl, P. (1978). Theoretical risks and tabular asterisks: Sir Karl. Sir Ronald. and the slow progress of soft psychology. *Journal of Consulting and Clinical Psychology, 46*, 806–834.

Morrison, D. E., & Henkel, R. E. (Eds.). (1970). *The significance test controversy*. Aldine.

Nickerson, R. S. (2000). Null hypothesis significance testing: A review of an old and continuing controversy. *Psychological Methods, 5*(2), 241–301.

O'Connor, B. P. (2017). A first steps guide to the transition from null hypothesis significance testing to more accurate and informative Bayesian analyses. *Canadian Journal of Behavioral Science, 49*(3), 166–182.

O'Connor, B. P. (2021). NO.PING.PONG: Incorporating previous findings when evaluating new data [Computer software manual]. https://cran.r-project.org/web/packages/NO.PING.PONG/index.html. (R package version 0.1.4).

O'Connor, B. P., & Ermacora, D. (2021). Illustrations of why previous findings should be taken into account when evaluating new datasets. *Canadian Journal of Behavioral Science, 53*(3), 328–341. https://doi.org/10.1037/cbs0000259

Oakes, M. L. (1986). *Statistical inference: A commentary for the social and behavioral sciences*. John Wiley.

Pollard, P., & Richardson, J. T. E. (1987). On the probability of making type I errors. *Psychological Bulletin, 102*, 159–163. https://doi.org/10.1037/0033-2909.102.1.159

Pruzek, R. M. (2016). An introduction to Bayesian Inference and its applications. In L. L. Harlow, S. A. Mulaik, & J. H. Steiger (Eds.), *What if there were no significance tests?* (2nd ed.). Routledge.

Quintana, M., Viele, K., & Lewis, R. J. (2017). Bayesian analysis: Using prior information to interpret the results of clinical trials. *Journal of the American Medical Association, 318*(16), 1605–1606.

Rindskopf, D. M. (2016). Testing "small," not null, hypotheses: Classical and Bayesian approaches. In L. L. Harlow, S. A. Mulaik, & J. H. Steiger (Eds.), *What if there were no significance tests?* (2nd ed.). Routledge.

Salsburg, D. (2001). *The lady tasting tea: How statistics revolutionized science in the twentieth century*. W. H. Freeman.

Schmidt, F. L. (1996). Significance testing and cumulative knowledge in psychology: Implications for training of researchers. *Psychological Methods, 1*, 115–129.
Schmidt, F. L., & Raju, N. S. (2007). Updating meta-analytic research findings: Bayesian approaches versus the medical model. *Journal of Applied Psychology, 92*(2), 297–308.
Sharpe, D. (2013). Why the resistance to statistical innovations? Bridging the communication gap. *Psychological Methods, 18*, 572–582.
Sowislo, J. F., &Orth, U. R. (2013). Does low self-esteem predict depression and anxiety? A meta-analysis of longitudinal studies. *Psychological Bulletin, 139*(1), 213–240.
Spence, J. R., & Stanley, D. J. (2018). Concise, simple, and not wrong: In search of a short-hand interpretation of statistical significance. *Frontiers in Psychology, 9*, Article 2185.
Tryon, W. W. (2016). Replication is about effect size: Comment on Maxwell, Lau, and Howard (2015). *American Psychologist, 71*(3), 236–237.
Viechtbauer, W. (2010). Conducting meta-analyses in R with the metafor package. *Journal of Statistical Software, 36*(3), 1–48. URL: http://www.jstatsoft.org/v36/i03/
Wagenmakers, E. J., Morey, R. D., & Lee, M. D. (2016a). Bayesian benefits for the pragmatic researcher. *Current Directions in Psychological Science, 25*(3), 169–176.
Wagenmakers, E. J., Verhagen, A. J., & Ly, A. (2016b). How to quantify the evidence for the absence of a correlation. *Behavior Research Methods, 48*, 413–426.
Wasserstein, R. L., & Lazar, N. A. (2016). The ASA's statement on *p* values: Context, process, and purpose. *American Statistician, 70*(2), 129–133.
Weston, L., Hodgekins, J., & Langdon, P. E. (2016). Effectiveness of cognitive behavioural therapy with people who have autistic spectrum disorders: A systematic review and metaanalysis. *Clinical psychology review, 49*, 41–54. https://doi.org/10.1016/j.cpr.2016.08.001
Wetzels, R., Matzke, D., Lee, M. D., Rouder, J. N., Iverson, G. J., & Wagenmakers, E.-J. (2011). Statistical evidence in experimental psychology: An empirical comparison using 855 *t* tests. *Perspectives on Psychological Science, 6*, 291–298.
Wilkinson, L., & Task Force on Statistical Inference. (1999). Statistical methods in psychology journals: Guidelines and explanations. *American Psychologist, 54*, 594–604.
Ziliak, S. T., & McCloskey, D. N. (2008). *The cult of statistical significance: How the standard error costs usjobs, justice, and lives*. University of Michigan Press.
Zyphur, M., & Oswald, F.(2015). Bayesian estimation and inference: A user's guide. *Journal of Management, 41*(2), 390–420.

Chapter 8
Hypothesizing After Results Are Known: HARKing

Ana J. Bridges

Abstract The scientific process rests on a set of core values, including objectivity, honesty, openness, accountability, fairness, and stewardship. Hypothesizing After Results Are Known (HARKing) threatens all of these values, and therefore is a threat to science itself. A commitment to reducing the prevalence of HARKing does not mean a stifling of exploration or creativity; indeed, these are at the core of discovery. Instead, that commitment is about presenting discovery and exploration honestly and not disguising it as confirmation. A final note on this topic is that unpredicted findings can be powerful scientific narratives that really engage audiences. Rather than seeing post-hoc findings as a pox on the study or investigator, as scholars and humans we should embrace the twists and turns our scholarship takes us, realizing that doing so is both more honest and more interesting.

Keywords HARKing · Questionable research practices · Clinical science · Psychological science

Introduction

The field of psychological science has found itself in the midst of a troubling situation known as a replication crisis (Pashler & Wagenmakers, 2012). In brief, efforts to replicate previously established psychological findings, such as currency primes make us more selfish (i.e., that being reminded of money, such as through subtle background images of $100 bills, makes people more likely to say they prefer solitary activities and to endorse a greater sense of self-sufficiency; Caruso et al., 2017), have more often failed than held up (Klein et al., 2014). Many explanations exist for

A. J. Bridges (✉)
University of Arkansas, Fayetteville, AR, USA
e-mail: abridges@uark.edu

W. O'Donohue et al. (eds.), *Avoiding Questionable Research Practices in Applied Psychology*, https://doi.org/10.1007/978-3-031-04968-2_8

this replication crisis—some researchers have focused on the changing nature of psychological phenomena (e.g., views of the American flag changing over time; Ferguson et al., 2014), some note the complex nature of what psychologists study that would lead to an expectation of failure to replicate across specific subsamples (Schimmack, 2020), and some note replications themselves are just as prone to inaccuracies as the original studies they are attempting to reproduce (Bryan et al., 2019). However, the replication crisis is largely understood through the lens of questionable research practices (John et al., 2012) that call into question the accuracy of the original results due to problematic practices in designing and conducting the original research. Among these questionable practices is hypothesizing after results are known, or HARKing (Kerr, 1998). In this chapter, I take a deeper look into HARKing—situating it within research design, describing why it is problematic, and suggesting solutions.

Confirmatory Versus Exploratory Research

Research can largely be categorized as falling under two domains: exploratory techniques and confirmatory techniques (de Groot, 2014). Exploratory research does not make a priori assumptions about findings; instead, as the name implies, it is concerned with describing a phenomenon or generating hypotheses about the relations among constructs that can later be tested in a more rigorous manner. Statistics can be used to describe findings in exploratory studies, but they cannot be used as evidence to adjudicate a particular hypothesis since no a priori hypotheses were offered. Good examples of exploratory research questions include how do homeless youth make meaning of their experiences (Toolis & Hammack, 2015) and what aspects of female governors' communication styles were associated with fewer deaths in their state residents during the COVID-19 pandemic (Sergent & Stajkovic, 2020).

Confirmatory research, in contrast, is concerned with evaluating the accuracy of a hypothesis. This kind of research does involve making an educated prediction regarding the relations among constructs that can be examined in a testable hypothesis. Most often, these hypotheses are subject to tests that allow researchers to *disconfirm* the claim, using rules of scientific logic well-articulated by Popper (1959). [In brief, Popper argues that in principle it would take an infinite number of observations to *confirm* a hypothesis but only one exception to *disconfirm* a hypothesis, and therefore scientific studies should be set up in such a manner that a researcher can eliminate one potential explanation for the observed effect or associations among constructs in favor of an alternative explanation.] Confirmatory techniques, therefore, are more powerful scientific tools because they permit, at least in the theory, the elimination of some explanations. Therefore, researchers can hone in on more likely explanations and discard those that are not useful or supported by evidence.

Because science requires confirmatory techniques to advance our understanding of a phenomenon, good research design emphasizes approaches that can lead to discarding certain theories or explanations. If scholars do not conduct a rigorous test of an empirically or theoretically informed a priori hypothesis about a phenomenon, then all they can say is that "this is surprising," or "this was unexpected," but they cannot advance the metaphoric scientific ball down the field. The ability to falsify a hypothesis is a basic motivation of scientific research; without falsifiable research claims, a theory is nothing more than faith. This sets up a powerful motivating context for scholars whose careers and livelihoods, reputations, and prestige depend on their ability to articulate, investigate, and falsify scientific claims (Kiai, 2019).

Defining HARKing

HARKing, hypothesizing after results are known, is the presentation of a posteriori hypothesis as an a priori hypothesis (Kerr, 1998). An a priori hypothesis is a hunch or prediction, based on scientific theory, about the outcome of an experiment or study that has not yet been conducted. It is considered the "cornerstone of the scientific method" (Erren, 2007). Training in psychological science is grounded in a priori hypotheses—we consider these so central to the proper conduction of scientific investigation that our institutions require students conducting research (e.g., honor's theses, master's theses, doctoral dissertations, and so forth) to formally propose their studies to a committee of scientists in advance of collecting (or analyzing, for already collected) data. In these proposals, young scholars are being asked to puzzle through theories and prior literature in order to formulate educated guesses about the outcomes of a particular study, and to explain their logic (the deductive reasoning they used) in order to arrive at these hypotheses.[1] A committee is there to ensure that the deductive logic used by the emerging scholar is sound; that the coverage of the background literature and content domain is sufficient; and that the methods proposed to evaluate or test the a priori hypothesis are adequate (e.g., measures are reliable and valid indicators of the constructs; analytic tests correspond with the research question; participants are sufficient in number and likely to yield variable responses on study measures).

The improper framing of exploratory research findings as confirmatory is equivalent to claiming "I knew it all along." It is a revision of history. Kerr (1998) explains that HARKing is *not* about including unexpected findings or results of post hoc analyses in a scientific paper. Indeed, it is wise for scholars to explore their data and describe to others what associations emerged. However, in hypothetico-deductive research, such findings must be clearly noted as having been exploratory or post hoc, rather than falsely being presented as a priori. Similarly, inductive reasoning in response to exploratory findings is appropriate—new insights about psychological phenomena can be gained from such exploratory endeavors. Here again, the primary concern is about improperly claiming to have foreseen the outcome when it

[1] Of course, there are other approaches to science, including inductive approaches that use empirical data to develop or arrive at theories; however, HARKing as a questionable research practice occurs in the context of hypothetico-deductive approaches to science.

was, in fact, not foreseen. Scientific papers that include exploratory findings post hoc and puzzle in the Discussion section about what these findings might mean can readily spur new, possibly fruitful lines of research (Erren, 2007). This clear generation of a hypothesis in the post hoc phase of study design is also called THARKing (transparent hypothesizing after results are known) and is not a questionable research practice (Vancouver, 2018)—it is the proper way for possible scientific discoveries to be acknowledged and tested in future confirmatory studies.

Within the realm of questionable research practices, some consider HARKing to be a rather minor offense (John et al., 2012). And in comparison to outright fabrication of data, of course, it probably is. However, HARKing is rather insidious in its harm because, even for well-intentioned researchers, it can misrepresent the probabilistic association between a theory and a predicted outcome (Kerr, 1998). The probability of an observation given a theory are not the same as the probability of a theory given an observation (this is also known as the fallacy of the transposed conditional or affirming the consequent; Evett, 1995). To illustrate, a scientist uses Theory A to derive a set of predictions. Based on that theory (and the body of empirical work done before this scientist's study), some outcomes are judged to be more probable or likely than other outcomes. The scientist makes a prediction that is most likely to occur, given the theory (theory ➔ prediction). However, when working backward (outcome ➔ theory), the probabilities are different. Confirmatory research asks "Given theory A, what is the most likely outcome?" while exploratory research asks "What is the most likely theory, given an outcome?" These conditional probabilities are typically *not* equivalent, but HARKing equates the two sets of probabilities. Perhaps a clear example of this fallacy is one familiar to most of us: using the Internet to search for the meaning of medical symptoms. The theory ➔ prediction might be that nearly everyone who has bacterial meningitis experiences headaches (the probability of having a headache given meningitis is approximately 90%; van de Beek et al., 2004). However, if someone has a headache, the likelihood that this headache is an indication of meningitis is nowhere close to 100%; instead, it is much closer to 0% (CDC, 2019).

There are numerous methods used to HARK. In its most prototypical form, researchers construct a hypothesis that fits the results after they have conducted analyses. However, HARKing can also include suppressing actual a priori hypotheses ("I never believed that to be the case," Kerr, 1998; Rubin, 2017), revising the introduction or background literature to a study in order to make a post-hoc hypothesis appear to have been well-reasoned and suspected (Rubin, 2017), presenting a not-previously considered theoretical model to justify a revised hypothesis (Vancouver, 2018), and even revising a manuscript to change commitments to hypotheses but only after reviewer or editor feedback (called passive HARKing; Rubin, 2017).

It is rather daunting to imagine how one might go about detecting HARKing after the fact. If authors are prone to do it, perhaps even unconsciously (e.g., failing to accurately recall what were a priori hypotheses and which ones were concocted post hoc—part of recall bias), then even asking people to reflect on their thoughts at the time they were designing a study will yield unreliable and unsatisfactory evidence. One way to detect this is to examine dissertation or thesis proposals and compare them to published papers resulting from these studies. Discrepancies between these two can yield strong evidence of HARKing (e.g., O'Donohue et al., 2016). Nevertheless, approximately 25–35% of research psychologists *do* in fact openly admit to HARKing and, based on these self-reported admission rates, authors estimate between 54 and 90% have engaged in this practice; John et al., 2012. Some potential indicators of HARKing are: weak prior evidence in support of specific parameters in a model (e.g., weak theoretical or prior empirical evidence to suggest an effect may only be observed in a specific subgroup of participants or in a specific context, given most psychological theories are vague; Eronen & Bringmann, 2021; Lishner, 2021), no prior studies with similar variables showing the association, a selective literature review that fails to consider what the "bulk" of the evidence is and instead appears to cherry-pick studies that support a particular directional hypothesis, exclusion of participants without clear and theoretically derived justification, measuring many variables and only reporting those that support the hypotheses (suppressing variables that failed to support the hypothesis), and *p*-values that are just under conventional significance. For instance, Masicampo and Lalande (2012) demonstrated that published work contains a rather unusually large number of *p*-values that fall just under the 0.05 threshold compared to what might be expected by chance alone.

I would argue that while it may be easiest to think of HARKing as a change in the presentation of a predictor (X) and a criterion (Y) relation (or, said differently, to think of HARKing as involving a study's main effect of X on Y), my personal experience in reading scholarly articles is that it tends to crop up more for mediators and moderators of the X ➔ Y relationship. For instance, a study may examine whether a pornography prime leads to sexually aggressive behavior in a laboratory interaction. The researchers may a priori hypothesize that a pornography prime (X) will result in sexually aggressive behavior (Y). However, after collecting data, the researchers may find that the X ➔ Y relationship is only observed for a subgroup of their sample (for instance, only in male participants). In this case, it may be tempting for researchers to think, "Of course- this makes perfect sense!" and then write up the study as though a primary study aim was to examine the moderating effect of gender on the X ➔ Y relationship. Certainly there is plenty of research available that would support such an a priori hypothesis. However, the researchers may not have considered this moderator ahead of time—either they failed to do due diligence when reviewing prior research or they considered the prior research not to be sufficiently compelling so as to hypothesize a moderating effect ahead of time. The researchers therefore likely did not power their study for testing moderation, did not intentionally recruit people who varied on the moderator, and did not intend to analyze their data with interaction terms. That in the end they find moderation does not somehow justify their design decisions (the study may still have been inappropriately designed and powered for moderation), nor does it honestly represent the state of the science at the time the study was designed. What would have been considered a weak hypothesis prior to knowing the outcome of a study is seen as much more probable or likely when the outcome of a study is known (something called the hindsight bias; Hawkins & Hastie, 1990; Slovic & Fischhoff, 1977). And because there may be a nearly limitless number of mediating and moderating variables that are covaried with X and Y, presenting any one of these as having been foreseen is especially problematic (Agler & De Boeck, 2017).

In summary, HARKing involves presenting known but unforeseen findings from a study as the *raison d'etre* for conducting the study—as the central parameters of interest that were under investigation. Consequently, the reader of a HARKed study assumes that all study design and analytic decisions were guided by this central question or hypothesis. It is one of a long set of decisions researchers make about how to conduct psychological science, all of which can influence the outcome of a study (Wicherts et al., 2016).

Consequences of HARKing

Like most behavior, HARKing creates a range of consequences to the individual, scientific community, and society at large. Like many questionable research practices, I consider HARKing to create a social dilemma. That is: what is good for the individual researcher in the short term creates long-standing problems for the collective (Table 8.1). Failing to acknowledge the benefits of HARKing means we will be ineffective at creating solutions.

HARKing occurs because it benefits the individual scientist who does this (or the research team—I do not mean to imply that HARKing is only occurring at the individual level). Whether intentional or not (and much of HARKing may be unintentional; Kerr, 1998), the scholars who do this benefit in the short term with clear tangibles: more publications, which in turn lead to higher success in their careers,

Table 8.1 Consequences of HARKing for individuals and the collective

	Individual	Collective
Short-term consequence	Higher rate of publication Promotion/advancement Scientific prestige (e.g., higher *h*-index) Media attention Increased income	Fast proliferation of scientific papers Difficulties keeping up with the rapid branching of scientific findings or facts Overconfidence in what is "established" scientific fact Demoralization for young scholars who may attribute their lack of significance in their studies a sign of personal shortcomings
Long-term consequence	Possibly tarnished reputation (if studies fail to replicate) Modeling of poor scientific practices for the next generation of scholars Existential crisis	Waste of limited resources (e.g., federal grant funding, time, talent) on potentially spurious findings Population mistrust of science and scientists Inflated effect sizes Inaccurate "facts"- a weak and poorly established knowledge basis in an area Less effective treatments delivered

job promotions, higher visibility as a scholar (e.g., higher *h*-index, a metric of a scholar's impact on the field), media attention for "splashy" findings, and so forth. Journal editors and reviewers may contribute to this inadvertently too—by pointing to alternative theories, suggesting alternative strategies for analyzing data, and even directly suggesting authors revise their introduction and hypotheses, they may be helping mold a paper that they anticipate will be highly cited, thereby raising the impact factor of that publication outlet. These are indeed powerful incentives that may drive scientists to HARK.

However, all other quadrants of Table 8.1 show the problems with HARKing. In the short term, the scientific community experiences a bit of a "mixed bag" of consequences. On one hand, scientific findings are proliferating at a rapid rate (if every study can be HARKed so that something interesting comes of it that is considered "publishable," then all studies will result in deliverables and these must be placed somewhere... which leads to a tangential but related problem with predatory journals whose purpose is to give outlet to oftentimes weak scientific papers. But this is a topic for another time.). However, the rapid proliferation of publications means it is difficult for someone to keep track of what all is known about a topic. And since HARKing has a high risk of capitalizing on spurious statistical findings (that is, findings that occur by chance alone and not because of some underlying, consistent, stable or causal link between the variables), we may be overconfident in what is "established" scientific fact in our field. Because journals historically have been reluctant to publish findings that are replications, novel discoveries quickly become canonized with limited opportunities for building consensus in published work. (Just recently, I received feedback on a manuscript I submitted. One reviewer noted "Although the article is well written and the literature is relevant, the study does not contribute anything new or unique to the literature. The authors are replicating a study already conducted." The manuscript was rejected.) If someone's experiment turns out opposite of an "established" (but HARKed) finding, that person may be reluctant to even submit the paper for publication—*if my study didn't replicate this well-established association between X and Y, clearly there is something wrong with my study.* That every study is flawed in some way means scholars may first make attributions that their own studies were erroneous, given the prior published studies showing a clear link between the variables. And so again, opportunities for science to either build scientific consensus with multiple observations of a phenomenon or self-correct with contradicting observations of a phenomenon are thwarted.

An insidious consequence of HARKing that I and others involved in education and training have observed is that young scholars develop an unrealistic expectation of what it means to be a good scientist (Kerr, 1998). When their own theses and dissertations (more often than not) fail to provide support for their hypotheses, they feel science is hard (it is!) and they aren't good at it (they are!). Even worse—for those who quickly understand that HARKing can lead to publication success, being trained to adopt questionable research practices such as HARKing means that the problem is prolonged for another generation.

In the long term, the consequences of HARKing tend to look dim for both the individual researcher and the collective. As illustrated by a recent set of studies attempting to replicate previously established psychological findings (Open Science Collaboration, 2015), scholars who HARK risk having their reputations tarnished.[2] Perhaps the most famous case of this is Dr. Daryl Bem, who claimed to have discovered evidence for extrasensory perception or, as he called it, precognition. Across nine studies that seemed methodologically rigorous, Bem (2011) appeared to show evidence that people could intuit the future. And yet, upon closer examination, there were numerous ways in which the decisions Dr. Bem had made about what studies to continue and which ones to stop prematurely, how many possible comparison conditions to include (and which ones to analyze, and which ones to report) were post hoc despite having been presented in his seminal paper as a priori or confirmatory (Engber, 2017). The outlandish findings from this set of nine studies are considered to be so fundamentally outside of the rigorous and committed scholar Dr. Bem was known to be that some have even suggested his famous publication was an attempt to illustrate precisely the problems with the field that we grapple with now, and that are the topic of this entire book (O'Donohue et al., n.d.). That empirically rigorous studies could provide evidence for extrasensory perception was so outlandish on the face of it that it threw psychology into an existential crisis.

Less dramatically, but no less important, HARKing's consequences include wasting valuable and limited resources such as federal grant dollars, time, and talent on the pursuit of what appear to be promising avenues of discovery but in fact may have been simply statistical artifacts. The resource issue is further amplified by a focus on innovation that is often part of grant evaluations.

A bias toward publishing significant effects (Lishner, 2021) means HARKing can result in an inflated estimate of effect sizes since contrary findings in the null direction are less likely to be disseminated. With inflated effect sizes, our confidence or certainty about what is "known" in our field or well-established is skewed. In clinical psychology, medicine, and other health disciplines, this can be especially problematic if it means that the efficacy of therapies or other treatments are inaccurately presented—perhaps a therapeutic approach with evidence largely driven by proponents who engage in HARKing may appear more effective and therefore deployed more than a therapy with less impressive effect sizes but whose effect sizes are honest, rather than inflated in magnitude. Well-intentioned clinicians and policy makers may be choosing a less effective treatment because they need to discern which treatment's evidence is most compelling without the benefit of full, accurate knowledge to assess bias in research findings. Finally, and not insignificantly, HARKing and other questionable research practices foment the public's mistrust of science (Kerr, 1998).

[2]There can *clearly* be cases of excellent, confirmatory research findings failing to replicate over time because circumstances change—early studies of sexual assault potential, for instance, often asked people if they would rape someone if they could be assured they would not get caught. While such a question might have resulted in variable responses in the past and were useful for predicting future behavior, for many years now this item no longer is useful and no longer can predict future behavior. Greater awareness of sexual assault and changing norms about sexually imposing behavior have shifted what was a real and established finding in the past.

Remedies

As a behaviorist, I think it is fundamental to shift the incentives for HARKing. While these incentives typically occur for the individual and in the short term (Table 8.1), they are supported by organizational and institutional practices and so remedies must be targeted at multiple levels. In particular, strategies can be implemented in the following domains: education and training, study design and preparation, scholarly products, publication outlets and grant funding agencies, and in high-stakes decisions such as promotion and tenure. Each is described in turn.

In Education and Training Although knowledge alone is insufficient to motivate change (Rothman, 2006), it is a critical first step in remedying questionable research practices like HARKing. Direct discussion of HARKing is helpful, of course, but it may be especially helpful in the context of our typical instruction in how to disseminate research findings, which is often when HARKing rears its head. The practice of good story-telling has been actively encouraged in books on academic writing (Olson, 2016) and many offer the advice to write articles starting with Method and Results so that you know your story and can clearly set the stage for the hypotheses. Indeed, some books suggest working backward from the results in order to deduce your research question and hypotheses (Bem, 2004). While this advice is well-intentioned and can help scholars avoid lengthy discussion of tangential topics in their literature review, it can also appear to be an implicit endorsement of HARKing by the scientific community. As such, discussions of good story telling and a proper narrative in scholarly work *must* include discussions of questionable research practices, including HARKing, to draw distinctions between style and substance. Narrative structure is style, but pretending that you "knew it all along" is a substantive matter (i.e., dishonest presentation of content matter).

Education in questionable research practices need not be the sole responsibility of lab principal investigators, professors, or advisors. In fact, it is possible these experts themselves may not be well educated on the topic (I know I was not until years after I became a professor). *Everyone,* students especially, is encouraged to access the numerous resources available in print, in video, and online. Self-directed learning has never been easier than at this point in our history. Reading books such as this one, conceptual and empirical articles tackling these issues, listening to podcasts and Ted Talks addressing questionable research practices, and attending conferences such as the annual meeting of the Society for the Improvement of Psychological Science are all good sources of education.

In addition to educating oneself about questionable research practices, better education about philosophy of science, logic, and strategies for comprehensive literature reviews would be beneficial (O'Donohue, 2013; Vancouver, 2018). The hypothetico-deductive approach to science relies on logic: using theory and/or a rendering of prior empirical findings and observations in order to deduce a precise, falsifiable prediction about the outcome of a study. However, graduate programs in science may not explicitly teach rules of logic or have courses dedicated to the philosophy of science. Direct instruction in scholarly tasks such as how to evaluate the soundness of a theory, how to render a literature in a manner that is fair rather than selectively citing studies in support of one's assumptions, how to use the rules of logic to derive predictions, and the importance of honesty for furthering science can help emerging researchers to recognize the diverse forms of HARKing and other questionable research practices and return to the core values of science (Kerr, 1998). (This seems particularly important if we assume that questionable practices are dynamic and new forms that we have not yet articulated are likely to emerge over time.)

In Study Design and Preparation Almost by definition, the sine qua non of remedying HARKing is pre-registration (Bergkvist, 2020; Wagenmakers et al., 2012). Because of issues with hindsight bias and inaccuracies in recall, the ability to reconstruct with accuracy prior hypotheses or beliefs about one's data should not be the foundation for a study. By committing ahead of time to the study purpose, hypotheses, measures, methods, and analyses, there is a clear record of the investigators' understanding of the theory, the hypotheses that were deemed most probable/likely from the theory and literature at the time the study was designed, how the variables were best operationalized, and how the data would best be analyzed to evaluate the hypotheses. There is no revision of history. Furthermore, pre-registration on web platforms such as Open Science Framework (https://osf.io; for a description, see Foster & Deardorff, 2017) allow that record to be tamper-proof and, should the investigators want it, available to others.

In order to combat one of the temptations of HARKing (my results were null; null findings have a lower chance of being published; I need to find something interesting in these data and restructure my manuscript to focus on that one interesting finding), the Center for Open Science (https://cos.io) has worked with journals to publish registered reports. Registered reports are peer-reviewed manuscripts where the review happens in two stages. At stage one, the investigators submit to the journal their study idea, including the introduction, method, hypotheses, proposed analyses, and any pilot data they have. This is functionally the equivalent of a master's thesis or doctoral dissertation proposal. The editor (and reviewers, if it is not desk rejected) provides feedback to the investigators, including possibly suggesting revisions. The process, if successful, results in an *in-principle acceptance* by the journal. After the investigators conduct the study, they may submit again to the journal for a second phase of peer review that includes the full manuscript (introduction, method, results, discussion). Importantly, the in-principle acceptance status of the manuscript means that the research will not be rejected at this second phase of review on the basis of the study outcomes, only on the basis of failing quality checks, failing to follow the original procedures, and so forth. The journal publishes the article with a "registered report" badge, which is designed to convey clearly to the reader that this study was confirmatory in nature and therefore was not HARKed. At the time of this chapter's writing, over 300 journals were allowing authors to submit registered reports.

Outside of pre-registration, HARKing may be combatted by having collaborative research teams comprised of people with diverse (and perhaps competing or contrary) perspectives. These adversarial research collaborations can mitigate HARKing because members of the team are hypothesizing different, even opposite, results when designing an experiment (Cowan et al., 2020). They can help reduce biases in all aspects of scholarship, from the framing of the study question to issues of design, measurement, analysis, and dissemination.

In Scholarly Products At times HARKing may occur because, as scholars, we are curious to understand and explore our data. Our motivation to do so likely increases in cases where the outcome of our study was not as predicted. If, in our exploratory journey, we find something interesting in our data that may help us make sense of our findings, of course we will want to share that with others.[3] Unfortunately, if we feel constrained by our circumstances (i.e., we must present these new findings as having been part of the plan all along; we must preview them in our introduction), then we will find ourselves HARKing.

In order to decrease the temptation to HARK, it would be wonderful to normalize having a subsection with the Results section of a manuscript that is labeled "Post-Hoc Analyses" or "Exploratory Analyses" (some have gone so far as to suggest that such a section be required; Erren, 2007). The ability to describe within scholarly products a clear demarcation of confirmatory versus exploratory findings would help reduce some of the temptations to HARK and it would help disseminate interesting scientific findings that can be subject to new, confirmatory tests. Without the ability to invite and encourage the labeling of exploratory hypotheses as a posteriori or post hoc, we risk a continued culture of HARKing. As Erren (2007) notes:

> While I would not think that anyone will need convincing that a posteriori hypotheses exist and that they may be even common…an examination of the numerous abstracts available in MEDLINE suggests that a posteriori hypotheses either do not exist or are disguised…[They] have such a negative connotation that scientists do not try their luck and appropriately document such modified thinking in a manuscript submission. Thus, there may be no a posteriori hypotheses published as such but I think that we must take seriously the possibility that there are quite a few concealed or obscured via 'politically correct' wording. (p. 450).

[3] I was attending a training recently and the lead of the training was describing a great study they and their team had conducted on substance use in youth. They explained how they had analyzed their data and found none of their predictions had panned out. When discussing this with their lab mates, the lab mates asked whether youth who were incarcerated (and therefore, presumably would not have had access to substances during the study time frame) had been filtered out of the data set. When they re-ran their data excluding these youth, their predictions were supported. How should the team have handled this? If they reported in the manuscript that they excluded these youth, would they be HARKing? Should they report that they first analyzed the data with all youth, and only afterward did they exclude some participants? I am not certain what they did, but it is easy to see how one might slap one's head, say "of course! That makes sense. I should have excluded them all along" and then proceed to write the paper as though one had always planned on such exclusions.

If he is correct and there is some sort of stigma attached to exploratory analyses that motivate their being presented as confirmatory, then we can bring to bear the science of stigma reduction to solving this problem. One of the most effective methods for reducing stigma is to have contact with people who do the stigmatizing thing *and* who are held in high regard or are of equal status (Corrigan & Penn, 1999). Therefore, having journal editors, senior authors, and thought leaders in the field model this behavior in their own manuscripts and research talks would go a long way toward creating conditions that promote greater honesty in research reporting.

Because detection of HARKing can be challenging, researchers should also consider: (a) making their data available (e.g., through data repositories); (b) including correlation matrices in their manuscripts or in online supplemental materials; (c) noting how many models were run on the data before the final model(s) that were presented in the manuscript; and (d) conducting sensitivity analyses (or testing the robustness of an effect). Data repositories and correlation tables allow other researchers to verify the results of a study. Robustness or sensitivity analyses allow readers to see whether the findings the authors report hold up across various constraints or conditions, such as when all or only a subset of people are included in the analyses; with and without controlling for sociodemographic or other covariates; when using one versus another form of model estimation, and so forth. One can be more confident of an effect if it is consistently present rather than an effect that seems only to arise under very specific and narrow circumstances (I am not suggesting this is not a "real" effect, only that it is not robust and likely would not be easy to confirm in future studies or to have been predicted in the current study).

On the issue of consistency, replication is also important to reduce the influence of HARKed findings on the body of scientific knowledge in a particular content domain. While individual researchers can do this within a single study (e.g., conducting exploratory analyses on some of the data and confirmatory analyses on a hold-out sample), replication is also important to do with newly collected samples/data and by independent research teams (Shrout & Rodgers, 2018).

In Publication Outlets and Grant Funding Agencies Journals and grant funding agencies have the ability to change incentives to HARK and thereby change individuals' behavior. Above I described the use of registered reports as a way for journals to clearly denote which published studies were fully confirmatory and underwent a pre-study peer review process. Other things journals can do include devoting more space explicitly for replication and null findings (e.g., earmarking 33% of each new issue to replication and null studies or including a special section in each journal for these kinds of studies—much like journals currently include special sections on book reviews, commentary, brief reports, and so forth). If scholars had more confidence that the likelihood of a null finding from a study had a good chance of being published, the desire to HARK in order to gain a publication is lowered. Journals may also want to leverage machine learning capabilities, which are being used to develop algorithms that create "credibility scores" for scientific studies (see https://www.darpa.mil/program/systematizing-confidence-in-open-research-and-evidence). It is too early to determine whether these efforts will be successful, but the promise to use computational power to help detect HARKing is exciting.

Grant reviews often assess the quality of an investigator when determining whether to award funds for a proposal. Quality metrics tend to be heavily influenced by rate of publication (and, of course, publication rate may be in part a function of how success a researcher has been at HARKing). Others have discussed the flaws with researcher quality metrics like the *h*-index and even proposed alternatives (Schimmack, 2020). One possibility is to divide the grant review process into two phases: in phase 1, the quality of the research design/proposal is assessed without knowledge of the investigators. If a proposal is deemed of high quality after this first round of review, a second round can then consider whether the investigators and the environment are adequate to supporting the project.[4]

In Promotion and Tenure A "publish or perish" environment has certainly fueled the use of questionable research practices, even from very well-intentioned and respected scholars (Kiai, 2019). Counting scholarly products like publications is easy and therefore many committees making important evaluative decisions (for awarding admissions, scholarships, internships, jobs, promotions, and so forth) will use quantity metrics to assist in their determinations. Considerable efforts should be devoted to either replacing or supplementing such easy-to-use metrics in high-stakes evaluative situations. Until such time, the demand for publishing felt by young scholars will be met with an increase in predatory journals that can provide an outlet for all kinds of scientific products of questionable quality. Perhaps the only good news about this vicious cycle is that it will likely accelerated the timeline for when committees and organizations decide they *must* consider more carefully the quality of research being conducted.

Conclusion

The scientific process rests on a set of core values, including objectivity, honesty, openness, accountability, fairness, and stewardship (National Academies of Sciences, Engineering, and Medicine, 2017). HARKing threatens all of these values, and therefore is a threat to science itself. A commitment to reducing the prevalence of HARKing does not mean a stifling of exploration or creativity; indeed, these are at the core of discovery. Instead, that commitment is about presenting discovery and exploration honestly and not disguising it as confirmation. A final note on this topic: unpredicted findings can be powerful scientific narratives that really engage audiences (Olson, 2016). Rather than seeing post-hoc findings as a pox on the study or investigator, as scholars and humans we should embrace the twists and turns our scholarship takes us, realizing that doing so is both more honest and more interesting.

[4] This may also have the benefit of reducing gender and racial biases in grant reviews.

References

Agler, R., & De Boeck, P. (2017). On the interpretation and use of mediation: Multiple perspectives on mediation analysis. *Frontiers in Psychology, 8*, 1984. https://doi.org/10.3389/fpsyg.2017.01984

Bem, D. J. (2004). Writing the empirical journal article. In J. M. Darley, M. P. Zanna, & H. L. Roediger (Eds.), *The compleat academic: A career guide* (pp. 185–219). American Psychological Association.

Bem, D. J. (2011). Feeling the future: Experimental evidence for anomalous retroactive influences on cognition and affect. *Journal of Personality and Social Psychology, 100*, 407–425. https://doi.org/10.1037/a0021524

Bergkvist, L. (2020). Preregistration as a way to limit questionable research practice in advertising research. *International Journal of Advertising, 39*, 1172–1180. https://doi.org/10.1080/02650487.2020.1753441

Bryan, C. J., Yeager, D. S., & O'Brien, J. M. (2019). Replicator degrees of freedom allow publication of misleading failures to replicate. *Proceedings of the National Academy of Sciences of the United States of America (PNAS), 116*, 25535–25545. https://doi.org/10.1073/pnas.1910951116

Caruso, E. M., Shapira, O., & Landy, J. F. (2017). Show me the money: A systematic exploration of manipulations, moderators, and mechanisms of priming effects. *Psychological Science, 8*, 1148–1159. https://doi.org/10.1177/0956797617706161

Centers for Disease Control and Prevention [CDC]. (2019). *Enhanced meningococcal disease surveillance report, 2019.* Retrieved from https://www.cdc.gov/meningococcal/downloads/NCIRD-EMS-Report-2019.pdf

Corrigan, P. W., & Penn, D. L. (1999). Lessons from social psychology on discreditingpsychiatric stigma. *American Psychologist, 54*(9), 765–776. https://doi.org/10.1037/0003-066X.54.9.765

Cowan, N., Belletier, C., Doherty, J. M., Jaroslawska, A. J., Rhodes, S., Forsberg, A., Naveh-Benjamin, M., Barrouillet, P., Camos, V., & Logie, R. H. (2020). How do scientific views change? Notes from an extended adversarial collaboration. *Perspectives on Psychological Science, 15*, 1011–1025. https://doi.org/10.1177/1745691620906415

de Groot, A. D. (2014). The meaning of "significance" for different types of research [translated and annotated by E. Wagenmakers, D. Borsboom, J. Verhagen, R. Kievit, M. Bakker, A. Cramer, D. Matzke, D. Mellenbergh, & H.L.J. van der Maas]. 1969. *Acta Psychologica, 148*, 188–194. https://doi.org/10.1016/j.actpsy.2014.02.001

Engber, D. (2017, June). Daryl Bem proved ESP is real: Which means science is broken. *Slate.* Retrieved from www.slate.com/health-and-science/2017/06/daryl-bem-proved-esp-is-real-showed-science-is-broken.html

Eronen, M. I., & Bringmann, L. F. (2021). The theory crisis in psychology: How to move forward. *Perspectives on Psychological Science, 16*, 779–788. https://doi.org/10.1177/1745691620970586

Erren, T. C. (2007). The case for a posteriori hypotheses to fuel scientific progress. *Medical Hypotheses, 69*, 448–453. https://doi.org/10.1016/j.mehy.2006.12.026

Evett, I. W. (1995). Avoiding the transposed conditional. *Science & Justice, 35*, 127–131. https://doi.org/10.1016/S1355-0306(95)72645-4

Ferguson, M. J., Carter, T. J., & Hassin, R. R. (2014). Commentary on the attempt to replicate the effect of the American flag on increased Republican attitudes. *Social Psychology, 45*, 299–311.

Foster, E. D., & Deardorff, A. (2017). Open Science Framework (OSF). *Journal of the Medical Library Association, 105*, 203–206. https://doi.org/10.5195/jmla.2017.88

Hawkins, S. A., & Hastie, R. (1990). Hindsight: Biased judgments of past events after the outcomes are known. *Psychological Bulletin, 107*, 311–327. https://doi.org/10.1037/0033-2909.107.3.311

John, L. K., Loewenstein, G., & Prelec, D. (2012). Measuring the prevalence of questionable research practices with incentives for truth telling. *Psychological Science, 23*, 524–532. https://doi.org/10.1177/0956797611430953

Kerr, N. L. (1998). HARKing: Hypothesizing after the results are known. *Personality and Social Psychology Review, 2*, 196–217. https://doi.org/10.1207/s15327957pspr0203_4

Kiai, A. (2019). To protect credibility in science, banish "publish or perish". *Nature Human Behaviour, 3*, 1017–1018. https://doi.org/10.1038/s41562-019-0741-0

Klein, R. A., Ratliff, K. A., Vianello, M., Adams, R. B., Jr., Bahnik, S., Bernstein, M. J., Bocian, K., Brandt, M. J., Brooks, B., Brumbaugh, C. C., Cemalcilar, Z., Chandler, J., Cheong, W., Davis, W. E., Devos, T., Eisner, M., Frankowska, N., Furrow, D., Galliani, E. M., … Nosek, B. A. (2014). Investigating variation in replicability: A "many labs" replication project. *Social Psychology, 45*, 142–152. https://doi.org/10.1027/1865-9335/a000178

Lishner, D. A. (2021). HARKing: Conceptualizations, harms, and two fundamental remedies. *Journal of Theoretical and Philosophical Psychology*. https://doi.org/10.1037/teo0000182

Masicampo, E. J., & Lalande, D. R. (2012). A peculiar prevalence of *p* values just below .05. *Quarterly Journal of Experimental Psychology, 65*, 2271–2279. https://doi.org/10.1080/17470218.2012.711335

National Academies of Sciences, Engineering, and Medicine. (2017). *Fostering integrity in research*. The National Academies Press. https://doi.org/10.17226/21896

O'Donohue, W. (2013). *Clinical psychology and the philosophy of science*. Springer International Publishing.

O'Donohue, W., Masuda, A., & Lilienfeld, S. O. (Eds.). (n.d.). *Questionable research practices: Designing, conducting, and reporting sound research in clinical psychology*. Springer Publication.

O'Donohue, W., Snipes, C., & Soto, C. (2016). A case study of overselling psychotherapy: An ACT intervention for diabetes management. *Journal of Contemporary Psychotherapy, 46*, 15–25. https://doi.org/10.1007/s10879-015-9308-1

Olson, R. (2016). *Houston, we have a narrative. While science needs story*. University of Chicago Press. https://doi.org/10.7208/chicago/9780226270982.001.0001

Open Science Collaboration. (2015). Estimating the reproducibility of psychological science. *Science, 349*, aac4716. https://doi.org/10.1126/science.aaac4716

Pashler, H., & Wagenmakers, E. (2012). Editors' introduction to the special section on replicability in psychological science: A crisis of confidence? *Perspectives on Psychological Science, 7*, 528–530. https://doi.org/10.1177/1745691612465253

Popper, K. (1959). *The logic of scientific discovery*. Routledge.

Rothman, A. J. (2006). Initiatives to motivate change: A review of theory and practice and their implications for older adults. In *In the National Research Council's When I'm 64*. National Academies Press.

Rubin, M. (2017). When does HARKing hurt? Identifying when different types of undisclosed post hoc hypothesizing harm scientific progress. *Review of General Psychology, 21*, 308–320. https://doi.org/10.1037/gpr0000128

Schimmack, U. (2020). A meta-psychological perspective on the decade of replication failures in social psychology. *Canadian Psychology, 61*, 364–376. https://doi.org/10.1037/cap0000246

Sergent, K., & Stajkovic, A. D. (2020). Women's leadership is associated with fewer deaths during the COVID-19 crisis: Quantitative and qualitative analyses of United States governors. *Journal of Applied Psychology*. https://doi.org/10.1037/apl0000577

Shrout, P. E., & Rodgers, J. L. (2018). Psychology, science, and knowledge construction: Broadening perspectives from the replication crisis. *Annual Review of Psychology, 69*, 487–510. https://doi.org/10.1146/annurev-psych-122216-011845

Slovic, P., & Fischhoff, B. (1977). On the psychology of experimental surprises. *Journal of Experimental Psychology, 3*, 544–551. https://doi.org/10.1037/0096-1523.3.4.544

Toolis, E. E., & Hammack, P. L. (2015). The lived experience of homeless youth: A narrative approach. *Qualitative Psychology, 2*, 50–68. https://doi.org/10.1037/qup0000019

van de Beek, D., de Gans, J., Spanjaard, L., Weisfelt, M., Reitsma, J. B., & Vermeulen, M. (2004). Clinical features and prognostic factors in adults with bacterial meningitis. *New England Journal of Medicine, 351*, 1849–1859. https://doi.org/10.1056/NEJMoa040845

Vancouver, J. B. (2018). In defense of HARKing. *Industrial and Organizational Psychology, 11*, 73–80. https://doi.org/10.1017/iop.2017.89

Wagenmakers, E. J., Wetzels, R., Borsboom, D., van der Maas, H. L. J., & Kievit, R. A. (2012). An agenda for purely confirmatory research. *Perspectives on Psychological Science, 7*, 632–638. https://doi.org/10.1177/1745691612463078

Wicherts, J. M., Veldkamp, C. L. S., Augusteijn, H. E. M., Bakker, M., van Aert, R. C. M., & van Assen, M. A. L. M. (2016). Degrees of freedom in planning, running, analyzing, and reporting psychological studies: A checklist to avoid *p*-hacking. *Frontiers in Psychology, 7*, 1832. https://doi.org/10.3389/psyg.2016.01832

Chapter 9
Statistical Controversies in Psychological Science

Andrew H. Hales and Natasha R. Wood

Abstract In this chapter, we provide an overview of some of the major historic and contemporary statistical controversies, including the use of qualitative versus quantitative methods, the role of description/exploration in research, and the nature of hypothesis testing. We also consider a number of statistical non-controversies that we believe are generally agreed upon, yet still worthy of consideration in the current overview, including the condemnation of fraud, the value of sharing data, and the use of broader/more diverse samples. Finally, we consider reasons why statistical debates can be surprisingly heated and conclude that—regardless of the reasons for controversy, or the tone of these debates—impressive progress has been made in the last decade. Given the tools that researchers now have to avoid the mistakes that led to the replication crisis, we expect the quality of research to improve. There will undoubtedly continue to be statistical controversy, but as new practices take hold, we may see a shift in the tone of these debates to being more civil.

Keywords Statistics · Quantitative · Qualitative · Hypothesis testing · Bayesian statistics · WEIRD samples

Background

Are humans blank slates, or do we have an essential nature? If humans have a nature, what characterizes that nature? Are people generally good and trustworthy? Can they change and improve? Do feelings, choices, and behaviors originate within a person, or do we mechanistically respond to our environment? These questions are at the very heart of ideological divides—both contemporary and classic. These questions are also at the very heart of psychological science. With such a polarizing and complicated subject matter, it is not surprising that the field often encounters

A. H. Hales (✉) · N. R. Wood
University of Mississippi, University, MS, USA
e-mail: ahales@olemiss.edu

W. O'Donohue et al. (eds.), *Avoiding Questionable Research Practices in Applied Psychology*, https://doi.org/10.1007/978-3-031-04968-2_9

controversies both in its approaches to questions about people and in the methods it uses to answer those questions. Given psychology's strong quantitative orientation, these are very often *statistical controversies* pertaining to the ways in which conclusions should be drawn from data. In recent years, the amount of attention given to statistical best practices has ballooned in response to the replication crisis—itself a massive controversy composited of many specific statistical realizations, developments, and, of course, disagreements.

In this chapter, we provide an overview of some of the major statistical controversies, with an emphasis on best research practices. If controversy is defined, simply, as a matter on which people strongly disagree, then the universe of statistical controversies ranges from the very broad (e.g., *is it possible to better understand human nature through quantitative analysis?*) to the very narrow (e.g., *which post-hoc correction is most appropriate when group variances are unequal and sample sizes are balanced?*). We will bounce around this broad-narrow continuum, but focus on the controversies that are most fundamental to the decisions that researchers make when planning and conducting their analyses, and the conclusions consumers should draw when reading reports of others' research.

We will also consider a number of statistical non-controversies. These are areas that we really do believe are uncontroversial, yet still worthy of consideration in the current overview—either because people may incorrectly assume there is controversy where none exists or simply to celebrate that progress is being made in areas with surprisingly little institutional/systemic resistance.

This chapter does not directly address statistical controversies surrounding particular findings of substance. These controversies certainly abound: Does exerting effort on one activity deplete one's ability to perform another (Carter et al., 2015; Hagger et al., 2016; Vohs et al., 2020)? Does contemplating one's own death alter their worldview (Greenberg et al., 1994; Klein et al., 2019)? Does standing in an expansive super-heroesque pose change one's physiology and performance (Credé & Phillips, 2017; Cuddy et al., 2018; Simmons & Simonsohn, 2017; Ranehill et al., 2015)? These questions—and others—have stirred up their fair share of controversy. It appears that in many cases these sorts of questions generated controversy not because of the actual empirical claims being promoted or rebutted (though this certainly happens; e.g., Bauer, 2020; AlShebli et al., 2020). Rather, these controversies appeared to be expressions of deep underlying statistical disagreements. Exactly how strong does evidence need to be in order to endorse a research claim? Why does it seem that stronger evidence is required to refute an already-published research claim than to establish that claim (Ferguson & Heene, 2012; Gelman, 2016)? What should become of a research claim that was introduced in a zeitgeist of looser statistical standards than what the field currently observes? If nothing else, controversies of substantive research claims remind us that there are stakes to the controversies of statistical practices that are the focus of this chapter, sometimes with serious policy implications (e.g., IJzerman et al., 2020; Van Bavel et al., 2020).

Controversies

Quantitative Versus Qualitative Methods

Before considering controversies of how statistics should be applied, it is necessary to consider the most fundamental statistical controversy of all. Namely, whether statistics should be used in the first place. There has long been tension between quantitative research methods—those focusing on numerical summaries of observations—and qualitative research methods—those focusing on narrative and linguistic accounts of observations—a disagreement dubbed "the paradigm wars" (Gage, 1989).

Despite the sharp contrast and apparent incompatibility between qualitative and quantitative methods (Jackson, 2015), it is easy to see how the two approaches are not only compatible, but symbiotic within a program of research (Landrum & Garza, 2015; Willig, 2019). Quantitative methods can provide somewhat objective and precise answers to specific questions. But insights from quantitative research are only as good as the questions being asked. Qualitative methods can provide rich insight and thick description (Ponterotto, 2006), while deftly capitalizing on unexpected insights as they arise in an investigation and through interaction with participants. With this approach, the answer to a research question is not necessarily bound by choices in experimental design or in survey content. This makes rigorous qualitative studies well-suited for the generation of meaningful hypotheses, which can subsequently be confirmed or refuted through rigorous quantitative studies. Such triangulation of methods, especially with qualitative research preceding quantitative research, is a powerful recipe for building strong theories (and is analogous to the prescription in quantitative research to "explore small, confirm big"; Sakaluk, 2016).

To illustrate, this pattern appears to have played out in the scientific investigation of ostracism. Early inquiries embraced qualitative approaches in seeking to understand the phenomenological experience of ostracism (Williams et al., 2000; Zadro, 2004), and valued open-ended reports of when, why, and how people use and receive ostracism (Sommer et al., 2001; Williams et al., 1998). These investigations helped shape modern ostracism theory (Williams, 2009), and also informed the development of quantitative experiments (e.g., Goodwin et al., 2010), and eventually meta-analysis (Hartgerink et al., 2015). An analogous trajectory characterizes the theory of cognitive dissonance, which began with the iconic qualitative case-study of the doomsday cult, the Seekers (Festinger et al., 1957; incidentally, this report was the first known use of the term "qualitative"; Jackson, 2015). This vivid, memorable, and richly described qualitative study initiated decades of quantitative experimental research on cognitive dissonance and related consistency theories. In short, qualitative and quantitative methods can not only be compatible, but actually quite complementary, resulting ultimately in a stronger scientific understanding than either approach would allow on its own.

Descriptive Analysis Versus Hypothesis Testing

> Both exploratory and confirmatory data analysis deserve our attention. Both detection and adjudication play crucial roles - in the progress of science as in the control of crime. To concentrate on confirmation, to the exclusion of exploration, is an obvious mistake. Where does new knowledge come from? How can an undetected criminal be put on trial? … There really seems to be no substitute for "looking at the data." (Tukey, 1969, p. 83)

A commitment to quantitative methods is not necessarily a commitment to conducting hypothesis tests with *p*-values and the other machinery that we often automatically associate with statistics. Conceptually prior to this formal testing stage is an entire world of description, exploration, visualization, and understanding that many have long plead for researchers to take more seriously (Meehl, 1978; Rozin, 2001; Scheel et al., 2020; Tukey, 1969, 1977).

Kerr (1998, p. 201), for example, observed how it is common for mentors to ask a student, *what are your hypotheses?*, but rare for them to ask *do you have any hypotheses?* Psychologists are deeply—and often implicitly—entrenched in the hypothetico-deductive tradition of first positing a hypothesis and subsequently subjecting it to confirmatory test. As a result, researchers reflexively employ hypothesis tests (usually null hypothesis tests—discussed below), even in situations where it is unnecessarily, or even silly to do so. This might happen, for example, when one conducts a t-test to show that two groups that differ by several standard deviations on a manipulation check are statistically significantly different, or when two groups created through median-split on a continuous variable are significantly different (Abelson, 1995, p. 76). If inferential statistics are a tool to help advance argument (Abelson, 1995), then invoking them in situations when no reasonable person would disagree dilutes their meaning, and has the potential to create an artificial precision to a claim (Gigerenzer, 2018).

Descriptive statistical techniques are routinely taught in undergraduate and graduate statistics courses, but very often as a steppingstone on the way to the (presumed) more relevant and useful inferential statistics employed to test hypotheses. Despite the ubiquity and knee-jerk use of hypothesis tests (Gigerenzer, 2004), there have been vocal and persuasive calls for a less rigid approach that is more exploratory and descriptive. For example, Rozin (2001) argued that psychology (particularly social psychology) is a relatively young science, and that it is prematurely conducting experiments and hypothesis tests. Instead, psychology should follow the trajectory of other more mature sciences and first spend time fully describing phenomena under investigation. Psychologists are often so interested in detecting significant differences between groups on various dimensions, that they forget to identify the *absolute* values that typically characterize the groups being studied (i.e., important *invariances*—something hypothesis testing is not well-suited to do).

One area of psychology in particular, behavior analysis, has embraced descriptive and graphical analyses and has largely eschewed hypothesis testing altogether (though see Fox, 2018 for a potential shift in this trend). The historical suspicion toward hypothesis testing dates back to B. F. Skinner who regarded large-group analysis, and statistics more generally, with some apparent strong dislike (Skinner, 1963, pp. 507–508). Modern behavior analysts favor descriptive and graphical analysis of a few individuals—ideally replicated across organisms—and reject statistical testing (e.g., Branch, 2014; Perone, 1999). Under this approach, the idea is to conduct an experiment involving only 3 subjects—where the 2 and 3rd are essentially replications of the single subject experimental design and observe a graph of the results (perhaps on an ongoing basis) and interpret it intelligently. "What is preferred [to numerical statistical analysis] is an experimental analysis so thorough, so powerful in its control over the subject matter of interested, that cause-effect relations are plain to see" (Perone, 1999, p. 114; also called the "inter-ocular traumatic test" because the result "hits you between the eyes"; Edwards et al., 1963).

This strategy is great when it works (i.e., when the experimental result is glaring), and indeed, other areas of psychology could improve their visualization practices. Ideally this would involve greater emphasis on showing raw data points rather than bar graph summaries of results (e.g., McCabe et al., 2018). This would serve the dual benefit of (1) maintaining emphasis on the *individuals* rather than groups as the unit that psychologists typically care about (Branch, 2014), and (2) avoiding obscuring important trends and irregularities that may be present in the data (e.g., nonlinear patterns or unduly influential outliers; Anscombe, 1973). However, this strategy also assumes that the graphical display is honestly arranged, the subjects were representative of the populations, and accurately represents the magnitude of effects (e.g., with choices in the y-axis that neither artificially magnify trivial findings, nor trivialize meaningful findings; Witt, 2019).

All of this assumes that researchers are bothering to look at any graph of their data before running hypothesis tests. Yanai and Lercher (2020) showed, amusingly, that when given a dataset and asked to answer a correlational question, several analysts advanced straight to computing a coefficient, and failed to notice that the dataset contained an "invisible gorilla." That is, had the researchers produced a scatterplot, they would have seen dots producing an image of a friendly gorilla waving at the researchers (in a nod to the iconic "gorilla" used to document the change-blindness phenomenon).

A final important message to take from the various discussions of how much emphasis to give to exploratory/non-hypothesis driven analysis concerns the extent to which group-level findings can meaningfully characterize individuals. Branch (2014) observed that statistical hypothesis testing is essentially *actuarial*. These analyses can reveal trends and patterns in groups, but there is no guarantee that those group-level differences generalize to specific individuals. Just as the "average" family has 1.93 children, yet no *actual* family has 1.93 children, so too do the mean descriptions of groups not necessarily characterize the individuals within those groups (Grice et al., 2020). It is entirely possible to use hypothesis tests to draw conclusions about groups of individuals that do not actually apply to the individuals within those groups (a phenomenon sometimes referred to as "Simpson's paradox" or the "ecological fallacy"; Robinson, 1950; Simpson, 1951). In fact, early indications worryingly suggest this may be the case for typical psychology findings (Fisher et al., 2018). This is not a trivial limitation of traditional group-level hypothesis testing. In fact, comparing only groups that differ on average, while remaining agnostic as to processes for any given case, represents a major retreat from the assumed goal of psychology—to explain the behavior of an individual. It is worthwhile for researchers to regard hypothesis testing as one tool—of many—to be used when the time is right.

Fisher Versus Neyman-Pearson

The currently ubiquitous system of testing psychological theories with *p*-values—null-hypothesis statistical testing—has its historical origins in two competing systems (as discussed by Gigerenzer, 2004; Salsburg, 2002). The first, developed by R.A. Fisher, introduced the *p*-value as the probability that results as or more extreme than that which was observed, *under the assumption that a null hypothesis is true*. The second, developed by Jerzy Neyman and Egon Pearson, also involved testing the plausibility of a null hypothesis, and introduced the presence of an alternative hypothesis, as well as the concepts of power and alpha levels, to control long-run error rates (see Perezgonzalez, 2015 for a comprehensive comparison of the two systems). Fisher vigorously opposed the Neyman-Pearson approach leading to longstanding and acrimonious disagreement (Salsburg, 2002). Today's commonly taught and practiced system of null hypothesis testing is a merging together of elements and interpretational practices from both systems.

While Fisher's model and the Neyman-Pearson model are based on fundamentally different assumptions about the mathematical nature of probability (Schneider, 2015), the most important consequence for the application of their models is that Fisher's system treats *p*-values as providing gradations of evidence against a null hypothesis; a *p*-value of 0.04 is stronger than a *p*-value of 0.05, but not *that much* stronger. In contrast, the Neyman-Pearson approach is concerned with controlling error rates in the long run. This necessitates treating a pre-determined alpha level, as a hard cutoff. In this model, a decision must be made, and the threshold must be determined a priori. Evidence either meets the standard or it doesn't. This approach has the advantage of putting null hypothesis testing on more solid mathematical grounding, by explicitly treating probability in the frequentist sense, the long-run frequency of events (Salsburg, 2002). In contrast, Fisher's system is vague in regard to its handling of probability, treating it more as a subjective degree of confidence in a hypothesis (Perezgonzalez, 2015). While the Neyman-Pearson approach brought mathematical coherence to hypothesis testing, it can reasonably be blamed for the widely-recognized practice of regarding *p*-values below a specific threshold has qualitatively more convincing than those just above that threshold, which itself is thought to be the very source of questionable research practices to begin with (Giner-Sorolla, 2012; Nosek et al., 2012). Given that null hypothesis testing emerged out of two contradictory frameworks, it is not surprising that it has been the target of fierce criticism for decades (e.g., Bakan, 1966; Cohen, 1994; Lykken, 1968; Nickerson, 2000).

Null Hypothesis Statistical Testing Versus the World

In 1997, weary of the debate on the merits of null hypothesis statistical testing, Robert Abelson titled his defense of the practice, "On the surprising longevity of flogged horses." The controversy has not calmed since. In fact, it has been revived with renewed urgency as the replication crisis revealed that the abuse of null hypothesis testing leads not only to theoretically-prophesied false positives (Ioannidis, 2005; Kerr, 1998; Simmons et al., 2011), but actually flesh-and-bone verification that rates of replicability in psychology are disappointing at best (Open Science Collaboration, 2015).

So what exactly is the problem with traditional null hypothesis testing? For one thing, people don't seem to understand it. This is predictable, given that the system itself is an amalgam of two opposing and incompatible systems (Schneider, 2015). Numerous commenters have catalogued the many misunderstandings that are common in the null-hypothesis testing framework (Branch, 2014; Goodman, 2008; Greenland et al., 2016). Chief among these is the extraordinary difficulty with conveying the correct meaning of a *p*-value (Anderson, 2020; namely, the likelihood of the given results, *given that the null hypothesis is true*). This confusion appears to be tracible back to the original incompatibilities between Fisher's original concept of the *p*-value as an index of the implausibility of the null hypothesis, and Neyman-Pearson's competing concept of the alpha level, or long-run rate of false positives given properties of the test situation (their system does not accommodate *p*-values). Today researchers commonly confuse one for the other (Hubbard, 2004).

But, even when properly understood, criticisms of null hypothesis testing abound. For example, it's been observed that the null hypothesis is never *actually* true (Lykken, 1968), at least not when comparing two groups in a population. If one were omnisciently able to know the value of every unit in a population, it's exceedingly unlikely that two groups being compared would have the *exact same* mean. So, the argument goes, it is pointless to test a null hypothesis to begin with because it is already known to be false. There are solutions to this criticism that involve recasting hypothesis tests as giving information about how confident one can be that they have correctly identified *the direction* of an effect rather than just its presence (Jones & Tukey, 2000). Even critics grant that null hypothesis testing can be useful for this purpose (e.g., Cohen, 1995). It is also worthwhile to note that there are in fact situations in which the null hypothesis is a tenable starting point—among them, the research claim that sparked the replication crisis: Bem's (2011) claim that people can "respond" to future events at above-chance levels.

Null hypothesis testing has also been blamed for focusing attention on statistical significance to the exclusion of *practical* significance. By anointing *p*-values above the common—yet arbitrary—threshold of 0.05 as significant, researchers often overlook the question of *how big* an effect is (Cumming, 2014a, 2014b). This is unfortunate not only because it incites the motivation for p-hacking, but also because it creates difficulty for policy makers who need to know not only whether an effect exists, but also whether it is large enough to justify the expense of implementation.

And indeed, defenders of null hypothesis testing loudly acknowledge the need to pair *p*-values with indices of effect size (e.g., Abelson, 1997; Lakens, 2020).

A final criticism of null hypothesis testing worth mentioning here is the continued misinterpretation of many researchers that a *p*-value greater than 0.05 represents evidence in favor of a null hypothesis (Goodman, 2008), and, more generally, that null hypothesis testing provides no ready way to provide evidence for a null hypothesis. Happily, the first issue is a matter of better statistical education (Lakens, 2020), which is difficult but possible (e.g., Nisbett, 2015). And the second issue actually can be addressed within the usual null hypothesis testing framework (Lakens, 2017). One simply has to designate as a null hypothesis an effect size that would be considered unmeaningful, and show that the true effect is smaller than this. In essence, one can't use *p*-values to show a "significant null effect," but one can use *p*-values to show that an effect is significantly smaller than "small."

Bayesian Statistics Versus Null Hypothesis Testing

One of the harshest complaints about null hypothesis testing is that people mistakenly take *p*-values to represent the probability of the null hypothesis. People fail to appreciate that the *p*-value is the probability of the observed data, *given that the null hypothesis is true.* Since we don't (and can't) know whether the null hypothesis is true, this is a strange thing on which to condition our test. So, many have argued, a better framework would be one that conditions our test on something we *do* have: our data. The Bayesian statistical framework does just this. Rather than telling the analyst the probability of their results, given a hypothesis, it does the reverse, and indicates the probability of a hypothesis, *given the results that were observed.* Based on this apparently more logical approach, many have argued that Bayesian analyses should be used as a default rather than the classical null hypothesis testing approach.

The Bayesian approach treats probability not as the hypothetical long run frequency of events (as in the Neyman-Pearson framework), but something more like a well-informed personally held subjective degree of credence given to a hypothesis (Edwards et al., 1963; in this sense the Bayesian view is closer to Fisher's treatment of *p*-values than Neyman-Fisher). As Edwards and colleagues put it, "The Bayesian approach is a common-sense approach. It is simply a set of techniques for orderly expression and revision of your opinions with due regard for internal consistency among their various aspects and for the data" (Edwards et al., 1963, p. 195).

An appealing feature of Bayesian analysis is its emphasis on the cumulative updating of beliefs as more and more data become available on a given issue. Because this is embedded into the nature of the framework, Bayesian analysts are relatively free to collect data and stop when satisfied (Rouder, 2014)—a practice that is highly problematic in the traditional frequentist framework (Wagenmakers, 2007). The Bayesian framework also has the advantage of allowing the researcher to directly assess evidence *in support* of the null hypothesis that there is no difference or relationship (Rouder et al., 2009), and to do so without the awkward step of identifying the smallest effect size of interest mentioned above (Dienes, 2014).

Given these advantages, one might wonder why Bayesian analysis is not ubiquitous. Aside from the usual inertial/sociological forces that make change slow, the Bayesian approach has a key limitation: "priors." In Bayesian analysis, the final outcome of a hypothesis test is highly contingent on the presumed *prior* probability of that hypothesis (i.e., the researcher's belief in the hypothesis prior to seeing the data). This itself is contingent on the researcher's beliefs or analytic choices. Daryl Bem may well have assigned a modest prior probability to the existence of psi/extrasensory perception. Other researchers would have put extremely small prior probability on this possibility, defensibly even zero, fating the posterior probability to also be zero (Abelson, 1995, p. 44; Wagenmakers et al., 2011). This subjectivity seems undesirable in a framework for making statistical decisions, especially considering that one of the main advantages of quantitative over qualitative methods is relatively greater objectivity. And indeed, there does seem to be evidence that when researchers employ Bayesian methods their conclusions vary as a function of individual characteristics, such as confidence in oneself and potentially even gender (Dunn et al., 2020). Despite all of the problems of null hypothesis testing and its over-use, the "sharp null hypothesis" (Edwards et al., 1963) starts to look like an appealing starting point in comparison to prior odds, which differ from researcher to researcher.

Non-Controversies

There is danger in declaring any issue of non-controversy; one only needs to identify a single dissenting voice to create an impression that a given position is meaningfully in-question. For the topics that follow, we do not deny that such dissenting voices may exist. But we were surprised, and sometimes pleased, to see that these topics have received relatively little pushback and in many cases are now taken as simple common practice.

Fraud

First and foremost, fraud is fraud. It is a serious concern, but one that is separable from the more ubiquitous problem of questionable research practices. Because a highly-publicized case of fraud coincided with the beginning of the replication crisis (Levelt committee, 2012), there was some possibility that people might conflate questionable research practices with fraud, and fail to distinguish major malfeasance from common and well-intentioned practices that nevertheless cause problems (i.e., undisclosed researcher degrees of freedom). Part of what made the original false-positive discovery so impactful was the recognition that the practices described in the paper (Simmons et al., 2011) really were widespread, and not something that people considered fraudulent.

Sometimes fraud is categorized as a questionable research practice. In our view this is a mistake. No reasonable person would question whether fraud is an acceptable practice in science. As the replication crisis has emerged, researchers have generally been restrained in reserving accusations of fraud for truly fraudulent behavior. This is good because it is true/promotes clear thinking and preserves the strength of the "fraud" concept (Haslam, 2016) by reserving its use for truly fraudulent cases. It is also good, because for a topic that is already rife with moralizing, it is wise to assure people that you are not accusing them of fraud when you are actually persuading them to take up practices to increase replicability.

Data Sharing

In 2011 it was rare to publicize the data corresponding to a research report for a published empirical article. Today it is entirely common, and we predict that in a few short years it will be strange for a paper to be published without a link to materials including an accompanying datafile. In fact, some journals now require posted data for publication (Grahe, 2021), while many others recognize this and other desiderata with badges. No doubt, a big reason for this shift in expectations is that new resources such as the Open Science Framework (osf.io) and ResearchBox (researchbox.org) have made it trivially easy for researchers to post a datafile and link to it in an accompanying researcher report.

This is a good thing, because making data available to other researchers promotes transparency, allows for quicker detection of errors, accelerates the pace of science, and can increase the knowledge-yield from a given study (Perrino et al., 2013; Simonsohn, 2013; Wicherts & Bakker, 2012). However, at the beginning of the replication crisis it was not at all obvious that researchers would take heed of the call to post their data. Journals did not require it, the infrastructure didn't exist to accommodate it, and it seemed quite effortful. Moreover, some expressed reasonable reservations that privacy concerns would not make it possible for all areas of social science to comply (Finkel et al., 2015). But researchers soon learned that these logistical issues, while present, are easily navigable and the transparency is worth the effort (Meyer, 2018). Even when researchers post data, it's not guaranteed to be in a form that allows for immediate reproduction of analyses by others (Obels et al., 2020), but the fact that people are routinely preemptively posting data is itself major progress. Also, readers will have to trust that this was a controversial proposal at the time. Today it seems hard to imagine anyone objecting to this simple prescription.

Non-WEIRD Samples

In 2010 Henrich and colleagues published a seminal article discussing social scientists' overreliance on WEIRD (Western, Educated, Industrialized, Rich, Democratic) samples in research. The authors argued that, modeling the physical sciences, psychologists attempt to explain universals—define psychological phenomena that describe all of humanity—but do so with data from WEIRD people, a narrow and odd sample of the world population (see also Norenzayan & Heine, 2005). To be even more specific, most empirical work uses undergraduate subject pools from United States universities (Peterson, 2001; Wintre et al., 2001). However, people from WEIRD societies tend to be at the outlying end of the distribution on a variety of measures, suggesting they are highly distinguishable from other people, and thus findings from studies using these samples cannot, and should not, be generalized to humans at large (Henrich et al., 2010). The argument implies that instead of studying human nature, we study the psychological processes of only WEIRD people. We miss important variation when samples are restricted to only WEIRD societies and thus limits our understanding of psychological phenomena.

Accuracy issues arise when researchers claim their findings from WEIRD samples are universal principles that generalize to a global population. Additionally, for applied research, it is problematic if policies that affect a diverse group of people are enacted based on the results of a series of studies using only WEIRD individuals. The overuse of WEIRD samples and the tendency to generalize from narrow populations is non-controversial and most social scientists—including not only psychologists, but also anthropologists, economists, and sociologists—would agree that our WEIRD-dominated data is a crisis. In the same issue of *Behavioral and Brain Sciences* as the original Henrich et al. (2010) article, dozens of commentators concurred and further elaborated with their argument. They suggest that in addition to using WEIRD samples of odd people, social scientists also rely on experimental designs that are culture-specifically contrived (Baumard & Sperber, 2010) and lack correlation with real-world situations (Rai & Fiske, 2010). It is important to note that some researchers disagree with the claim that WEIRD samples are problematic—suggesting that while behavior might differ, humans are all the same species thus WEIRD samples can represent universal human processes (Gaertner et al., 2010)—though this argument is in the minority.

The WEIRD sample problem is exacerbated by the over-representation of WEIRD researchers within the field (Meadon & Spurrett, 2010). These researchers share cultural similarities with their participants, which hinder their ability to break from their intuition when theorizing and choosing research questions (Fessler, 2010). Additionally, WEIRD researchers have a "home culture bias" of methods and result interpretation within cross-culture comparisons (Bennis & Medin, 2010). Many commentators suggest that a potential solution of the WEIRD sample reliance is expanding research capabilities in non-WEIRD societies.

While the Henrich et al. (2010) article shook the psychological world, it should not have come as a surprise that researchers were vocalizing the issue of making broad generalizations based on narrow samples. Psychologists have been commenting on this problem for decades (Rozin, 2001; Smart, 1966). The gap in income, education, and physical health (which all contribute to psychological processes) between WEIRD and non-WEIRD societies is widening, in turn exacerbating the crisis. Thus, the need to use diverse samples in psychological research is at an all-time high, a conclusion with which most psychologists would agree (Arnett, 2008).

Interestingly, though generalizing from narrow samples is non-controversial, there seems to be a disconnect between admitting psychology has a problem and the application of solutions (Rad et al., 2018). We have known for a long time that psychological research relies too heavily on WEIRD samples (and US undergraduates, in particular; Sears, 1986), yet decades later this issue continues (Arnett, 2008; Henrich et al., 2010), and abounds further even after highly cited articles kickstart the discussion again (Rad et al., 2018). Analyses of the top journals in the psychological subdisciplines suggest that most authors and samples are based in the United States (73% and 68% respectively in the mid-2000s; Arnett, 2008). The use of WEIRD samples is so pervasive that our implicit assumption is that research findings result from a US or WEIRD sample (titles and abstracts typically only mention sample characteristics if the sample is non-WEIRD; Cheon et al., 2020).

While the overabundance of WEIRD samples is not a controversy among psychologists, it seems that pressures from the field prohibit researchers from implementing solutions to the problem. WEIRD samples are convenient, which allows for a greater volume of research to be produced. Due to favoritism toward multistudy articles in high-impact journals and publication pressure needed for job procurement and career advancement, we continue to publish papers and award grants that allow WEIRDness to prosper in psychology (Rozin, 2009). However, many researchers have suggested ways to alleviate the over-representation of WEIRD samples. Authors should always report sample characteristics, WEIRD and non-WEIRD samples alike (Rad et al., 2018). Similarly, others suggest including a "Constraints on Generality" statement in the discussion section that emphasizes why the sample was chosen and justifies the generalizability of findings to the target population (Simons et al., 2017). Like the 21-word solution for data collection and analyses (Simmons et al., 2012), a Constraints on Generality statement normalizes the recognition of WEIRD sample limitations. One of the frequently mentioned solutions to expand sampling is to use internet-based data collection (Gosling et al., 2010), though this recommendation should be taken with caution as psychology is experiencing an influx of online studies which limits the scope and real-world similarity of experimental designs (Anderson et al., 2019). At the journal level, special issues focused on studies using methods with diverse samples written and edited by non-WEIRD researchers should become more regular (Arnett, 2008).

In sum, we should obviously be cautious about generalizing findings from a narrow sample, but that does not mean that studies conducted using a US undergraduate subject pool or WEIRD participants are useless. To the contrary, the convenience provided by WEIRD sampling can allow researchers to explore new theories and draw tentative conclusion (Khemlani et al., 2010). However, it is important to recognize the limitation of narrow samples to universal generalizability and thus even robust findings should continue to be explored in diverse populations.

One-Tailed Tests

In the past—prior to the ability to preregister an analysis plan—a one-tailed hypothesis test could be viewed with skepticism. Is the researcher just trying to scoot an inconvenient "marginal" *p*-value below 0.05? How can we know that they *really* intended to perform a one-tailed test? The choice of one- versus two-tailed tests is a prototypical researcher degree of freedom, and skeptics would be entirely justified in wondering if the result of the test affected the decision to report it as one-tailed instead of two-tailed.

In recent years however, many have noticed that (1) one-tailed tests are a free and effortless way to increase power, and (2) preregistration makes it possible and easy to certify that the decision to use a one-tailed test preceded the data (Hales, 2016; Hales et al., 2019; Lakens, 2016; Maner, 2014). We are not aware of anyone who has argued (at all, let along convincingly) that researchers should continue to be compelled to run two-tailed tests, even when they are willing to perform a risky preregistered one-tailed test. Our view on this matter is one of statistical libertarianism; researchers who want to risk a one-tailed test should be permitted to do so. Reasons to do this include: a study being a direct replication (in which case, a significant effect in the unexpected direction would be so confusing it still would probably not lead to a rejection of the null hypothesis), a study testing an intervention against another that is already known to be effective (in which case the decision is simply whether the new intervention is better than the old one), or simple confidence in one's hypothesis. Whatever the reason for the researcher's decision, it is not controversial to say that a researcher who preregisters and then properly conducts a one-tailed hypothesis test is playing by the rules of null hypothesis testing, and has not inappropriately inflated their chance of a false positive. Moreover, they've probably run a more powerful test.

Even before the widespread adoption of preregistration, there were cogent arguments for one-tailed tests (Cho & Abe, 2013; Jones, 1952). Now that preregistration is commonplace, one-tailed tests should be as well (provided that is how a researcher elects to distribute their alpha, in the spirit of statistical libertarianism). While it is still not common to see one-tailed tests in the literature, when we do encounter preregistered one-tailed tests, it seems to be an unremarkable and clearly justified analytic decision (e.g., Effron, 2018). We expect to see more of these in the future.

Conclusion

Statistical disagreements have been surprisingly contentious in psychology, especially in recent years (in fact, more so than this chapter has conveyed; skeptical readers can google the term "methodological terrorists" for evidence). Perhaps this is surprising, given statistic's reputation for being dry and mathematical. So why are statistical issues so controversial?

One reason for the contention relates to the unique position that statistics and methodology hold in the psychology curriculum. Psychologists are well-aware of the naturalistic fallacy (Hume, 1969/1739; Moore, 1903/1996), and are proscribed from directly drawing any moral conclusions from their empirically descriptive research, at least not in heavily-policed peer-reviewed outlets. Statistical methods represent an exception to this ban on prescriptive language. Psychologists writing on this topic are free to say that one *ought* to analyze their data a certain way, or that one *ought not* to engage in certain research practices. Of course, these statements are based on the (often) unstated premise that doing so will lead to unreliable conclusions which—assuming one values reliable research—"ought" to be avoided. Relative to other topics in psychology, in statistical debates, the taboo against "should" and "ought" statements is relatively thin. This has led to some unhelpful moralizing at times (e.g., causing people to think of preregistration as a morally virtuous thing to do, rather than just one way to rule out analytic flexibility as one potential pesky alternative explanation for findings; see Simmons et al., 2017 for this alternative-explanation perspective). The freedom to make prescriptive statements has also likely contributed to the heated nature of debates on this topic, making statistics, surprisingly, one of the more controversial areas of psychology.

A second potential reason for the contentiousness concerns the stakes of statistical practices. Controversies of substantial research findings are local, in that they affect only the theories and topics that they touch. Statistical controversies, on the other hand, are global, in that they affect quite literally the entire field, and raise the possibility that the whole enterprise could be "rotten to the core" (Motyl et al., 2017). This helps explain why there is much hand-wringing about the implications of the replication crisis not only in-house but also for how psychology is viewed by the public and by policy-makers (e.g., Mede et al., 2020).

Regardless of the reasons for controversy, or the tone of the debate, it is hard to deny that impressive progress has been made in the last decade, and this is certainly cause for optimism. We believe that informed researchers are now armed with the tools to avoid the mistakes that led to the replication crisis (Hales et al., 2019). There will undoubtedly continue to be statistical controversy. But as these new practices take hold, we may see a shift in the tone of these debates to being more civil. Either way, scientific progress will not only continue, but, we predict, accelerate.

References

Abelson, R. P. (1995). *Statistics as principled argument*. Lawrence Erlbaum Associates.

Abelson, R. P. (1997). On the surprising longevity of flogged horses: Why there is a case for the significance test. *Psychological Science, 8*(1), 12–15. https://doi.org/10.1111/j.1467-9280.1997.tb00536.x

AlShebli, B., Makovi, K., & Rahwan, T. (2020). Retraction note: The association between early career informal mentorship in academic collaborations and junior author performance. *Nature Communications, 11*(1), 1–8.

Anderson, S. F. (2020). Misinterpreting p: The discrepancy between p values and the probability the null hypothesis is true, the influence of multiple testing, and implications for the replication crisis. *Psychological Methods, 25*(5), 596–609. https://doi.org/10.1037/met0000248

Anderson, C. A., Allen, J. J., Plante, C., Quigley-McBride, A., Lovett, A., & Rokkum, J. N. (2019). The MTurkification of social and personality psychology. *Personality and Social Psychology Bulletin, 45*(6), 842–850. https://doi.org/10.1177/0146167218798821

Anscombe, F. J. (1973). Graphs in statistical analysis. *American Statistician, 27*(1), 17–21.

Arnett, J. J. (2008). The neglected 95%: Why American psychology needs to become less American. *American Psychologist, 63*(7), 602–614. https://doi.org/10.1037/0003-066X.63.7.602

Bauer, P. J. (2020). A call for greater sensitivity in the wake of a publication controversy. *Psychological Science, 31*(7), 767–769. https://doi.org/10.1177/0956797620941482

Bakan, D. (1966). The test of significance in psychological research. *Psychological Bulletin, 66*(6), 423–437. https://doi.org/10.1037/h0020412

Baumard, N., & Sperber, D. (2010). Weird people, yes, but also weird experiments. *Behavioral and Brain Sciences, 33*(2–3), 84–85. https://doi.org/10.1017/S0140525X10000038

Bem, D. J. (2011). Feeling the future: Experimental evidence for anomalous retroactive influences on cognition and affect. *Journal of Personality and Social Psychology, 100*(3), 407–425. https://doi.org/10.1037/a0021524

Bennis, W. M., & Medin, D. L. (2010). Weirdness is in the eye of the beholder. *Behavioral and Brain Sciences, 33*(2–3), 85–86. https://doi.org/10.1017/S0140525X1000004X

Branch, M. (2014). Malignant side effects of null-hypothesis significance testing. *Theory & Psychology, 24*(2), 256–277. https://doi.org/10.1177/0959354314525282

Carter, E. C., Kofler, L. M., Forster, D. E., & McCullough, M. E. (2015). A series of metaanalytic tests of the depletion effect: Self-control does not seem to rely on a limited resource. *Journal of Experimental Psychology: General, 144*(4), 796–815. https://doi.org/10.1037/xge0000083

Cheon, B. K., Melani, I., & Hong, Y. Y. (2020). How USA-centric is psychology? An archival study of implicit assumptions of generalizability of findings to human nature based on origins of study samples. *Social Psychological and Personality Science, 11*(7), 928–937. https://doi.org/10.1177/1948550620927269

Cho, H.-C., & Abe, S. (2013). Is two-tailed testing for directional research hypotheses tests legitimate? *Journal of Business Research, 66*(9), 1261–1266. https://doi.org/10.1016/j.jbusres.2012.02.023

Cohen, J. (1994). The earth is round ($p < .05$). *American Psychologist, 49*(12), 997–1003. https://doi.org/10.1037/0003-066X.49.12.997

Cohen, J. (1995). The earth is round ($p < .05$): Rejoinder. *American Psychologist, 50*(12), 1103. https://doi.org/10.1037/0003-066X.50.12.1103

Cuddy, A. J. C., Schultz, S. J., & Fosse, N. E. (2018). P-curving a more comprehensive body of research on postural feedback reveals clear evidential value for power-posing effects: Reply to Simmons and Simonsohn (2017). *Psychological Science, 29*(4), 656–666. https://doi.org/10.1177/0956797617746749

Cumming, G. (2014a). The new statistics: Why and how. *Psychological Science, 25*(1), 7–29. https://doi.org/10.1177/0956797613504966

Credé, M., & Phillips, L. A. (2017). Revisiting the power pose effect: How robust are the results reported by Carney, Cuddy, and Yap (2010) to data analytic decisions? *Social Psychological and Personality Science, 8*(5), 493–499. https://doi.org/10.1177/1948550617714584

Cumming, G. (2014b). The new statistics: Why and how. *Psychological Science, 25*(1), 7–29. https://doi.org/10.1177/0956797613504966

Dienes, Z. (2014). Using Bayes to get the most out of non-significant results. *Frontiers in Psychology, 5*, 781. https://doi.org/10.3389/fpsyg.2014.00781

Dunn, E. W., Chen, L., Proulx, J. D. E., Ehrlinger, J., & Savalei, V. (2020). Can researchers' personal characteristics shape their statistical inferences? *Personality and Social Psychology Bulletin, 47*(6), 969–984. https://doi.org/10.1177/0146167220950522

Edwards, W., Lindman, H., & Savage, L. J. (1963). Bayesian statistical inference for psychological research. *Psychological Review, 70*(3), 193–242. https://doi.org/10.1037/h0044139

Effron, D. A. (2018). It could have been true: How counterfactual thoughts reduce condemnation of falsehoods and increase political polarization. *Personality and Social Psychology Bulletin, 44*(5), 729–745. https://doi.org/10.1177/0146167217746152

Ferguson, C. J., & Heene, M. (2012). A vast graveyard of undead theories: Publication bias and psychological science's aversion to the null. *Perspectives on Psychological Science, 7*(6), 555–561. https://doi.org/10.1177/1745691612459059

Fessler, D. M. (2010). Cultural congruence between investigators and participants masks the unknown unknowns: Shame research as an example. *Behavioral and Brain Sciences, 33*(2–3), 92. https://doi.org/10.1017/S0140525X10000087

Festinger, L., Riecken, H., & Schachter, S. (1957). *When prophecy fails*. University of Minnesota Press.

Finkel, E. J., Eastwick, P. W., & Reis, H. T. (2015). Best research practices in psychology: Illustrating epistemological and pragmatic considerations with the case of relationship science. *Journal of Personality and Social Psychology, 108*(2), 275–297. https://doi.org/10.1037/pspi0000007

Fisher, A. J., Medaglia, J. D., & Jeronimus, B. F. (2018). Lack of group-to-individual generalizability is a threat to human subjects research. *Proceedings of the National Academy of Sciences of the United States of America, 115*(27), E6106–E6115. https://doi.org/10.1073/pnas.1711978115

Fox, A. E. (2018). The future is upon us. *Behavior Analysis: Research & Practice, 18*(2), 144–150. https://doi.org/10.1037/bar0000106

Gage, N. L. (1989). The paradigm wars and their aftermath. A "historical" sketch of research on teaching since 1989. *Educational Researcher, 18*(7), 4–10. https://doi.org/10.3102/0013189X018007004

Gaertner, L., Sedikides, C., Cai, H., & Brown, J. D. (2010). It's not WEIRD, it's WRONG: When Researchers Overlook uNderlying Genotypes, they will not detect universal processes. *Behavioral and Brain Sciences, 33*(2–3), 93–94. https://doi.org/10.1017/S0140525X10000105

Gelman. (2016). *The time-reversal heuristic – a new way to think about a published finding that is followed up by a large, preregistered replication (in context of claims about power pose)*. https://statmodeling.stat.columbia.edu/2016/01/26/more-power-posing/

Gigerenzer, G. (2004). Mindless statistics. *The Journal of Socio-Economics, 33*(5), 587–606. https://doi.org/10.1016/j.socec.2004.09.033

Gigerenzer, G. (2018). Statistical rituals: The replication delusion and how we got there. *Advances in Methods and Practices in Psychological Science, 1*(2), 198–218. https://doi.org/10.1177/2515245918771329

Giner-Sorolla, R. (2012). Science or art? How aesthetic standards grease the way through the publication bottleneck but undermine science. *Perspectives on Psychological Science, 7*(6), 562–571. https://doi.org/10.1177/1745691612457576

Goodman, S. (2008). A dirty dozen: Twelve p-value misconceptions. *Seminars in Hematology, 45*(3), 135–140. https://doi.org/10.1053/j.seminhematol.2008.04.003

Goodwin, S. A., Williams, K. D., & Carter-Sowell, A. R. (2010). The psychological sting of stigma: The costs of attributing ostracism to racism. *Journal of Experimental Social Psychology, 46*(4), 612–618. https://doi.org/10.1016/j.jesp.2010.02.002

Gosling, S. D., Sandy, C. J., John, O. P., & Potter, J. (2010). Wired but not WEIRD: The promise of the Internet in reaching more diverse samples. *Behavioral and Brain Sciences, 33*(2-3), 94–95. https://doi.org/10.1017/S0140525X10000300

Grahe, J. (2021). The necessity of data transparency to publish. *The Journal of Social Psychology, 161*(1), 1–4. https://doi.org/10.1080/00224545.2020.1847950

Greenberg, J., Pyszczynski, T., Solomon, S., Simon, L., & Breus, M. (1994). Role of consciousness and accessibility of death-related thoughts in mortality salience effects. *Journal of Personality and Social Psychology, 67*(4), 627–637. https://doi.org/10.1037/0022-3514.67.4.627

Greenland, S., Senn, S. J., Rothman, K. J., Carlin, J. B., Poole, C., Goodman, S. N., & Altman, D. G. (2016). Statistical tests, P values, confidence intervals, and power: A guide to misinterpretations. *European journal of epidemiology, 31*(4), 337–350. https://doi.org/10.1007/s10654-016-0149-3

Grice, J. W., Medellin, E., Jones, I., Horvath, S., McDaniel, H., O'lansen, C., & Baker, M. (2020). Persons as effect sizes. *Advances in Methods and Practices in Psychological Science, 3*(4), 443–455. https://doi.org/10.1177/2515245920922982

Hagger, M. S., Chatzisarantis, N. L. D., Alberts, H., Anggono, C. O., Batailler, C., Birt, A. R., Brand, R., Brandt, M. J., Brewer, G., Bruyneel, S., Calvillo, D. P., Campbell, W. K., Cannon, P. R., Carlucci, M., Carruth, N. P., Cheung, T., Crowell, A., De Ridder, D. T. D., Dewitte, S., … Zwienenberg, M. (2016). A multilab preregistered replication of the ego-depletion effect. *Perspectives on Psychological Science, 11*(4), 546–573. https://doi.org/10.1177/1745691616652873

Hales, A. H. (2016). Does the conclusion follow from the evidence? Recommendations for improving research. *Journal of Experimental Social Psychology, 66*, 39–46. https://doi.org/10.1016/j.jesp.2015.09.011

Hales, A. H., Wesselmann, E. D., & Hilgard, J. (2019). Improving psychological science through transparency and openness: An overview. *Perspectives on Behavior Science, 42*(1), 13–31. https://doi.org/10.1007/s40614-018-00186-8

Hartgerink, C. J., van Beest, I., Wicherts, J. M., & Williams, K. D. (2015). The ordinal effects of ostracism: A meta-analysis of 120 cyberball studies. *PLoS One, 10*(5), e0127002. https://doi.org/10.1371/journal.pone.0127002

Haslam, N. (2016). Concept creep: Psychology's expanding concepts of harm and pathology. *Psychological Inquiry, 27*, 1–17. https://doi.org/10.1080/1047840X.2016.1082418

Henrich, J., Heine, S. J., & Norenzayan, A. (2010). The weirdest people in the world? *Behavioral and Brain Sciences, 33*(2-3), 61–83. https://doi.org/10.1017/S0140525X0999152X

Hubbard, R. (2004). Alphabet soup: Blurring the distinctions between p's and α's in psychological research. *Theory & Psychology, 14*(3), 295–327. https://doi.org/10.1177/0959354304043638

Hume, D. (1969/1739). *A Treatise on Human Nature*. Penguin.

Ioannidis, J. P. A. (2005). Why most published research findings are false. *PLoS Medicine, 2*(8), e124. https://doi.org/10.1371/journal.pmed.0020124

IJzerman, H., Lewis, N. A., Przybylski, A. K., Weinstein, N., DeBruine, L., Ritchie, S. J., Vazire, S., Forscher, P. S., Morey, R. D., Ivory, J. D., & Anvari, F. (2020). Use caution when applying behavioural science to policy. *Nature Human Behavior, 4*, 1092–1094.

Jackson, M. R. (2015). Resistance to qual/quant parity: Why the "paradigm" discussion can't be avoided. *Qualitative Psychology, 2*(2), 181–198. https://doi.org/10.1037/qup0000031

Jones, L. V. (1952). Test of hypotheses: one-sided vs two-sided alternatives. *Psychological Bulletin, 49*(1), 43–46. https://doi.org/10.1037/h0056832

Jones, L. V., & Tukey, J. W. (2000). A sensible formulation of the significance test. *Psychological Methods, 5*(4), 411–414. https://doi.org/10.1037/1082-989X.5.4.411

Kerr, N. L. (1998). HARKing: Hypothesizing after the results are known. *Personality and Social Psychology Review, 2*(3), 196–217. https://doi.org/10.1207/s15327957pspr0203_4

Khemlani, S. S., Lee, N. Y., & Bucciarelli, M. (2010). Determinants of cognitive variability. *Behavioral and Brain Sciences, 33*(2-3), 97–98. https://doi.org/10.1017/S0140525X10000130

Klein, R. A., Cook, C. L., Ebersole, C. R., Vitiello, C. A., Nosek, B. A., Chartier, C. R., … Ratliff, K. A. (2019). Many Labs 4: Failure to replicate mortality salience effect with and without original author involvement. *Collabra: Psychology, 8*(1). https://doi.org/10.31234/osf.io/vef2c

Lakens, D. (2020). The practical alternative to the p-value is the correctly used p-value. *Perspectives on Psychological Science, 16*(3), 639–648. https://doi.org/10.31234/osf.io/shm8v

Lakens, D. (2017). Equivalence tests: A practical primer for t tests, correlations, and meta-analyses. *Social Psychological and Personality Science, 8*(4), 355–362. https://doi.org/10.1177/1948550617697177

Landrum, B., & Garza, G. (2015). Mending fences: Defining the domains and approaches of quantitative and qualitative research. *Qualitative Psychology, 2*(2), 199–209. https://doi.org/10.1037/qup0000030

Levelt Committee. (2012). *Flawed science: The fraudulent research practices of social psycholo- gist Diederik Stapel.* Retrieved from https://www.rug.nl/about-ug/latest-news/news/archief2012/nieuwsberichten/stapel-eindrapport-eng.pdf

Lakens, D. (2016). *One-sided tests: Efficient and underused.* http://daniellakens.blogspot.com/2016/03/one-sided-tests-efficient-and-underused.html

Lykken, D. T. (1968). Statistical significance in psychological research. *Psychological Bulletin, 70*(3), 151–159. https://doi.org/10.1037/h0026141

Meadon, M., & Spurrett, D. (2010). It's not just the subjects–there are too many WEIRD researchers. *Behavioral and Brain Sciences, 33*(2-3), 104–105. https://doi.org/10.1017/S0140525X10000208

Maner, J. K. (2014). Let's put our money where our mouth is: If authors are to change their ways, reviewers (and editors) must change with them. *Perspectives on Psychological Science, 9*(3), 343–351. https://doi.org/10.1177/1745691614528215

McCabe, C. J., Kim, D. S., & King, K. M. (2018). Improving present practices in the visual display of interactions. *Advances in Methods & Practices in Psychological Science, 1*(2), 47–165. https://doi.org/10.1177/2515245917746792

Mede, N. G., Schäfer, M. S., Ziegler, R., & Weißkopf, M. (2020). The "replication crisis" in the public eye: Germans' awareness and perceptions of the (ir)reproducibility of scientific research. *Public Understanding of Science, 30*(1), 91–102. https://doi.org/10.1177/0963662520954370

Meehl, P. E. (1978). Theoretical risks and tabular asterisks: Sir Karl, Sir Ronald, and the slow progress of soft psychology. *Journal of Consulting and Clinical Psychology, 46*(4), 806–834. https://doi.org/10.1037/0022-006X.46.4.806

Meyer, M. N. (2018). Practical tips for ethical data sharing. *Advances in Methods and Practices in Psychological Science, 1*(1), 131–144. https://doi.org/10.1177/2515245917747656

Moore, G. E. (1903/1996). *Principia ethica.* Cambridge University Press.

Motyl, M., Demos, A. P., Carsel, T. S., Hanson, B. E., Melton, Z. J., Mueller, A. B., Prims, J. P., Sun, J., Washburn, A. N., Wong, K. M., Yantis, C., & Skitka, L. J. (2017). The state of social and personality science: Rotten to the core, not so bad, getting better, or getting worse? *Journal of Personality and Social Psychology, 113*(1), 34–58. https://doi.org/10.1037/pspa0000084

Nickerson, R. S. (2000). Null hypothesis significance testing: A review of an old and continuing controversy. *Psychological Methods, 5*(2), 241–301. https://doi.org/10.1037/1082-989X.5.2.241

Nisbett, R. E. (2015). *Mindware: Tools for smart thinking*. Farrar, Straus and Giroux.

Nosek, B. A., Spies, J. R., & Motyl, M. (2012). Scientific utopia: II. Restructuring incentives and practices to promote truth over publishability. *Perspectives on Psychological Science, 7*(6), 615–631. https://doi.org/10.1177/1745691612459058

Norenzayan, A., & Heine, S. J. (2005). Psychological universals: What are they and how can we know? *Psychological Bulletin, 131*(5), 763–784. https://doi.org/10.1037/0033-2909.131.5.763

Obels, P., Lakens, D., Coles, N. A., Gottfried, J., & Green, S. A. (2020). Analysis of open data and computational reproducibility in registered reports in psychology. *Advances in Methods and Practices in Psychological Science, 3*(2), 229–237. https://doi.org/10.1177/2515245920918872

Open Science Collaboration. (2015). Estimating the reproducibility of psychological science. *Science, 346*(6251), aac4716. https://doi.org/10.1126/science.aac4716

Perezgonzalez, J. D. (2015). Fisher, Neyman-Pearson or NHST? A tutorial for teaching data testing. *Frontiers in Psychology, 6*, 223. https://doi.org/10.3389/fpsyg.2015.00223

Perone, M. (1999). Statistical inference in behavior analysis: Experimental control is better. *The Behavior Analyst, 22*(2), 109–116. https://doi.org/10.1007/BF03391988

Perrino, T., Howe, G., Sperling, A., Beardslee, W., Sandler, I., Shern, D., … Brown, C. (2013). Advancing science through collaborative data sharing and synthesis. *Perspectives on Psychological Science, 8*(4), 433–444. https://doi.org/10.1177/1745691613491579

Peterson, R. A. (2001). On the use of college students in social science research: Insights from a second-order meta-analysis. *Journal of Consumer Research, 28*(3), 450–461. https://doi.org/10.1086/323732

Ponterotto, J. G. (2006). Brief note on the origins, evolution, and meaning of the qualitative research concept thick description. *The Qualitative Report, 11*(3), 538–549. Retrieved from https://nsuworks.nova.edu/tqr/vol11/iss3/6

Rad, M. S., Martingano, A. J., & Ginges, J. (2018). Toward a psychology of Homo sapiens: Making psychological science more representative of the human population. *Proceedings of the National Academy of Sciences, 115*(45), 11401–11405. https://doi.org/10.1073/pnas.1721165115

Ranehill, E., Dreber, A., Johannesson, M., Leiberg, S., Sul, S., & Weber, R. A. (2015). Assessing the robustness of power posing: No effect on hormones and risk tolerance in a large sample of men and women. *Psychological Science, 26*(5), 653–656. https://doi.org/10.1177/0956797614553946

Rai, T. S., & Fiske, A. (2010). ODD (observation-and description-deprived) psychological research. *Behavioral and Brain Sciences, 33*(2-3), 106–107. https://doi.org/10.1017/S0140525X10000221

Robinson, W. S. (1950). Ecological correlations and the behavior of individuals. *American Sociological Review, 15*, 351–357. https://doi.org/10.2307/2087176

Rouder, J. N. (2014). Optional stopping: No problem for Bayesians. *Psychonomic Bulletin & Review, 21*(2), 301–308. https://doi.org/10.3758/s13423-014-0595-4

Rouder, J. N., Speckman, P. L., Sun, D., Morey, R. D., & Iverson, G. (2009). Bayesian t tests for accepting and rejecting the null hypothesis. *Psychonomic Bulletin & Review, 16*(2), 225–237. https://doi.org/10.3758/PBR.16.2.225

Rozin, P. (2001). Social psychology and science: Some lessons from Solomon Asch. *Personality and Social Psychology Review, 5*(1), 2–14. https://doi.org/10.1207/S15327957PSPR0501_1

Rozin, P. (2009). What kind of empirical research should we publish, fund, and reward? A different perspective. *Perspectives on Psychological Science, 4*(4), 435–439. https://doi.org/10.1111/j.1745-6924.2009.01151.x

Sakaluk, J. K. (2016). Exploring Small, Confirming Big: An alternative system to The New Statistics for advancing cumulative and replicable psychological research. *Journal of Experimental Social Psychology, 66*, 47–54. https://doi.org/10.1016/j.jesp.2015.09.013

Salsburg, D. (2002). *The lady tasting tea: How statistics revolutionized science in the twentieth century*. Owl Books.

Scheel, A. M., Tiokhin, L., Isager, P. M., & Lakens, D. (2020). Why hypothesis testers should spend less time testing hypotheses. *Perspectives on Psychological Science, 16*(4), 744–755. https://doi.org/10.1177/1745691620966795

Schneider, J. (2015). Null hypothesis significance tests. A mix-up of two different theories: The basis for widespread confusion and numerous misinterpretations. *Scientometrics, 102*, 411–432. https://doi.org/10.1007/s11192-014-1251-5

Sears, D. O. (1986). College sophomores in the laboratory: Influences of a narrow data base on social psychology's view of human nature. *Journal of Personality and Social Psychology, 51*(3), 515–530. https://doi.org/10.1037/0022-3514.51.3.515

Simmons, J. P., Nelson, L. D., & Simonsohn, U. (2011). False-positive psychology: Undisclosed flexibility in data collection and analysis allows presenting anything as significant. *Psychological Science, 22*(11), 1359–1366. https://doi.org/10.1177/0956797611417632

Simmons, J. P., Nelson, L. D., & Simonsohn, U. (2012). A 21 word solution. *SSRN*. https://doi.org/10.2139/ssrn.2160588

Simmons, J. P., Nelson, L. D., & Simonsohn, U. (2017). *How to properly preregister a study*. http://datacolada.org/64

Simmons, J. P., & Simonsohn, U. (2017). Power posing: P-curving the evidence. *Psychological Science, 28*(5), 687–693. https://doi.org/10.1177/0956797616658563

Simons, D. J., Shoda, Y., & Lindsay, D. S. (2017). Constraints on generality (COG): A proposed addition to all empirical papers. *Perspectives on Psychological Science, 12*(6), 1123–1128. https://doi.org/10.1177/1745691617708630

Simonsohn, U. (2013). Just post it: The lesson from two cases of fabricated data detected by statistics alone. *Psychological Science, 24*(10), 1875–1888. https://doi.org/10.1177/0956797613480366

Simpson, E. H. (1951). The interpretation of interaction in contingency tables. *Journal of the Royal Statistical Society, 13*(2), 238–241. https://doi.org/10.1111/j.2517-6161.1951.tb00088.x

Skinner, B. F. (1963). Operant behavior. *American Psychologist, 18*(8), 503–515. https://doi.org/10.1037/h0045185

Smart, R. G. (1966). Subject selection bias in psychological research. *Canadian Psychologist/Psychologie canadienne, 7*(2), 115–121. https://doi.org/10.1037/h0083096

Sommer, K. L., Williams, K. D., Ciarocco, N. J., & Baumeister, R. F. (2001). When silence speaks louder than words: Explorations into the intrapsychic and interpersonal consequences of social ostracism. *Basic and Applied Social Psychology, 23*(4), 225–243. https://doi.org/10.1207/153248301753225694

Tukey, J. W. (1969). Analyzing data: Sanctification or detective work? *American Psychologist, 24*(2), 83–91. https://doi.org/10.1037/h0027108

Tukey, J. W. (1977). *Exploratory data analysis*. Addison-Wesley.

Van Bavel, J. J., Baiker, K., Boggio, P. S., Valerio, C., Cichocka, A., Cikara, M., Crockett, M. J., Crum, A. J., Douglas, K. M., Druckman, J. N., Drury, J., Oeindrila, D., Ellemers, N., Finkel, E. F., Fowler, J. H., Gelfand, M. J., Shihui, H., Haslam, A., Jetten, J., … Willer, R. (2020). Using social and behavioral science to support COVID-19 pandemic response. *Nature Human Behavior, 4*, 460–471. https://doi.org/10.1038/s41562-020-0884-z

Vohs, K. D., Schmeichel, B. J., Lohmann, S., Gronau, Q., Finley, A. J., Wagenmakers, E.-J., & Albarracín, D. (2020). A multi-site preregistered paradigmatic test of the ego depletion effect. *Psychological Science, 32*(10), 1566–1581.

Wagenmakers, E.-J. (2007). A practical solution to the pervasive problems of p values. *Psychonomic Bulletin & Review, 14*(5), 779–804. https://doi.org/10.3758/BF03194105

Wagenmakers, E.-J., Wetzels, R., Borsboom, D., & van der Maas, H. L. J. (2011). Why psychologists must change the way they analyze their data: The case of psi: Comment on Bem (2011). *Journal of Personality and Social Psychology, 100*(3), 426–432. https://doi.org/10.1037/a0022790

Wicherts, J. M., & Bakker, M. (2012). Publish (your data) or (let the data) perish! Why not publish your data too? *Intelligence, 40*(2), 73–76. https://doi.org/10.1016/j.intell.2012.01.004

Williams, K. D. (2009). Ostracism: Effects of being excluded and ignored. In M. P. Zanna (Ed.), *Advances in experimental social psychology* (Vol. 41, pp. 275–314). Academic Press.

Williams, K. D., Bernieri, F. J., Faulkner, S. L., Gada-Jain, N., & Grahe, J. E. (2000). The scarlet letter study: Five days of social ostracism. *Journal of Personal and Interpersonal Loss, 5*(1), 19–63. https://doi.org/10.1080/10811440008407846

Williams, K. D., Shore, W. J., & Grahe, J. E. (1998). The silent treatment: Perceptions of its behaviors and associated feelings. *Group Processes and Intergroup Relations, 1*(2), 117–141. https://doi.org/10.1177/1368430298012002

Willig, C. (2019). What can qualitative psychology contribute to psychological knowledge? *Psychological Methods, 24*(6), 796–804. https://doi.org/10.1037/met0000218

Wintre, M., North, C., & Sugar, L. A. (2001). Psychologists' response to criticisms about research based on undergraduate participants: A developmental perspective. *Canadian Psychology/Psychologie Canadienne, 42*(3), 216–225. https://doi.org/10.1037/h0086893

Witt, J. K. (2019). Graph construction: An empirical investigation on setting the range of the Y-axis. *Meta-Psychology, 3*. https://doi.org/10.5626/MP.2018.895

Yanai, I., & Lercher, M. A. (2020). A hypothesis is a liability. *Genome Biology, 21*, 1–5. https://doi.org/10.1186/s13059-020-02133-w

Zadro, L. (2004). *Ostracism: Empirical studies inspired by real-world experiences of silence and exclusion* (Unpublished doctoral dissertation). University of New South Wales, Sydney, NSW.

Chapter 10
Publication Bias

Robbie C. M. van Aert and Helen Niemeyer

Abstract Meta-analysis is the statistical method for synthesizing studies on the same topic and is often used in clinical psychology to quantify the efficacy of treatments. A major threat to the validity of meta-analysis is publication bias, which implies that some studies are less likely to be published and are therefore less often included in a meta-analysis. A consequence of publication bias is the overestimation of the meta-analytic effect size that may give a false impression with respect to the efficacy of a treatment, which might result in (avoidable) suffering of patients and waste of resources. Guidelines recommend to routinely assess publication bias in meta-analyses, but this is currently not common practice. This chapter describes popular and state-of-the-art methods to assess publication bias in a meta-analysis and summarizes recommendations for applying these methods. We also illustrate how these methods can be applied to two meta-analyses that are typical for clinical psychology such that psychologists can readily apply the methods in their own meta-analyses.

Keywords Publication bias · Questionable research practices · Methods to assess publication biases

Introduction

A meta-analysis provides a quantitative summary of studies on the same topic, and its results are seen as the best available evidence (Aguinis et al., 2011; Head et al., 2015). However, the quality of a meta-analysis fully depends on the quality of the included studies, and an important threat for the validity of a meta-analysis arises if the included studies are not representative for all studies conducted on

R. C. M. van Aert (✉)
Department of Methodology and Statistics, Tilburg University, Tilburg, The Netherlands
e-mail: R.C.M.vanAert@tilburguniversity.edu

H. Niemeyer
Department of Clinical Psychological Intervention, Freie Universität Berlin, Berlin, Germany

W. O'Donohue et al. (eds.), *Avoiding Questionable Research Practices in Applied Psychology*, https://doi.org/10.1007/978-3-031-04968-2_10

this topic. Publication bias is one of the possible causes of a meta-analysis containing an unrepresentative set of studies (Rothstein et al., 2005), which means that statistically nonsignificant studies have a lower probability of being published than significant studies. Publication bias may be caused by editors and reviewers who are more reluctant to positively evaluate statistically nonsignificant compared to significant studies or by authors who do not submit nonsignificant studies for publication (Cooper et al., 1997; Coursol & Wagner, 1986). The consequences of publication bias are severe and hamper the progress of science, because it yields overestimated effect size in the individual studies and when combining these studies in a meta-analysis (e.g., Kraemer et al., 1998; Lane & Dunlap, 1978). For this reason, guidelines on how to conduct a meta-analysis such as the Meta-Analytic Reporting Standards (MARS, Appelbaum et al., 2018), Preferred Reporting Items for Systematic Reviews and Meta-Analyses (PRISMA, Moher et al., 2009), and the Cochrane Handbook for Systematic Reviews of Interventions (Page et al., 2019) all encourage meta-analysts to routinely assess publication bias in their meta-analysis.

There is strong evidence for the presence of publication bias in the psychological literature. For example, Fanelli (2012) showed that 90% of a random sample of studies published in the psychological literature found support for their main hypothesis. This large percentage is in disagreement with the on average low statistical power of studies in psychology (Bakker et al., 2012; Ellis, 2010), which is too low to find support for the hypothesis this often. More direct evidence of publication bias has been observed in Franco et al. (2014) who determined whether the publication status of studies that were awarded a grant depended on their results. They concluded that studies with null or mixed results remained more often unpublished than studies with strong results (i.e., predominantly statistically significant results). Publication bias has also been studied in clinical psychology. For example, Driessen et al. (2015) compared the publication status of studies awarded with a grant focusing on research studying the efficacy of psychological treatments for patients with major depressive disorder. They showed that 13 out of 55 (23.6%) studies that were awarded with a grant were not published in the literature. Adding these unpublished studies to the meta-analysis of published studies resulted in a reduction in effect size estimate of 0.13 standardized mean difference.

If the efficacy of interventions is overestimated due to publication bias and publication bias remains undetected, this can have severe consequences. Taking the example of depression, efficacious treatments are essential to reduce impaired functioning and risk of suicide that are caused by depression (Holma et al., 2010). If publication bias is present, clinical guidelines may prompt psychotherapists to apply interventions in routine care that may be less efficacious than assumed. This would not only prevent individuals from receiving the best possible treatment but also result in unnecessarily high costs for the health care system (Jaycox & Foa, 1999; Maljanen et al., 2016; Margraf, 2009). Moreover, publication bias in research on etiological assumptions such as genetic predispositions, biological mechanisms, detrimental environmental exposures, cognitive distortions, or attentional biases would also be hampering knowledge accumulation about the underlying mechanisms that contribute to the onset and maintenance of mental disorders. Thus, assessing publication bias in all fields of clinical psychology is strongly recommended, but it has not been routinely done in meta-analyses. For example, Niemeyer et al. (2013) found that in the majority (82%) of meta-analyses on the efficacy of (psychotherapeutic) interventions for depression publication bias was not considered in the statistical analyses. In addition, 81.2% of the meta-analyses explicitly did not include unpublished studies. Publication bias is also not routinely assessed in education research where 44% (Banks et al., 2012) did not assess publication bias, and industrial and organizational psychology where publication bias was not assessed in 92.7% (Aguinis et al., 2010) and 82.3% (Aytug et al., 2012) of a large number of meta-analyses.

Of note is that evidence for publication bias in psychology is, however, not always observed when published meta-analyses are reanalyzed using publication bias methods in so-called meta-meta-analyses (i.e., meta-analysis of meta-analyses). For example, publication bias was detected in approximately 15% of reanalyzed meta-analyses published on psychotherapeutic interventions for schizophrenia and depression (Niemeyer et al., 2012, 2013). Another study also observed only weak evidence for publication bias in reanalyzed meta-analyses published in Psychological Bulletin and the Cochrane Database of Systematic Reviews (Van Aert et al., 2019). A possible reason for not observing strong evidence for publication bias are the challenging conditions of the meta-analyses under study for publication bias methods. The publication bias methods that were available at that time could only be applied to a small subset of meta-analyses in these studies due to strong assumptions of the methods. For example, the applied publication bias methods assume each study in the meta-analysis to estimate the same true effect size. This implies that no heterogeneity in true effect size is allowed, which is especially uncommon in clinical psychology research where studies in psychotherapy research are, for instance, administered at different locations and by different therapists. Moreover, many disorders are heterogeneous in their symptom presentation and comorbidity is frequent (e.g., Deisenhofer et al., 2018).

Another complicating factor that is common for meta-analyses in clinical psychology research are the small number of studies included in meta-analyses. Meta-analyses containing less than five studies are not uncommon in medical research (e.g., Rhodes et al., 2015; Turner et al., 2015) in general and clinical psychology research in particular (Niemeyer et al., 2020). Examining publication bias based on such a small number of studies is challenging, because the number of data points in the analysis equals the number of studies in the meta-analysis. The two complicating factors (heterogeneity and small number of studies) are also not unrelated. For example, Schumacher et al. (2018) meta-analyzed hormonal dysregulation in post-traumatic stress disorder (PTSD), but these data of 108 studies and more than 6000 participants were very heterogeneous. An option was to create subgroups of more homogeneous studies and assessing publication bias in these subgroups, but these subgroups comprised a very small number of studies.

Simulation studies tailored to characteristics of meta-analyses on clinical psychology research also confirmed that the conditions were unfavorable for the available publication bias methods (Niemeyer et al., 2020). However, newly developed publication bias methods are better equipped to be applied to meta-analyses that are typical for research in clinical psychology. A clear overview of the existing methods and software on how to apply these methods is currently lacking in the literature. The goal of this chapter is to provide such an overview together with summarizing recommendations for applying these methods. Many different publication bias methods have been developed, so we focus in this chapter on the most popular methods and state-of-the-art methods that have shown to outperform these most popular methods. Methods to investigate publication bias can serve two different purposes: first to estimate an effect size in the presence of publication bias, and second to assess the degree of publication bias. Publication bias methods for both purposes will be illustrated using the statistical software R (R Core Team, 2020) and by applying these to two examples that are typical for meta-analyses in clinical psychology.

We continue this chapter by introducing the statistical software R. Subsequently, we will describe graphical methods to assess publication bias, methods to correct effect size estimates for publication bias, and methods to assess the presence of publication bias in a meta-analysis. These methods will be applied to a meta-analysis on the efficacy of cognitive-behavior therapy (CBT) for treating pathological and problem gambling (Cowlishaw et al., 2012) and a meta-analysis on the added value of collaborative care for patients with depression or anxiety problems (Archer et al., 2012). Both meta-analyses provide paradigmatic examples, because CBT is a guideline-recommended treatment for most disorders (David et al., 2018), and second, depression and anxiety are among the most prevalent disorders (Alonso et al., 2004). The chapter ends with recommendations for clinical psychologists on how to deal with publication bias in meta-analyses.

Software

The publication bias methods that are discussed in this chapter are illustrated using the statistical software R (Version 4.0.3; R Core Team, 2020). R is free and open-source programming software with a primary focus on statistical computing and creating graphics. An important feature of R is that researchers can contribute to the software by developing so-called packages that can easily be loaded in R. Packages contain all sorts of functions to, for example, run statistical analyses and visualize data. After downloading R via https://cran.r-project.org/ and installing it, packages can be downloaded and installed by running the R code

```
install.packages("PACKAGE")
```

where PACKAGE needs to be replaced by the name of the package you want to download and install. The functions in a package become available by loading it using the R code

```
library("PACKAGE")
```

A popular R package for conducting meta-analyses is metafor (Viechtbauer, 2010). This package (Version 2.5.60) will be used throughout this chapter, because it contains besides functions for conducting meta-analyses also functions for applying a large number of publication bias methods. However, we sometimes have to rely on other packages if a particular method is not included in the metafor package, which will be introduced when explaining these methods.

Note that we make excessive use of R for applying publication bias methods in this chapter, but familiarity with R or programming experience is not a prerequisite. All R code will be provided for applying the publication bias methods such that this code can be easily used by interested readers who want to apply these methods to their own data. An annotated version of all the codes used in this chapter is also available at https://osf.io/qjk9b/. Readers who want to learn more about R are referred to https://cran.r-project.org/doc/manuals/R-intro.pdf for an elaborate introduction or introductory books on R such as Matloff (2011) and Teetor (2011).

Examples

Example 1: Cowlishaw et al. (2012)

The publication bias methods will be applied to two meta-analyses that are typical for meta-analyses in clinical psychology research. The first meta-analysis synthesizes seven studies on the efficacy of CBT for treating pathological and problem gambling (analysis 1.2 in Cowlishaw et al., 2012). For each study, a standardized mean difference (i.e., Hedges' g) is computed that compares the difference in financial loss of patients who received CBT in the last three months with a control group. A positive standardized mean difference indicates that the financial loss was smaller in the group of patients who received CBT compared to those in the control group.

Cowlishaw et al. (2012) fitted a random-effects model to the included studies in the meta-analysis and, therefore, assumed that each study had its own unique true effect size (for an elaborate description of the random-effects model see Borenstein et al., 2010). This random-effects model can also be fitted to the data using the metafor package after creating two vectors[1] containing the standardized mean

[1]A vector is R terminology for a particular data structure that contains in our case seven numeric values with the studies' standardized mean difference (yi) and corresponding sampling variance (vi).

differences and corresponding sampling variances (i.e., squared standard errors). The vectors are named `yi` and `vi` and can be created using

```
yi <- c(0.587, 0.706, 0.552, 0.515, 0.566, 0.291, 0.989)
vi <- c(0.076, 0.067, 0.074, 0.217, 0.047, 0.028, 0.157)
```

the vectors are subsequently be used in the `rma()` function of the `metafor` package to fit the random-effects model,

```
res <- rma(yi = yi, vi = vi)
```

where the results are stored in the R object `res`. The average effect size in this meta-analysis was 0.519 with 95% confidence interval (CI) equal to (0.332; 0.706), and the null-hypothesis of no effect is rejected (z = 5.432, two-tailed p-value <0.001). The estimated between-study variance in true effect size is 0 with 95% CI equal to (0; 0.125). The Q-test (Cochran, 1954) for testing the null-hypothesis of no heterogeneity is not statistically significant ($Q = 3.897$, one-tailed p-value is 0.691). To conclude, the financial loss of the group of patients who received CBT was smaller than in the control group, and the difference between both groups was of medium size according to the rules of thumb by Cohen (1988). The between-study variance in true effect size was estimated as zero indicating that the studies' true effect size was homogeneous. However, estimation of the between-study variance was imprecise due to the small number of studies in the meta-analysis, which is apparent in the wide CI.

Example 2: Archer et al. (2012)

The second example used in this chapter is the meta-analysis by Archer et al. (2012) on the added value of collaborative care measured by patient satisfaction for patients with depression or anxiety problems. This meta-analysis consists of 24 studies and patient satisfaction was reported with a dichotomous variable in each study. The effect size measure of interest was a risk ratio (a.k.a. relative risk). The risk ratios were first transformed to log risk ratios before synthesizing these, because an assumption of common meta-analysis models is that the effect size measure follows a normal distribution. This is approximately the case for log risk ratios but not for risk ratios.

We follow Archer et al. (2012) by also fitting a random-effects model to these data. The estimated average risk ratio was 1.271 (95% CI (1.180; 1.368)), and the null-hypothesis of no effect was rejected ($z = 6.347$, two-tailed p-value <0.001). The between-study variance was estimated as 0.021 (95% CI (0.009; 0.070)), and the null-hypothesis of no heterogeneity was rejected (Q = 83.580, one-tailed p-value

<0.001). These results show that patients receiving collaborative care were more satisfied than patients receiving the usual care. The true effect sizes were heterogeneous, so the effectiveness of collaborative care varied across studies.

We have presented the results of the two meta-analyses when using conventional meta-analysis methods that do not correct for publication bias in this section. We will compare these results to those obtained with publication bias methods later in this chapter. We continue by explaining the publication bias methods that are also summarized in Table 10.1.

Graphical Methods to Assess Publication Bias

Funnel Plot

A regularly reported figure for assessing publication bias in a meta-analysis is the funnel plot (Light & Pillemer, 1984). A funnel plot of the meta-analysis by Cowlishaw et a. (2012) is presented in the left panel of Fig. 10.1.[2] The *x*-axis of a funnel plot shows the observed effect sizes of the studies included in the meta-analysis, and a measure of the studies' precision is depicted on the *y*-axis. The standard error is displayed on the *y*-axis of the funnel plot in Fig. 10.1, but other measures of a study's precision can also be displayed (e.g., sampling variance, sample size, or the inverse of the standard error). A funnel plot can be created using the funnel () function incorporated in the metafor package by using the code

```
funnel(res)
```

where `res` is the object that was created earlier when conducting the random-effects meta-analysis.

Publication bias can be assessed using a funnel plot by examining whether the studies resemble the shape of an inverted funnel. Some studies in the left bottom corner are missing in the funnel plot in the left panel of Fig. 10.1 to closely resemble an inverted funnel. This implies that studies with a negative observed effect size might be suppressed from being published in the literature, and therefore could not be included in the meta-analysis. It is important to emphasize that funnel plots not resembling an inverted funnel can also be caused by other factors than publication bias. An asymmetric funnel plot is indicative for larger observed effect sizes going along with larger imprecision (i.e., larger standard errors) of studies. These so-called small-study effects (Egger et al., 1997) may be caused by publication bias but also by other factors such as heterogeneity in true effect size. Heterogeneity is common for meta-analyses in clinical psychology, so prudence is in order when

[2] The funnel plot based on the data of the meta-analysis by Archer et al. (2012) is available in the annotated R codes (https://osf.io/qjk9b/)

Table 10.1 Summary of the methods described in this chapter

	Description	Characteristics/ Recommendations	R function
Graphical methods:			
Funnel plot	Figure displaying the relation between effect size and their precision (so-called small-study effects).	Small-study effects can be caused by publication bias but also by other factors. Eyeballing a funnel plot is subjective, so funnel plot asymmetry tests are recommended instead.	`funnel()` in `metafor`
Meta-plot	Figure displaying the results of cumulative meta-analysis with studies ordered by their precision.	The meta-plot can be used for assessing small-study effects and publication bias, and it is an improvement over the funnel plot.	`meta_plot()` in `puniform`
Correcting effect size for publication bias:			
Top 10% and WAAP	Meta-analysis based on the 10% most precise and adequately powered studies.	Methods only perform well if there is no heterogeneity and many studies may be discarded from the meta-analysis.	`rma()` in `metafor` after selecting studies
Trim-and-fill	Corrects for small-study effects by imputing studies in the funnel plot until symmetry is reached.	Method is discouraged to be used, because it falsely imputes studies if heterogeneity is present and is outperformed by other methods.	`trimfill()` in `metafor`
PET-PEESE	Estimate corrected for small-study effects is the intercept of regressing the effect size on either the standard error or sampling variance.	Method is discouraged to be applied in case of less than 10 studies and similar precisions of the studies.	Regression model fitted with `lm()` depending on whether true effect is zero
p-uniform and *p*-curve	Estimate equals the value where the *p*-value distribution of only the significant studies is uniform.	Methods recommended to be applied when heterogeneity is less than moderate.	`puniform()` in `puniform` for *p*-uniform
p-uniform*	Extension of *p*-uniform that does not discard nonsignificant studies and allows heterogeneous effects.	Method is discouraged to be applied if publication bias is extreme and there are only significant studies.	`puni_star()` in `puniform`
Weight-fun.	Corrected estimates obtained by estimating and incorporating weights of studies that reflect the extent of publication bias.	Method is discouraged to be applied if publication bias is extreme and there are only significant studies. Convergence problems may arise in case of a small number of studies.	`weightfunct()` in `weightr`

(continued)

Table 10.1 (continued)

	Description	Characteristics/ Recommendations	R function
Assessment of publication bias:			
Fail-safe *N*	Computes the number of studies that are needed to make the null-hypothesis of no meta-analytic effect nonsignificant.	Method is discouraged to be used due to, for example, the assumptions of no heterogeneity and missing studies having an effect of zero.	`fsn()` in `metafor`
Funnel plot asymmetry tests	Rank-correlation and Egger's regression test for small-study effects in a funnel plot.	Tests for small-study effects rather than publication bias. Methods are recommended to be applied with at least 10 studies in the meta-analysis.	`ranktest()` and `regtest()` in `metafor`
Test of excess significance (TES)	Tests whether more statistical significant studies are observed than expected based on their power.	Method is discouraged to be applied in case of heterogeneity and is known to be conservative.	`tes()` in `metafor`
Publication bias tests selection models	*p*-uniform and weight-function model test difference between models corrected and not corrected for publication bias.	*p*-uniform's test is conservative if true effect size is large. Properties of the test of the weight-function model are currently unknown.	`puniform()` in `puniform` and `weightfunct()` in `weightr`

Note: WAAP weighted average of the adequately powered studies, *PET* precision-effect test, *PEESE* precision-effect estimate with standard error, *Weight-fun.* weight-function model

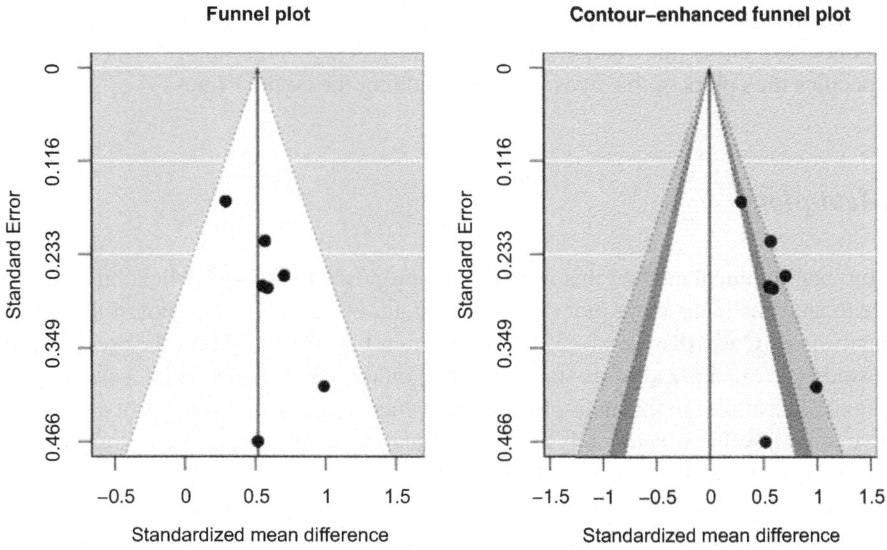

Fig. 10.1 Funnel plot (left panel) and contour-enhanced funnel plot (right panel) for the meta-analysis by Cowlishaw et al. (2012)

concluding that publication bias is present solely based on visually inspecting a funnel plot.

Another reason why meta-analysts should be cautious when drawing conclusions by inspecting funnel plots is that funnel plots can be misleading. Based on a large number of funnel plots, researchers correctly identified publication bias in only 52.5% of the funnel plots (Terrin et al., 2005). Moreover, changing the study's precision on the *y*-axis may also have a major impact on the on the shape of the funnel plot. The contour-enhanced funnel plot (Peters et al., 2008) was proposed to counteract the drawbacks of the funnel plot. The contour-enhanced funnel plot of the meta-analysis by Cowlishaw et al. (2012) is presented in Fig. 10.1 and modifies the funnel plot in two important ways. First, the contour-enhanced funnel plot is always centered at an effect size of zero, whereas the funnel plot is centered at the meta-analytic effect size estimate. Second, contour lines are added to the plot reflecting the *p*-values of studies. That is, studies in the white area of the contour-enhanced funnel plot have two-tailed *p*-values between 0.1 and 1, whereas studies in the dark gray, gray, and outside the funnel have two-tailed *p*-values in the intervals 0.05 and 0.1, 0.01 and 0.05, and 0 and 0.01, respectively. These contour lines help evaluating whether publication bias is the cause of funnel plot asymmetry, because they show whether statistically nonsignificant studies are missing in the meta-analysis. A contour-enhanced funnel plot can also be created using the `funnel()` function,

```
funnel(res, refline = 0, level = c(90, 95, 99),
       shade = c("white", "gray55", "gray75"))
```

where `refline = 0` is the center of the funnel, `level = c(90, 95, 99)` defines the contour lines, and `shade = c("white," "gray55," "gray75")` specifies the colors of the areas created by adding the contour lines.

Meta-plot

Another graphical method that was recently proposed to assess publication bias in a meta-analysis is the meta-plot (Van Assen et al., 2022). The meta-plot of the meta-analysis by Cowlishaw et al. (2012) is shown in Fig. 10.2. It shows the precision of a study (i.e., reciprocal of its standard error) on the *x*-axis and the effect size on the *y*-axis. The circles in the meta-plot are the average effect size estimates of a cumulative random-effects meta-analysis. In a cumulative meta-analysis (Lau et al., 1992), multiple meta-analyses are conducted where the first meta-analysis is based on a single study and in each subsequent meta-analysis a study is added. The order of the studies being added to the cumulative meta-analysis in the meta-plot is based on studies' precision. That is, the rightmost dot is the meta-analysis based on only the

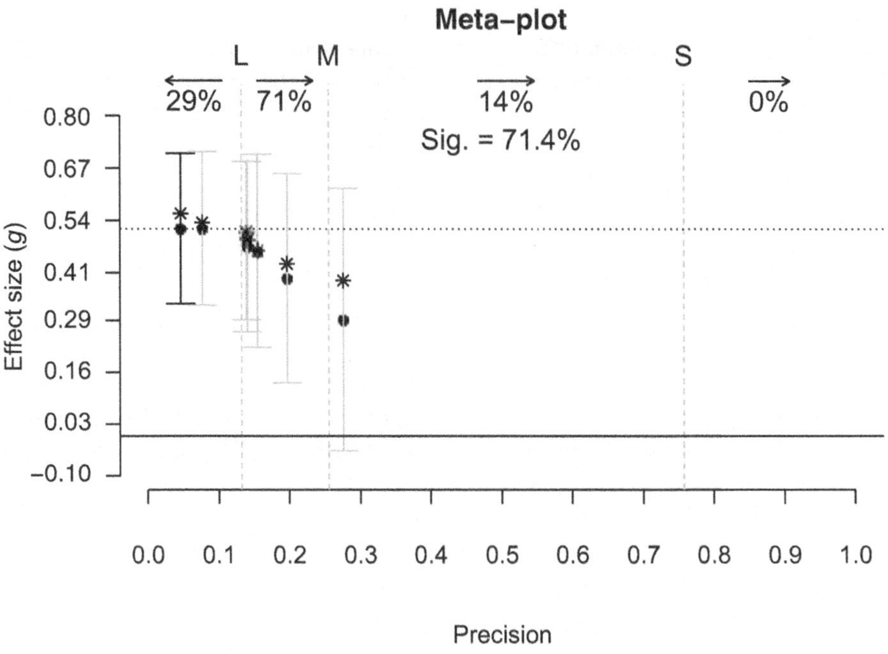

Fig. 10.2 Meta-plot of the meta-analysis by Cowlishaw et al. (2012)

study that is most precise and the leftmost dot is the meta-analysis based on all studies. Each dot is accompanied by its 95% CI. The meta-plot in Fig. 10.2 shows a decreasing trend in the cumulative meta-analysis from left to right. This is indicative for small-study effects, because the average effect size estimate of the meta-analysis based on all studies is larger than meta-analyses based on more precise studies. An advantage of the meta-plot over the funnel plot is that small-study effects are more visible as the effect size in the plot refers to the results of meta-analyses rather than individual studies.

The meta-plot also contains other relevant information for meta-analysts. First, it states the percentage of statistically significant results in the meta-analysis (71.4% in the meta-analysis of Cowlishaw et al. (2012)). Second, it shows information about the statistical power of the studies in the meta-analysis at the top of the plot. The leftmost percentage indicates the percentage of studies whose statistical power was insufficient (less than 80%) to detect a large population effect. The remaining three percentages at the top of the plot describe the percentages of studies with sufficient statistical power to detect a large (L), medium (M), and small (S) effect, respectively. Finally, the asterisks in the meta-plot refer to the expected estimates in the cumulative meta-analysis if the population effect size is zero in combination with extreme publication bias (i.e., only statistically significant studies get published). Asterisks that are larger than the dots imply that the results of the meta-analysis can also be explained by extreme publication bias in combination with no effect. This is the case for the meta-plot in Fig. 10.2, so authors are recommended to be cautious when interpreting the results of this meta-analysis.

Functions for creating the meta-plot are available in the R package `puniform` (Version 0.2.3; Van Aert, 2020). After installing and loading this package as described above, the meta-plot can be created using the code

```
meta_plot(m1i = m1i, m2i = m2i, n1i = n1i, n2i = n2i, sd1i = sd1i,
      sd2i = sd2i, pub_bias = TRUE)
```

where `m1i, n1i`, and `sd1i` are the study's mean, sample size, and standard deviation of patients receiving usual care and `m2i, n2i`, and `sd2i` are the study's mean, sample size, and standard deviation of patients receiving collaborative care.[3] Setting the argument `pub_bias` to `TRUE` makes sure that the asterisks are plotted.

The above introduced funnel plot and meta-plot enable to visually inspect whether small-study effects or publication bias are present in a meta-analysis. For an applied researcher, it is usually more of interest what the impact is of these biases on the results of a meta-analysis. In the next section, we will introduce methods that can be used for this purpose.

Methods to Estimate Effect Size in the Presence of Publication Bias

WAAP and Top 10%

Two intuitive approaches to estimate the effect size in the presence of publication bias are the weighted average of the adequately powered (WAAP) studies (Ioannidis et al., 2017) and the Top 10% approach (Stanley et al., 2010). Both approaches rest on the idea that the effect sizes of the most precise studies (i.e., studies with the largest sample size) in a meta-analysis are less overestimated due to publication bias. Less precise studies are more vulnerable to publication bias, because overestimation of effect size needs to be larger in these studies in order to be statistically significant. The WAAP uses this idea by meta-analyzing only the studies whose statistical power to reject the null hypothesis of no effect is larger than 80%.[4] The Top 10% does not take statistical power into account, but meta-analyzes only the 10% most precise studies. Others have argued to not focus on the 10% most precise studies but interpret the study with the largest precision as the best effect size estimate if publication bias is present (Ioannidis, 2013). Although, the intuition of these approaches

[3] The study's mean, sample size, and standard deviation of both groups are available on page 73 of Cowlishaw et al. (2012).

[4] Statistical power of the studies is computed using the estimate of the fixed-effect model as proxy for the true effect size and a two-tailed hypothesis with significance level 0.05 (Stanley et al., 2017).

is appealing, they should only be used if there is no heterogeneity in the meta-analysis. Drawing conclusions based on only a subset of studies is ill-advised in case of heterogeneity, because the true effect size of studies is different, and a subset of studies is not a good representation of all studies in the meta-analysis.

Trim-and-fill

The trim-and-fill method (Duval & Tweedie, 2000a, 2000b) is the most often used method to correct effect size for publication bias. The trim-and-fill method is an iterative procedure that *trims* the most extreme effect sizes from the right hand side of the funnel plot and *fills* these in the funnel plot until it is symmetric. The meta-analytic estimate corrected for bias is the estimate based on the observed studies as well as the imputed studies. The left panel of Fig. 10.3 visually shows the procedure for the meta-analysis of Cowlishaw et al. (2012) where the solid and open circles are the observed and filled studies, respectively.

Multiple researchers have criticized the trim-and-fill method and discourage meta-analysts to use the method. A prevalent issue with the trim-and-fill method is that it is based on the funnel plot and therefore actually corrects for small-study effects rather than publication bias. Simulation studies have confirmed that the trim-and-fill method yields misleading results if heterogeneity is present in a meta-analysis (Peters et al., 2007; Terrin et al., 2003). Moreover, the trim-and-fill method

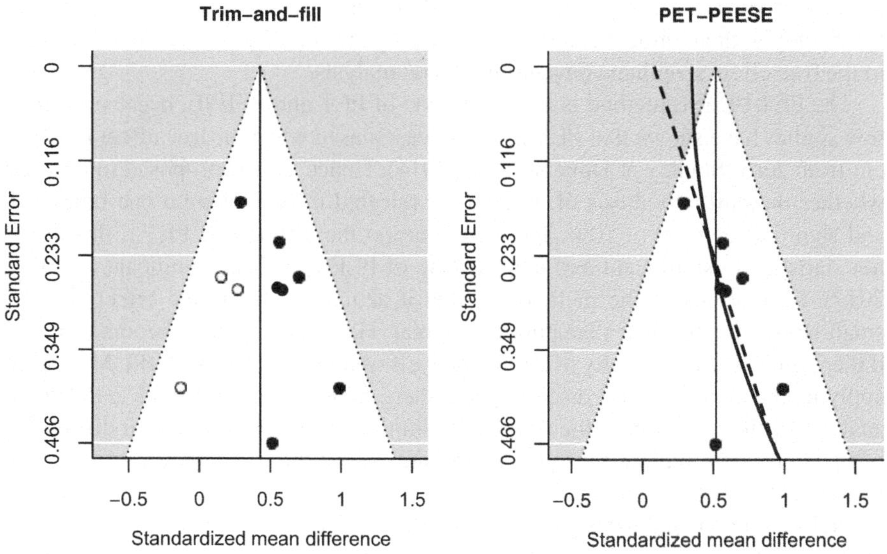

Fig. 10.3 Illustration of the trim-and-fill method (left panel) and the PET-PEESE method (right panel) when applied to the meta-analysis by Cowlishaw et al. (2012). The dashed line in the right panel refers to the PET analysis and the solid line to the PEESE analysis

is outperformed by other methods (e.g., Van Assen et al., 2015; Moreno et al., 2009; Simonsohn et al., 2014) that will be discussed next making it a method that should better be avoided. Nevertheless, if researchers want to report the results of the trim-and-fill method in their meta-analysis, they can apply the method using this line of code

```
trimfill(res)
```

PET-PEESE

Another method that uses the relationship between studies' effect size and precision is the PET-PEESE method (Moreno et al., 2009; Stanley & Doucouliagos, 2014). PET-PEESE is a combination of two distinct methods: the *precision-effect test* (PET) and the *precision-effect estimate with standard error* (PEESE). The rationale of this method can best be explained by the funnel plot based on the meta-analysis by Cowlishaw et al. (2012) in the right panel of Fig. 10.3. PET and PEESE both fit a regression line through the points in the funnel plot. The lines of PET (dashed line) and PEESE (solid line) in Fig. 10.3 are based on a linear regression with the study's standard error and sampling variance as predictor, respectively. The effect size estimates of PET and PEESE are the values where the slope of the regression line is 0 (i.e., the estimate of the intercept). This occurs in the right panel of Fig. 10.3 at the top of the funnel plot where the lines end, because this is the point where the standard error equals zero. The rationale of both methods is that these estimates resemble a study with an infinite sample size, and they are therefore expected to be closer to the true effect size than conventional meta-analysis.

The PET-PEESE method is a combination of PET and PEESE, because simulation studies have shown that PEESE is the least biased when the true effect is different from zero (Stanley & Doucouliagos, 2014). Hence, it was proposed to first test whether the null-hypothesis of no effect is rejected in PET using a one-tailed test and significance level of 10%, and then interpret the estimate of PET if this test is not statistically significant and the estimate of PEESE if it is significant (Stanley, 2017). Limitations of the method are that it actually corrects the effect size for small-study effects rather than publication bias. Hence, the method becomes biased if there is large heterogeneity in a meta-analysis (Alinaghi & Reed, 2018). Moreover, applying the method is also discouraged if there are less than 10 studies in the meta-analysis or the precision of the studies are similar, because this makes it difficult to fit the regression lines and results in an imprecise estimate (Niemeyer et al., 2020; Stanley et al., 2017; Stanley, 2017).

PET can be applied using the following line of code

```
lm(yi ~ I(sqrt(vi)), weights = 1/vi)
```

where `yi ~ I(sqrt(vi))` specifies that the effect size is regressed on the standard error and `weights = 1/vi` make sure that studies in the analysis are weighted by the reciprocal of their sampling variance. If the null hypothesis of no effect is statistically significant in PET, PEESE can be applied using the code

```
lm(yi ~ vi, weights = 1/vi)
```

Selection Model Approaches

Selection model approaches are nowadays seen as the state-of-the-art methods to correct for publication bias in a meta-analysis (McShane et al., 2016). These selection model approaches assign weights to studies to take into account that some studies are less likely to be published than others. For example, statistical nonsignificant studies will most likely receive a larger weight than significant studies to compensate for nonsignificant studies being less likely to be published. These weights are then taken into account when meta-analyzing studies using a conventional meta-analysis model such as the random-effects model that does not correct for publication bias.

Selection model approaches were known to suffer from convergence problems if less than 100 studies are included in a meta-analysis (e.g., Borenstein et al., 2009; Terrin et al., 2003). However, these convergence problems were less of an issue in recent studies (Carter et al., 2019; McShane et al., 2016; Van Aert & Van Assen, 2022), which was probably caused by the development of new selection model approaches in combination with improved software implementation. Many different selection model approaches exist (for an overview see Marks-Anglin and Chen (2020), Jin et al. (2014), and the supplements of Van Aert and Van Assen et al. (2022)) that mainly differ on how the weights of the studies are computed. We will focus in this book chapter on three selection model approaches that do not require the meta-analyst to make sophisticated choices and are therefore easy to implement, have shown to outperform the existing methods that were introduced above, or are regularly used in practice.

P-*uniform and* p-*curve*

P-uniform (Van Assen et al., 2015) and *p*-curve (Simonsohn et al., 2014) are two methods based on the same methodology that slightly differ in how they are implemented (for a description of the differences see Van Aert et al., 2016). Both methods correct for publication bias in a meta-analysis by only focusing on the statistically significant studies and discarding the nonsignificant studies. The methods use the distribution of statistically significant *p*-values for effect size estimation. The estimate of both methods equals zero if this *p*-value distribution is uniformly distributed under the null-hypothesis. A *p*-value distribution with small *p*-values being overrepresented is indicative for an effect larger than zero, whereas a distribution with an overrepresentation of *p*-values close to the significance level is evidence for an effect smaller than zero. The effect size estimate of *p*-uniform and *p*-curve is obtained by means of an iterative procedure to find the effect size where the *p*-values are uniformly distributed. The methods assume that each statistically significant study is equally likely to be published (i.e., the same weight for each study).

P-uniform and *p*-curve have shown to yield accurate estimates in the presence of publication bias and homogeneous true effect size and outperformed the trim-and-fill method (Simonsohn et al., 2014; Van Assen et al., 2015). However, the methods overestimate effect size if a meta-analysis is heterogeneous (Carter et al., 2019; McShane et al., 2016; Van Aert et al., 2016). For that reason, Van Aert et al. (2016) recommended to only interpret the effect size estimate of both methods as the estimate of the population effect if the true effect sizes are homogeneous or if heterogeneity is less than moderate.[5] Another limitation of the methods is that effect size estimates may become unrealistically low in case of *p*-uniform or peculiar in case of *p*-curve if a preponderance of studies has *p*-values just under the significance level (Van Aert et al., 2016). This may be caused by researchers having used questionable research practices (a.k.a. *p*-hacking or researcher degrees of freedom, Simmons et al., 2011; Wicherts et al., 2016) in the studies to get *p*-values below the threshold of statistical significance.

We only show how *p*-uniform can be applied, because there is no R package that contains functions for applying *p*-curve and, in contrast to *p*-curve, a publication bias test and 95% CIs have been developed for *p*-uniform. *P*-uniform can be applied by using the puniform() function in the puniform package,

```
puniform(yi = yi, vi = vi, side = "right")
```

where `side = "right"` specifies that the method should be applied to the studies that are statistically significant based on a right-tailed test. Specifying `side = "left"` allows applying *p*-uniform to studies that are based on a left-tailed test.

[5]Moderate heterogeneity is defined in terms of the I^2-statistic that is commonly used in meta-analysis to quantify the heterogeneity. The I^2-statistic (Higgins & Thompson, 2002) indicates the proportion of total variance that can be attributed to heterogeneity in true effect size. Moderate heterogeneity is $I^2 = 0.5$ according to the rules-of-thumb proposed in Higgins et al. (2003).

P-*uniform**

The *p*-uniform* method (Van Aert & Van Assen, 2022) is an extension of *p*-uniform that solves the problem of overestimation of effect size if there is heterogeneity in a meta-analysis. Furthermore, it also enables, in contrast to *p*-uniform and also *p*-curve, estimation of heterogeneity and testing the null-hypothesis of no heterogeneity. *P*-uniform* is based on the same rationale as *p*-uniform and *p*-curve, but also includes statistically nonsignificant studies. That is, the method implicitly assigns different weights to statistically significant and nonsignificant studies by taking into account the likelihood of a study getting published given its statistical (non)significance (for technical details see Van Aert & Van Assen, 2022). An important assumption of *p*-uniform* is that all statistically significant studies are assumed to be equally likely published and the same holds for all statistically nonsignificant studies. This implies that studies with statistically nonsignificant *p*-values of, for instance, 0.1 and 0.9 are assumed to be published with the same probability, but that this probability might differ for a study with a statistically significant *p*-value of 0.04.[6]

A recent simulation study (Van Aert & Van Assen, 2022) has shown that *p*-uniform* is indeed an improvement over *p*-uniform if heterogeneity is present and both statistically significant and nonsignificant studies are included in a meta-analysis. Researchers should, however, be cautious when interpreting the results of *p*-uniform* when publication bias is expected to be extreme in combination with only statistically significant studies in a meta-analysis. *P*-uniform*'s performance was not good in this condition and was outperformed by *p*-uniform if there was no heterogeneity. *P*-uniform* might also yield a very negative effect size estimate if many studies with *p*-values just below the significance threshold are included, but this was less of a problem than with *p*-uniform due to the inclusion of also statistically nonsignificant studies in *p*-uniform*.

P-uniform* can be applied by using the `puni_star()` function included in the `puniform` package,

```
puni_star(yi = yi, vi = vi, side = "right")
```

[6] Research is currently ongoing to study whether this assumption can be relaxed by not only weighing statistically significant and nonsignificant studies differently in *p*-uniform* but also allow more complex weighting schemes. For example, marginally significant studies (i.e., studies with *p*-values just above the significance threshold) may have a different probability of being published than other nonsignificant studies. Weighing these studies differently may improve estimation and drawing inferences.

Weight-function Model

The weight-function model (Hedges, 1992; Vevea et al., 1993) also enables estimation of the average effect size as well as between-study variance in a meta-analysis. The method creates intervals based on *p*-values, and then estimates the weights for the studies with *p*-values belonging to these intervals. Studies in the same interval get the same weight in the weight-function model. The intervals have to be specified by the meta-analyst and a reasonable choice is to create two intervals such that statistically significant and nonsignificant studies are treated differently. This model with two intervals is sometimes also referred to as the three-parameter selection model, because three parameters are estimated: the average effect size, between-study variance in true effect size (i.e., heterogeneity), and the relative weight specifying how much less likely a statistically nonsignificant study is published compared to a significant study.

The weight-function model outperformed the trim-and-fill method, *p*-uniform, and *p*-curve in simulation studies (Carter et al., 2019; McShane et al., 2016). A recent study (Van Aert & Van Assen, 2022) comparing the weight-function model to *p*-uniform* revealed that the performance of both methods was comparable. Performance of the weight-function model was, just as of *p*-uniform*, not good in case of extreme publication bias in combination with only statistically significant studies in a meta-analysis, so the method is not recommended to be applied in meta-analyses with these characteristics. The weight-function model requires, in contrast to *p*-uniform, *p*-curve, and *p*-uniform*, estimation of the weights of the studies. This may cause convergence problems if a small number of studies is included in some of the intervals. Furthermore, Hedges and Vevea (1996) showed that estimation of the weights is often inaccurate, but that this hardly affected estimation of the average effect size and heterogeneity.

The weight-function model can be applied by using the `weightfunct()` function in the `weightr` package (Version 2.0.2, Coburn & Vevea, 2016),

```
weightfunct(effect = yi, v = vi)
```

where the study's effect sizes and corresponding sampling variances can be supplied using the arguments `effect` and `v`, respectively.

Assessment of Publication Bias

We focused in the previous section on methods to correct for bias in the meta-analysis. Meta-analysts might, however, also want to quantify whether publication bias is likely present in their meta-analysis or test whether the hypothesis of no publication bias is rejected. We discuss methods for these purposes in this section.

Fail-safe N

The most popular method to study the impact of publication bias in a meta-analysis is the fail-safe *N* method (Rosenthal, 1979). This method quantifies how many studies with an effect size of zero need to be added to a meta-analysis such that the meta-analytic effect size changes from being statistically significant to nonsignificant. Publication bias is unlikely if the fail-safe *N* is large, because many studies with an effect size of zero are then needed to no longer reject the null-hypothesis of no effect in the meta-analysis.

The fail-safe *N* method has been heavily criticized (e.g., Becker, 2005; Iyengar & Greenhouse, 1988; Orwin, 1983; Scargle, 2000; Schonemann & Scargle, 2008) for multiple reasons. First, the method does not take the sample size of studies into account by treating all studies as if they are equally precise. Second, there is no clear criterion defining what a large fail-safe *N* is. Third, only studies with an effect size of zero are assumed to be missing.

For this reason, Orwin (1983) extended the fail-safe *N* method by allowing meta-analysts to specify an average effect size of the missing studies that may differ from zero, and allowing computing the number of studies needed to get a meta-analytic estimate smaller than a user-specified effect size. A drawback of the originally proposed fail-safe *N* method as well as Orwin's extension is that heterogeneity in the meta-analysis is not taken into account, because all missing studies are assumed to have a common effect size. Due to these limitations, the fail-safe *N* method and Orwin's extension are discouraged to be used (Becker, 2005; Jin et al., 2014; Vevea & Woods, 2005), and meta-analysts are referred to other methods that will be discussed next. Nevertheless, the fail-safe *N* can be computed using

```
fsn(yi = yi, vi = vi)
```

Funnel Plot Asymmetry Tests

The funnel plot introduced earlier can be used to examine visually whether small-study effects are present in a meta-analysis. However, eyeballing a funnel plot to assess small-study effects is known to be difficult (Terrin et al., 2005). Hence, hypothesis tests were developed to test whether a funnel plot is asymmetric and thus small-study effects are present in a meta-analysis. The rank-correlation test (Begg & Mazumdar, 1994) tests whether the Kendall's rank correlation between the studies' effect sizes and sampling variances differs from zero after first stabilizing the sampling variances by standardizing the effect sizes (for technical details see Begg & Mazumdar, 1994). A positive correlation implies that large effect sizes go along with large sampling variances and is indicative for small-study effects.

Another funnel plot asymmetry test is Egger's regression test (Egger et al., 1997) that actually formed the basis of the PET-PEESE method to correct effect size estimates. In Egger's regression test, the slope of the regression line fitted by applying PET is tested for statistical significance, and evidence for small-study effects is observed if this slope is significantly larger than zero. Egger's regression test has been modified in various ways where especially other predictors than the studies' standard error are used as predictor (for an overview see Jin et al., 2014).

Simulation studies have shown that statistical power of Egger's regression test is generally larger than of the rank-correlation test (Sterne et al., 2000). However, statistical power of both methods is low when a small number of studies are included in the meta-analysis (Deeks et al., 2005; Macaskill et al., 2001). Hence, both methods are recommended to be only applied if a meta-analysis contains more than ten studies (Sterne et al., 2011), and a significance level of 0.1 is recommended to be used for hypothesis testing (Egger et al., 1997). Another limitation of funnel plot asymmetry tests is that these, just as the funnel plot itself and other methods based on the funnel plot, test whether small-study effects are present and not explicitly test for publication bias.

The rank-correlation test can be applied using the following code

```
ranktest(res)
```

Egger's regression test is incorporated in the PET analysis when testing whether the slop coefficient is statistically significant and can also be obtained using the code

```
regtest(res)
```

Test of Excess Significance

The test of excess significance (TES, Ioannidis & Trikalinos, 2007) tests whether more studies in a meta-analysis are statistically significant than expected. The expected number of statistically significant studies is obtained by taking the sum of each study's statistical power given that the meta-analytic effect size estimate is the true effect size. A hypothesis test (e.g., an exact, binomial, or Pearson's χ^2-test) can subsequently be used to test whether the observed number of statistically significant studies is larger than expected.

A problem with the TES is that the expected number of statistically significant studies is based on the meta-analytic effect size estimate that is likely to be overestimated if publication bias is present. Consequently, the statistical power of the studies and, in turn, also the expected number of statistically significant studies will be overestimated. This has also been observed in simulation studies where the TES was conservative (Francis, 2013; Van Assen et al., 2015; Vandekerckhove et al., 2013). Hence, it is recommended to apply the TES using 0.1 as significance level (Ioannidis & Trikalinos, 2007). It is important to emphasize that publication bias is not the only cause of an excess of significant studies. Another reason is considerable heterogeneity, and the TES is therefore advised to be not applied when this is present in a meta-analysis (Ioannidis & Trikalinos, 2007).

The TES can be applied using the code

```
tes(res)
```

Publication Bias Tests Based on Selection Model Approaches

The selection model approaches *p*-uniform and the weight-function model also implemented publication bias tests. In these methods, the estimated model that corrects for publication bias is compared with the conventional meta-analysis model that does not correct for bias. A statistically significant difference between these two models indicates that a selection model approach better fits the data, and that publication bias might be present.

Simulation studies have shown that *p*-uniform's publication bias test is conservative if the true effect size is large, and that statistical power of *p*-uniform's test was generally higher than of TES except for meta-analyses with a large true effect and more than 30 studies in the meta-analysis (Renkewitz & Keiner, 2019; Van Assen et al., 2015). The properties of the publication bias test of the weight-function model are unknown and are therefore topic for future research. These publication bias tests are reported in the output of *p*-uniform and the weight-function model that can be obtained by applying these methods as described in the section on correcting effect size estimation corrected for bias.

Applying Methods to Examples

We apply the described methods to the earlier introduced meta-analyses of Cowlishaw et al. (2012) and Archer et al. (2012). Annotated R code of all analyses is available at to facilitate the application of these methods.

Example 1: Cowlishaw et al. (2012)

Table 10.2 shows the earlier described results of applying the random-effects meta-analysis to the data of Cowlishaw et al. (2012), and the results of the methods that correct for bias. This meta-analysis only contains seven studies and is therefore typical for meta-analyses in clinical psychology. All methods that estimate the between-study variance in true effects estimate it as zero and testing the null-hypothesis of homogeneity is for none of the methods statistically significant.

Hence, the results of the methods that require homogeneous true effect size in the meta-analysis (WAAP, Top 10%, trim-and-fill, and *p*-uniform) can also be safely interpreted. Note that some results regarding estimation and testing the between-study variance are missing in Table 10.2 and denoted by "-," because these results could not be computed or are not reported by the methods.

The average effect size estimate of all methods was closer to zero than of the random-effects model. The smallest correction was by trim-and-fill that imputed three missing studies and the largest correction was by PET-PEESE that yielded an estimate close to zero. The results of WAAP and Top 10% have to be interpreted with caution, because estimates of these methods were only based on the most precise study in the meta-analysis. For this reason, the between-study variance in true effect size could also not be estimated for these methods. Only trim-and-fill, *p*-uniform*, and the weight-function model rejected the null-hypothesis of no effect and corroborated the hypothesis test of the random-effects model. Table 10.3 shows in the first column the results of the tests for small-study effects and publication bias. No method rejected the null hypothesis of no bias in this meta-analysis.

Table 10.2 Results of applying random-effects meta-analysis and methods to correct for bias to the meta-analysis by Cowlishaw et al. (2012)

Overall mean					Between-study variance		
	k	Estimate (SE)	(95% CI)	Test of no effect	Estimate (SE)	(95% CI)	Test of homogeneity
RE	7	0.519 (0.096)	(0.332;0.706)	$z = 5.432$, $p < .001$	0 (0.035)	(0;0.125)	$Q = 3.897$, $p = .691$
WAAP	1	0.291 (0.168)	(−0.037;0.619)	$z = 1.737$, $p = .082$	–[b]	–[b]	$Q = 0, p = 1$
Top 10%	1	0.291 (0.168)	(−0.037;0.619)	$z = 1.737$, $p = .082$	–[b]	–[b]	$Q = 0, p = 1$
Trim-and-fill	10	0.430 (0.083)	(0.267;0.593)	$z = 5.162$, $p < .001$	0 (0.030)	(0;0.202)	$Q = 8.204$, $p = .514$
PET-PEESE[a]	7	0.084 (0.195)	(−0.418;0.586)	$t = 0.430$, $p = .685$	–[c]	–[c]	–[c]
p-uniform	5	0.218 (–)	(−0.787;0.656)	$L_0 = -0.672$, $p = .251$	–[c]	–[c]	–[c]
p-uniform*	7	0.394 (–)	(0.059;0.721)	$L_0 = 5.414$, $p = .020$	0 (–)	(0;0.064)	$L_{het} = 0, p = 1$
Weight-fun.	7	0.328 (0.156)	(0.022;0.634)	$z = 2.100$, $p = .036$	0 (–[b])	–[b]	–[c]

Note: For the random-effects model and Trim-and-fill, between-study variance is estimated with the restricted maximum likelihood estimator (Raudenbush, 2009) and corresponding confidence intervals are created using the *Q*-profile method (Viechtbauer, 2007). *RE* random-effects model, *WAAP* weighted average of the adequately powered studies, *PET* precision-effect test, *PEESE* precision-effect estimate with standard error, *Weight-fun.* weight-function model, *k* number of studies in the analysis, *SE* standard error, *CI* confidence interval

[a]Results of PET analysis

–[b]Could not be computed by the method

–[c]Estimation or testing of the between-study variance is not included by the method

However, this may be caused by the small number of studies resulting in low statistical power of these tests. To conclude, correcting for bias yielded estimates closer to zero of all methods, and the null hypothesis of no effect was not rejected by some methods. Although the tests for bias were not statistically significant, we argue that the evidence for CBT resulting in less financial loss of patients is weak at best.

Example 2: Archer et al. (2012)

Table 10.4 shows the results of effect size estimation and drawing inferences for the meta-analysis by Archer et al. (2012). This meta-analysis is typical for clinical psychology, because there is a large amount of heterogeneity in the meta-analysis. All methods estimated the between-study variance as positive and rejected the null-hypothesis of homogeneity. Hence, interpreting the results of the methods that do not perform well if large heterogeneity is present should best be avoided (WAAP, Top 10%, trim-and-fill, PET-PEESE, and *p*-uniform) and are only reported for completeness. The methods that allow large heterogeneity (*p*-uniform* and the weight-function model) estimated a lower average effect size than the random-effects model that was statistically significant. Estimates of the between-study variance were similar of the random-effects model and *p*-uniform* and the weight-function model. The rank-correlation test, Egger's test, and the publication bias test of the weight-function model were statistically significant (second column of Table 10.3). This suggests that small-study effects or publication bias were present and might be the cause of the large effect size of the random-effects model compared to the other methods. To conclude, there is evidence for bias in the meta-analysis by Archer et al. (2012), because tests for small-study effects and publication bias were statistically significant and the corrected average effect size for bias was smaller than the one of the random-effects 611 meta-analysis. However, the effect was larger than zero after correcting for bias, so collaborative care appeared to be beneficial for patients with depression or anxiety problems.

Table 10.3 Results of applying tests for small-study effects and publication bias to the meta-analyses of Cowlishaw et al. (2012) and Archer et al. (2012)

	Cowlishaw et al. (2012)	Archer et al. (2012)
Fail-safe *N*	$N = 75$	$N = 1216$
Rank-cor. test	$\tau = 0.238, p = 0.562$	$\tau = 0.391, p = 0.007$
Egger's test	$z = 1.426, p = 0.154$	$z = 3.17, p = 0.002$
TES[a]	Exact $p = 0.192$	$\chi^2 = 1.545, p = 0.107$
p-uniform	$L_{pb} = 1.284, p = 0.100$	$L_{pb} = -0.552, p = 0.709$
Weight-fun.	$\chi^2 = 3.292, p = 0.070$	$\chi^2 = 4.687, p = 0.030$

Note: [a]The default implementation of the Test of Excess Significance (TES) in the tes() function was used. Using this implementation an exact test was conducted for the meta-analysis by Cowlishaw et al. (2012) and a Pearson's χ^2-test for the meta-analysis by Archer et al. (2012)

Table 10.4 Results of applying random-effects meta-analysis and methods to correct for bias to the meta-analysis by Archer et al. (2012)

			Overall mean			Between-study variance	
	k	Estimate (SE)	(95% CI)	Test of no effect	Estimate (SE)	(95% CI)	Test of homogeneity
RE	24	0.240 (0.038)	(0.166;0.314)	$z = 6.347$, $p < 0.001$	0.021 (0.010)	(0.009;0.070)	$Q = 83.580$, $p < 0.001$
WAAP	6	0.155 (0.068)	(0.022;0.289)	$z = 2.285$, $p = 0.022$	0.024 (0.018)	(0.008;0.164)	$Q = 47.556$, $p < 0.001$
Top 10%	2	0.287 (0.135)	(0.023;0.550)	$z = 2.131$, $p = 0.033$	0.034 (0.051)	(0.005;36.862)	$Q = 16.416$, $p < 0.001$
Trim-and-fill	27	0.210 (0.041)	(0.130;0.290)	$z = 5.136$, $p < 0.001$	0.029 (0.012)	(0.016;0.105)	$Q = 100.725$, $p < 0.001$
PET-PEESE[a]	24	0.160 (0.042)	(0.073;0.247)	$t = 3.835$, $p = 0.001$	–[c]	–[c]	–[c]
p-uniform	16	0.240 (–)	(0.154;0.374)	$L_0 = -4.309$, $p < 0.001$	–[c]	–[c]	–[c]
p-uniform*	24	0.175 (–)	(0.067;0.280)	$L_0 = 9.913$, $p = 0.002$	0.015 (–)	(0.005;0.040)	$L_{het} = 29.502$, $p < 0.001$
Weight-fun.	24	0.148 (0.057)	(0.036;0.261)	$z = 2.593$, $p = 0.010$	0.017 (0.009)	(0;0.035)	–[b]

Note: Estimates and confidence intervals are log-transformed risk ratios. For the random-effects model, WAAP, Top 10%, and Trim-and-fill, between-study variance is estimated with the restricted maximum likelihood estimator (Raudenbush, 2009) and corresponding confidence intervals are created using the *Q*-profile method (Viechtbauer, 2007). *RE* random-effects model, *WAAP* weighted average of the adequately powered studies, *PET* precision-effect test, *PEESE* precision-effect estimate with standard error, *Weight-fun.* weight-function model, *k* number of studies in the analysis, *SE* standard error, *CI* confidence interval

[a]Results of PEESE analysis

[b]Could not be computed by the method

[c]Estimation or testing of the between-study variance is not included by the method

Summary

It is of utmost importance to address publication bias in every meta-analysis, which has also been advised by MARS (Appelbaum et al., 2018), PRISMA (Moher et al., 2009), and the Cochrane Collaboration (Page et al., 2019). We believe that publication bias should also be routinely assessed when developing and revising evidence-based clinical guidelines, such as the NICE guidelines in the UK or the AWMF guidelines in Germany, and when identifying empirically supported treatments (ESTs) by the American Psychological Association's (APA) Division 12 (Tolin et al., 2015). In this chapter, we have described methods that can be applied for this purpose and summarized recommendations on when to apply each method (see Table 10.1).

Clinical psychologists who conduct a meta-analysis often encounter difficulties when addressing publication bias, because meta-analyses in clinical psychology are usually heterogeneous and contain a small number of studies, which are unfavorable conditions for the vast majority of publication bias methods (Niemeyer et al., 2020). However, recent research has shown that selection model approaches perform reasonably well when the number of studies in the meta-analysis is at least ten (Van Aert et al., 2019). Despite the promising results of selection model approaches, it is important that meta-analysts apply multiple publication bias methods in a so-called triangulation approach (Coburn & Vevea, 2015; Kepes et al., 2012), because there is no publication bias method that outperformed all other methods in all conditions (Carter et al., 2019; Renkewitz & Keiner, 2019). Such a triangulation approach should be preceded by a performance check to assess which methods perform well for the characteristics of the meta-analysis under study (Carter et al., 2019; Niemeyer et al., 2020). A performance check can be conducted by scrutinizing the literature 636 on publication bias methods or assessing the performance of publication bias methods in a simulation study that resembles the characteristics of the meta-analysis as closely as possible.

We hope that this chapter helps clinical psychologists to apply state-of-the-art publication bias methods in their meta-analyses. Application of these publication bias methods has high potential for yielding relevant scientific insights, and will benefit policy-making and treatment of patients that is commonly based on the conclusions of meta-analyses.

Author Note We would like to thank Claudia Kapp, Manuel Heinrich, and Johannes Heekerens for commenting on a previous version of this chapter.

The authors made the following contributions. Robbie C.M. van Aert: Conceptualization, Formal analysis, Writing—Original Draft Preparation, Writing—Review & Editing; Helen Niemeyer: Conceptualization, Writing—Review & Editing.

References

Aguinis, H., Dalton, D. R., Bosco, F. A., Pierce, C. A., & Dalton, C. M. (2010). Meta-analytic choices and judgment calls: Implications for theory building and testing, obtained effect sizes, and scholarly impact. *Journal of Management, 37*(1), 5–38. https://doi.org/10.1177/0149206310377113

Aguinis, H., Gottfredson, R. K., & Wright, T. A. (2011). Best-practice recommendations for estimating interaction effects using meta-analysis. *Journal of Organizational Behavior, 32*(8), 1033–1043. https://doi.org/10.1002/job.719

Alinaghi, N., & Reed, W. R. (2018). Meta-analysis and publication bias: How well does the FAT-PET-PEESE procedure work? *Research Synthesis Methods, 9*(2), 285–311. https://doi.org/10.1002/jrsm.1298

Alonso, J., Angermeyer, M. C., Bernert, S., Bruffaerts, R., Brugha, T. S., Bryson, H., Girolamo, G., Graaf, R., Demyttenaere, K., Gasquet, I., Haro, J. M., Katz, S. J., Kessler, R. C., Kovess, V., Lépine, J. P., Ormel, J., Polidori, G., Russo, L. J., Vilagut, G., … Vollebergh, W. A. (2004). Prevalence of mental disorders in Europe: Results from the European Study of the Epidemiology of Mental Disorders (ESEMeD) project. *Acta Psychiatrica Scandinavica. Supplementum, 420*, 21–27. https://doi.org/10.1111/j.1600-0047.2004.00327.x

Appelbaum, M., Cooper, H., Kline, R. B., Mayo-Wilson, E., Nezu, A. M., & Rao, S. M. (2018). Journal article reporting standards for quantitative research in psychology: The APA Publications and Communications Board task force report. *The American Psychologist, 73*(1), 3–25. https://doi.org/10.1037/amp0000191

Archer, J., Bower, P., Gilbody, S., Lovell, K., Richards, D., Gask, L., Dickens, C., & Coventry, P. (2012). Collaborative care for depression and anxiety problems. *Cochrane Database of Systematic Reviews, 10*. https://doi.org/10.1002/14651858.CD006525.pub2

Aytug, Z. G., Rothstein, H. R., Zhou, W., & Kern, M. C. (2012). Revealed or concealed? Transparency of procedures, decisions, and judgment calls in meta-analyses. *Organizational Research Methods, 15*(1), 103–133. https://doi.org/10.1177/1094428111403495

Bakker, M., Van Dijk, A., & Wicherts, J. M. (2012). The rules of the game called psychological science. *Perspectives on Psychological Science, 7*(6), 543–554. https://doi.org/10.1177/1745691612459060

Banks, G. C., Kepes, S., & Banks, K. P. (2012). Publication bias: The antagonist of meta-analytic reviews and effective policymaking. *Educational Evaluation and Policy Analysis, 34*(3), 259–277. https://doi.org/10.3102/0162373712446144

Becker, B. J. (2005). Failsafe N or file-drawer number. In H. R. Rothstein, A. J. Sutton, & M. Borenstein (Eds.), *Publication bias in meta-analysis: Prevention, assessment and adjustments* (pp. 111–125). Wiley.

Begg, C. B., & Mazumdar, M. (1994). Operating characteristics of a rank correlation test for publication bias. *Biometrics, 50*(4), 1088–1101.

Borenstein, M., Hedges, L. V., Higgins, J. P. T., & Rothstein, H. R. (2009). *Introduction to meta-analysis*. Wiley.

Borenstein, M., Hedges, L. V., Higgins, J. P. T., & Rothstein, H. R. (2010). A basic introduction to fixed-effect and random-effects models for meta-analysis. *Research Synthesis Methods, 1*(2), 97–111. https://doi.org/10.1002/jrsm.12

Carter, E. C., Schönbrodt, F. D., Gervais, W. M., & Hilgard, J. (2019). Correcting for bias in psychology: A comparison of meta-analytic methods. *Advances in Methods and Practices in Psychological Science, 2*(2), 115–144. https://doi.org/10.1177/2515245919847196

Coburn, K. M., & Vevea, J. L. (2015). Publication bias as a function of study characteristics. *Psychological Methods, 20*(3), 310–330. https://doi.org/10.1037/met0000046

Coburn, K. M., & Vevea, J. L. (2016). *weightr: Estimating weight-function models for publication bias*. https://cran.r-project.org/package=weightr

Cochran, W. G. (1954). The combination of estimates from different experiments. *Biometrics, 10*(1), 101–129.

Cohen, J. (1988). *Statistical power analysis for the behavioral sciences* (2nd ed.).

Cooper, H., DeNeve, K., & Charlton, K. (1997). Finding the missing science: The fate of studies submitted for review by a human subjects committee. *Psychological Methods, 2*(4), 447–452. https://doi.org/10.1037/1082-989X.2.4.447

Coursol, A., & Wagner, E. E. (1986). Effect of positive findings on submission and acceptance rates: A note on meta-analysis bias. *Professional Psychology: Research and Practice, 17*(2), 136–137. https://doi.org/10.1037/0735-7028.17.2.136

Cowlishaw, S., Merkouris, S., Dowling, N., Anderson, C., Jackson, A., & Thomas, S. (2012). Psychological therapies for pathological and problem gambling. *Cochrane Database of Systematic Reviews, 11*. https://doi.org/10.1002/14651858.CD008937.pub2

David, D., Cotet, C., Matu, S., Mogoase, C., & Stefan, S. (2018). 50 years of rational-emotive and cognitive-behavioral therapy: A systematic review and meta-analysis. *Journal of Clinical Psychology, 74*(3), 304–318. https://doi.org/10.1002/jclp.22514

Deeks, J. J., Macaskill, P., & Irwig, L. (2005). The performance of tests of publication bias and other sample size effects in systematic reviews of diagnostic test accuracy was assessed. *Journal of Clinical Epidemiology, 58*(9), 882–893. https://doi.org/10.1016/j.jclinepi.2005.01.016

Deisenhofer, A., Delgadillo, J., Rubel, J. A., Böhnke, J. R., Zimmermann, D., Schwartz, B., & Lutz, W. (2018). Individual treatment selection for patients with posttraumatic stress disorder. *Depression and Anxiety, 35*(6), 541–550. https://doi.org/10.1002/da.22755

Driessen, E., Hollon, S. D., Bockting, C. L. H., Cuijpers, P., & Turner, E. H. (2015). Does publication bias inflate the apparent efficacy of psychological treatment for major depressive disorder? A systematic review and meta-analysis of US National Institutes of Health-Funded Trials. *PLoS One, 10*(9), e0137864. https://doi.org/10.1371/journal.pone.0137864

Duval, S., & Tweedie, R. L. (2000a). A nonparametric "trim and fill" method of accounting for publication bias in meta-analysis. *Journal of the American Statistical Association, 95*(449), 89–98. https://doi.org/10.1080/01621459.2000.10473905

Duval, S., & Tweedie, R. L. (2000b). Trim and fill: A simple funnel-plot-based method of testing and adjusting for publication bias in meta-analysis. *Biometrics, 56*(2), 455–463.

Egger, M., Smith, G. D., Schneider, M., & Minder, C. (1997). Bias in meta-analysis detected by a simple, graphical test. *British Medical Journal, 315*, 629–634.

Ellis, P. D. (2010). *The essential guide to effect sizes: An introduction to statistical power, meta-analysis, and the interpretation of research results*. Cambridge University Press. https://doi.org/10.1017/CBO9780511761676

Fanelli, D. (2012). Negative results are disappearing from most disciplines and countries. *Scientometrics, 90*(3), 891–904. https://doi.org/10.1007/s11192-011-0494-7

Francis, G. (2013). Replication, statistical consistency, and publication bias. *Journal of Mathematical Psychology, 57*(5), 153–169. https://doi.org/10.1016/j.jmp.2013.02.003

Franco, A., Malhotra, N., & Simonovits, G. (2014). Publication bias in the social sciences: Unlocking the file drawer. *Science, 345*(6203), 1502–1505. https://doi.org/10.1126/science.1255484

Head, M. L., Holman, L., Lanfear, R., Kahn, A. T., & Jennions, M. D. (2015). The extent and consequences of p-hacking in science. *PLoS Biology, 13*(3), e1002106. https://doi.org/10.1371/journal.pbio.1002106

Hedges, L. V. (1992). Modeling publication selection effects in meta-analysis. *Statistical Science, 7*(2), 246–255.

Hedges, L. V., & Vevea, J. L. (1996). Estimating effect size under publication bias: Small sample properties and robustness of a random effects selection model. *Journal of Educational and Behavioral Statistics, 21*(4), 299–332.

Higgins, J. P. T., & Thompson, S. G. (2002). Quantifying heterogeneity in a meta-analysis. *Statistics in Medicine, 21*(11), 1539–1558. https://doi.org/10.1002/sim.1186

Higgins, J. P. T., Thompson, S. G., Deeks, J. J., & Altman, D. G. (2003). Measuring inconsistency in meta-analyses. *British Medical Journal, 327*(7414), 557–560. https://doi.org/10.1136/bmj.327.7414.557

Holma, K. M., Melartin, T. K., Haukka, J., Holma, I. A. K., Sokero, T. P., & Isometsä, E. T. (2010). Incidence and predictors of suicide attempts in DSM–IV major depressive disorder: A five-year prospective study. *American Journal of Psychiatry, 167*(7), 801–808. https://doi.org/10.1176/appi.ajp.2010.09050627

Ioannidis, J. P. A. (2013). Clarifications on the application and interpretation of the test for excess significance and its extensions. *Journal of Mathematical Psychology, 57*(5), 184–187. https://doi.org/10.1016/j.jmp.2013.03.002

Ioannidis, J. P. A., Stanley, T. D., & Doucouliagos, H. (2017). The power of bias in economics research. *The Economic Journal, 127*(605), F236–F265. https://doi.org/10.1111/ecoj.12461

Ioannidis, J. P., & Trikalinos, T. A. (2007). An exploratory test for an excess of significant findings. *Clinical Trials, 4*(3), 245–253. https://doi.org/10.1177/1740774507079441

Iyengar, S., & Greenhouse, J. B. (1988). Selection models and the file drawer problem: Rejoinder. *Statistical Science, 3*(1), 133–135. http://www.jstor.org/stable/2245932

Jaycox, L. H., & Foa, E. B. (1999). Cost-effectiveness issues in the treatment of posttraumatic stress disorder. In *Cost-effectiveness of psychotherapy: A guide for practitioners, researchers, and policymakers* (pp. 259–269). Oxford University Press.

Jin, Z. C., Zhou, X. H., & He, J. (2014). Statistical methods for dealing with publication bias in meta-analysis. *Statistics in Medicine, 34*(2), 343–360. https://doi.org/10.1002/sim.6342

Kepes, S., Banks, G. C., McDaniel, M., & Whetzel, D. L. (2012). Publication bias in the organizational sciences. *Organizational Research Methods, 15*(4), 624–662. https://doi.org/10.1177/1094428112452760

Kraemer, H. C., Gardner, C., Brooks, J., & Yesavage, J. A. (1998). Advantages of excluding underpowered studies in meta-analysis: Inclusionist versus exclusionist viewpoints. *Psychological Methods, 3*(1), 23–31. https://doi.org/10.1037/1082-989X.3.1.23

Lane, D. M., & Dunlap, W. P. (1978). Estimating effect size: Bias resulting from the significance criterion in editorial decisions. *British Journal of Mathematical & Statistical Psychology, 31*, 107–112.

Lau, J., Antman, E. M., Jimenez-Silva, J., Kupelnick, B., Mosteller, F., & Chalmers, T. C. (1992). Cumulative meta-analysis of therapeutic trials for myocardial infarction. *New England Journal of Medicine, 327*(4), 248–254. https://doi.org/10.1056/nejm199207233270406

Light, R. J., & Pillemer, D. B. (1984). *Summing up: The science of reviewing research.* Harvard University Press.

Macaskill, P., Walter, S. D., & Irwig, L. (2001). A comparison of methods to detect publication bias in meta-analysis. *Statistics in Medicine, 20*(4), 641–654.

Maljanen, T., Knekt, P., Lindfors, O., Virtala, E., Tillman, P., & Härkänen, T. (2016). The cost-effectiveness of short-term and long-term psychotherapy in the treatment of depressive and anxiety disorders during a 5-year follow-up. *Journal of Affective Disorders, 190*, 254–263. https://doi.org/10.1016/j.jad.2015.09.065

Margraf, J. (2009). *Kosten und Nutzen der Psychotherapie.* Springer Medizin. http://public.ebookcentral.proquest.com/choice/publicfullrecord.aspx?p=450836

Marks-Anglin, A., & Chen, Y. (2020). A historical review of publication bias. *Research Synthesis Methods, 11*(6), 725–742. https://doi.org/10.1002/jrsm.1452

Matloff, N. S. (2011). *The art of R programming: Tour of statistical software design.* No Starch Press. http://site.ebrary.com/id/10513550

McShane, B. B., Böckenholt, U., & Hansen, K. T. (2016). Adjusting for publication bias in meta-analysis: An evaluation of selection methods and some cautionary notes. *Perspectives on Psychological Science, 11*(5), 730–749. https://doi.org/10.1177/1745691616662243

Moher, D., Liberati, A., Tetzlaff, J., Altman, D. G., & The Prisma Group. (2009). Preferred reporting items for systematic reviews and meta-analyses: The PRISMA statement. *PLoS Medicine, 6*(7), e1000097. https://doi.org/10.1371/journal.pmed.1000097

Moreno, S. G., Sutton, A. J., Ades, A. E., Stanley, T. D., Abrams, K. R., Peters, J. L., & Cooper, N. J. (2009). Assessment of regression-based methods to adjust for publication bias through a comprehensive simulation study. *BMC Medical Research Methodology, 9*(2), 1–17. https://doi.org/10.1186/1471-2288-9-2

Niemeyer, H., Musch, J., & Pietrowsky, R. (2012). Publication bias in meta-analyses of the efficacy of psychotherapeutic interventions for schizophrenia. *Schizophrenia Research, 138*(2), 103–112. https://doi.org/10.1016/j.schres.2012.03.023

Niemeyer, H., Musch, J., & Pietrowsky, R. (2013). Publication bias in meta-analyses of the efficacy of psychotherapeutic interventions for depression. *Journal of Consulting and Clinical Psychology, 81*(1), 58–74. https://doi.org/10.1037/a0031152

Niemeyer, H., Van Aert, R. C. M., Schmid, S., Uelsmann, D., Knaevelsrud, C., & Schulte-Herbrueggen, O. (2020). Publication bias in meta-analyses of posttraumatic stress disorder interventions. *Meta-Psychology, 4*, 31.

Orwin, R. G. (1983). A fail-safe N for effect size in meta-analysis. *Journal of Educational Statistics, 8*(2), 157–159.

Page, M. J., Higgins, J. P. T., & Sterne, J. A. C. (2019). Chapter 13: Assessing risk of bias due to missing results in a synthesis. In J. P. T. Higgins, J. Thomas, J. Chandler, M. Cumpston, T. Li, M. J. Page, & V. A. Welch (Eds.), *Cochrane handbook for systematic reviews of interventions version 6.0.*

Peters, J. L., Sutton, A. J., Jones, D. R., Abrams, K. R., & Rushton, L. (2007). Performance of the trim and fill method in the presence of publication bias and between-study heterogeneity. *Statistics in Medicine, 26*(25), 4544–4562. https://doi.org/10.1002/sim.2889

Peters, J. L., Sutton, A. J., Jones, D. R., Abrams, K. R., & Rushton, L. (2008). Contour-enhanced meta-analysis funnel plots help distinguish publication bias from other causes of

asymmetry. *Journal of Clinical Epidemiology, 61*(10), 991–996. https://doi.org/10.1016/j.jclinepi.2007.11.010
Raudenbush, S. W. (2009). Analyzing effect sizes: Random-effects models. In H. Cooper, L. V. Hedges, & J. C. Valentine (Eds.), *The handbook of research synthesis and meta-analysis* (pp. 295–315). Russell Sage Foundation.
R Core Team. (2020). R: A language and environment for statistical computing. http://www.r--project.org/
Renkewitz, F., & Keiner, M. (2019). How to detect publication bias in psychological research. *Zeitschrift Für Psychologie*. https://doi.org/10.1027/2151-2604/a000386
Rhodes, K. M., Turner, R. M., & Higgins, J. P. (2015). Predictive distributions were developed for the extent of heterogeneity in meta-analyses of continuous outcome data. *Journal of Clinical Epidemiology, 68*(1), 52–60. https://doi.org/10.1016/j.jclinepi.2014.08.012
Rosenthal, R. (1979). The file drawer problem and tolerance for null results. *Psychological Bulletin, 86*(3), 638–641.
Rothstein, H. R., Sutton, A. J., & Borenstein, M. (2005). *Publication bias in meta-analysis: Prevention, assessment and adjustments*. Wiley.
Scargle, J. D. (2000). Publication bias: The "file-drawer problem" in scientific inference. *Journal of Scientific Exploration, 14*(1), 91–106. http://arxiv.org/abs/physics/9909033
Schonemann, P. H., & Scargle, J. D. (2008). A generalized publication bias model. *Chinese Journal of Psychology, 50*(1), 21–29.
Schumacher, S., Niemeyer, H., Engel, S., Cwik, J. C., & Knaevelsrud, C. (2018). Psychotherapeutic treatment and HPA axis regulation in posttraumatic stress disorder: A systematic review and meta-analysis. *Psychoneuroendocrinology, 98*, 186–201. https://doi.org/10.1016/j.psyneuen.2018.08.006
Simmons, J. P., Nelson, L. D., & Simonsohn, U. (2011). False-positive psychology: Undisclosed flexibility in data collection and analysis allows presenting anything as significant. *Psychological Science, 22*(11), 1359–1366. https://doi.org/10.1177/0956797611417632
Simonsohn, U., Nelson, L. D., & Simmons, J. P. (2014). P-curve and effect size: Correcting for publication bias using only significant results. *Perspectives on Psychological Science, 9*(6), 666–681. https://doi.org/10.1177/1745691614553988
Stanley, T. D. (2017). Limitations of PET-PEESE and other meta-analysis methods. *Social Psychological and Personality Science, 8*(5), 581–591. https://doi.org/10.1177/1948550617693062
Stanley, T. D., & Doucouliagos, H. (2014). Meta-regression approximations to reduce publication selection bias. *Research Synthesis Methods, 5*(1), 60–78.
Stanley, T. D., Doucouliagos, H., & Ioannidis, J. P. (2017). Finding the power to reduce publication bias. *Statistics in Medicine, 36*(10), 1580–1598. https://doi.org/10.1002/sim.7228
Stanley, T. D., Jarrell, S. B., & Doucouliagos, H. (2010). Could it be better to discard 90% of the data? A statistical paradox. *The American Statistician, 64*(1), 70–77. https://doi.org/10.1198/tast.2009.08205
Sterne, J. A. C., Gavaghan, D., & Egger, M. (2000). Publication and related bias in meta-analysis: Power of statistical tests and prevalence in the literature. *Journal of Clinical Epidemiology, 53*(11), 1119–1129. https://doi.org/10.1016/S0895-4356(00)00242-0
Sterne, J. A. C., Harbord, R. M., Sutton, A. J., Jones, D. R., Ioannidis, J. P., Terrin, N., Lau, J., Schmid, C. H., Carpenter, J., Rucker, G., Schwarzer, G., Tetzlaff, J., Moher, D., Deeks, J. J., Peters, J., Macaskill, P., Duval, S., Altman, D. G., & Higgins, J. P. T. (2011). Recommendations for examining and interpreting funnel plot asymmetry in meta-analyses of randomised controlled trials. *British Medical Journal, 343*(7818), 1–8. https://doi.org/10.1136/bmj.d4002
Teetor, P. (2011). *R cookbook*. O'Reilly.
Terrin, N., Schmid, C. H., & Lau, J. (2005). In an empirical evaluation of the funnel plot, researchers could not visually identify publication bias. *Journal of Clinical Epidemiology, 58*(9), 894–901. https://doi.org/10.1016/j.jclinepi.2005.01.006

Terrin, N., Schmid, C. H., Lau, J., & Olkin, I. (2003). Adjusting for publication bias in the presence of heterogeneity. *Statistics in Medicine, 22*(13), 2113–2126. https://doi.org/10.1002/sim.1461

Tolin, D. F., McKay, D., Forman, E. M., Klonsky, E. D., & Thombs, B. D. (2015). Empirically supported treatment: Recommendations for a new model. *Clinical Psychology: Science and Practice, 22*(4), 317–338. https://doi.org/10.1111/cpsp.12122

Turner, R. M., Jackson, D., Wei, Y., Thompson, S. G., & Higgins, J. P. T. (2015). Predictive distributions for between-study heterogeneity and simple methods for their application in Bayesian meta-analysis. *Statistics in Medicine, 34*(6), 984–998. https://doi.org/10.1002/sim.6381

Van Aert, R. C. M. (2020). *Puniform: Meta-analysis methods correcting for publication bias.* https://cran.r-project.org/package=puniform

Van Aert, R. C. M., & Van Assen, M. A. L. M. (2022). Correcting for publication bias in a meta-analysis with the p-uniform* method. Manuscript submitted for publication. https://doi.org/10.31222/osf.io/zqjr9

Van Aert, R. C. M., Wicherts, J. M., & Van Assen, M. A. L. M. (2016). Conducting meta-analyses on p-values: Reservations and recommendations for applying p-uniform and p-curve. *Perspectives on Psychological Science, 11*(5), 713–729. https://doi.org/10.1177/1745691616650874

Van Aert, R. C. M., Wicherts, J. M., & Van Assen, M. A. L. M. (2019). Publication bias examined in meta-analyses from psychology and medicine: A meta-meta-analysis. *PLoS One, 14*(4), e0215052. https://doi.org/10.1371/journal.pone.0215052

Van Assen, M. A. L. M., Van Aert, R. C. M., & Wicherts, J. M. (2015). Meta-analysis using effect size distributions of only statistically significant studies. *Psychological Methods, 20*(3), 293–309. https://doi.org/10.1037/met0000025

Van Assen, M. A. L. M., Van Den Akker, O. R., Augusteijn, H. E. M., Bakker, M., Nuijten, M. B., Olsson-Collentine, A., Stoevenbelt, A. H., Wicherts, J. M., & Van Aert, R. C. M. (2022). The meta-plot: A graphical tool for interpreting the results of a meta-analysis. Manuscript submitted for publication. https://doi.org/10.31234/osf.io/cwhnq

Vandekerckhove, J., Guan, M., & Styrcula, S. A. (2013). The consistency test may be too weak to be useful: Its systematic application would not improve effect size estimation in meta-analyses. *Journal of Mathematical Psychology, 57*(5), 170–173. https://doi.org/10.1016/j.jmp.2013.03.007

Vevea, J. L., Clements, N. C., & Hedges, L. V. (1993). Assessing the effects of selection bias on validity data for the General Aptitude Test Battery. *Journal of Applied Psychology, 78*(6), 981–987. https://doi.org/10.1037/0021-9010.78.6.981

Vevea, J. L., & Woods, C. M. (2005). Publication bias in research synthesis: Sensitivity analysis using a priori weight functions. *Psychological Methods, 10*(4), 428–443. https://doi.org/10.1037/1082-989X.10.4.428

Viechtbauer, W. (2007). Confidence intervals for the amount of heterogeneity in meta-analysis. *Statistics in Medicine, 26*(1), 37–52. https://doi.org/10.1002/sim.2514

Viechtbauer, W. (2010). Conducting meta-analyses in R with the metafor package. *Journal of Statistical Software, 36*(3), 1–48. https://doi.org/10.18637/jss.v036.i03

Wicherts, J. M., Veldkamp, C. L. S., Augusteijn, H. E. M., Bakker, M., van Aert, R. C. M., & van Assen, M. A. L. M. (2016). Degrees of freedom in planning, running, analyzing, and reporting psychological studies: A checklist to avoid p-hacking. *Frontiers in Psychology, 7*, 1832. https://doi.org/10.3389/fpsyg.2016.01832

Chapter 11
Avoiding Questionable Research Practices Surrounding Statistical Power Analysis

Jolynn Pek, Kathryn J. Hoisington-Shaw, and Duane T. Wegener

Abstract The purposes of this chapter are to provide statistical justifications and illustrations of whether and when statistical power can be used to improve the conduct of psychological science, reduce questionable research practices (QRPs), and perhaps even detect QRPs. In general, the utility of power analysis in countering QRPs is narrower than commonly believed. We begin by reviewing concepts that lead to a formal definition of statistical power. We highlight the meaning of the probability that a power value quantifies and identify assumptions necessary for calculated power values to be accurate. Next, we describe how power analysis can be fruitfully applied in the study design phase, emphasizing that power is valid only as a *pre-study* concept (i.e., before a study is implemented). We then examine uses of power in the post-study phase, where power values are calculated from collected data to aid in the interpretation and evaluation of results. Because of inherent ontological inconsistencies, post-study applications of power are unjustified and misplaced. Finally, we briefly note that other design features play essential roles in enhancing the credibility of research results.

Keywords Questionable research practice · Psychological science · Clinical science · Power analysis

Statistical power analysis is regarded as one of several means to reduce questionable research practices (QRPs; e.g., see Appelbaum et al., 2018; Cooper, 2016; Funder et al., 2013; Simmons et al., 2012). According to the frequentist significance testing paradigm, power for a procedure is the probability of rejecting the null hypothesis (H_0) over repeated samples, assuming that some specific effect size value under the alternative hypothesis (H_1) is true (for a particular sample size N and Type I error rate, α). Despite its straightforward definition, power as a concept continues to be

J. Pek (✉) · K. J. Hoisington-Shaw · D. T. Wegener
Department of Psychology, The Ohio State University, Columbus, OH, USA
e-mail: pek.5@osu.edu

W. O'Donohue et al. (eds.), *Avoiding Questionable Research Practices in Applied Psychology*, https://doi.org/10.1007/978-3-031-04968-2_11

misunderstood and misapplied in practice (e.g., see Hoenig & Heisey, 2001; Lenth, 2001; McShane et al., 2020). Therefore, applying power analysis to research is not as simple as it may seem.

Applications of power have spanned the entire research process: from designing studies, to the interpretation of published results. Some applications are justified by statistical theory, whereas others are not and can result in practices that are theoretically questionable. The purposes of this chapter are to provide statistical justifications and illustrations of whether and when statistical power can be used to improve the conduct of psychological science, reduce QRPs, and perhaps even detect QRPs. In general, the utility of power analysis in countering QRPs is narrower than commonly believed. We begin by reviewing concepts that lead to a formal definition of statistical power. We highlight the meaning of the probability that a power value quantifies and identify assumptions necessary for calculated power values to be accurate. Next, we describe how power analysis can be fruitfully applied in the study design phase, emphasizing that power is valid only as a *pre-study* concept (i.e., before a study is implemented). We then examine uses of power in the post-study phase, where power values are calculated from collected data to aid in the interpretation and evaluation of results (e.g., see Appelbaum et al., 2018; Funder et al., 2013; Giner-Sorolla et al., 2019). Because of inherent ontological inconsistencies, post-study applications of power are unjustified and misplaced (Hoenig & Heisey, 2001; Lenth, 2001; McShane et al., 2020; Yuan & Maxwell, 2005). Finally, we briefly note that other design features play essential roles in enhancing the credibility of research results.

Preliminaries

Population, Model, and Data

Research can be defined as a process of discovering or understanding phenomena (Fox, 1958). A *population* and its unknown effect size formally represent the phenomenon of interest. This phenomenon occurs in reality and is *not* theoretical. After *data* are sampled from the target population, a theoretical statistical *model* that serves as an approximation to the population (Box, 1976; MacCallum, 2003) is employed to summarize the data. Technically, the statistical model is an abstraction that parsimoniously represents the population. For example, assuming that the population is normally distributed with unknown effect size μ and nuisance parameter σ, a statistical model with the same structure is fit to sampled data. Here, the mean and standard deviation of the data, $\overline{X}$ and s_X, respectively, estimate the unknown population effect, μ, and the standard deviation, σ. In general, we denote the parameters of the model with $\boldsymbol{\theta}$, and the effect of interest as the focal parameter, θ_f. Parameters that are not the effect of interest but complete the model are called nuisance parameters, θ_n. Thus, for the normal distribution, $\boldsymbol{\theta} = (\mu, \sigma)'$, $\theta_f = \mu$ and $\theta_n = \sigma$. In more complex models, there could be multiple effects and nuisance parameters.

To make an inference about the population using collected data (i.e., to make a claim about μ using $\bar{X}$), a significance test can be conducted in which a null hypothesis (H_0) about the effect is rejected if the obtained data are sufficiently unlikely to be produced by the distribution implied by that null hypothesis. Following the logic of falsification (Popper, 1959), the null hypothesis typically specifies lack of an effect (e.g., $H_0 : \mu = 0$) in which the effect of interest would have a non-zero value. When a significance test rejects H_0 based on data sufficiently inconsistent with H_0, it suggests that the data support an effect (i.e., a non-zero value) in the direction of the obtained data. Significance tests are outputs of a statistical model fit to the collected data, and these tests are the very devices that link sampled data to the population.

Consider influences of position justification on perceptions of source bias, where inferences of source bias are thought to be more likely when a position is justified by weak, specious arguments rather than strong, compelling arguments (Wallace et al., 2021). We assume that the population of differences in perceived source bias between strong and weak argument conditions follows a normal distribution (either because the bias scores are normally distributed or because sufficient data are collected and the central limit theorem justifies the assumption). Thus, the specified model will compare two groups with normal distributions in which the focal parameter is the group differences in perceived source bias between strong and weak argument conditions. Parameters of this model that approximate the population (composed of two groups) are then estimated from data. Specifically, the mean and standard deviation parameters for the two groups (μ_S and σ_S versus μ_W and σ_W, with S and W denoting strong and weak arguments, respectively) are estimated by their sample means and standard deviations ($\bar{X}_S$ and s_S versus $\bar{X}_W$ and s_W, respectively); $\boldsymbol{\theta} = (\mu_S, \mu_W, \sigma_S, \sigma_W)'$ and $\hat{\theta} = \left(\bar{X}_S, \bar{X}_W, s_S, s_W\right)'$.

An appropriate statistical test of a null hypothesis of no difference between the two groups uses the t-test, with $t = \dfrac{\bar{X}_S - \bar{X}_W}{SE_{\bar{X}_S - \bar{X}_W}}$, where $\bar{X}_S - \bar{X}_W$ is the observed difference in sample means between the strong and weak argument conditions and $SE_{\bar{X}_S - \bar{X}_W}$ is the standard error of the difference in means estimated using the observed standard deviation of the bias scores within each cell and the respective cell sizes (n_S and n_W). The test statistic is deemed significant and H_0 is rejected when the p-value is below some arbitrary but agreed-upon level of significance, say $\alpha = .05$. α is also the Type I error rate, which will be defined in the power analysis section. Note that the p-value relies not only on the value of the statistic that tests $H_0 : \mu_S - \mu_W = 0$, but also on *every assumption necessary* to compute the test statistic (e.g., the assumption of random sampling and normality for the t-test; Greenland & Poole, 2013). Stated differently, the hypothesis H_0 about θ_f is embedded within a statistical model with parameters $\boldsymbol{\theta}$. These parameters are estimated as $\hat{\theta}$ from the data, which give rise to the p-value.

Taken together, the population represents the target phenomenon. The sample is an instance drawn from the population that provides imperfect information about the phenomenon of interest. Finally, the statistical model is an abstract and inexact representation of the population and is a structure imposed onto the data for the purpose of obtaining estimates to infer the nature of the population. In our example, the population, sample, and model are *assumed* to have normal distributions. Assuming normality allows for easy derivation of the sampling distribution of the t-statistic. The distributions of the population, model, and sample, however, need not be the same. In practice, the structure (distribution) of the population is unknown, samples are often nonrandom (e.g., drawn with convenience sampling) and may not necessarily reflect the distribution of the population, and the proposed statistical model may not fit the data well or represent the population accurately. It has to be emphasized that any separation of population, model, and data has different implications for the application of statistical concepts before data are collected (in the pre-study design phase) versus after data have been collected (in the post-study phase).

Pre-study Design versus Post-study Analysis

Pre-study Design When designing a study, decisions are made about the features of the study to be implemented. Study features include manipulations, measures of variables, random assignment, type of sampling, sample size, type of design (e.g., between-subjects, within-subjects, cross-sectional, and longitudinal), and planned statistical analyses (e.g., equal versus unequal variance t-test; see also Trochim & Land, 1982). It could also be argued that part of design includes the specification of the Type I error, α, to be used in the statistical tests (see Benjamin et al., 2018; Lakens et al., 2018; Neyman, 1957). Pre-study design is distinct from the post-study phase because the data considered during the design are hypothetical and treated as random (i.e., they are not yet fixed but will vary across repeated samples).

In the pre-study phase, the underlying structure of the population, statistical model, and theoretical data are generally treated as equivalent in structure (e.g., all follow a normal distribution). Assuming equivalence of the population, model, and data vastly simplifies statistical derivations (e.g., formulas to compute power). To illustrate this equivalence, we review the decision probabilities of the Type I error rate (α), Type II error rate (β), and power ($1 - \beta$) below. These probabilities are theoretical abstractions because they are based on hypothetical random data. The key takeaway is to recognize that statistical power is derived from the unrealistic but simplifying assumption that the population, model, and data are equivalent in structure. When the population, model, and data are different, which routinely occurs in the post-study phase of research (MacCallum, 2003), calculated power values for a design cannot be considered accurate in relation to a completed study using the design.

Decision Probabilities

The decision probabilities of the Type I error rate (α), Type II error rate (β), and power ($1 - \beta$) are illustrated in Fig. 11.1. Let the effect size be scaled to follow the Cohen's d metric ($d = \left[\overline{x}_S - \overline{x}_W\right] / s_{\text{pooled}}$, where $\overline{x}_s - \overline{x}_W$ is the observed mean difference between two groups and s_{pooled} is the estimated pooled standard deviation of the two groups) and let the sample size in each cell be $n = 60$. We use δ to denote the population effect size and d to denote the effect size estimated from data. Note that δ is a combination of the parameters from the normal distribution, $\delta = (\mu_S - \mu_W)/\sigma_{\text{pooled}}$. In significance testing, decisions are made based on the p-value associated with the t-test, and the p-value is computed on the assumption that $H_0 : \delta = 0$ is true. When the p-value is lower than some specified significance level (i.e., the Type I error, α), the decision is made to reject H_0. When binary decisions are made about a continuous measure such as the p-value, there is some non-zero probability of making decisional errors over repeated samples and their respective tests. These probabilities quantify how frequently decisional errors occur *over the long run* (i.e., over repeated and randomly drawn samples from the same population).

A Type I error occurs when $H_0 : \delta = 0$ is true in the population but the significance test leads to a decision to reject H_0. In Fig. 11.1a, $H_0 : \delta = 0$ is assumed to be true in

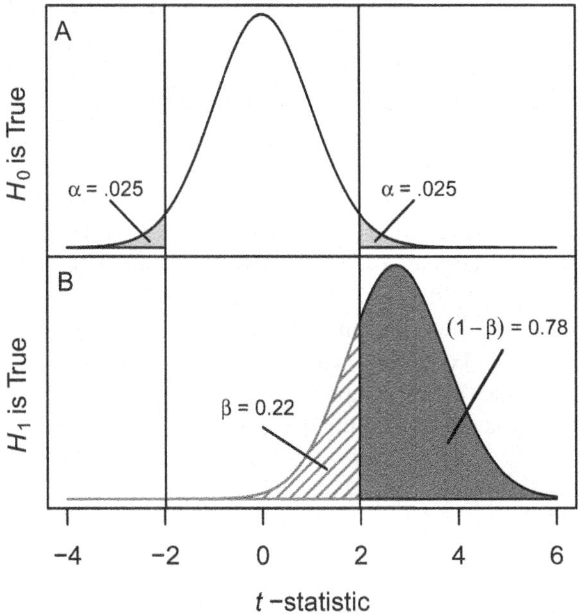

Fig. 11.1 Theoretical decision probabilities. A: Sampling distribution of d under the null hypothesis where $\delta = 0$. B: Sampling distribution of d under the alternative hypothesis where $\delta = 0.5$. *Note*. The t-distribution represents the sampling distribution of theoretically observed t-test statistic values that are sampled from a population with Cohen's $\delta = 0$ under H_0 and Cohen's $\delta = 0.5$ under H_1. The Type I error rate, $\alpha = .05$, and sample size $n = 60$ for each group (total $N = 120$). For a two-sample, two-sided t-test of $H_0 : \delta = 0$, power is $(1 - \beta) = 0.78$ and the Type II error rate, $\beta = 0.22$

the population.[1] When a sample of size $n = 60$ from each group is drawn from the population with $\delta = 0$ (assuming H_0 is true), and Cohen's d statistics are computed for each sample, these Cohen's d values will form a sampling distribution (i.e., a distribution of random samples of the same size drawn from the same population). The curve[2] in Fig. 11.1a represents the sampling distribution of hypothetically observed t-test statistics translated from Cohen's d under these assumptions. Because the samples are drawn from a population with $\delta = 0$, the sampling distribution is centered about $d = 0$ (translated to $t = 0$). This sampling distribution quantifies sampling variability (i.e., variation from sample to sample) inherent in Cohen's d or t-statistics that summarize random samples drawn from the population.

The vertical lines in Fig. 11.1 represent the threshold values that hypothetical t-test statistics need to exceed such that their commensurate p-values will be deemed significant in the correct tail of a two-tailed test with 5% level of significance, $\alpha = .05$. Stated differently, values to the right (or left) of the vertical reference lines result in a decision to reject $H_0 : \delta = 0$. These thresholds are related to the critical value of the t-distribution that is used to compute the p-value [where the critical value is the $\left(1 - \frac{\alpha}{2}\right) * 100$ th percentile of the t-distribution].[3] The light grey area inFig. 11.1a, located to the right of the vertical reference line, represents 2.5% of the sampling distribution associated with $H_0 : \delta = 0$, which is half of the Type I error rate, α. This area is also the probability over random data (represented by the sampling distribution) of making the incorrect decision of rejecting H_0 based on extreme t-statistic values even though the population is consistent with $H_0 : \delta = 0$. The entire Type I error rate (α) is the probability of rejecting H_0 based on either positive or negative t-test statistics over repeated samples when H_0 is true. Importantly, the uncertainty quantified by the probability α is over random data. Stated differently, α is a frequentist probability that measures how often a decisional error of Type I is made in the long run over repeated samples and is thus a pre-data concept.

A Type II error occurs when H_1 is true in the population but the significance test leads to a decision not to reject H_0. In Fig. 11.1b, $H_1 : \delta = 0.5$ (t-distribution noncentrality parameter $\lambda = 2.74$) is assumed true in the population. Under the assumption that the population $\delta = 0.5$, the curve centered about $t = 2.74$ (noncentrality parameter λ) is the sampling distribution of hypothetical t-test statistics over repeated samples of data for $n = 60$ per group. The shaded area to the left of the vertical reference line ($t = 1.98$) represents the Type II error probability, β. This probability quantifies the rate of making the incorrect decision (over repeated samples; i.e., random

[1] Decision probabilities in the frequentist perspective are not conditional in that there is no probability attached to the occurrence of H_0 or H_1. Instead, these probabilities are computed assuming either H_0 is true or H_1 is true.

[2] Although the curve looks normal, this is technically the t-distribution that underlies the t-test with $n = 60$ for each group that is applied to hypothetical data.

[3] In the context of power (correct rejections), we are considering only one tail of the distribution, though the complete α is distributed across both tails of the relevant distribution of t (or Cohen's d) values.

data) of not rejecting H_0 even when $H_1 : \delta = 0.5$ is true in the population. For this example, 22% of decisions will be to retain H_0 and will, therefore, be incorrect if $H_1 : \delta = 0.5$.

Power, $(1 - \beta)$, is the mathematical complement of the Type II error, β. When $\beta = 0.22$, $(1 - \beta) = 1 - 0.22 = 0.78$. In Fig. 11.1b, power is represented by the dark shaded area of the sampling distribution (to the right of the vertical reference line, $t = 1.98$) associated with a population where $\delta = 0.5$. Statistical power is the frequentist probability over random data that the test correctly rejects H_0, assuming $H_1 : \delta = 0.5$ is true in the population. Note that the concept of power is relevant only when the population effect size $\delta \neq 0$. Although we depict power as the area under the curve (sampling distribution) for a single point value of δ, power is better expressed as a function across a range of nonzero δ values (see Morey, 2020). Let power be expressed as a function of its inputs: $(1 - \beta) = f(\alpha, N, \boldsymbol{\theta})$, where $f(\cdot)$ is tied to the form of the statistical model that yields a test. From this expression, power is regarded as a function of varying inputs (α, N, and $\boldsymbol{\theta}$) and is a property of a design. In this vein, power represents how well the procedure (test), that is tied to unique forms of $f(\cdot)$, performs under different combinations of α, N, and $\boldsymbol{\theta}$. We elaborate on this use of power in the section on "Uses of pre-study power for design" below. Taken together, the classical approach of presenting the concept of power as a single value obfuscates the nature of power as a function of design that applies over different values of α, N, and $\boldsymbol{\theta}$ (Morey, 2020; cf. Cohen, 1988). Because power is a property of pre-study design, we term such pre-study calculations *power for design* (see Fig. 11.2).

Post-study Analysis The application of a model to empirical data occurs in the post-data collection phase of research in which collected data are analyzed (see right column of Fig. 11.2). This phase of research begins after study implementation; that is, after data have been collected such that data are fixed and no longer considered random. The statistical model that is inspired by substantive theory about the unknown effect, θ_f, is fit to collected data such that a significance test is conducted to draw an inference about θ_f. The p-value that is used to make an inference about θ_f is computed using the estimated effect size, $\hat{\theta}_f$. For the example on the normal distribution, $\theta_f = \mu$ and $\hat{\theta}_f = \overline{X}$. There are clear separations among the three entities in the post-study phase: (a) the population represents reality (where the effect size θ_f is unknown), (b) the statistical model (with parameters in $\boldsymbol{\theta}$, including θ_f) is a formalization of an aspect of substantive theory, and (c) collected data are sampled from the target population (where the model parameters are estimated by $\hat{\theta}$, including the effect size $\hat{\theta}_f$). Given that statistical models are imperfect representations of the population (Greenland, 2017), where model assumptions are not entirely met, calculated statistics (e.g., effect size $\hat{\theta}_f$ and its accompanying p-value) cannot be exactly correct. According to Box (1976), useful results come from a statistical model that approximates the population well-enough, even though the model is expected not to be precisely correct (see also Cudeck & Henly, 1991; MacCallum, 2003).

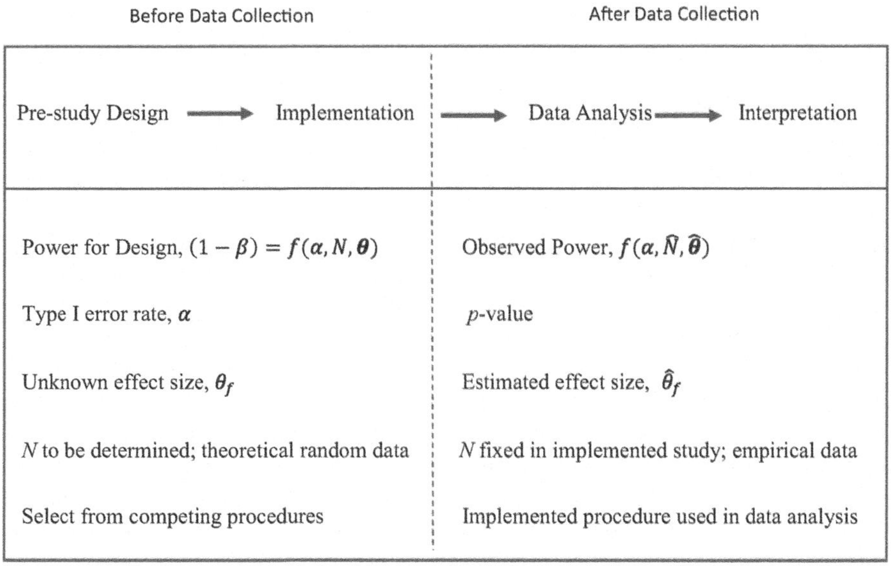

Fig. 11.2 Phases of research. *Note.* During the theoretical pre-study design phase, where data have yet to be collected, the population, statistical model, and data are assumed identical. However, after data are collected, it is assumed that the population, the statistical model, and the data are distinct from one another

Pre-study versus Post-study Concepts

It is essential to recognize that the error probabilities (α and β) and power ($[1 - \beta]$) are theoretical entities that are independent of empirical data (Cohen, 1973; Cox, 1958; Senn, 2002). Figure 11.1 was constructed by making the strong assumption that the population of differences follows a normal distribution. This assumption results in randomly sampled data ($N = 60$) that are normally distributed and Cohen's d values calculated from repeated samples that follow a t-distribution. Further, the statistical model underlying the t-test is specified to match the assumed population structure and resulting sampled data so that α, β, and $(1 - \beta)$ are exactly correct. The conceptualizations of α, β, and $(1 - \beta)$ are based on speculative assumptions about the population, collected data, and planned analyses, and the numerical values assigned to these concepts are correct only when these underlying assumptions are met. These assumptions are treated as being fully met in the (theoretical) pre-study design phase.

In the post-study phase, however, there is an undeniable separation between the distributional forms of the population, statistical model, and collected data (Box, 1976; MacCallum, 2003; Rodgers, 2010; Tukey, 1969). The assumptions required for the values carried by α, β, and $(1 - \beta)$ to be accurate (i.e., equivalent distributions among the population, model, and sample) are likely unmet in the post-study phase. Unfortunately, there are several challenges to extending the pre-study design concept of power to the post-study phase. These challenges rest directly on the assumptions underlying power analysis. First, in pre-study power for design, every data set is a *random* sample from the same population. However, most completed studies are based on convenience samples that differ from one another in various ways. Even if the repeated samples participate in the same lab and with the same materials, they almost certainly differ in time (along with any changes in current events encountered outside the lab). Study samples also often differ on other characteristics such as different labs, different materials, different geographical locations within a country, different cultures, or simply different characteristics of the participants themselves (e.g., in demographic terms or across psychological variables related to the phenomenon of interest). Any of these differences across samples create a distance between the assumption of random sampling from the same population and the reality of how research is conducted. Second, the specified population is represented by a single value of the population parameter underlying the sampled data (i.e., there is a point value for the target effect size). However, large replication studies of the same phenomenon have consistently observed effect size heterogeneity (e.g., see Hedges & Schauer, 2019; Kenny & Judd, 2019; results from the Many Labs project by Klein et al., 2014, 2018). That heterogeneity might ultimately be related to the same variables that differ across samples and make them come from potentially different populations rather than one common population. Third, sampling distributions are made up of samples of the same size N, thus all studies should be of the same size N for downstream calculations such as power (see also definition of sampling distribution in previous section). However, replication efforts and meta-analyses typically contain studies with different N (e.g., see Many Labs Project [Klein et al., 2014] and Registered Replication Reports [Simons et al., 2014]). Fourth, the statistical model would have to be a perfect representation of complex reality (i.e., the population), but it is not (see Box, 1976; MacCallum, 2003). Taken together, assumptions underlying values of power calculated during the pre-study design stage are thus likely violated in the post-study phase of research. Because of the distance between assumptions made to compute pre-study power for design values and the likely violation of these assumptions in the post-study phase of research, pre-study power cannot be directly applied to results of a completed study based on the same design.

Furthermore, recall that power is a probability over random data that is a performance measure attached to a design $f(\cdot)$. When analyzing data collected in a realized study, the data are fixed (not random), so power as a concept can no longer be directly applied to the realized study's (fixed) results. This is an ontological impossibility (McShane, Böckenholt, & Hansen, 2020). Consider the distinction between Type I error rate, α, and the p-value in that the Type I error rate applies to theoretical random data across samples, whereas the p-value applies to the observed data from a single sample relative to a sampling distribution. The level of α set before the data were collected does not influence the p-value that is obtained. Similarly, power, $(1 - \beta)$, applies to theoretical random data, whereas its observed counterpart, typically called observed power (see O'Keefe, 2007), would potentially apply to fixed observed data. Yet, observed power is isomorphic with the p-value and no longer provides information about the actual power for design before the study was conducted (Hoenig & Heisey, 2001; Lenth, 2001). Thus, there is an *ontological inconsistency* between the meaning of probability that is attached to pre-study power for design (uncertainty over random data) and that of observed power (a function of the *p*-value; see Greenwald et al., 1996; Hoenig & Heisey, 2001; Lenth, 2001). The concepts of pre-study power for design and post-study observed power are fundamentally distinct. In a nutshell, as emphasized by methodologists and statisticians alike, statistical power is only relevant during the pre-data design phase (Cohen, 1973; Cox, 1958; Senn, 2002). Indeed, after outlining pre-study power for design considerations, Cohen (1973, p. 226) wrote, "… nothing said thus far has had any bearing on the results of the research when it is completed; all of the above reasoning is complete prior to the examination and, in principle, prior even to the collection of the sample data."

In the sections to follow, we describe how statistical power can be employed to facilitate the development of quantitative methods and the design of studies during the pre-study stage. Next, we address popular misconceptions pertaining to the application of the concept of power to post-study results where power calculations are often used to suggest QRPs (cf. statistical forensics, Morey, 2019). Finally, we end with a discussion on the appropriate uses of power with respect to QRPs and highlight the role of other design features (e.g., experimental control, construct validity, measurement reliability, accumulation of information across studies) that are important for bolstering the credibility of statistical results.

Uses of Pre-study Power for Design

As a Sharp Measure

Statistical power was originally formulated as a measure of performance to evaluate competing statistical procedures (cf. the Neyman-Pearson's [1933] lemma). Suppose that a researcher is interested in a difference between reaction times of two conditions, where reaction times are usually non-normally distributed (e.g., see Van Zandt, 2002). This difference can be evaluated using a *t*-test with equal variances, a *t*-test with unequal variances, a sign-test, and a likelihood ratio test (LRT), where the fit of each test's underlying model to the collected data is likely imperfect. The relative performance of each of these tests can be compared to one another by computing power for the same N and α over the same repeated samples of data while varying $f(\cdot)$ in $(1 - \beta) = f(\alpha, N, \boldsymbol{\theta})$. Here, different $f(\cdot)$ map onto different statistical procedures and the inputs to $f(\cdot)$ are held constant. Thus, differences in $(1 - \beta)$ are unequivocally attributed to differences between the competing tests, and it matters not whether the hypothetical data align with the population in reality. Any procedure that preserves the nominal Type I error rate, α, and has higher power even by a small percentage, is quantitatively better than lower powered tests. Thus, power is useful for *sharply distinguishing* which tests are best applied in particular settings, allowing the researcher to select the best test within the context of their planned study (Neyman, 1957). In developing quantitative methods, power is an essential performance measure used to evaluate the robustness of statistical models and methods across different data conditions (e.g., assumption violation, missing data). For example, MacKinnon, Lockwood, Hoffman, West, and Sheets (MacKinnon et al., 2002) evaluated the performance of 14 different approaches to testing simple mediation and ultimately recommended the bootstrap of the product of coefficients because of its superior performance in terms of preserving α and optimizing power.

As a Blunt Measure

It is common practice to use power calculations to determine N (e.g., Cohen, 1988), where the focus is on determining a particular N to achieve a target level of power, expressed as $N = f(\alpha, [1 - \beta], \boldsymbol{\theta})$. Recall, however, that calculated pre-study power for design values (i.e., values applied across random data) does not directly translate to completed studies (where data are fixed). The distance between a calculated power for design value and any notion of power applied to a completed study using the same design reflects not only the departure from the frequentist concepts underlying power for design but also because of the distance between the theorized pre-study population, model, and data and the *actual* (post-study) population, model, and data. Thus, recommendations for designs to achieve particular levels of power by varying N (e.g., 80% by Cohen, 1988; 90% by Beribisky et al., 2019) do not translate to a single completed study using that design. Instead, power can be used as a blunt measure to determine ballpark requirements of sample size (i.e., it can be used by researchers as a guide to help determine feasibility of a study in terms of N requirements).

In fact, the discipline of psychology supports the use and reporting of statistical power as a blunt measure and a way to emphasize the importance of careful, thoughtful study design. Wilkinson and the Task Force for Statistical Inference (TFSI, 1999, p. 596) recommended reporting power because "[t]he intellectual exercise required to do this stimulates authors to take seriously prior research and theory in their field, and it gives an opportunity, with incumbent risk, for a few to offer the challenge that there is no applicable research behind a given study." Similarly, Simmons, Nelson, and Simonsohn (2011, 2012) recommended reporting how a fixed sample size N was determined prior to data collection (cf. N-determination via power analysis) to potentially curb the QRP of conducting multiple tests on data as they stream in without error control. These authors did not focus much on specifying numerical values of pre-study power for design but pointed to the act of conducting power analysis as evidence against disguising exploratory research as hypothetico-deductive or confirmatory research. Pek and Park (2019) forwarded a similar argument in the context of incorporating additional sources of uncertainty in the calculation of pre-study power for design. These authors demonstrated that taking into account realistic sources of uncertainty in power analysis only tends to increase the uncertainty of power estimates whereby the range of power estimates tend to cover most of the space between zero and 1. Therefore, the real value in engaging in power analysis is that it provides *a systematic framework for researchers to consider possible effect sizes associated with possible design features and to consider issues related to potential statistical models and tests.*

As an Index to Interpret Results

Recent arguments that promote power analysis as part of good scientific practice, however, seem to imply that the value of calculated pre-study power for design is informative for interpreting results of completed studies. For instance, Cooper's (2016) *Journal of Personality and Social Psychology* editorial emphasized the problems of publishing underpowered studies, seemingly attaching the concept of power to fixed data in the post-study phase. Similarly, the APA Publications and Communications Board Task Force Report recommended reporting pre-study power analysis not only in the methods (pre-study) section but also in the analysis (post-study) section (Appelbaum et al., 2018). Additionally, the *Social and Personality and Social Psychology* Task Force on Publication and Research Practices emphasized the importance of sufficiently powered studies and "recommend that a priori statistical power be reported whenever possible and considered as one factor among many when interpreting results" (Funder et al., 2013, p. 7). Inherent in these recommendations is the implicit (mis)belief that pre-study power for design transfers onto a realized study that used this design. Yet, because power is a pre-study concept applied to random data across samples, use of power for result interpretation is logically inconsistent with the statistical theory underlying power (Cohen, 1973; Cox, 1958; Senn, 2002). Recognizing that pre-study power does not determine the strength of the results from a completed study, Fisher (1947, p. 24) wrote, "[power] contribute[s] nothing to the validity of the experiment and of the test of significance by which we determine its result."

Misuses of Power for Evaluating Completed Studies

Use of power to evaluate completed studies ignores the distance between the pre-study design and post-study phase and treats calculated values of power as sharp measures even though they are blunt (cf. Fisher, 1947). Unfortunately, the popularity of this use of power has also perpetuated other misconceptions that have been counterproductive to progress in psychological science. We describe some of these in the sections to follow.

Power of Published Studies Cannot Be Reasonably Estimated

Many reviews report low power of published studies in the literature (e.g., median power of .21 as reported by Button et al., 2013; see also Chan & Altman, 2005; Cohen, 1962; Dumas-Mallet et al., 2017; Fraley & Vazire, 2014; Freiman et al., 1978; Rossi, 1990; Sedlmeier & Gigerenzer, 1989; Smaldino & McElreath, 2016; Stanley et al., 2018; Szucs & Ioannidis, 2017). Given that power is a pre-study concept, one should question how accurate estimates of power for published studies are. Power for published studies is computed using the same expression as pre-study power for design, $(1 - \beta) = f(N, \alpha, \boldsymbol{\theta})$. However, instead of using hypothetical values for N and $\boldsymbol{\theta}$ while holding α at the nominal .05 level, these reviews compute power with N and $\boldsymbol{\theta}$ observed from the literature. Following the convention of using the caret symbol "^" to represent values obtained from collected data, let $\hat{N}$ and $\hat{\theta}$ denote sample-based values of N and $\boldsymbol{\theta}$, respectively. By using $\hat{N}$ and $\hat{\theta}$ to compute power, the obtained value would seem to reflect a feature of the completed study or studies that gave rise to $\hat{N}$ and $\hat{\theta}$.

There are two competing interpretations tied to values computed using $f\left(\alpha,\hat{N},\hat{\theta}\right)$. The first is that the calculated value is a re-expression of the p-value and does not quantify the same uncertainty over random data as in pre-study power for design (cf. later discussion of observed power). The second is that the value $f\left(\alpha,\hat{N},\hat{\theta}\right)$ is as an estimate of pre-study power for design of the studies that produced $\hat{N}$ and $\hat{\theta}$. Treating $f\left(\alpha,\hat{N},\hat{\theta}\right)$ as an estimate of $f(\alpha, N, \boldsymbol{\theta})$ implies a belief that the calculated value quantifies the same probability over random data as the concept of pre-study power for design. However, as discussed in the following sections, methodological research reveals that $f\left(\alpha,\hat{N},\hat{\theta}\right)$ as an estimate of $f(\alpha, N, \boldsymbol{\theta})$ cannot be interpreted with confidence because of its high imprecision (i.e., the estimate is highly variable and therefore unreliable).

When $\hat{N}$ and $\hat{\theta}$ come from a single completed study, calculated power has been called observed power (O'Keefe, 2007). This value has been used to provide a reason why statistical significance was not achieved in a completed study, which we term power for evaluation (e.g., the p-value is larger than α because the completed study was underpowered). However, power for evaluation as a concept has been debunked because it is a mere re-expression of the p-value (Hoenig & Heisey, 2001, Lenth, 2001; see also p_{rep} by Killeen [2005] as debunked by Iverson, Lee, Zhang, & Wagenmakers [2009] and Maraun and Gabriel [2010]). The use of power for evaluation to explain statistical nonsignificance is tautological. Recall that the distinction between α and the p-value extends to $(1 - \beta)$ and power for evaluation. Whereas $(1 - \beta)$ is the probability over random samples that the test rejects H_0 if H_1 is true for some N and α (cf. interpretation of α), power for evaluation is a statement about the observed (fixed) data and more extreme data relative to H_1 where $\theta_f = \hat{\theta}_f$ for $\hat{N}$ and some level of α (cf., interpretation of p-value). Thus, pre-study power for design is a statement about random (hypothetical) data whereas power for evaluation is a statement about fixed (observed) data.

For the value calculated with $f\left(\alpha,\hat{N},\hat{\theta}\right)$ to be conceptually distinct from power for evaluation and to quantify the same concept of probability over random data in pre-study power for design, $f\left(\alpha,\hat{N},\hat{\theta}\right)$ must adequately estimate $f(\alpha, N, \boldsymbol{\theta})$ (target power of the original design). Yuan and Maxwell (2005) examined how well observed power $f\left(\alpha,\hat{N},\hat{\theta}\right)$ based on data from a single study estimates pre-study target power for design behind the completed study, $f(\alpha,N,\theta)$. From analytics and simulation studies, they concluded that $f\left(\alpha,\hat{N},\hat{\theta}\right)$ is an extremely imprecise estimate of $f(\alpha, N, \boldsymbol{\theta})$ due to the power estimate having high variability. Stated differently, confidence intervals (CIs) around $f\left(\alpha,\hat{N},\hat{\theta}\right)$ are extremely wide when its value is not .05 or 1. An estimate is statistically consistent when $N\rightarrow\infty$ results in the estimate (e.g., $\overline{x}$) approaching its target—usually a model parameter (e.g., μ)—in probability. Statistical consistency means that when $N\rightarrow\infty$, $\overline{x}\xrightarrow{p}\mu$, where the p indicates that the convergence is in probability. Such convergence occurs when sample means are used to estimate the population mean, but part of the reason for this is that the value of the population mean does not depend on N (whereas the variability of the sample means does depend on N). In the case of estimating power for published studies, however, as $N\rightarrow\infty$, the target pre-study power for design, $f[\alpha, N, \boldsymbol{\theta}]$, is not independent of N. Rather, as N increases, the target power value also changes, and the sampling variability of its estimate $f\left(\alpha,\hat{N},\hat{\theta}\right)$ hardly decreases; as $N\rightarrow\infty$, $f\left(\alpha,\hat{N},\hat{\theta}\right)\overset{p}{\nrightarrow}f(\alpha,N,\theta)$. Thus, estimates of power have much poorer statistical properties than estimates of population parameters, such as means or regression slopes.

Consider pre-study power for design for a paired-samples two-sided t-test. We increase N to increase pre-study power for design while holding $\alpha = .05$ and $\delta = 0.3$.

Table 11.1 Pre-data power for design for $\delta = 0.3$ and $\alpha = .05$ for varying levels of N

		Sampling distribution of estimate $f\left(\alpha,\hat{N},d\right)$		
N	Pre-study power for design $f(\alpha, N, \delta)$	Mean	Median	95% CI
25	.30	.37	.30	[.05, .96]
56	.60	.57	.59	[.06, .99]
89	.80	.73	.80	[.13, 1.00]
117	.90	.81	.90	[.24, 1.00]
192	.99	.94	.98	[.58, 1.00]
250	≈1	.97	1.00	[.79, 1.00]

Note. N = sample size, 95% CI = empirical confidence interval over 1000 Monte Carlo samples. The sampling distribution of the estimate $f\left(\alpha,\hat{N},d\right)$ is skewed as seen by the distance between the mean and median values of $f\left(\alpha,\hat{N},d\right)$, and has a very wide range. Increasing N does not greatly shrink the variance of the sampling distribution of $f\left(\alpha,\hat{N},d\right)$, pointing to high variability in the estimate

Table 11.1 presents different levels of pre-study power for design against N and summary statistics of the simulated sampling distribution of the estimate $f\left(\alpha,\hat{N},d\right)$ computed over 1000 Monte Carlo samples. From Table 11.1, increasing N systematically increases pre-study power for design $f(\alpha, N, \delta)$. Given a design, 1000 Monte Carlo samples of size N were drawn from a standard normal population with mean $\delta = 0.3$ and a Cohen's d estimate was computed for each sample. Using these sample estimates of d, 1000 values of $f\left(\alpha,\hat{N},d\right)$ were computed to construct an empirical distribution of this estimate. With increasing N, the mean of the distribution of $f\left(\alpha,\hat{N},d\right)$ approaches the pre-study power for design target value, but reveals bias even when pre-study power for design is ≈ 1.00 at $N = 250$. Instead, the median of the sampling distribution of $f\left(\alpha,\hat{N},d\right)$, reproduces pre-study power for design. The separation between the mean and median implies that the sampling distribution of $f\left(\alpha,\hat{N},d\right)$ is skewed. More importantly, high variability is evident in the very wide 95% empirical CIs (covering most of the possible range from 0 to 1) for most values of pre-study power for design. When N increases from 25 to 192, the variability in $f\left(\alpha,\hat{N},d\right)$ remains high, implying that this estimate of pre-study power for design cannot be interpreted with confidence because of its imprecision. An R Shiny application that dynamically presents the sampling distribution of d and $f\left(\alpha,\hat{N},d\right)$ in relation to specified δ and N is available online.[4]

One might imagine that the high variability in the estimate of pre-study power for design is due to using $\hat{N}$ and $\hat{\theta}$ from a single study where information is limited. To distinguish the estimate $f\left(\alpha,\hat{N},\hat{\theta}\right)$ from observed power, researchers have also used meta-analytic values of $\hat{N}$ and $\hat{\theta}$ to estimate pre-study power behind a collection of studies in an area of scholarship (e.g., Button et al., 2013; Dumas-Mallet et al., 2017; Stanley et al., 2018). There are several advantages of using information from a meta-analysis in place of a single study, including the ability to incorporate heterogeneous effect sizes across studies, study-level moderators, publication bias, and other factors. The meta-analytic effect size is also likely more precise than that from a single study. McShane, Böckenholt, and Hansen (2020) reported on the properties of $f\left(\alpha,\hat{N},\hat{\theta}\right)$ as an estimate of pre-study power for design that underlie the studies in the meta-analysis. Compared to single study $f\left(\alpha,\hat{N},\hat{\theta}\right)$, meta-analytic $f\left(\alpha,\hat{N},\hat{\theta}\right)$ has a sampling distribution that is less skewed and closer to the normal distribution (see Fig. 11.1 in McShane et al., 2020). However, similar to a single study $f\left(\alpha,\hat{N},\hat{\theta}\right)$, meta-analytic $f\left(\alpha,\hat{N},\hat{\theta}\right)$ estimates continue to be highly variable and imprecise (as communicated by large CIs), and attempts to correct for publication bias only increase this imprecision (McShane et al., 2020). Thus, regardless of whether single or multiple collected studies inform $f\left(\alpha,\hat{N},\hat{\theta}\right)$, this value remains an imprecise estimate of the pre-study target power for design $f(\alpha, N, \boldsymbol{\theta})$ behind the completed study, and therefore cannot be treated as a precise estimate of pre-study

[4] https://seeing-statistics.shinyapps.io/EstimatedPower/

power. Thus, it is unfortunate that researchers have treated values computed with $f\left(\alpha,\hat{N},\hat{\theta}\right)$ as though they accurately reflect the power of completed studies.[5] Methodological research reveals that $f\left(\alpha,\hat{N},\hat{\theta}\right)$ is an estimate of $f(\alpha, N, \boldsymbol{\theta})$ that cannot be interpreted with confidence because of its high imprecision (i.e., the estimate is highly variable with large CI). Obtaining values of the power of completed studies thus remains elusive, and the application of the power concept to completed data remains contrary to the notion that power has no influence on the strength (or weakness) of the evidence conveyed by collected data (e.g., see Cohen, 1973; Cox, 1958; Senn, 2002).

Because power of completed studies cannot be calculated with reasonable precision, arguments that make use of values of $f\left(\alpha,\hat{N},\hat{\theta}\right)$ in power for evaluation cannot be interpreted with confidence. One such argument links low statistical power to false findings[6] (e.g., see Christley, 2010; Colquhoun, 2014; Ioannidis, 2005; Pashler & Harris, 2012). These researchers treat decision probabilities (α, β, and $[1 - \beta]$) as conditional probabilities (even though they are *not* conditional probabilities under the frequentist framework; see Morey & Lakens, 2016). They regard the Type I error rate (α) as the probability of rejecting H_0 conditioned on H_0 true; that is, P[reject H_0| H_0 true]. Similarly, they treat the Type II error rate (β) as the probability of not rejecting H_0 conditioned on H_1 is true; that is, P[not reject H_0| H_1 true]. Accordingly, they treat power, $(1 - \beta)$, as the probability of correctly rejecting H_0 conditioned on H_1 is true; that is, P[reject H_0| H_1 true]. Key to this argument is the treatment of decision probabilities as conditional, such that there is a probability attached to the occurrence of H_0 and H_1. Let π denote the probability of H_0. Then, $\pi = P(H_0)$ and $1 - \pi = P(H_1)$. Given a value of π, use of Bayes' Theorem to invert the Type I error rate P(reject H_0| H_0 true) produces the false finding rate (FFR).

$$\text{FFR} = P\left(H_0 \text{ true}|\text{reject } H_0\right) = \frac{\pi\alpha}{\pi\alpha + (1-\pi)(1-\beta)}. \tag{11.1}$$

The conditional probability in Eq. 11.1 is interpreted as the proportion of false findings among all rejections of H_0. For example, if $\pi = P(H_0) = .90$, $(1 - \pi) = P(H_1) = .10$, $\alpha = .05$ and $(1 - \beta) = .35$, $\text{FFR} = \frac{.90 \times .05}{.90 \times .05 + .10 \times .35} = .56$ (e.g., see Pashler & Harris, 2012). With 35% power, an alarming 56% of H_0 rejections with $\alpha = .05$ and $\pi = .90$ are false findings. Increasing power to 80% reduces the FFR to 36% (a 20%

[5] Power for completed studies have also been computed by using Cohen's t-shirt effect sizes (small, medium, large) and sample sizes from observed studies, $f\left(\alpha,\hat{N},\theta\right)$. These values also cannot accurately reflect power for a specific area of study because the input of $\boldsymbol{\theta}$ is even less likely to reflect the phenomenon under study.

[6] We use the term false finding rate (FFR) as coined by Mayo and Morey (2017) instead of the false discovery rate (e.g., see Ioannidis, 2005) to distinguish the target concept from corrections to multiple testing (e.g., see Benjamini & Hochberg, 1995). The FFR is the probability of H_0 being true given a decision to reject H_0; FFR = $P(H_0$ true | reject $H_0)$.

reduction), ceteris paribus. These values are taken to support the argument that increasing power will have a strong effect of reducing the FFR, thereby buttressing the importance of implementing highly powered studies. The theoretical relation between power and the FFR is undeniably formalized in Eq. 11.1, but for it to be applicable to published research, a number of assumptions inherent in those calculations must also accurately reflect conducted research. First, the value of power used in Eq. 11.1 must reflect the power of published studies. However, recall that methodological work confirms that the power for published studies is estimated with high imprecision. Second, reasonable values for the prior probability of the null hypothesis must be established. However, little justification (and no empirical basis) for the prior probabilities has been given, and the relation between power and the FFR becomes quite weak for most plausible values of $P(H_0)$. Thus, the relation between power and the FFR remains a theoretical abstraction. For other criticisms about the relations between power and FFR made by statisticians, methodologists, and philosophers, see Goodman and Greenland (2007a, 2007b), Mayo and Morey (2017), and Wegener et al. (2022).

Statistical Forensics

The use of statistics to hunt for anomalies, termed "statistical forensics" by Morey (2019), has a long history that includes Fisher's (1936) use of the χ^2 test to raise suspicions that Mendel's (1886) results on pea plants were "too good." In brief, Mendel's results did not exhibit enough variability relative to predictions from a statistical model, *suggesting* but not confirming that the results could have been selectively reported. Mendel has been deemed innocent of Fisher's suspicions (Franklin et al., 2008; Hartl & Fiarbanks, 2007). However, note that, opposite from the usual perspective taken during data analysis in which the model is assumed to be in error relative to the data (cf. model diagnostics), statistical forensics assumes that the reported data are potentially erroneous relative to a statistical model.

With similar motivations to the Mendel-Fisher controversy, power calculations have been used in statistical forensic methods to uncover plausible instances of QRPs (i.e., publication bias, selective reporting, and *p*-hacking) in modern psychological research. These methods similarly assume that the investigated data or results are potentially erroneous relative to the model used to evaluate these data. Here, a statistical model is used to represent characteristics of data that have been analyzed without using QRPs. Then, results from analyzed data are compared to this model. When the distance between the model and data is large, QRPs are suspected. Example methods are Ioannidis and Trinkalinos' (2007) exploratory excessive significance test,[7] Francis' (2013) consistency test, Schimmack's incredibility index

[7] Ioannidis and Trikalinos (2007) were extremely careful in labeling their test as an exploratory method. They explicitly acknowledged the possibility of assumption violation that would invalidate their test's results and emphasized caution in broadly applying it.

(2012), and Brunner and Schimmack's *z*-curve method (2020). In general, these methods incorporate statistical power to evaluate whether QRPs have occurred. These methods have been theoretically derived using the concept of pre-study power for design, $f(\alpha, N, \boldsymbol{\theta})$ and evaluated using Monte Carlo simulations that assume equivalence between the population, model, and data. However, when these methods are applied to published research, the population, model, and data are undeniably distinct. Currently, there are no model diagnostics to evaluate how closely these assumed models approximate empirical data analyzed without QRPs, leaving unanswered the question of whether these models are reasonable benchmarks to use in the post-study phase of research (cf. Box's, 1976 position that all models are wrong and some are useful; see also MacCallum, 2003; Rodgers, 2010; Tukey, 1969). Furthermore, these methods make use of power calculations from observed data, $f\left(\alpha,\hat{N},\hat{\theta}\right)$, that tend to be highly imprecise estimates (cf. Table 11.1). In the following paragraphs, we explicate these issues using one such statistical forensic method.

Consider Ioannidis and Trinkalinos' (2007) exploratory χ^2 test to detect excess significance among a collection of studies. Let S be the total number of completed studies indexed by s such that $s = 1, \cdots, S$. Let O be the count of observed statistically significant findings (e.g., $p < .05$) and E be the expected count of statistically significant findings. E is calculated based on power for each study

$$E = \sum_{s=1}^{S} f_s\left(\alpha,N,\theta\right), \tag{11.2}$$

where $f_s(\alpha, N, \boldsymbol{\theta})$ is power calculated for the sth study. Given that power is the probability over random data that the test will reject H_0 under the assumption that H_1 is true, the sum of $(1-\beta)_s = f_s(\alpha, N, \boldsymbol{\theta})$ across the S studies will give rise to the average number of significant findings based on the design of these studies. The test statistic is $X^2 = \dfrac{\left(O-E\right)^2/E}{\left(O-E\right)^2/\left(S-E\right)}$ and is assumed to follow a χ^2 distribution with 1 degree of freedom. Note that O represents the data under evaluation and E represents the statistical model that is assumed to reflect data analyzed without QRPs (e.g., without selectivity in reporting results). A statistically significant test result indicates a discrepancy between O and E, which is taken to suggest that the collection of S studies has an excess of statistically significant results that imply the use of QRPs. Monte Carlo simulations that assume equivalence between the population, model, and data show that this test performs according to expectations in theory (Francis, 2013).

However, when this test is applied to collected data, its performance remains unknown because the population, model, and data are distinct. How well the test performs under Monte Carlo simulation is unlikely to reflect how well the test performs with empirical data because of the separation between (pre-study) theory and (post-study) reality. We list four places of potential separation between theory and practice in Ioannidis and Trinkalinos' (2007) test.

First, O is derived from the p-value that unrealistically assumes that the model fit to the data is a perfect representation of the population (e.g., see Box, 1976; Greenland & Poole, 2013; MacCallum, 2003). When the model is an imperfect representation of the population, the calculated p-values will not be accurate to some degree. The accuracy of the p-value is moderated by how well the statistical model fits the sample data and distally approximates the population.

Second, power calculated using $f_s\left(\alpha,\hat{N},\hat{\theta}\right)$ is a re-expression of the p-value for the sth study (cf. observed power; Greenwald et al., 1996; Hoenig & Heisey, 2001; Lenth, 2001). When $f_s\left(\alpha,\hat{N},\hat{\theta}\right)$ is treated as a re-expression of the p-value, O would also be a function of E because both values are based on the same p-value. Then, the comparison between O and E would technically be comparing variations of the same statistic (i.e., the p-value). In this vein, the X^2 test statistic does not compare the data (O) against a model assuming no QRPs (E), but instead compares the data (O) to a function of the same data (E as a re-expression of O). If $f_s\left(\alpha,\hat{N},\hat{\theta}\right)$ is instead treated as an estimate of pre-study target power for design underlying the sth study, $f_s(\alpha, N, \boldsymbol{\theta})$, then the values of E are highly imprecise (cf. Yuan & Maxwell, 2005, discussed earlier, and Table 11.1), and results of the χ^2 test cannot be interpreted with confidence.

Third, the summation in Eq. 11.2 to obtain values of E assumes that the S studies all come from the same population (i.e., the studies are independently and identically distributed); if the studies are from different populations, then their summation is not justified. In empirical research, there is heterogeneity of results even in direct replications (Hedges & Schauer, 2019; Kenny & Judd, 2019; results from the Many Labs project by Klein et al., 2014, 2018), pointing to potential violation of this essential assumption.

Fourth, there is currently no way to check whether the model assuming no QRPs approximates empirical data well (cf. model diagnostics under data analysis) because the data are assumed to have undergone QRPs. Thus, conclusions from this test are based on the strong, likely unrealistic, and a currently untestable assumption that the model is a perfect representation of data free from QRPs. Because it is unclear how tests of QRPs perform under assumption violation, conclusions made from results of such tests would carry much uncertainty.

Although we focus on Ioannidis and Trinkalinos' (2007) exploratory test, the other tests purported to uncover QRPs (Brunner & Schimmack, 2020; Francis, 2013; Schimmack, 2012) rely on similar assumptions and calculations of power for completed studies. Simulation studies validate the performance of these methods on hypothetical data, but the separation between theoretical and empirical data raises questions about the valid performance of these methods on collected data. Similar to Ioannidis and Trinkalinos' (2007) position of caution, we encourage much care in the application of statistical forensics that rely on power values calculated from collected data, $f\left(\alpha,\hat{N},\hat{\theta}\right)$, because of the unbridgeable separation between the pre-study and post-study phases of research. Results from such tests for QRPs cannot be definitive and remain, at best, exploratory (see also Fabrigar & Wegener, 2016, 2017).

Summary and Discussion

The concept of statistical power is seemingly simple. Pre-study power for design is just the probability (over random data) of rejecting H_0 if some H_1 value for θ_f is true for specific α, sample size, and $\boldsymbol{\theta}_n$; $(1 - \beta) = f(\alpha, N, \boldsymbol{\theta})$. However, in the post-study phase of research, when data are no longer random but fixed, the concept of power is irrelevant to determining the strength or weakness of the obtained results (Cohen, 1973; Cox, 1958; Senn, 2002). That is, just as the pre-study α value does not add to the informativeness of the p-value associated with obtained data, the pre-study power for design does not change the informativeness of the obtained data. The population, statistical model, and data are conveniently assumed to be equivalent in the pre-study phase of research to allow for easy analytical derivations (e.g., α and β). However, these three entities are distinct in the post-study phase of research. Thus, power calculations that are based on the assumed equivalence of the population, model, and data in the pre-study phase of research cannot be automatically generalized to the post-study phase of research where the population, model, and data are recognized to be distinct (Box, 1976; MacCallum, 2003). To do so is, in fact, highly questionable from a statistical point of view.

Even if the population, model, and data were highly similar, there remains ontological inconsistency in the meaning of the probability quantified by a value of power in the pre- versus the post-study phase. Pre-study power for design is a probability over random data (similar to the Type I error rate, α). In contrast, power calculated from collected (fixed) data, $f\left(\alpha,\hat{N},\hat{\theta}\right)$, is a probability about the observed data and more extreme (future and past) data relative to H_1 given an observed effect size, $\hat{\theta}_f$, for observed $\hat{N}$, and some level of α (similar to the p-value). Attempts to address this ontological inconsistency use $f\left(\alpha,\hat{N},\hat{\theta}\right)$ as an estimate of pre-study target power for design, $f(\alpha, N, \boldsymbol{\theta})$. Yet, this estimate has high variability in that its values tend to be imprecise and relatively uninformative (McShane et al., 2020; Yuan & Maxwell, 2005).

Because power calculated from collected data, $f\left(\alpha,\hat{N},\hat{\theta}\right)$, has high variability that undermine its informativeness, claims that make use of its numerical value should also be interpreted with the expectation that downstream numerical values are equally uninformative (i.e., imprecise). Contrary to reviews on the power of studies in the literature, power for completed studies cannot be estimated with precision. Although power might, in concept, be linked to the FFR in the pre-study design phase, the strength of this link remains unknown in the post-design phase (see also Goodman & Greenland, 2007a, 2007b; Mayo & Morey, 2017). Finally, methods to detect QRPs that rely on power calculated from collected data should be cautiously utilized with an appreciation that many assumptions underlying these methods cannot be empirically evaluated and are surely violated in the post-study phase of research. In sum, power calculations based on collected data have limited use to interpret results because this value is simply a re-expression of the *p*-value (Greenwald et al., 1996; Hoenig & Heisey, 2001; Lenth, 2001). Alternatively, power calculations based on collected data as an estimate of power for design tend to be highly imprecise and should be used with caution. As a result, any inferences about completed studies that rely on power calculations using the collected data should also be recognized as questionable because such inferences often entail heavy reliance on estimated power values that have high variability.

In the pre-study phase of research, calculated values of pre-study power for design are useful for evaluating the performance of procedures under different data conditions (cf. Neyman-Pearson [1933] lemma). Such tools are often used in quantitative methodological research. Calculated values of pre-study power for design are irrelevant to collected data (Fisher, 1947). However, power analysis remains useful for assessing design feasibility under resource constraints as long as uncertainty surrounding the calculations is taken into account. The main benefits of power analysis lie in the *act* of considering different designs and downstream data analytic plans (cf. Pek & Park, 2019). Initial recommendations for reporting power analysis emphasized the value of relating a study to previous research as evidence that the reported study was not presented as confirmatory if it was exploratory (Wilkinson & TFSI, 1999). Similarly, reporting a power analysis was recommended to justify the collected sample size N to discourage multiple testing without corrections (Simmons et al., 2012). These positive outcomes of power analysis need not be tied to specific values of pre-study power for design (e.g., 80% by Cohen, 1988). Rather, these considerations fall under the larger umbrella of research methods that is often tied to a much broader validity framework (Shadish et al., 2002; see also Fabrigar et al., 2020; Finkel et al., 2017).

Acknowledgement of that broader framework brings into sharp relief the multifaceted nature of research evaluation and the credibility of claims based on research results. That is, rather than focusing merely on the statistical results of reported studies, a broader evaluation of the research claims based on the results must go far beyond an index of research credibility based on estimated power values. Such evaluations also take into account the goals of the research, the design features of the research—including qualities of the manipulation and measurement of relevant variables, the use of appropriate analyses, and the fit of each of those elements to the claims that are ultimately advanced (for additional discussion, see Fabrigar & Wegener, 2016; Fabrigar et al., 2020). To put too much emphasis on estimates of power in such evaluations is to accept risks associated with departures from the statistical theory underlying the very index on which one is basing one's claims. Such emphasis also incurs risks associated with reliance on estimates with high variability and risks associated with unknown (and potentially unbridgeable) differences between the assumed statistical model and the realities of the research setting. Indeed, power analysis is not so simple as it often might seem.

References

Appelbaum, M., Cooper, H., Kline, R. B., Mayo-Wilson, E., Nezu, A. M., & Rao, S. M. (2018). Journal article reporting standards for quantitative research in psychology: The APA publications and communications board task force report. *American Psychologist, 73*(1), 3–25. https://doi.org/10.1037/amp0000191

Benjamin, D. J., Berger, J. O., Johannesson, M., Nosek, B. A., Wagenmakers, E. J., Berk, R., … Cesarini, D. (2018). Redefine statistical significance. *Nature Human Behaviour, 2*(1), 6–10. https://doi.org/10.1038/s41562-017-0189-z

Benjamini, Y., & Hochberg, Y. (1995). Controlling the false discovery rate: A practical and powerful approach to multiple testing. *Journal of the Royal Statistical Society, Series B, 57*(1), 289–300.

Beribisky, N., Alter, U., & Cribbie, R. A. (2019). A multi-faceted mess: A review of statistical power analysis. *Psychology Journal Articles* [preprint]. https://doi.org/10.31234/osf.io/3bdfu.

Box, G. E. (1976). Science and statistics. *Journal of the American Statistical Association, 71*(356), 791–799. https://doi.org/10.1080/01621459.1976.10480949

Brunner, J., & Schimmack, U. (2020). Estimating population mean power under conditions of heterogeneity and selection for significance. *Meta-Psychology, 4*. https://doi.org/10.15626/mp.2018.874.

Button, K. S., Ioannidis, J. P., Mokrysz, C., Nosek, B. A., Flint, J., Robinson, E. S., & Munafò, M. R. (2013). Power failure: Why small sample size undermines the reliability of neuroscience. *Nature Reviews Neuroscience, 14*(5), 365–376. https://doi.org/10.1038/nrn3475

Chan, A.-W., & Altman, D. G. (2005). Epidemiology and reporting of randomised trials published in PubMed journals. *The Lancet, 365*(9465), 1159–1162. https://doi.org/10.1016/s0140-6736(05)71879-1

Christley, R. (2010). Power and error: Increased risk of false positive results in underpowered studies. *The Open Epidemiology Journal, 3*(1), 16–19. https://doi.org/10.2174/1874297101003010016

Cohen, J. (1973). Brief notes: Statistical power analysis and research results. *American Educational Research Journal, 10*(3), 225–229. https://doi.org/10.3102/00028312010003225

Cohen, J. (1962). The statistical power of abnormal-social psychological research: A review. *The Journal of Abnormal and Social Psychology, 65*(3), 145–153. https://doi.org/10.1037/h0045186

Cohen, J. (1988). *Statistical power analysis for the behavioral sciences*. Academic Press.

Colquhoun, D. (2014). An investigation of the false discovery rate and the misinterpretation of *p*-values. *Royal Society Open Science, 1*(3), 140216. https://doi.org/10.1098/rsos.140216

Cooper, M. L. (2016). Editorial. *Journal of Personality and Social Psychology, 110*(3), 431–434. https://doi.org/10.1037/pspp0000033

Cox, D. R. (1958). *Planning of experiments*. Wiley.

Cudeck, R., & Henly, S. J. (1991). Model selection in covariance structures analysis and the "problem" of sample size: A clarification. *Psychological Bulletin, 109*(3), 512–519. https://doi.org/10.1037/0033-2909.109.3.512

Dumas-Mallet, E., Button, K. S., Boraud, T., Gonon, F., & Munafò, M. R. (2017). Low statistical power in biomedical science: A review of three human research domains. *Royal Society Open Science, 4*(2), 160254. https://doi.org/10.1098/rsos.160254

Fabrigar, L. R., & Wegener, D. T. (2016). Conceptualizing and evaluating the replication of research results. *Journal of Experimental Social Psychology, 66*, 68–80. https://doi.org/10.1016/j.jesp.2015.07.009

Fabrigar, L. R., & Wegener, D. T. (2017). Further considerations on conceptualizing and evaluating the replication of research results. *Journal of Experimental Social Psychology, 69*, 241–243. https://doi.org/10.1016/j.jesp.2016.09.003

Fabrigar, L. R., Wegener, D. T., & Petty, R. E. (2020). A validity-based framework for understanding replication in psychology. *Personality and Social Psychology Review, 24*(4), 316–344. https://doi.org/10.1177/1088868320931366

Finkel, E. J., Eastwick, P. W., & Reis, H. T. (2017). Replicability and other features of a high-quality science: Toward a balanced and empirical approach. *Journal of Personality and Social Psychology, 113*(2), 244–253. https://doi.org/10.1037/pspi0000075

Fisher, R. A. (1936). Has Medel's work been rediscovered? *Annals of Science, 1*(2), 115–137. https://doi.org/10.1080/00033793600200111

Fisher, R. A. (1947). *The design of experiments* (4th ed.). Hafner Press.

Fox, J. H. (1958). Criteria of good research. *The Phi Delta Kappa, 39*(6), 284–286.

Fraley, R. C., & Vazire, S. (2014). The N-pact factor: Evaluating the quality of empirical journals with respect to sample size and statistical power. *PLoS One, 9*(10), e109019. https://doi.org/10.1371/journal.pone.0109019

Francis, G. (2013). Replication, statistical consistency, and publication bias. *Journal of Mathematical Psychology, 57*(5), 153–169. https://doi.org/10.1016/j.jmp.2013.02.003

Franklin, A., Edwards, A. W. F., Fairbanks, D. J., Hartl, D. L., & Seidenfeld, T. (2008). *Ending the Mendel-Fisher controversy*. University of Pittsburg Press.

Freiman, J. A., Chalmers, T. C., Smith, H., & Kuebler, R. R. (1978). The importance of beta, the type II error, and sample size in the design and interpretation of the randomized controlled trial. *New England Journal of Medicine, 299*(13), 690–694. https://doi.org/10.1201/9780429187445-19

Funder, D. C., Levine, J. M., Mackie, D. M., Morf, C. C., Vazire, S., & West, S. G. (2013). Improving the dependability of research in personality and social psychology: Recommendations for research and educational practice. *Personality and Social Psychology Review, 18*(1), 3–12. https://doi.org/10.1177/1088868313507536

Giner-Sorolla, R., Aberson, C. L., Bostyn, D. H., Carpenter, T., Conrique, B. G., A Lewis Jr., Montoya, N., Ng, A. K., Reifman, B. W., Schoemann, A.M., & Soderberg, C. (2019). Power to detect what? Considerations for planning and evaluating sample size. [preprint]. osf.io/9bt5s.

Goodman, S. N., & Greenland, S. (2007a). Why most published research findings are false: Problems in the analysis. *PLoS Medicine, 4*(4), e168. https://doi.org/10.1371/journal.pmed.0040168

Goodman, S. N., & Greenland, S. (2007b). *Assessing the unreliability of the medical literature: A response to "why most published research findings are false" [working paper 135]*. Johns Hopkins University, Department of Biostatistics. Retrieved from: https://biostats.bepress.com/jhubiostat/paper135

Greenland, S. (2017). Invited commentary: The need for cognitive science in methodology. *American Journal of Epidemiology, 186*(6), 639–645. https://doi.org/10.1093/aje/kwx259

Greenland, S., & Poole, C. (2013). Rejoinder: Living with statistics in observational research. *Epidemiology, 24*(1), 73–78. https://doi.org/10.1097/EDE.0b013e3182785a49

Greenwald, A. G., Gonzalez, R., Harris, R. J., & Guthrie, D. (1996). Effect sizes and p values: What should be reported and what should be replicated? *Psychophysiology, 33*(2), 175–183. https://doi.org/10.1111/j.1469-8986.1996.tb02121.x

Hartl, D. J., & Fiarbanks, D. J. (2007). Mud sticks: On the alleged falsification of Mendel's data. *Genetics, 175*(2), 975–979.

Hedges, L. V., & Schauer, J. M. (2019). Statistical analyses for studying replication: Meta-analytic perspectives. *Psychological Methods, 24*, 557–570. https://doi.org/10.1037/met0000189

Hoenig, J. M., & Heisey, D. M. (2001). The abuse of power: The pervasive fallacy of power calculations for data analysis. *The American Statistician, 55*(1), 19–24. https://doi.org/10.1198/000313001300339897

Ioannidis, J. P. (2005). Why most published research findings are false. *PLoS: Medicine, 2*(8), e124. https://doi.org/10.1371/journal.pmed.0020124

Ioannidis, J. P., & Trikalinos, T. A. (2007). An exploratory test for an excess of significant findings. *Clinical Trials, 4*(3), 245–253. https://doi.org/10.1177/1740774507079441

Iverson, G. J., Lee, M. D., Zhang, S., & Wagenmakers, E.-J. (2009). Prep: An agony in five fits. *Journal of Mathematical Psychology, 53*(4), 195–202. https://doi.org/10.1016/j.jmp.2008.09.004

Kenny, D. A., & Judd, C. M. (2019). The unappreciated heterogeneity of effect sizes: Implications for power, precision, planning of research, and replication. *Psychological Methods, 24*(5), 578. https://doi.org/10.1037/met0000209

Killeen, P. R. (2005). An alternative to null-hypothesis significance tests. *Psychological Science, 16*(5), 345–353. https://doi.org/10.1111/j.0956-7976.2005.01538.x

Klein, R. A., Ratliff, K. A., Vianello, M., Adams, R. B., Bahník, Š., Bernstein, M. J., Bocian, K., Brandt, M. J., Brooks, B., Brumbaugh, C. C., Cemalcilar, Z., Chandler, J., Cheong, W., Davis, W. E., Devos, T., Eisner, M., Frankowska, N., Furrow, D., Galliani, E. M., … Nosek, B. A. (2014). Investigating variation in replicability: A "many labs" replication project. *Social Psychology, 45*(3), 142–152. https://doi.org/10.1027/1864-9335/a000178

Klein, R. A., Vianello, M., Hasselman, F., Adams, B. G., Adams, B. R., Jr., Alper, S., Aveyard, M., Axt, J. R., Babalola, M. T., Bahník, Š., Batra, R., Berkics, M., Bernstein, M. J., Berry, D. R., Bialobrzeska, O., Binan, E. D., Bocian, K., Brandt, M. J., Busching, R., … Nosek, B. A. (2018). Many labs 2: Investigating variation in replicability across samples and settings. *Advances in Methods and Practices in Psychological Science, 1*(4), 443–490. https://doi.org/10.1177/2515245918810225

Lakens, D., Adolfi, F. G., Albers, C. J., Anvari, F., Apps, M. A. J., Argamon, S. E., Baguley, T., Becker, R. B., Benning, S. D., Bradford, D. E., Buchanan, E. M., Caldwell, A. R., Calster, B. V., Carlsson, R., Chen, S.-C., Chung, B., Colling, L. J., Collins, G. S., Crook, Z., … Zwaan, R. A. (2018). Justify your alpha. *Nature Human Behaviour, 2*(3), 168–171. https://doi.org/10.1038/s41562-018-0311-xL

Lenth, R. V. (2001). Some practical guidelines for effective sample size determination. *The American Statistician, 55*(3), 187–193. https://doi.org/10.1198/000313001317098149

MacCallum, R. C. (2003). 2001 presidential address: Working with imperfect models. *Multivariate Behavioral Research, 38*(1), 113–139. https://doi.org/10.1207/S15327906MBR3801_5

MacKinnon, D. P., Lockwood, C. M., Hoffman, J. M., West, S. G., & Sheets, V. (2002). A comparison of methods to test mediation and other intervening variable effects. *Psychological Methods, 7*(1), 83–104. https://doi.org/10.1037/1082-989X.7.1.83

Maraun, M., & Gabriel, S. (2010). Killeen's (2005) P_{rep} coefficient: Logical and mathematical problems. *Psychological Methods, 15*(2), 182. https://doi.org/10.1037/a0016955

Mayo, D. G., & Morey, R. D. (2017). *A poor prognosis for the diagnostic screening critique of statistical tests*. [preprint]. https://doi.org/10.31219/osf.io/ps38b.

McShane, B. B., Böckenholt, U., & Hansen, K. T. (2020). Average power: A cautionary note. *Advances in Methods and Practices in Psychological Science, 3*(2), 185–199. https://doi.org/10.1177/2515245920902370.

Mendel, G. J. (1886). *Versuche über Pflanzen-Hybriden [Experiments Concerning Plant Hybrids]. In Verhandlungen des naturforschenden Vereines in Brünn [Proceedings of the Natural History Society of Brünn]*, 4, 3–47.

Morey, R. D. (2019, July 31). *Statistical forensics* [Paper presentation]. *Summer seminar: Philosophy of statistics,* Virginia Polytechnic Institute and State University, Virginia.

Morey, R. D. (2020, June 12). *Power and precision: Why the push for replacing "power" with precision" is misguided* [Blog post]. Retrieved from https://medium.com/@richarddmorey/power-and-precision-47f644ddea5e

Morey, R. D., & Lakens, D. (2016). *Why most of psychology is statistically unfalsifiable.* Unpublished manuscript.

Neyman, J. (1957). The use of the concept of power in agricultural experimentation. *Journal of the Indian Society of Agricultural Statistics, 9*(1), 9–17.

Neyman, J., & Pearson, E. S. (1933). IX. On the problem of the most efficient tests of statistical hypotheses. *Philosophical Transactions of the Royal Society of London, Series A, 231*(694–706), 289–337.

O'Keefe, D. J. (2007). Post hoc power, observed power, a priori power, retrospective power, prospective power, achieved power: Sorting out appropriate uses of statistical power analyses. *Communication Methods and Measures, 1*(4), 291–299. https://doi.org/10.1080/19312450701641375

Pashler, H., & Harris, C. R. (2012). Is the replicability crisis overblown? Three arguments examined. *Perspectives on Psychological Science, 7*(6), 531–536. https://doi.org/10.1177/1745691612463401

Pek, J., & Park, J. (2019). Complexities in power analysis: Quantifying uncertainties with a Bayesian-classical hybrid approach. *Psychological Methods, 24*(5), 590–605. https://doi.org/10.1037/met0000208

Popper, K. (1959). *The logic of scientific discovery*. Routledge. https://doi.org/10.4324/9780203994627

Rodgers, J. L. (2010). The epistemology of mathematical and statistical modeling: A quiet methodological revolution. *American Psychologist, 65*(1), 1–12. https://doi.org/10.1037/a0018326

Rossi, J. S. (1990). Statistical power of psychological research: What have we gained in 20 years? *Journal of Consulting and Clinical Psychology, 58*(5), 646–656. https://doi.org/10.1037//0022-006x.58.5.646

Schimmack, U. (2012). The ironic effect of significant results on the credibility of multiple-study articles. *Psychological Methods, 17*(4), 551–566. https://doi.org/10.1037/a0029487

Sedlmeier, P., & Gigerenzer, G. (1989). Do studies of statistical power have an effect on the power of studies? *Psychological Bulletin, 105*(2), 309–316. https://doi.org/10.1037//0033-2909.105.2.309

Senn, S. J. (2002). Power is indeed irrelevant in interpreting completed studies. *BMJ, 325*(7375), 1304. https://doi.org/10.1136/bmj.325.7375.1304

Shadish, W. R., Cook, T. D., & Campbell, D. T. (2002). *Experimental and quasi-experimental designs for generalized causal inference*. Houghton Mifflin.

Simmons, J. P., Nelson, L. D., & Simonsohn, U. (2012). A 21 word solution. *SSRN*. https://doi.org/10.2139/ssrn.2160588

Simmons, J. P., Nelson, L. D., & Simonsohn, U. (2011). False-positive psychology undisclosed flexibility in data collection and analysis allows presenting anything as significant. *Psychological Science, 22*(11), 1359–1366. https://doi.org/10.1177/0956797611417632

Simons, D. J.,Holcombe, A. O., & Spellman, B. A. (2014). An introduction to registered replication reports at perspectives on psychological science. *Perspectives on Psychological Science, 9*(5), 552–555. doi: https://doi.org/10.1177/1745691614543974.

Smaldino, P. E., & McElreath, R. (2016). The natural selection of bad science. *Royal Society Open Science, 3*(9), 160384. https://doi.org/10.1098/rsos.160384

Stanley, T., Carter, E. C., & Doucouliagos, H. (2018). What meta-analyses reveal about the replicability of psychological research. *Psychological Bulletin, 144*(12), 1325–1346. https://doi.org/10.1037/bul0000169

Szucs, D., & Ioannidis, J. P. (2017). Empirical assessment of published effect sizes and power in the recent cognitive neuroscience and psychology literature. *PLoS Biology, 15*(3), e2000797. https://doi.org/10.1371/journal.pbio.2000797

Trochim, W., & Land, D. (1982). Designing designs for research. *The Researcher, 1*(1), 1–6.

Tukey, J. W. (1969). Analyzing data: Sanctification or detective work? *American Psychologist, 24*(2), 83–91. https://doi.org/10.1037/h0027108

Van Zandt, T. (2002). Analysis of response time distributions. In J. T. Wixted (Ed.), *Stevens' handbook of experimental psychology: Vol. 4: Methodology in experimental psychology* (3rd ed, pp. 461–516). Wiley Online Library.

Wallace, L. E., Wegener, D. T., Quinn, M., & Ross, A. (2021). Influences of position justification on perceived bias: Carry over across persuasive messages. *Personality and Social Psychology Bulletin, 47(7)*, 1188-1204. https://doi.org/10.1177/0146167220963678

Wegener, D. T., Fabrigar, L. R., Pek, J., & Hoisington-Shaw, K. (2022). Evaluating Research in Personality andSocial Psychology: Considerations of Statistical Power and Concerns About False Findings. *Personality and Social PsychologyBulletin, 48*(7),1105-11117. https://doi.org/10.1177/01461672211030811

Wilkinson, L., & the Task Force on Statistical Inference. (1999). Statistical methods in psychology journals: Guidelines and explanations. *American Psychologist, 54*(8), 594–604. https://doi.org/10.1037/0003-066X.54.8.594

Yuan, K.-H., & Maxwell, S. (2005). On the post hoc power in testing mean differences. *Journal of Educational and Behavioral Statistics, 30*(2), 141–167. https://doi.org/10.3102/10769986030002141

Chapter 12
Questionable Research Practices in Single-Case Experimental Designs: Examples and Possible Solutions

Matt Tincani and Jason Travers

Abstract Questionable research practices (QRPs) are a variety of research choices that introduce bias into the body of scientific literature. Researchers have documented widespread presence of QRPs across disciplines and promoted practices aimed at preventing them. More recently, Single-Case Experimental Design (SCED) researchers have explored how QRPs could manifest in SCED research. In the chapter, we describe QRPs in participant selection, independent variable selection, procedural fidelity documentation, graphical depictions of behavior, and effect size measures and statistics. We also discuss QRPs in relation to the file drawer effect, publication bias, and meta-analyses of SCED research. We provide recommendations for researchers and the research community to promote practices for preventing QRPs in SCED.

Keywords Single-case experimental design · Questionable research practices · Publication bias · Open science

Overview of Single-Case Experimental Designs

Single-case experimental designs (SCED) are used by social and behavioral sciences researchers to evaluate effects of clinical interventions on therapeutic outcomes. Rooted in B.F. Skinner's (1953) radical behaviorism, these designs began to appear widely in journals in the 1960s–1980s (Barlow & Hersen, 1984), and remain popular today. SCED are found in a broad array of clinical fields and specialties (e.g., Krasny-Pacini & Evans, 2018; Lobo et al., 2017; Tanious & Onghena, 2019). Whereas group designs rely on inferential statistics to determine average experimental effect between groups, SCED employ visual analysis of individual

M. Tincani (✉) · J. Travers
Department of Teaching and Learning, Temple University, Philadelphia, PA, USA
e-mail: tincani@temple.edu

W. O'Donohue et al. (eds.), *Avoiding Questionable Research Practices in Applied Psychology*, https://doi.org/10.1007/978-3-031-04968-2_12

responding between conditions, with each subject serving as their own control. SCED require fewer subjects to demonstrate experimental control, typically 3–5, as statistical power requirements for group design samples are not applicable.

SCED employ inductive reasoning to determine functional relationships between independent and dependent variables. Often, following a period of baseline data (A condition), in which an individual's responding is measured repeatedly in the absence of the intervention, an intervention (B condition) is applied, and repeated measurement continues. If application of the intervention corresponds with therapeutic changes in level, trend, and variability of targeted responses, the researcher concludes with some degree of confidence the intervention was effective (i.e., there was a functional relationship between the intervention and observed outcomes). The researcher's confidence in a functional relationship increases with replications of experimental effect. SCED researchers employ a variety of designs that permit both within and between subject replication (Ledford & Gast, 2018). Two SCED are most common. In the reversal design, following baseline (A), an intervention (B) is applied, removed, and then reapplied. The multiple baseline (MB) design entails multiple A-B series where baseline condition lengths are staggered and intervention is introduced at different points in time (across different behaviors, participants, and settings). There are numerous variations in SCED that include combined designs like, for example, the MB design with a reversal. Figures 12.1 and 12.2 show hypothetical examples of reversal and multiple baseline designs demonstrating experimental effect as evidenced through visual inspection.

As inductive research designs, a noteworthy feature of SCED is flexibility for adaptation according to each subject's response to an intervention. For example, if

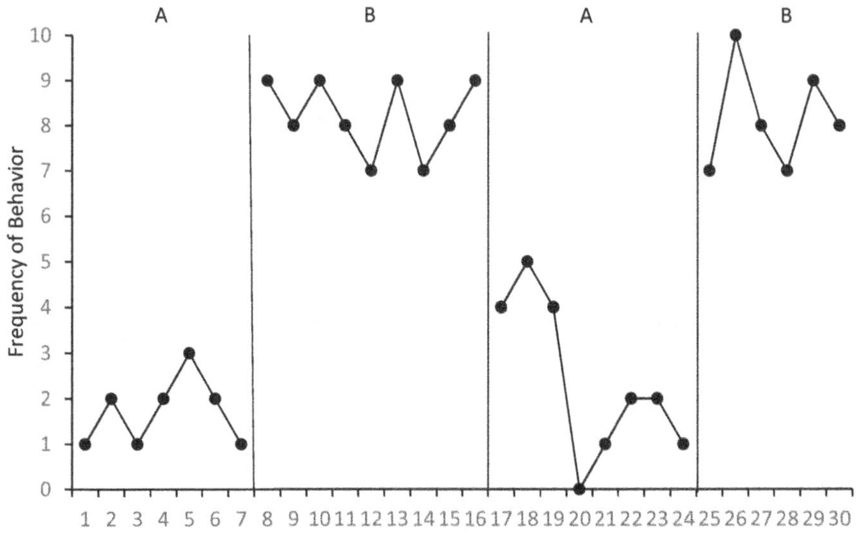

Fig. 12.1 Reversal design graph with hypothetical data

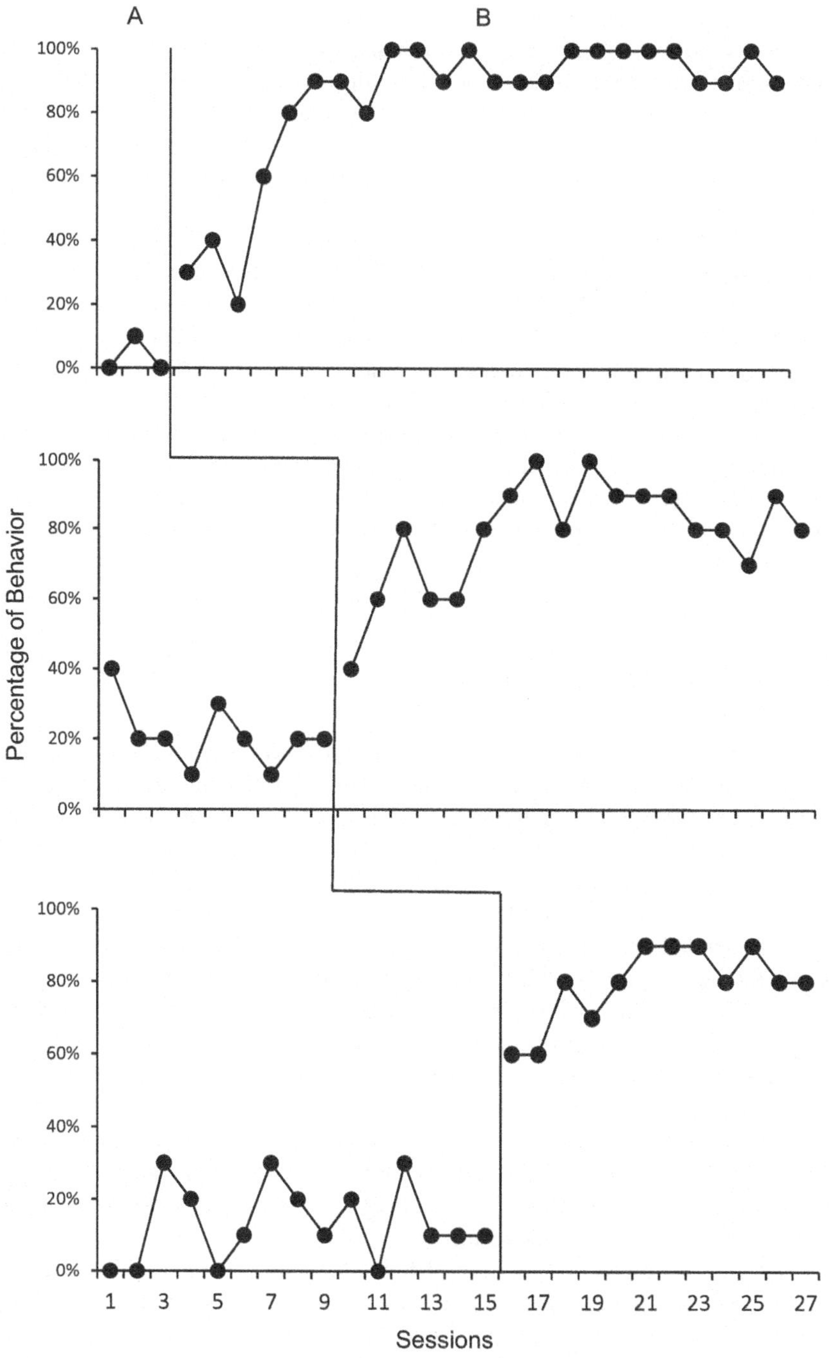

Fig. 12.2 Multiple baseline graph with hypothetical data

a subject's improvement with an intervention (B^1) is insufficient to meet desired clinical thresholds, a therapeutic modification to the intervention (B^2), or a new intervention (C), can be introduced, and any improvements can be evaluated accordingly (e.g., Crozier & Tincani, 2005). Such inductive design adaptations mitigate risk of clinical treatment failure associated with more traditional group designs, and permit a closer examination of factors that appear to influence treatment responsiveness.

As with other research design traditions, contemporary SCED researchers have established consensus on what constitutes rigorous study features in line with key design elements (Horner et al., 2005; Kratochwill et al., 2010; Ledford et al., 2018). These include selection of target behaviors and interventions that align with consumer preferences and values (i.e., social validity; Schwartz & Baer, 1991); precise operationalization of outcome variables with measures to establish reliability; complete and replicable descriptions of treatments with documented fidelity; sufficient data points within conditions to establish functional relations (or non-responsiveness); and design features that permit replication of experimental effect. Outcomes of studies reflecting these collective elements are considered more trustworthy than those that do not.

Questionable Research Practices

Questionable research practices (QRPs) are a variety of research choices that introduce bias into the body of research literature. The most blatant QRPs include undisclosed researcher conflicts of interest; post-hoc selection of statistical techniques to support a priori hypotheses (p-hacking; Head et al., 2015); post-hoc creation of hypotheses in accordance with observed results (harking; Kerr, 1998); selectively reporting data in published articles (Piggot et al., 2013); and outright suppression of results (Rosenthal, 1979). However, more subtle forms of QRPs are also likely quite common (Gerrits et al., 2019). Writing an article abstract to downplay non-significant findings, interpreting findings in the results to overplay significant findings, or changing the order of findings in the discussion from the aims of the study are examples of these more subtle QRPs. In the context of our current discussion, we conceive QRPs in the broadest sense to include both researcher and editorial behavior. For instance, a manuscript author might selectively report data in a paper submitted for publication to make results appear more favorable, or an editor might request an author to omit data in a revision of a paper, or otherwise reject a paper, because it contains unfavorable results.

Scientific researchers in a variety of fields have documented widespread presence of QRPs and have promoted practices aimed at preventing them. One of the most prominent efforts is the open science movement, which includes calls for preregistering studies and making all study data publicly available (Nosek et al., 2015). Open science practices prevent more blatant QRPs when researchers declare publicly their study hypotheses, data collection procedures, and data analysis techniques in a transparent manner prior to conducting a study. Similarly, publicly sharing of datasets prevents omission of data in research reports and enables outside researchers to replicate and conduct their own independent analyses of the data. Other measures aimed at preventing QRPs include editorial practices that promote publication of studies with non-significant results, and calls for publication of registered reports, a type of peer-review whereby research papers are accepted for publication with the introduction and methods only (Scheel et al., 2021). The latter practice ensures authors follow methods and analysis procedures they committed to prior to conducting the study and prevents journal editors and reviewers from biased editorial decisions based on study results.

QRPs in SCED

While QRPs have been documented in a variety of research fields, examination of QRPs has until recently tended to focus exclusively on mainstream group design approaches. Only recently have SCED researchers focused any significant attention on whether QRPs are present with these designs, and if so, what should be done about it (Tincani & Travers, 2018, 2019). Clearly, there is potential overlap between many QRPs in the broader researcher community and QRPs as they could manifest in SCED. Given both SCED and group design studies involve formulation of research questions and collection of data, biases related to selective reporting of results or post hoc alterations of research questions could occur regardless of design. Nonetheless, unique features of SCED suggest certain QRPs are likely to appear differently within these designs, and other QRPs not present in group designs might appear uniquely in SCED. What follows is a description of several key QRPs that could manifest in SCED, and steps SCED researchers and the SCED research community can take to mitigate them. Where useful, we include hypothetical examples of QRPs in SCED. We acknowledge examination of QRPs in SCED is a relatively new area of study and this is not necessarily an exhaustive list of practices. We hope our tentative discussion and recommendations will inspire future research aimed at examining these important concerns.

Participant Selection

Because SCED studies typically employ a small number of participants who serve as their own controls, often fewer than five in a given study, participant selection bias is a potential concern. This QRP occurs when experimenters selectively recruit participants whose behavior is more likely to change therapeutically with application of an intervention (i.e., selection bias; Ledford & Gast, 2018). Of course, this is a potential problem with any study of a therapeutic procedure, but the threat is mitigated to a degree in group designs, as outcomes of an intervention are aggregated across a larger number of participants within a treatment group, and one or more comparison groups are not exposed to the intervention.

It is typical with any SCED study for a therapeutic intervention to target a population with a common set of characteristics. For example, interventions that improve functional communication skills should target participants who would most likely benefit from the intervention due to clinical diagnoses associated with poor functional communication, such as severe autism spectrum disorder (ASD; Ganz, 2015). Selecting study participants from a population most likely to benefit from successful intervention is an important aspect of social validity and definitely not a QRP. However, intentionally selecting a subset of participants within a population because the experimenter knows, or at least suspects they will perform better than the target population at large, is a QRP.

The following example illustrates participant selection bias. A hypothetical research team is interested in evaluating an intervention to teach basic communication skills using speech generating devices (SGDs) to children with severe ASD (Tincani et al., 2020). The specific intervention involves teaching children to select picture icons from a computer screen which activates corresponding voice output. As part of their participant recruitment procedures, the team interviews parents about their children's current communicative skills, and while none of the children is currently using an SGD, they learn that one child has a history using a different albeit similar type of picture-based communication system. The team encourages the parents to consent for participation because they suspect this child will more easily acquire communication once the intervention is introduced, enhancing the appearance of experimental control (i.e., treatment effect). Consequently, it is impossible to tell how much of the observed results are due to the intervention effects or due to the participant's reported history of exposure prior to the study.

Participant selection bias in SCED can be prevented in the following ways. First, the description of recruitment procedures for a study should clearly detail all criteria for participation, along with a thorough description of recruitment procedures, to make clear participants were not selectively recruited based on potential intervention responsiveness. Considering the previous example, this would include screening questions that exclude participants for previous exposure to the intervention or highly similar interventions, or other characteristics that could selectively enhance intervention responsiveness. Second, researchers could employ a randomization procedure to recruit participants. This would begin with the researchers identifying a pool of participants whose characteristics match those of the study, and then randomly selecting a subset of individuals for participation. A limitation of the latter approach is the researcher must have a sufficient participant pool for randomization, which may not always be the case, especially if the target population has a rare or low incidence condition and few potential participants are readily available.

Independent Variable Selection

This QRP is a type of data falsification that occurs when researchers intentionally select or modify independent variables to enhance the appearance of intervention responsiveness, or to support their a priori hypotheses (Tijdink et al., 2014). This includes simplifying operational definitions of target behaviors on one or more salient dimensions, effectively lowering the bar for a successful outcome. This also includes omitting key response dimensions from operational definitions to enhance the appearance of success.

Consider a hypothetical research team evaluating an intervention to teach vocational skills to adults with intellectual and developmental disabilities in a community work setting. After observing participants in their work setting, they select a work task consisting of 15 steps completed in sequential order; the task is considered learned when participants complete all 15 steps with 100% accuracy. During the baseline condition, researchers note that while participants complete most steps with poor and inconsistent accuracy, steps 1 and 2 are especially challenging and are almost never performed correctly. Anticipating these steps will be especially difficult to teach with intervention, the team omits them from the task analysis, so participants must complete only 13 steps with 100% accuracy. Participant data during baseline and intervention are graphed as a percentage of steps completed correctly; therefore, omission of the two steps is not reflected in the data. A different problematic example would be if researchers selected a different and easier to perform target behavior altogether, and then began from scratch collecting new baseline data on the easier-to-perform behavior.

The potential for target behavior selection bias can be mitigated as follows. First, researchers should provide evidence that operational definitions of target behaviors reflect all key dimensions of meaningful dependent variables. In the previous example, this would include evidence that participants' work supervisors (or other secondary consumers, as appropriate) were consulted on target behavior selection prior to the baseline condition, and they verified all relevant steps of the target behavior were included in the task analysis, which remained consistent during baseline and intervention. Pre-registration of the study prior to initiation, including complete operational definitions of each target behavior, enhance research transparency and prevent researchers from modifying target behaviors in ways that work against an unbiased approach. Finally, avoidance of graphing dependent responses using less dimensional measurement systems (e.g., percentage correct) in favor of more dimensional measurement systems (e.g., frequency correct) lessens the likelihood that data are depicted in a deceptive way (Johnston et al., 2010).

Procedural Fidelity Documentation

Thus far, our discussion of QRPs has focused on ways researchers can manipulate dependent variables to produce the appearance of favorable findings. Data collection on independent variables can also be manipulated in ways that enhance appearance of a study's credibility. Procedural fidelity, the degree to which experimenters adhere to a specified experimental protocol, is a key part of SCED intervention research (Ledford & Wolery, 2013). Documenting strong procedural fidelity is critical for demonstrating that changes in the dependent variable are attributable to changes in the independent variable. Conversely, poor or inconsistent procedural fidelity reduces the researcher's confidence that observed outcomes are attributable to treatment.

In SCED, procedural fidelity is typically documented using a checklist of steps that represent adherence to the treatment protocol, often completed by a researcher implementing the procedure along with secondary observers to ensure reliability. If a treatment is particularly complex, the checklist may have many steps, including ones that are conditional based on participants' response to an intervention; a study may have multiple checklists representing different phases of intervention. The percent or number of treatment steps completed is often reported as a measure of procedural fidelity in SCED research reports. In reports published in refereed journals, the percent or number of completed steps reported typically is high. However, low fidelity to the treatment protocol usually is considered a threat to internal validity and reduces chances of the study's publication (Tincani & Travers, 2018).

Maintaining high procedural fidelity can be challenging, particularly if a treatment protocol is complex. In lieu of rigorous training to ensure close adherence to a stated protocol, researchers may simplify a protocol checklist to omit key steps of an intervention, or use generalized descriptions of procedural steps to increase the likelihood of yes responses on a checklist. For example, rather than describing the checklist steps consistent with the stated procedure, "The therapist delivers verbal praise immediately after a correct response or corrective feedback immediately after an incorrect response", the checklist simply states, "The therapist delivers feedback." The latter description would more readily engender a "yes" response regardless of whether the feedback was correctly given. In a different and more blatant example, if a treatment step requires particularly skillful implementation that is prone to error, researchers could simply omit the step when calculating fidelity and bolster the apparent adherence to the protocol and chances of publication.

To prevent QRPs in procedural fidelity documentation, researchers should make their procedural fidelity checklists or similar forms publicly available. This level of transparency allows readers to compare the stated experimental procedure with the method of procedural fidelity documentation and the reported procedural fidelity data. Sharing details of procedural fidelity documentation can be accomplished through inclusion in a published manuscript, in a publicly available data repository associated with the journal's publisher, or other public data repository like the open science framework (https://osf.io). Also, where possible, researchers should employ secondary, blinded observers who do not implement the treatment to complete procedural fidelity checklists.

Graphical Depictions of Behavior

SCED studies involve graphical depiction of data for visual analysis to determine experimental control. QRPs can manifest when researchers depict behavior on graphs to enhance the appearance of experimental effect. Specifically, researchers can modify dimensions of the Y-axis scale of a graph to enhance the appearance of a robust treatment effect where less treatment effect actually exists (Dart & Radley, 2017). Consider the three hypothetical graphs depicted in Fig. 12.3. Each graph depicts the same results, but the scale of the Y-axis scale varies between 10, 20, and 40. As the largest unit on the Y-axis increases, the appearance of treatment effect is lessened, since the vertical distance between the data paths and the floor of the graph decreases, and the vertical distance between the data paths in the two phases also decreases. Researchers may intentionally select a lower maximum Y-axis scale value, like the one shown in the bottom panel of Fig. 12.2, to enhance the appearance of treatment effect.

Some SCED researchers have recommended use of standard graphical displays, which would prevent misrepresentation of SCED data by varying the Y-axis scale (Calkin, 2005). Currently, such standard displays are not in wide usage. Alternatively, two rules of thumb in scaling the Y-axis should be followed. First, where multiple graphs representing the same dependent variable are presented (e.g., across different participants or settings), the Y-axis scaling should be consistent across the graphs. Second, the maximum value of the Y-axis should correspond with the optimal level of *socially significant behavior change*, rather than simply defaulting to the highest level of behavior observed during the experiment. For example, in a study of a reading intervention where the goal is to teach students to read 120 words-per-minute, the maximum Y-axis value should be *at least* 120 regardless of whether participants' maximum reading performance fell below 120 words-per-minute. In this way, readers can interpret results of the experiment in relation to the established goals of intervention.

Effect Size Measures and Statistics

SCED researchers rely primarily on visual analysis to determine whether changes in trend, level, and variability of behavior indicate experimental control for interventions. In addition to visual analysis, researchers have devised a variety of effect size (ES) measures and statistical tests to estimate treatment effect in SCED research (Parker et al., 2011; Parker & Brossart, 2003; Shadish et al., 2014). These techniques are specific to SCED since the assumptions of statistical tests used with

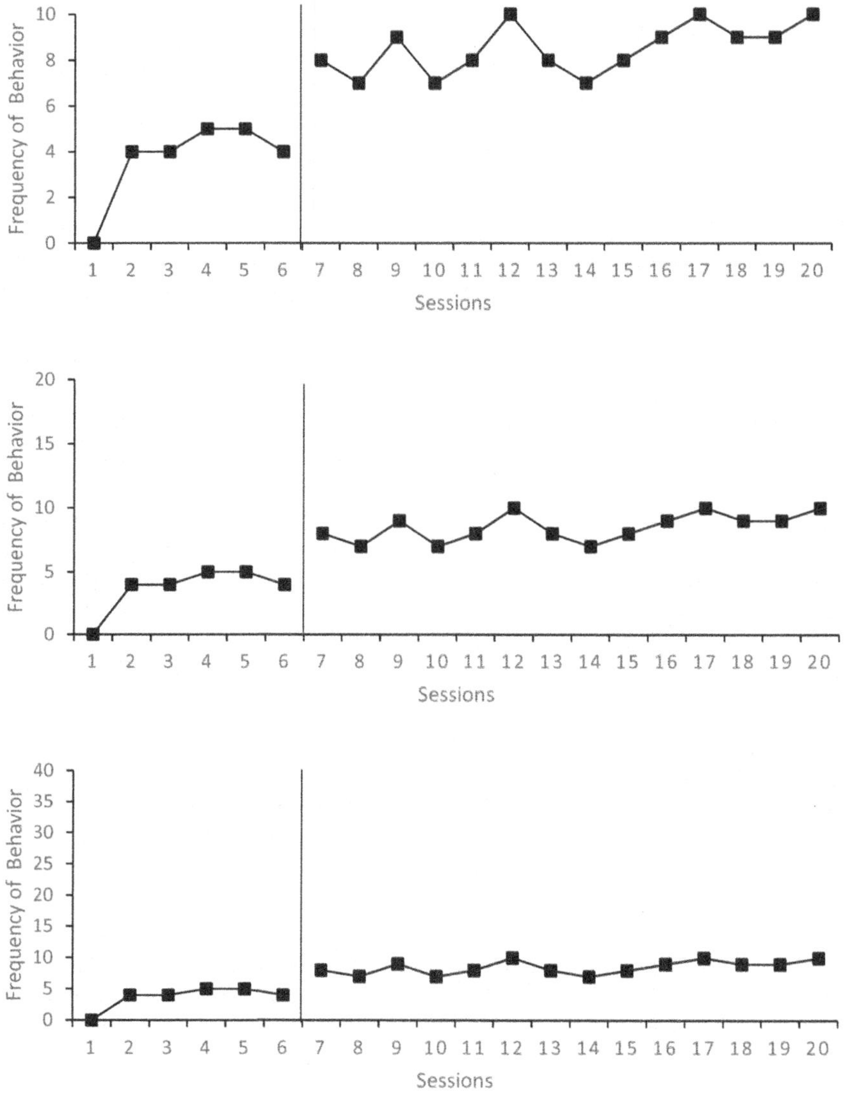

Fig. 12.3 Variations in Y-axis scaling to alter appearance of experimental effect

group design research are often violated or not applicable to SCED datasets. The most commonly used procedures evaluate non-overlap of data between baseline and intervention conditions. Additionally, a wide variety of parametric and non-parametric statistical techniques have been developed to quantify magnitude of treatment effect between baseline and intervention conditions. While these techniques have become increasingly commonplace in meta-analyses of SCED studies, they also frequently appear in single SCED study reports (Bondy & Tincani, 2018).

Use of statistics to quantity behavior change has long been an area of controversy and disagreement in the SCED research community, with some arguing these methods cannot adequately capture dynamic aspects of visual analysis and should be staunchly avoided (e.g., Baer, 1977). Others contend these methods are useful in addressing shortcomings of visual analysis (e.g., lack of precision, need for training), including when synthesizing a body of literature to determine whether an intervention is effective (Kratochwill & Levin, 2014). There is evidence to suggest inconsistent and, in some cases, poor correspondence when quantitative measures are compared to each other, and when quantitative measures are compared to visual analysis (Wolery et al., 2010; Yucesoy-Ozkan et al., 2020; Zimmerman et al., 2018). Furthermore, each available quantitative technique has salient limitations. For example, non-overlap methods provide no quantification of the magnitude of experimental effect, and statistical approaches may be compromised by certain properties of SCED data, such as autocorrelation and trend within conditions. Although these techniques are continually evolving and show promise for quantifying SCED data in more robust ways, currently there is no single statistical analysis technique that captures all aspects of visual analysis.

Incorporation of ES measures and statistics in SCED research is not itself a QRP. However, statistical methods of analysis create the possibility of QRPs. In the absence of robust treatment effect as evidenced by visual analysis, researchers may be tempted to include a quantitative measure to bolster the appearance of treatment effect. This seems especially likely when researchers can find one or more measures that will yield a moderate to large ES estimate. This QRP is similar to p-hacking in the group design research literature. Given myriad quantitative metrics available, different assumptions and calculation procedures, and varying results, this QRP is a very real possibility. To prevent this, researchers should publicly report and provide a rationale for specific ES metrics and statistical tests (along with visual analytic techniques) before the experiment. Importantly, given limitations of current quantitative techniques for SCED data, we strongly recommend researchers use these techniques for individual study datasets only after well-established visual analysis approaches have been used (Lane & Gast, 2014; Harrington & Velicer, 2015; Kratochwill & Levin, 2014). Finally, given the novelty of some techniques and lack of wide usage across the SCED literature, we recommend researchers provide an explanation for why and how a particular technique will be used, how the data assumptions for the technique are met with the current dataset, and any anticipated limitations with the technique.

The File Drawer Effect and Publication Bias

The file drawer effect is a longstanding problem in the social and behavioral sciences research literature (Rosenthal, 1979). It occurs when studies that fail to yield statistically significant results are not published in refereed journals. If the study is part of a policy report, doctoral dissertation, or master's thesis, it will forever remain as part of the grey literature, unless it is incorporated into a published research synthesis that includes grey literature (e.g., Dowdy et al., 2020). Non-significant findings are crucial in understanding how interventions can fail to achieve desired therapeutic effects (e.g., Lang et al., 2012). The file drawer effect is problematic as it results in a skewed research literature disproportionately represented by successful outcomes (Gage et al., 2017; Scheel et al., 2021).

The file drawer effect is thought to be a manifestation of publication bias (Tincani & Travers, 2019). Publication bias occurs when researchers selectively submit studies showing positive effects for publication, and/or when journal reviewers and editors favor such studies when rendering editorial recommendations and decisions. Publication bias also occurs when researchers selectively report study data showing positive effects to increase the likelihood of publication. As with other QRPs, publication bias can manifest with either group design or SCED research. For example, researchers using any type of design who have invested substantial time and resources to develop an intervention might be disinclined to publish data showing equivocal or negative effects of the intervention. Similarly, journal reviewers and editors may consider non-significant or negative results uninteresting or otherwise indicative of a poor study. One unique aspect of SCED research is the longstanding view that strong experimental control is the hallmark of a good study (Tincani & Travers, 2018). That is, a study is worthy of dissemination only when researchers demonstrate control over behavior through application of an intervention. Shadish et al. (2016) found that SCED researchers were more likely to rate a dataset as publishable if it demonstrated strong experimental control. In contrast, since contemporary SCED are used to evaluate a wide variety of behavioral, educational, psychosocial, and other therapeutic interventions, it is crucial that failures to produce therapeutic effects are documented in published research (Johnson & Cook, 2019). This information is critical in facilitating consumers' knowledge of the boundaries of interventions, including those in widespread usage (Leaf et al., 2021).

Journal editors can implement editorial practices that foster publication of non-effect studies. Stated editorial policies should explicitly encourage authors to submit for publication studies that fail to yield experimental control (Kittelman et al., 2018). Journal editors can provide instructions to reviewers as guidance on how to review these studies, outlining features of high quality SCED that fail to produce optimal therapeutic effects (Tincani & Travers, 2018). Registered reports also may increase publication of non-effect studies (Johnson & Cook, 2019). In registered reports, researchers submit the introduction (i.e., rationale) and method for conducting a study for publication prior to conducting the experiment. Assuming a solid rationale and adherence to registered experimental procedures, journals are obligated to publish the final manuscript (with results and discussion) regardless of the direction of findings. The acceptance for publication based on the proposed study and strict adherence to the accepted method precludes the possibility of editorial decisions biased by study findings. Scheel et al. (2021) found that registered reports included substantially more studies demonstrating non-statistically significant findings, compared to papers published through the typical peer review process. Registered reports show promise for disseminating non-effect findings of both group design and SCED studies, yet they are not currently in wide usage in journals publishing either types of design (c.f., Cook et al., 2021).

Research funding agencies can counter publication bias by encouraging researchers to incorporate open science principles into their research. This includes requiring research grant recipients to preregister their studies on open science platforms (e.g., U.S. Department of Education, 2020). Granting agencies should also encourage researchers to incorporate open science principles into their grant applications, and advise reviewers to score them accordingly. For instance, granting agencies can allow applicants to achieve maximum review scores only if they commit to submitting at least a portion of their research for publication as registered reports, and otherwise demonstrate they will take explicit steps to disseminate non-effect studies, along with other open science practices.

Meta-Analyses of SCED Research

Thus far we have restricted our discussion to QRPs to individual SCED studies, but QRPs can occur with syntheses and meta-analyses. Increasingly, SCED researchers have adapted meta-analysis techniques to aggregate findings across individual studies to specify what interventions work, for whom they are effective, and under what conditions effects are observed (Dowdy et al., in press; Maggin et al., 2011). Meta-analysis is an established research methodology in the social and behavioral sciences, and SCED researchers have been adopting these techniques in their work for decades (Allison & Gorman, 1993; Shadish et al., 2008; Vannest et al., 2018). However, many SCED researchers are unfamiliar with contemporary meta-analysis techniques as they apply to SCED, and lack training in requisite statistical techniques used in meta-analysis (Dowdy et al., in press). Given these concerns, QRPs can manifest when researchers employ less-than-rigorous procedures to conduct meta-analyses of individual SCED studies (Jamshidi et al., 2018).

As with any research methodology, there are established standards for conducting high quality meta-analyses to which all researchers should adhere (e.g., Moher et al., 2009; Shea et al., 2007). These include a priori formulation of research questions, rigorous procedures for extracting targeted studies, strategies for detecting bias within studies, and procedures for evaluating quality of studies. Many of these standards are the same regardless of whether researchers are synthesizing group design studies, SCED studies, or both together (e.g., Gage et al., 2017). However, there are at least two unique considerations for SCED meta-analyses. First, as discussed, ES estimation and statistical procedures for aggregating SCED data are different than those for aggregating group design data, with little consensus among SCED researchers for which is best given a particular dataset. Therefore, as with individual SCED studies, we strongly recommend SCED researchers preregister their meta-analysis studies on an open science platform, such as PROSPERO (Booth et al., 2012). They should provide a rationale for and associated assumptions for their selected ES metric. Researchers also should describe how the assumptions are likely to be met by the proposed dataset and explain the calculation procedures. Additionally, we strongly recommend researchers report multiple ES metrics and explain any differences between results obtained from the different metrics. In addition, researchers may wish to consider reporting structured visual analysis techniques in tandem with ES metrics (Lane & Gast, 2014).

Second, given the possibility of publication bias, we strongly encourage SCED researchers to include grey literature in their meta-analyses. We recommended that any meta-analysis include procedures for detecting publication bias, such as the funnel plot technique (Duval & Tweedie, 2000). However, given differing statistical assumptions of SCED experiments, procedures for detecting publication bias in group design studies may not be appropriate for SCED studies. Alternatively, to detect publication bias, SCED researchers can calculate ES metrics for published studies and grey studies separately and compare them (Dowdy et al., 2020; Sham & Smith, 2014).

Conclusion

We discussed several QRPs in SCED research and outlined tentative solutions for preventing them. In drawing attention to QRPs in SCED, it is not our intention to convey that SCED research is less rigorous or somehow more prone to these practices (though we do recognize that many researchers are unfamiliar with and have incorrect understandings of SCED research). In fact, some well-documented QRPs like p-hacking are less likely to occur in SCED given the traditional reliance on visual analysis over statistics to determine experimental effect. Nonetheless, QRPs can manifest in any research methodology or scientific discipline, and SCED researchers are not immune from them. Clearly, more research on questionable practices in SCED is needed to better understand the scope of these misguided research choices, and how they appear uniquely within SCED. For example, surveys of researchers in other fields have revealed that QRPs are commonplace (Gerrits et al., 2019; Loewenstein & Prelec, 2012; Tijdink et al., 2014). We hope our tentative discussion in this chapter sheds light on the possibility of QRPs in our research, and encourages researchers, journal editors, and other members of the research community to embrace practices to prevent them.

References

Allison, D. B., & Gorman, B. S. (1993). Calculating effect sizes for meta-analysis: The case of the single case. *Behaviour Research and Therapy, 31*(6), 621–631.

Baer, D. M. (1977). Perhaps it would be best not to know everything. *Journal of Applied Behavior Analysis, 10*(1), 167–172.

Barlow, D. H., & Hersen, M. (1984). *Single-case experimental designs: Strategies for studying behavior change* (2nd ed.). Pergamon Press.

Bondy, A. H., & Tincani, M. (2018). Effects of response cards on students with autism spectrum disorder or intellectual disability. *Education and Training in Autism and Developmental Disabilities, 53*(1), 59–72.

Booth, A., Clarke, M., Dooley, G., Ghersi, D., Moher, D., Petticrew, M., & Stewart, L. (2012). The nuts and bolts of PROSPERO: An international prospective register of systematic reviews. *Systematic Reviews, 1*(1), 1–9.

Calkin, A. B. (2005). Precision teaching: The standard celeration charts. *The Behavior Analyst Today, 6*(4), 207–215.

Cook, B. G., Maggin, D. M., & Robertson, R. E. (2021). Registered reports in special education: Introduction to the special series. *Remedial and Special Education.* https://doi.org/10.1177/0741932521996459

Crozier, S., & Tincani, M. J. (2005). Using a modified social story to decrease disruptive behavior of a child with autism. *Focus on Autism and Other Developmental Disabilities, 20*(3), 150–157.

Dart, E. H., & Radley, K. C. (2017). The impact of ordinate scaling on the visual analysis of single-case data. *Journal of School Psychology, 63*(1), 105–118.

Dowdy, A., Tincani, M., & Schneider, J. (2020). Evaluation of publication bias in response interruption and redirection: A meta-analysis. *Journal of Applied Behavior Analysis, 53*(4), 2151–2171.

Dowdy, A., Peltier, C., Tincani, M., Schneider, J., Hantula, D., & Travers, J. (in press). The utility of meta-analyses in applied behavior analysis: A discussion and review. *Journal of Applied Behavior Analysis.*

Duval, S., & Tweedie, R. (2000). Trim and fill: A simple funnel-plot–based method of testing and adjusting for publication bias in meta-analysis. *Biometrics, 56*(2), 455–463.

Gage, N. A., Cook, B. G., & Reichow, B. (2017). Publication bias in special education meta-analyses. *Exceptional Children, 83*(4), 428–445.

Ganz, J. B. (2015). AAC interventions for individuals with autism spectrum disorders: State of the science and future research directions. *Augmentative and Alternative Communication, 31*(3), 203–214.

Gerrits, R. G., Jansen, T., Mulyanto, J., van den Berg, M. J., Klazinga, N. S., & Kringos, D. S. (2019). Occurrence and nature of questionable research practices in the reporting of messages and conclusions in international scientific Health Services Research publications: A structured assessment of publications authored by researchers in the Netherlands. *BMJ Open, 9*(5), e027903.

Harrington, M., & Velicer, W. F. (2015). Comparing visual and statistical analysis in single-case studies using published studies. *Multivariate Behavioral Research, 50*(2), 162–183.

Head, M. L., Holman, L., Lanfear, R., Kahn, A. T., & Jennions, M. D. (2015). The extent and consequences of p-hacking in science. *PLoS Biology, 13*(3), e1002106.

Horner, R. H., Carr, E. G., Halle, J., McGee, G., Odom, S., & Wolery, M. (2005). The use of single-subject research to identify evidence-based practice in special education. *Exceptional children, 71*(2), 165–179.

Jamshidi, L., Heyvaert, M., Declercq, L., Fernández-Castilla, B., Ferron, J. M., Moeyaert, M., … Van den Noortgate, W. (2018). Methodological quality of meta-analyses of single-case experimental studies. *Research in Developmental Disabilities, 79*, 97–115.

Johnson, A. H., & Cook, B. G. (2019). Preregistration in single-case design research. *Exceptional Children, 86*(1), 95–112. https://doi.org/10.1177/0014402919868529

Johnston, J. M., Pennypacker, H. S., & Green, G. (2010). *Strategies and tactics of behavioral research.* Routledge.

Kerr, N. L. (1998). HARKing: Hypothesizing after the results are known. *Personality and Social Psychology Review, 2*(3), 196–217.

Kittelman, A., Gion, C., Horner, R. H., Levin, J. R., & Kratochwill, T. R. (2018). Establishing journalistic standards for the publication of negative results. *Remedial & Special Education, 39*(3), 171–176.

Kratochwill, T. R., Hitchcock, J., Horner, R. H., Levin, J. R., Odom, S. L., Rindskopf, D. M & Shadish, W. R. (2010). Single-case designs technical documentation. Retrieved from What Works Clearinghouse website: http://ies.ed.gov/ncee/wwc/pdf/wwc_scd.pdf

Kratochwill, T. R., & Levin, J. R. (Eds.). (2014). Enhancing the scientific credibility of single-case intervention research: Randomization to the rescue. In T. R. Kratochwill & J. R. Levin (Eds.), Single-case intervention research: Methodological and statistical advances (pp. 53–89). American Psychological Association. https://doi.org/10.1037/14376-003

Krasny-Pacini, A., & Evans, J. (2018). Single-case experimental designs to assess intervention effectiveness in rehabilitation: A practical guide. *Annals of Physical and Rehabilitation Medicine, 61*(3), 164–179.

Lane, J. D., & Gast, D. L. (2014). Visual analysis in single case experimental design studies: Brief review and guidelines. *Neuropsychological rehabilitation, 24*(3-4), 445–463.

Lang, R., O'Reilly, M., Healy, O., Rispoli, M., Lydon, H., Streusand, W., … Giesbers, S. (2012). Sensory integration therapy for autism spectrum disorders: A systematic review. *Research in Autism Spectrum Disorders, 6*(3), 1004–1018.

Leaf, J. B., Sato, S. K., Javed, A., Arthur, S. M., Creem, A. N., Cihon, J. H., … Oppenheim-Leaf, M. L. (2021). The evidence-based practices for children, youth, and young adults with autism report: Concerns and critiques. *Behavioral Interventions, 36*(2), 457–472.

Ledford, J. R., & Gast, D. L. (2018). *Single case research methodology*. Routledge.

Ledford, J. R., Lane, J. D., & Tate, R. (2018). Evaluating quality and rigor in single case research. In J. R. Ledford & D. L. Gast (Eds.), *Single case research methodology* (pp. 365–392). Routledge.

Ledford, J. R., & Wolery, M. (2013). Procedural fidelity: An analysis of measurement and reporting practices. *Journal of Early Intervention, 35*(2), 173–193.

Lobo, M. A., Moeyaert, M., Cunha, A. B., & Babik, I. (2017). Single-case design, analysis, and quality assessment for intervention research. *Journal of Neurologic Physical Therapy, 41*(3), 187–197.

Maggin, D. M., O'Keeffe, B. V., & Johnson, A. H. (2011). A quantitative synthesis of methodology in the meta-analysis of single-subject research for students with disabilities: 1985–2009. *Exceptionality, 19*(2), 109–135.

Moher, D., Liberati, A., Tetzlaff, J., Altman, D. G., & PRISMA Group*. (2009). Preferred reporting items for systematic reviews and meta-analyses: the PRISMA statement. *Annals of internal medicine, 151*(4), 264–269.

Nosek, B. A., Alter, G., Banks, G. C., Borsboom, D., Bowman, S. D., Breckler, S. J., … Yarkoni, T. (2015). Promoting an open research culture. *Science, 348*(6242), 1422–1425.

Parker, R. I., & Brossart, D. F. (2003). Evaluating single-case research data: A comparison of seven statistical methods. *Behavior Therapy, 34*(2), 189–211.

Parker, R. I., Vannest, K. J., & Davis, J. L. (2011). Effect size in single-case research: A review of nine nonoverlap techniques. *Behavior Modification, 35*(4), 303–322.

Piggott, T. D., Valentine, J. C., Polanin, J. R., Williams, R. T., & Canada, D. D. (2013). Outcome-reporting bias in education research. *Educational Researcher, 42*(8), 424–432.

Rosenthal, R. (1979). The file drawer problem and tolerance for null results. *Psychological Bulletin, 86*(3), 638–641.

Scheel, A. M., Schijen, M. R., & Lakens, D. (2021). An excess of positive results: Comparing the standard psychology literature with registered reports. *Advances in Methods and Practices in Psychological Science, 4*(2), 25152459211007467.

Schwartz, I. S., & Baer, D. M. (1991). Social validity assessments: Is current practice state of the art? *Journal of Applied Behavior Analysis, 24*(2), 189–204.

Shadish, W. R., Hedges, L. V., Pustejovsky, J. E., Boyajian, J. G., Sullivan, K. J., Andrade, A., & Barrientos, J. L. (2014). A d-statistic for single-case designs that is equivalent to the usual between-groups d-statistic. *Neuropsychological Rehabilitation, 24*(3–4), 528–553.

Shadish, W. R., Rindskopf, D. M., & Hedges, L. V. (2008). The state of the science in the meta-analysis of single-case experimental designs. *Evidence-Based Communication Assessment and Intervention, 2*(3), 188–196.

Shadish, W. R., Zelinsky, N. A., Vevea, J. L., & Kratochwill, T. R. (2016). A survey of publication practices of single-case design researchers when treatments have small or large effects. *Journal of Applied Behavior Analysis, 49*(3), 656–673.

Sham, E., & Smith, T. (2014). Publication bias in studies of an applied behavior-analytic intervention: An initial analysis. *Journal of Applied Behavior Analysis, 47*(3), 663–678.
Shea, B. J., Bouter, L. M., Peterson, J., Boers, M., Andersson, N., Ortiz, Z., … Grimshaw, J. M. (2007). External validation of a measurement tool to assess systematic reviews (AMSTAR). *PLoS One, 2*(12), e1350.
Skinner, B. F. (1953). *Science and human behavior*. Simon and Schuster.
Tanious, R., & Onghena, P. (2019). Randomized single-case experimental designs in healthcare research: What, Why, and How? *Healthcare, 7*, 143. https://doi.org/10.3390/healthcare7040143
Tincani, M., Miller, J., Nepo, K., & Lorah, E. R. (2020). Systematic review of verbal operants in speech generating device research from Skinner's analysis of verbal behavior. *Perspectives on Behavior Science, 43*, 387–413.
Tincani, M., & Travers, J. (2018). Publishing single-case research design studies that do not demonstrate experimental control. *Remedial and Special Education, 39*(2), 118–128.
Tincani, M., & Travers, J. (2019). Replication research, publication bias, and applied behavior analysis. *Perspectives on Behavior Science, 42*, 59–75.
Tijdink, J. K., Verbeke, R., & Smulders, Y. M. (2014). Publication pressure and scientific misconduct in medical scientists. *Journal of Empirical Research on Human Research Ethics, 9*(5), 64–71.
U.S. Department of Education. (2020). *Request for applications*. Special Education Research Grant Program.
Vannest, K. J., Peltier, C., & Haas, A. (2018). Results reporting in single case experiments and single case meta-analysis. *Research in Developmental Disabilities, 79*, 10–18.
Wolery, M., Busick, M., Reichow, B., & Barton, E. E. (2010). Comparison of overlap methods for quantitatively synthesizing single-subject data. *The Journal of Special Education, 44*(1), 18–28.
Yucesoy-Ozkan, S., Rakap, S., & Gulboy, E. (2020). Evaluation of treatment effect estimates in single-case experimental research: Comparison of twelve overlap methods and visual analysis. *British Journal of Special Education, 47*(1), 67–87.
Zimmerman, K. N., Pustejovsky, J. E., Ledford, J. R., Barton, E. E., Severini, K. E., & Lloyd, B. P. (2018). Single-case synthesis tools II: Comparing quantitative outcome measures. *Research in Developmental Disabilities, 79*, 65–76.

Chapter 13
Presenting the Psychometric Evidence for Psychological Measures: A Proposal and Thoughts on Questionable Research Practices

William O'Donohue, Akihiko Masuda, and Stephen N. Haynes

Abstract There is little guidance and much variability regarding the description of psychological measures and their associated psychometric evidence in scientific reports in psychology. This hinders the ability of the reader to properly evaluate the degree to which these measures actually measure what the authors intend to measure. This chapter advocates for a more standardized approach to reporting the psychometric evidence that is based on five key principles. In reference to the psychometric evidence of a measure used in a scientific report, the write-up should: (a) argue what are the most relevant psychometric dimensions; (b) acknowledge the multidimensional nature of psychometric evidence; (c) acknowledge any missing and conflicting psychometric data; (d) present quantified psychometric information; and (e) acknowledge the conditional nature of the psychometric evidence across sample characteristics, dimensions of individual differences, and assessment contexts. A case example is provided to exemplify the use of these principles in a scientific report in psychology.

Keywords Questionable research practices · Psychometric evidence · Reporting psychometric evidence

---

W. O'Donohue (✉)
University of Nevada, Reno, Reno, NV, USA
e-mail: wto@unr.edu

A. Masuda · S. N. Haynes
University of Hawai'i at Mānoa, Honolulu, HI, USA

W. O'Donohue et al. (eds.), *Avoiding Questionable Research Practices in Applied Psychology*, https://doi.org/10.1007/978-3-031-04968-2_13

Background

There are significant discrepancies among researchers about how measures are described in reports of psychological research. Part of this variability may emanate from the lack of specifics in the American Psychological Association's (APA's) Publication Manual (APA, 2011), which states:

> **Measures and covariates.** Include in the Method section information that provides definitions of all primary and secondary outcome measures and covariates, including measures collected but not included in the report. Describe the methods used to collect the data (e.g., written questionnaires, interviews, observations) as well as methods used to enhance the quality of the measurements (e.g., the training and reliability of assessors or the use of multiple observations). Provide information on instruments used including psychometric and biometric properties and evidence of cultural validity. (p. 31)

This statement provides some guidance for certain research and clinical contexts (e.g., treatment outcome studies), but is short on specific guidelines and is open to alternative interpretations along several dimensions. An important consideration is the goal of clearly, accurately, and comprehensively presenting psychometric evidence. What is the primary purpose of providing information about the psychometric properties of measures in a report? Is it to convince readers that the measure is "good enough," or perhaps "better" than alternative measures of the same construct? Or is it to give clear and precise information about a measure's strengths and weaknesses so that readers can better interpret results based on this information? How ought this psychometric information be described—with general qualitative terms like "good" or ought this psychometric evidence be quantified? Should the reader be presented with an argument regarding what are the most important dimensions of psychometric evidence for the particular purposes of the study? As discussed elsewhere (Hunsley & Mash, 2019), the relevance of the various dimensions of psychometric evidence about measures varies across the purposes for which the measures are being used and the inferences that are to be derived. For example, predictive validity evidence might be more important than internal consistency evidence when a measure is being used to make inferences about the future.

Discrepancies across psychology reports may also be based on confusion, uncertainty, and variability regarding the dimensions, applicability, and meaning of psychometric evidence. These problems can vary from elementary mistakes of confusing reliability information with validity information (Haynes et al., 2019), such as when authors present Cronbach alphas to support the validity of a measure. However, these confusions also can be based on an insufficient appreciation of the advantages of precise psychometric information and guidelines about how psychometric information should be presented.

Authors may also be insensitive to the conditional and multidimensional nature of psychometric information, and assume that psychometric information resides in, and thus is a stable trait of, a measure (Haynes et al., 2019). These assumptions about the nature and relevance of psychometric evidence are, of course, incorrect. The sample characteristics from which psychometric information about a given measure has been derived should be compared to the sample characteristics of those in the report, and statements ought to be made about the degree of generalizability of the evidence.

A final source of confusion may be due to the primary function of providing psychometric information in psychological reports. It appears that many authors emphasize one function of the psychometric evidence—to convince readers and reviewers that "this measure is good enough." That is, the measure and its attendant psychometric evidence (e.g., however these are described and however few or many of these psychometric dimensions are described) ought not to cause sufficient concern to call into question any conclusions, cast doubt on the quality of the measures reported, or certainly provide a reason to not publish this paper. Gross (1996) and other contemporary philosophers of science have written on the rhetorical functions of scientific writing—that is, there are persuasive burdens that must be met—and here the persuasive task might be to convince the reader that the measure is "good enough" or perhaps "worry free."

This chapter provides principles to resolve these problems and to assist writers and readers of a scientific paper to understand which psychometric evidence is relevant to a measure and its application. This approach allows the reader then to more fully understand the appropriateness of the study's inferences and conclusions and the writer to select the most relevant psychometric evidence to report. We regard problematically communicating or understanding key psychometric information on the measures used in a study as a questionable research practice, because this can intentionally or unintentionally create a false understanding or, at a minimum, a misleading impression regarding the quality of the measures used. In particular, when a report fails to explicate missing or poor psychometric data, the reader may gain the false impression that the measure is better than it actually is and thus not be in a position to properly evaluate the study's conclusions. The move toward evidence-based assessment is laudatory but some researchers may have measurement interests for which sound measures for the construct of interest simply do not exist. All measures contain error and communications about a measure should be constructed so that the reader has an accurate understanding of the measure's strengths and weaknesses.

Principles of Description of Psychometric Evidence in Psychology Reports

Given these areas of variability in describing measures and their psychometrics delineated above, we propose that when authors describe the measures that are used in a study, they should:

1. **Specify the relative importance among various dimensions of psychometric information for each measure used in a study.**

The relevance of different forms of psychometric evidence varies across the behaviors and events that are being measured, the strategies and methods of assessment, the goals of assessment, the inferences to be made in the study, and the characteristics of the study participants. Thus, there should be an explicit argument of what kinds of psychometric information are particularly important for the inferences to be made from the measures in the particular study.

Construct validity is almost always an important composite judgment of psychometric evidence, particularly when the intent is to measure higher-level constructs such as intelligence, depression, rape proclivity, and so on (Haynes et al., 2019). However, at other times the importance of other psychometric indices can be important. For example, in a study involving repeated measures over a 2-week interval, 2-week *test-retest reliability* may be particularly important. *Internal consistency* is an important source of psychometric evidence for many self-report instruments, such as a multi-item depression inventory, that provide measures based on the aggregation of scores from multiple elements that are presumed to be correlated.

On the other hand, internal consistency is a less important source of psychometric evidence for other purposes or methods, such as some direct observation methods of discrete behaviors in which elements are not theorized to co-occur. For example, for a rating scale that inquiries about a client's experience with traumatic life events, we would not necessarily expect experiences with events such as divorce/separation, sexual assault, and death of a loved one to be significantly correlated. The number and severity of a client's traumatic life events could be summed into an aggregate score as an index of life traumatic experiences, but an index of internal consistency would not be informative for evaluating its psychometric properties. In other words, a low coefficient would not necessarily mean that scores from the instrument are invalid. Thus, in reports describing the measures and their psychometric evidence, there should be explicit arguments about the relative importance of the different kinds of psychometric information to aid the reader in understanding the assessment instrument and its measures in the context of the particular study.

2. **Acknowledge the multidimensional nature of psychometric evidence.**

It is incorrect and misleading to talk about reliability and validity as if these are unidimensional. For example, the glossary in Haynes et al. (2019) mentions 13 forms of validity evidence. Communicating and understanding this multidimensional information are challenging. For example, a measure of children's attention deficit hyperactivity disorder (ADHD) symptoms that has demonstrated a high degree of *discriminative validity* may not demonstrate a high degree of *discriminant validity*. That is, it may accurately identify children who have attentional deficits but not discriminate well between these children and children with conduct disorders.

It cannot be assumed that a single validity index for a measure (e.g., content validity) is generalizable across other validity indices for that measure (e.g., convergent validity, divergent validity, predictive variability).

3. **Describe missing psychometric data as well as conflicting psychometric data of a measure in all relevant psychometric domains, and the implications of**

these gaps for understanding the measure as well as the interpretation of results.

Many studies do not present all relevant psychometric information that is known about a given measure used and few disclose important psychometric evidence that is missing. Additionally, some authors report only research that has generally positive psychometric information and fail to report research that has been unsupportive. Instead of just burying this information, these lacunae need to be explicated and their implications be made clear to the reader. Scientific methods are designed in part to safeguard against confirmation bias (e.g., Popper, 1957). Consistent with that goal, studies that provide less positive psychometric evidence or show measures with superior relevant psychometric evidence should be acknowledged.

Disregarding conflicting evidence about the internal structure of an assessment device is an example of this problem. Factor structures identified during the original development of an instrument often fail to be satisfactorily replicated in subsequent studies, especially those that involve participants who differ in important ways from participants in the original studies. An unreliable factor structure indicates either problems with the content of the instrument/scale or that there are real and important differences in the targeted construct across populations or assessment contexts. The unreliability of some factor structures also means that the clinician should carefully consider the characteristics of prior studies when making clinical judgments on the basis of scale scores derived from prior factor analyses.

4. **Present quantified psychometric information.**

Psychometric information ought to be presented in ways that are precise and quantified. Broad statements that "The reliability of the measure is (or has been found to be) good" are insufficiently helpful in evaluating the psychometric strength of a measure for its current application. Rather, it is essential to include quantified statements of psychometric evidence that allow more precise understanding of the magnitude of the potential error of measurement in the current study, which can then affect the study's inferences and conclusions.

5. **Avoid describing psychometric information as trait-like.**

Psychometric evidence can generalize across persons, settings, or assessment contexts and goals, and time, but does not necessarily do so (Haynes et al., 2018). Thus, psychometric evidence ought not to be presented as if it is a stable, trait-like characteristic of a measure. Validity evidence for a measure can vary depending on the particular judgment to which it is applied and the cultural characteristics and setting of a client. For example, the validity evidence for a measure of couple adjustment can vary depending on whether the measure is used for brief screening purposes or for clinical case formulation, and whether it is used to measure adjustment in younger or older couples.

As we have noted, inferences about the psychometric characteristics of measures from an instrument are always conditional and these conditions should be reported precisely. Psychometric data and inferences depend on sample composition, convergent measures used, and other conditions of the evaluation. As stated above, psychometric characteristics do not "reside" with an instrument—past indices of validity do not mean that the instrument provided valid inferences in the current study. When providing supporting evidence for the psychometric characteristics of measures, the past tense should be used and, when feasible, the characteristics of the psychometric study (especially sample composition and convergent measures) should be reported. Evidence should focus on measures rather than instruments because some instruments provide multiple measures that vary in terms of their psychometric support.

In sum, these five principles provide a template to more fully and clearly describe the goals, strengths, weaknesses, and unknowns regarding a measure. They provide a useful and a more comprehensive and valid basis upon which to make more accurate evaluations of the study's conclusions. These recommendations might also have the salutary effect of encouraging investigators to more carefully select measurement instruments if they knew they had to report more comprehensively on the measures selected.

A Case Illustration

The following example was drawn from Masuda et al. (2007), one of the second author's previously published papers. To exemplify the confusion, uncertainty, and variability relevant to the presentation of the psychometric information of a measure, we are going to focus on the Acceptance and Action Questionnaire (AAQ; Bond & Bunce, 2003; Hayes, Strosahl, et al., 2004), a self-report measure used in Masuda et al. (2007).

In the paper, Masuda et al. (2007) presented a randomized controlled trial that examined whether individuals high in the construct of psychological flexibility and those low in psychological flexibility responded differently to two types of psychosocial interventions designed to reduce mental-health-related stigma in a nonclinical sample of college students. Ninety-five participants (64 women; 2 participants failed to note their gender) attended the workshops; 52 (38 women; 2 unidentified) assigned to the Acceptance and Commitment Training (ACT; Hayes, Bissett, et al., 2004; Hayes et al., 2012) intervention and 43 (26 women) to the education intervention. The average age was 19.7 years. The majority of participants were non-Hispanic Caucasians (non-Hispanic Caucasian = 70, Asian/Pacific Islander = 6, Hispanic = 7, African American = 2, multiethnic/others = 8, and unidentified = 2). To accomplish this end, a 16-item AAQ scores at pre-intervention were used to dichotomize the study sample (i.e., high psychological flexibility vs. low psychological flexibility). The study participants were recruited from a state university in Reno, Nevada, in the United States. In the published paper, the AAQ was described as follows (p. 2767):

> The Acceptance and Action Questionnaire (AAQ; Bond & Bunce, 2003; Hayes, Strosahl, et al., 2004) was used to categorize participants by their degree of psychological inflexibility, cognitive fusion, and experiential avoidance [in this study, we will use the term "psychological inflexibility" to refer to these ACT processes]. The AAQ is a 7-point Likert scale with *adequate reliability* (α of .72–.79; Bond & Bunce, 2003; Hayes, Bissett, et al., 2004; Hayes, Strosahl, et al., 2004), and inquiries about avoidance of emotions, fusion with thoughts, and the inability to act in the presence of difficult thoughts and feelings. The 16-item version (Bond & Bunce, 2003) was used and scored so that higher scores correspond to higher levels of psychological flexibility.

In order to categorize participants, the mean score for clinical populations (Hayes, Strosahl, et al., 2004) was used as a cutoff. If a participant's pre-treatment AAQ score was 66 or lower, the participant was categorized as being psychologically inflexible; if the score was 67 or higher, the participant was categorized as being "*psychologically flexible*." (Italics added).

Critique

As the pre-treatment AAQ score was used to differentiate those high in psychological inflexibility from those low in psychological inflexibility, it is safe to say that Masuda et al. (2007) viewed psychological inflexibility as a relatively stable trait-like behavioral pattern. Masuda et al. (2007) used a 16-item version AAQ developed by Bond and Bunce (2003). It is also important to note that at the time of the study, there were several other versions of AAQ available, including another 16-item version of AAQ, and that psychometric indices of these alternative AAQs were reported in detail in Hayes, Strosahl, et al. (2004). As described in detail below, Masuda et al. (2007) violated a number of the proposed principles described earlier, including treating psychometric properties as sample independent and thus trait-like.

Inconsistent with Principle 1 (provide explicit arguments about relative importance of psychometric evidence), Masuda et al. (2007) did not present any argument about the relative importance of various psychometric indices for the 16-item version of AAQ in the context of the study. Thus, the reader had no guidance about which psychometric dimensions were particularly important to know. Given the purpose of AAQ in Masuda et al. (2007), there should have been an explicit argument about the importance of knowing its *content validity*, *convergent validity*, *divergent validity*, and *factor structure*. These arguments in this study needed to be nuanced because Masuda et al. (2007) implied that three interrelated constructs were purportedly measured by this measure: psychological inflexibility, cognitive fusion, and experiential avoidance. In addition, arguments should have been made (e.g., correlations among factor/scale scores) about how these three constructs can be combined into a measure of one superordinate construct *psychological in/flexibility*. Additionally, particularly relevant to the way that AAQ score was used in the study, arguments should have been made about the importance of the *discriminative validity* of scores on the AAQ to using a cut score of 66 as well as the temporal stability of AAQ score in form of *test-retest reliability*.

In regard to Principle 2 (i.e., psychometric evidence is multidimensional), Masuda et al. (2007) also failed to report the available psychometric data of the 16-item version of AAQ that were particularly relevant to their study. Instead, Masuda et al. (2007) implied the overall adequacy of this measure by reporting only its internal consistency (i.e., "α of .72–.79"). As discussed extensively elsewhere (Haynes et al., 2019), one should not assume other validity and reliability indices of a scale (e.g., convergent validity, divergent validity, and predicted validity; test-retest reliability) based on one index of that scale (e.g., internal consistency).

At the time of the study, results of internal consistency (Cronbach's alpha of .88 and .90), one-year test-retest reliability (r = .72), *confirmatory factor analysis* (CFA), and *convergent validity* (with mental ill-health measured by General Health Questionnaire-12 [GHQ-12; Goldberg, 1978]; r = −. 61) were available for the 16-item version of AAQ with a sample of English and Scottish employees who work in the customer service centers of a United Kingdom financial institution (N = 412, mean age = 30.87 years, 68% women, 66% working part-time; Bond & Bunce, 2003). The CFA found a two-factor solution was a good fit to the data of this study sample: $\chi^2(101, N = 412) = 233$, p = .02, comparative fit index (CFI) = .97; and root-mean-square error of approximation (RMSEA) = .05, with one factor appearing to represent one's "willingness to experience unwanted events" and the other appearing to reflect one's "ability to take action, even in the face of unwanted internal events." Masuda et al. (2007) could have reported these findings as well as provided how these two factors mapped onto the construct of *psychological flexibility*.

At the time of the study in Masuda et al. (2007), a comprehensive psychometric examination of the 16-item 2-factor version of AAQ had not been done. The only available psychometric evidence for this version of 16-item AAQ was that reported in Bond and Bunce (2003). More thorough psychometric evaluation had been done with one of the other versions of AAQ (i.e., 9-item single-factor version of AAQ; Hayes, Strosahl, et al., 2004) with multiple non-clinical and clinical samples in the United States, and, as described below, Masuda et al. (2007) appeared to *assume wrongly* that the 16-item version of AAQ was psychometrically equivalent to the 9-item version.

With regard to Principle 3 (i.e., discuss missing psychometric data, negative psychometric information, and the implications of these gaps for understanding the measure as well as the interpretation of results), Masuda et al. (2007) did not present any argument to help the reader understand the implications of key missing psychometric information. As stated above regarding Principle 1 there was missing psychometric information, particularly with regard to *content validity*, *convergent validity*, and *divergent validity*. As such, the readers are left to their own to understand implications of other psychometric information.

Furthermore, Masuda et al. (2007) also failed to report negative psychometric information of the 16-item version AAQ as well as other versions of AAQ. At the time of the study, the 16-item 2-factor version of AAQ used in Masuda et al. (2007) appeared to be internally consistent with the sample of English and Scottish employees who work in the customer service centers of a United Kingdom financial institution (Cronbach's alpha of .88 and .90; Bond & Bunce, 2003) more so than other versions of AAQ (e.g., Cronbach's alpha = .70 for the 9-item version, and Cronbach's alpha = .61 for the other 16-item version of AAQ with a sample of clients in a university counseling center in the United States; Hayes, Strosahl, et al., 2004). However, it is also important to note that since the publication of Masuda et al. (2007), the 16-item AAQ used in the study has been replaced with a revised 1-factor version of AAQ (i.e., AAQ-II) for the purpose of refining its construct validity. Furthermore, more recently, the measure of AAQ has come under attack for its questionable construct validity. Several authors have argued that the measure of even the most recent version of AAQ (i.e., AAQ-II) is flawed psychometrically, as the data indicate inadequate discriminant validity—the confounding of constructs, particularly with negative emotionality (Gámez et al., 2011; Tyndall et al., 2019; Wolgast, 2014).

Inconsistent with Principle 4 (i.e., present quantified psychometric information), additional quantitative psychometric information should have been presented. The only quantified psychometric presented was an alpha coefficient (i.e., "α of .72–.79") that does not provide sufficient information to fully understand the strengths and weaknesses of the measure, particularly in the context of what was argued as the most important psychometric information (e.g., content validity, convergent validity, divergent validity, and factor structure). Masuda et al. (2007) also failed to provide psychometric information regarding test-retest reliability but still described the measure as having "adequate reliability," treating reliability as a trait-like characteristic of the measure.

Furthermore, Masuda et al. (2007) failed to report that these numeric values of alphas as well as how the cutoff score of 66 were also drawn from the findings of other versions of AAQ that used quite different samples of subjects (i.e., 460 clients [mean age = 26; 63% women, 85% Caucasian] in an university counseling center; 419 clients receiving a service from a large Health Maintenance Organization in Seattle [mean age = 38.6; 65.4% women, no information of client's ethnic and racial background], and 41 adult treatment-seeking individuals [mean age = 38.0; 70.7% women, 100% Caucasian]). In addition, Masuda et al. (2007) did not present a quantified index of the internal consistency of AAQ that could have been computed in their study sample. This set of information is important given the conditional nature of psychometric information of a measure (Haynes et al., 2019).

Principle 5 (psychometric information is conditional and not trait-like) also was violated in that all information about the sample and testing characteristics of the studies that produced psychometric information were not described. Consequently, it is not possible to understand how to make a detailed comparison of characteristics of past studies with those of the present study. These include the characteristics of study sample of the reported psychometric information. For example, data from the Bond and Bunce (2003) study were derived from 412 full-time customer service center workers in the United Kingdom, compared to the college student sample in Masuda et al. Approximately 30% of study participants in Masuda et al. (2007) were ethnic minorities, but they failed to suggest the dissimilarities of their sample with the samples used to describe the psychometric properties of the scale. As Masuda et al. (2007) viewed psychological flexibility as a trait-like feature of an individual variable, data on test-retest reliability of the AAQ were particularly important, given the test-retest nature of the study.

Examples of an Improved Presentation

What follows is an example of how the measure of AAQ in Masuda et al. (2007) could have been described in a way that is more consistent with the five principles described above.

"The score of 16-item 2-factor version of Acceptance and Action Questionnaire (AAQ; Bond & Bunce, 2003) was used to categorize participants into two groups by their degree of psychological flexibility. The AAQ used in this study was a 16-item self-report measure, and each item is rated on a 7-point Likert-like scale ranging from 1 (*never true*) to 7 (*always true*). Scores of all items were summed to yield the total score of AAQ, ranging from 16 to 112. Higher scores are interpreted to indicate higher levels of psychological flexibility. For the purpose of the present study, the most important psychometric information of AAQ are internal consistency, test-retest reliability, convergent validity, discriminant validity, and discriminative validity. Internal consistency was regarded as an important psychometric property because behavioral phenomena described in the items of AAQ were theorized to co-occur. Test-retest reliability (i.e., one-month test-retest reliability) was important because the time period between pretest and follow-up test was 4 weeks so the stability of the measure in this time period would be important to allow a more valid interpretation of cutoff score. Convergent validity with conceptually relevant constructs, such as thought suppression, emotion regulation, and mindfulness, was thought to be particularly important in this study because the construct of psychological flexibility purported to be measured by the 16-item version of AAQ was relatively new and not subjected to extensive psychometric evaluation. Discriminant validity and discriminative validity were thought to be important because in this study, alternative interpretations based on social disability, therapeutic allegiance, and willingness to report negative emotions would need to be ruled out for valid interpretation.

At the time of the present study, the psychometric research of this version of AAQ was fairly primitive and few psychometric studies were available. To date, results of internal consistency (Cronbach's alpha of .88 and .90), one-year test-retest reliability (r = .72), *confirmatory factor analysis* (CFA), *convergent validity* (with mental ill-health; $r = -.61$), *and criterion-related validation* (e.g., predicting mental ill-health 1 year later) were available for the 16-item version of AAQ only with a sample of English and Scottish employees who work in the customer service centers of a United Kingdom financial institution. This sample differs in age, employment status, nationality, educational status, as well as other variables from the sample in this study. A confirmatory factor analysis found a two-factor solution was a good fit to the data of this study sample: $\chi^2(101, N = 412) = 233$, $p = .02$, comparative fit index (CFI) = .97; and root-mean-square error of approximation (RMSEA) = .05, with one factor appearing to represent one's *willingness to experience unwanted events* (e.g., "It's OK to feel depressed and anxious") and the other appearing to reflect one's *ability to take action, even in the face of unwanted internal events* (e.g., "Despite doubts, I feel as though I can set a course in my life and then stick to it"). However, the construct allegedly being measured in the paper was psychological flexibility and these two factors are not identical with this. Moreover, to date no psychometric study has been done to examine its psychometric properties with the US college students, the sample of present study, and it was unclear the extent to which the psychometric evidence of AAQ found in Bond and Bunce (2003) was applicable to the present sample.

Furthermore, in order to categorize participants, the mean score for clinical populations drawn from another version of AAQ (i.e., 9-item version; Hayes, Strosahl, et al., 2004) was used as a cutoff. More specifically, if a participant's pre-treatment AAQ score was 66 or lower, the participant was categorized as being psychologically inflexible; if the score was 67 or higher, the participant was categorized as being psychologically flexible." Psychometric evidence of using a score of AAQ as a clinical cutoff was missing.

Conclusions

It is the premise of this chapter that there has been too little guidance on how to report psychometric information in studies in psychology. This has led to a wide variability in reporting practices as well as reporting practices that do not allow the reader to fully understand the measures' strengths and weaknesses in the context of the study in which they are used. In an attempt to provide increased clarity regarding the quality of the measures used, this paper proposed five principles to present and organize complex psychometric information.

If adopted, this more standardized approach ought to both give scholars more guidance on writing about measures and help readers to understand and better interpret the inferences made in the study. This will increase the length of the Measurement section in journal articles but is justified by its importance. This proposal should be regarded as preliminary and it is hoped that others will provide improvements to these principles so that measures can be better understood and the data these measures produce are more validly interpreted.

References

American Psychological Association. (2011). *Publication manual of the American Psychological Association* (6th ed.). American Psychological Association.

Bond, F. W., & Bunce, D. (2003). The role of acceptance and job control in mental health, job satisfaction, and work performance. *Journal of Applied Psychology, 88*, 1057–1067.

Gámez, W., Chmielewski, M., Kotov, R., Ruggero, C., & Watson, C. (2011). Development of a measure of experiential avoidance: The multidimensional experiential avoidance questionnaire. *Psychological Assessment, 23*, 692–713. https://doi.org/10.1037/a0023242

Goldberg, D. (1978). Manual of the general health questionnaire. Windsor: National Foundation for Educational Research.

Gross, A. (1996). *The rhetoric of science*. Harvard University Press.

Hayes, S. C., Bissett, R., Roget, N., Padilla, M., Kohlenberg, B. S., Fisher, G., Masuda, A., Pistorello, J., Rye, A. K., Berry, K., & Niccolls, R. (2004). The impact of acceptance and commitment training and multicultural training on the stigmatizing attitudes and professional burnout of substance abuse counselors. *Behavior Therapy, 35*, 821–836.

Hayes, S. C., Strosahl, K. D., & Wilson, K. G. (2012). *Acceptance and commitment therapy: The process and practice of mindful change* (2nd ed.). Guilford Press.

Hayes, S. C., Strosahl, K., Wilson, K. G., Bissett, R. T., Pistorello, J., Toarmino, D., … Stewart, S. H. (2004). Measuring experiential avoidance: A preliminary test of a working model. *The Psychological Record, 54*(4), 553–578.

Haynes, S. N., Kaholokula, J. K. A., & Tanaka-Matsumi, J. (2018). Psychometric foundations of psychological assessment with diverse cultures: What are the concepts, methods, and evidence? In *Cultural competence in applied psychology* (pp. 441–472). Springer.

Haynes, S. N., Smith, G., & Hunsley, J. R. (2019). *Scientific foundations of clinical assessment* (2nd ed.). Taylor and Francis/Routledge.

Hunsley, J., & Mash, E. J. (Eds.). (2019). *A guide to assessments that work* (2nd ed.). Oxford University Press.

Masuda, A., Hayes, S. C., Fletcher, L. B., Seignourel, P. J., Bunting, K., Herbst, S. A., … Lillis, J. (2007). Impact of acceptance and commitment therapy versus education on stigma toward people with psychological disorders. *Behaviour Research and Therapy, 45*(11), 2764–2772.

Popper, K.R. (1957). The logic of scientific discovery. Open Court.

Tyndall, I., Waldeck, D., Pancani, L., Whelan, R., Roche, B., & Dawson, D. L. (2019). The Acceptance and Action Questionnaire-II (AAQ-II) as a measure of experiential avoidance: Concerns over discriminant validity. *Journal of Contextual Behavioral Science, 12*, 278–284.

Wolgast, M. (2014). What does the Acceptance and Action Questionnaire (AAQ-II) really measure? *Behavior Therapy, 45*(6), 831–839.

Part III
Possible Solutions

Chapter 14
Replicability and Meta-Analysis

Jacob M. Schauer

Abstract In this chapter, I will discuss statistical considerations for studying replication. More specifically, I will approach replication from a framework based on meta-analysis. To do so, I will focus on *direct* replications, where studies are designed to be as similar as possible, as opposed to *conceptual* replications that (systematically or haphazardly) vary in at least one aspect of an experiment. The chapter starts with a brief description of recent research on replication in psychology and uses examples from that research to highlight relevant considerations in defining and parametrizing "replication." It then outlines different ways to frame analyses of replication and provides examples. Finally, it takes one possible definition of replication—that effects found across studies involving the same phenomenon are consistent—and describes relevant analyses and their properties.

Keywords Replicability · Meta-analysis · Replication research · Direct replication

Introduction

In the first two decades of the twenty-first century, multiple research programs called into question the replicability of scientific findings in several fields, including psychology (e.g., Ioannidis, 2005; Open Science Collaboration, 2015; Camerer et al., 2016; Klein et al., 2014). These findings would seem to have serious implications for the evidence behind evidence-based practices, particularly in clinical and behavioral psychology. In response, various scientific bodies, such as the Association for Psychological Science (APS) and National Institute of Health (NIH), as well as individual researchers called for steps to improve the transparency and reproducibility of scientific research (see Pashler & Harris, 2012; Collins & Tabak, 2014; Perrin, 2014; Bollen et al., 2015; Head et al., 2015).

J. M. Schauer (✉)
Department of Preventive Medicine, Northwestern University, Evanston, IL, USA
e-mail: jms@u.northwestern.edu

W. O'Donohue et al. (eds.), *Avoiding Questionable Research Practices in Applied Psychology*, https://doi.org/10.1007/978-3-031-04968-2_14

Though enhanced transparency can improve the face validity and potential replicability of research results, a critical step in establishing scientific evidence involves actually conducting replication studies. Yet, until the mid to late 2010s, literature on methods for designing and analyzing replication studies was limited (Schmidt, 2009; Hedges & Schauer, 2019b). Methods for studying replication would seem to be simple: Just conduct an additional study (or several studies) and examine whether results are *the same*. But empirical research on replication has demonstrated that replication is anything but simple (see Bollen et al., 2015). It can be extremely difficult and time-consuming to standardize procedures to ensure that relevant factors are controlled across multiple studies. Moreover, emerging work on the statistical aspects of studying replication has revealed several key challenges for researchers.

These statistical challenges intersect everything from the definition of replication to analytic methods to the design and sample size considerations for replication studies. Precise definitions of replication, which are seldom directly specified, are required in order to identify a relevant analytic method for replication. Schauer and Hedges (2021) argue that there are several possible definitions of replication, including agreement in the direction, interpretation, and magnitude of effects across studies. Moreover, a definitive analytic method must be specified in order to determine the sample size required of replication studies, or indeed, to determine how many studies need to be conducted.

In this chapter, I will discuss statistical considerations for studying replication. More specifically, I will approach replication from a framework based on meta-analysis. To do so, I will focus on *direct* replications, where studies are designed to be as similar as possible, as opposed to *conceptual* replications that (systematically or haphazardly) vary in at least one aspect of an experiment (for discussion, see Collins, 1992; Schmidt, 2009). The chapter starts with a brief description of recent research on replication in psychology and uses examples from that research to highlight relevant considerations in defining and parametrizing "replication." It then outlines different ways to frame analyses of replication and provides examples. Finally, it takes one possible definition of replication—that effects found across studies involving the same phenomenon are consistent—and describes relevant analyses and their properties.

What Does Research on Replication Look Like?

To understand what replication research looks like, it helps to look at how researchers have approached the study of replication, including those in psychology. Perhaps the most high-profile replication research projects were the Replication Project: Psychology (RPP; Open Science Collaboration, 2015) and the Replication Project: Economics (RPE; Camerer et al., 2016). Both of these projects took a series of scientific findings and ran a single replication of each: The RPE focused on 18 different experiments in behavioral economics, while the RPP looked at 100 social and behavioral psychology experiments, 73 of which they identified as a "meta-analytic subset" for which meta-analysis methods would be appropriate. The lengths taken to standardize and register protocols, ensuring that the replications in question could be seen as direct replications (or as direct as possible), were documented by the RPP (Open Science Collaboration, 2012).

However, there is no reason that researchers need to stop at just one replication study. Depending on the finding in question, it may be possible to conduct multiple replication studies, as was the case with the Many Labs Replication Project (Klein et al., 2014). Many Labs recruited 36 laboratories to run the same set of experiments. In the same year Many Labs published their results, the APS announced a series on the Registered Replication Reports (Simons, Holcombe, & Spellman, 2014). These efforts have conducted several replications of a given finding, from as few as 13 to as many as 33 studies (see Alogna et al., 2014; Bouwmeester et al., 2017; Cheung et al., 2016; Eerland et al., 2016; Hagger et al., 2016; Wagenmakers et al., 2016). Subsequent projects, including various iterations of Many Labs (e.g., Ebersole et al., 2016; Klein et al., 2018, 2019) and the Pre-Publication Independent Replication (PPIR) project (Schweinsberg., 2016) have also approached the study of replication as one that relies on several replication studies. This approach has been adopted by the Psychological Science Accelerator, an international collaboration of over 500 laboratories across the globe dedicated to conducting simultaneous replications across several laboratories (see Moshontz et al., 2018). To date, large-scale programs devised to study replication have seldom attempted to replicate findings in clinical psychology.

To unpack what these programs imply about replication research, we can zoom in on a single experiment. For instance, the RPP (Open Science Collaboration, 2015) ran a replication of an experiment first described by Payne et al. (2008). The original study examined the correlation between time spent awake and participant's memory of negative objects or scenes. The experiment involved presenting participants with a series of negative and neutral images, randomizing participants to conditions that corresponded with different sleeping conditions, and then asking them to respond to a set of images similar to those they were shown previously. The RPP conducted a single replication of this experiment. In that sense, programs, such as the RPP and RPE, have taken an approach of studying replication by conducting a single replication study for a finding, typically with a larger sample size than the original study.

Contrast that with a program, such as Many Labs, which replicated experiments like the reverse gambler's fallacy (Oppenheimer & Monin, 2009). In the original experiment, participants were asked to imagine a man rolling dice at a casino. In the two arms of the study, participants imagined seeing the man roll three sixes versus seeing him rolling two sixes and a three. Participants were then asked how many times they thought the man had rolled the dice before they witnessed the result in their assigned condition. On average, participants who imagined seeing three sixes tended to estimate the man had rolled the dice more times than those who imagined seeing only two sixes. Many Labs ran this experiment 36 times across different laboratories at (roughly) the same time. Analyses used by Many Labs included comparing the original study to an average of the effects found in the replication studies, as well as examining variation across the replication studies. Viewed this way, the study of replication may require several replication studies conducted simultaneously, and potentially in different laboratories or settings.

Model and Notation

As noted above in this chapter, I adopt a model and notation to describe the data generated by replication studies that is commonly used in meta-analysis. Meta-analysis is particularly germane to discussions of replication as it concerns the statistical methodology for studying the results of multiple (i.e., two or more) studies (Hedges & Olkin, 1985; Cooper et al., 2019). Suppose there are k studies conducted; replication research involves $k \geq 2$ studies. Often, one of these studies is published, though it is neither infeasible nor without precedent that multiple replication studies could be conducted prior to publishing any result (see e.g., Schweinsberg et al., 2016; Moshontz, et al., 2018): For the Payne/RPP sleep-memory study, $k = 2$ (i.e., an original and a replication study), for Many Labs' gambler's fallacy study, $k = 36$.

In any single study, the focus of statistical analysis is a quantity known as the *estimand.* An estimand is a quantity to be estimated or evaluated in a statistical analysis. The term is used to more clearly distinguish the target of inference (i.e., the *estimand*) from the method used to make inferences about that target (i.e., the estimator) and the specific value obtained from a given method and dataset (i.e., the estimate; for discussion in participant or patient outcomes, see Lawrance et al., 2020). An estimand could be a treatment effect in a randomized trial, such as a standardized mean difference or log odds ratio, a population parameter, or some parameter in a statistical model. For the sake of simplicity, the language in this chapter will refer to *effects* or *treatment effects* and assume that effects are one of the standard effect size indices commonly used in meta-analysis, such as the mean difference, standardized mean difference (Cohen's *d*), log odds ratio, risk ratio, and correlation coefficient (see Cooper et al., 2019). I will present results on the scale of standardized mean differences; however, the statistical results presented largely hold for other quantities.

Within study $i = 1, \ldots, k$, let θ_i be the effect or estimand of interest. Note that it may be possible (even probable) that $\theta_i \neq \theta_j$ even among direct replications if there are any uncontrolled sources of variation between studies (e.g., samples derived from different populations, potentially unknown deviations in protocols; Hedges & Schauer, 2019b). In later sections I will discuss an important way to conceive of the θ_i as either fixed but unknown quantities, or as random variables (referred to as the random effects model in meta-analysis). When the θ_i are treated as random variables, their distribution is assumed to have a mean μ and variance τ^2.

In practice, we do not observe θ_i directly, but instead must estimate it from data collected within study i. Denote T_i as the estimate of θ_i and let v_i be the estimation variance of T_i. Thus, from each study, we obtain an effect estimate T_i, and a variance

v_i or standard error $\sqrt{v_i}$. The statistical results in this chapter make three assumptions about T_i. First, that T_i is an unbiased estimator of θ_i. Second, that T_i is normally distributed. Third, that v_i is known (or estimated with very little uncertainty). Taken together, these assumptions imply

$$T_i \sim N(\theta_i, v_i)$$

where v_i is known. This will be exactly or approximately true for estimates of most effect size indices, including standardized mean differences (with reasonably large sample sizes), mean differences, log odds ratios, or z-transformed correlation coefficients (Cooper et al., 2019; Borenstein et al., 2009). Note that in a two-armed experiment, the variance v_i of the standardized mean difference can be expressed as

$$v_i = \frac{n_i^T + n_i^C}{n_i^T n_i^C} + \frac{\theta_i^2}{2\left(n_i^T + n_i^C\right)} \tag{14.1}$$

where n_i^T and n_i^C are the sample size of a treatment and control groups, respectively (Hedges, 1982). In a balanced experiment, where $n_i^T = n_i^C = n_i/2$ (i.e., n_i is the total sample size), so long as effects are relatively small and sample sizes within groups $n_i/2$ are reasonably large, we can write

$$v_i \approx \frac{4}{n_i} \tag{14.2}$$

Additional Notation

A key attribute of the statistical model above is that it distinguishes between an effect parameter θ_i and effect estimate T_i. Understanding this distinction will be sufficient for unpacking most of the key considerations for defining and evaluating replicability. For readers interested in more technical aspects of analyses for replication, this section provides some other useful values that arise in analysis methods discussed in this chapter. These serve as a reference for subsequent equations in this chapter.

- The precision weighted average of effect parameters. This is one way to average the effects across replication studies, wherein effects that are more precisely estimated receive more weight.

$$\bar{\theta}_{\cdot} = \sum_{i=1}^{k} \frac{\theta_i}{v_i} \tag{14.3}$$

- The unweighted average of effect parameters. This is an alternative to the weighted average in (14.3).

$$\bar{\theta} = \sum_{i=1}^{k} \frac{\theta_i}{k} \tag{14.4}$$

  - Note that when all v_i are equal so that $v_i = v$, then $\bar{\theta}$ is equivalent to $\bar{\theta}_{\cdot}$. Unless there is substantial variation in sample sizes across studies, the averages in (14.3) and (14.4) will often be similar in value.

- The precision weighted average of effect estimates. This is typically used to summarize or average effect estimates in meta-analysis, and gives greater weight to studies with smaller variances (i.e., more weight is given to studies with bigger sample sizes).

$$\bar{T}_{\cdot} = \left(\sum_{i=1}^{k} \frac{T_i}{v_i}\right) \Big/ \left(\sum_{i=1}^{k} \frac{1}{v_i}\right) \tag{14.5}$$

  - Note that when all v_i are equal so that $v_i = v$, then $\bar{T}_{\cdot}$ is equivalent to the unweighted mean $\bar{T} = \sum_{k}^{i=1} T_i / k$.
  - Among the $k = 36$ effect estimates and variances reported by Many Labs' reverse gambler's fallacy experiments (see Table 14.3), the weighted mean of effects is 0.63 and the unweighted mean is 0.61.

- The Q statistic is used to test heterogeneity and estimate between-study variance:

$$Q = \sum_{i=1}^{k} \frac{\left(T_i - \bar{T}_{\cdot}\right)^2}{v_i} \tag{14.6}$$

  - For the Many Labs' reverse gambler's fallacy example, the Q statistics is 51.61.
  - For $k = 2$ studies, the Q statistic reduces to $(T_1 - T_2)^2/(v_1 + v_2)$.

- A sum of (powers of) precisions S_j is used in computing various quantities related to variation between studies and standard errors of meta-analytic estimates:

$$S_j = \sum_{i=1}^{k} \frac{1}{v_i^j} \tag{14.7}$$

  - Note that when all v_i are equal so that $v_i = v$, $S_j = k/v^j$.

- The constant S is a function of the estimation error variances v_i used in common estimators of between-study variation:

$$S = S_1 - \frac{S_2}{S_1} = \sum_{i=1}^{k} \frac{1}{v_i} - \frac{\sum_{i=1}^{k} \frac{1}{v_i^2}}{\sum_{i=1}^{k} \frac{1}{v_i}} \tag{14.8}$$

- Note that when all v_i are equal so that $v_i = v$, $S = (k - 1)/v$.

- An estimate of the variation between effect parameters is based on the Q statistic in (14.6) (DerSimonian & Laird, 1986):

$$\hat{\tau}^2_{DL} = \frac{Q-(k-1)}{S} \tag{14.9}$$

  - For the reverse gambler's fallacy example, the estimated between-study variance is $\hat{\tau}^2_{DL} = 0.01$.

- A random effects weighted average of effect estimates:

$$\bar{T}_{\cdot}^{*} = \left(\sum_{i=1}^{k} \frac{T_i}{v_i + \hat{\tau}^2_{DL}}\right) \Bigg/ \left(\sum_{i=1}^{k} \frac{1}{v_i + \hat{\tau}^2_{DL}}\right) \tag{14.10}$$

  - This is analogous to the weighted average in (14.5), except the weights in (14.10) involve the estimated between-study variation $\hat{\tau}^2_{DL}$. In the reverse gambler's fallacy example, the unweighted mean is $\bar{T} = 0.61$, the precision weighted mean is $\bar{T}_{\cdot} = 0.63$, and the random effects weighted mean is $\bar{T}_{\cdot}^{*} = 0.61$.

What We Mean When We Say "Replication"

Conventional understanding of successful replications is that they get "the same" result or outcome. Yet, when it comes to statistical analyses, "the same" has proven to be tricky to characterize precisely (see Valentine et al., 2011; Bollen et al., 2015; Hedges & Schauer 2019b; Schauer & Hedges, 2021; Schauer et al., 2021). Doing so requires some decision-making about the studies involved and how they pertain to the finding under scrutiny. Because of this, there are *several* possible definitions of replication success or failure, and different ways to quantify these definitions (Schauer & Hedges, 2021). As one might expect, an analytic method for one definition of replication may be wholly inappropriate for a different definition of replication. Thus, before conducting any analysis of replication, it is critical to formalize the relevant definition. In this section, I will discuss various ways in which definitions of replication can be structured, and why that can matter for making inferences about the replicability of scientific findings.

Definition of Replication Versus Analysis Methods

An important distinction to make in the context of replication research is between the underlying *definition* of replication and *analysis methods* for a given definition. A definition of replication ought to concern the effect parameters θ_i. The θ_i are the actual effects produced in an experiment or study, and hence reflect a study's true results (i.e., the true effect). An analysis method uses data (i.e., the effect estimates T_i and variances v_i) to infer something about the relationships between the θ_i. That is, an analysis method—which is a function of the T_i and v_i—concerns a specific formal definition of replication—which is a function of the θ_i.

To unpack this distinction, consider a research design with $k = 2$ studies: an original study (study 1) and a replication (study 2). There appear to be two commonly accepted definitions of replication success for two studies. First, effects could agree in sign/direction, so that both effects are positive or negative (e.g., a treatment improves outcomes in both studies). Second, effects could agree in magnitude, so that effects are the same size in each study. The first column of Table 14.1 shows a mathematical formalization of these two definitions. Note that the sign () and 1{} functions are as follows:

$$sign(x) = \begin{cases} 1 \text{ if } x > 0 \\ 0 \text{ if } x = 0 \\ -1 \text{ if } x < 0 \end{cases}, \qquad 1\{x\} = \begin{cases} 1 \text{ if } x \text{ is TRUE} \\ 0 \text{ if } x \text{ is FALSE} \end{cases} \tag{14.11}$$

Common analysis methods used to determine if study 2 failed to replicate study 1 include the statistical significance criterion, the confidence interval overlap (CIO) procedure (Brandt et al., 2014), and the prediction interval (PI) procedure (Patil, Peng, & Leek, 2016). Table 14.1 describes these approaches as statistical procedures.

Table 14.1 Some definitions of replication and commonly used methods to assess those definitions

Definition of replication	Some proposed analyses
Agreement in sign/*direction* of effects sign (θ_1) = sign (θ_2)	*Significance criterion:* sign (T_1) = sign (T_2) AND both are significant (or both nonsignificant)
Agreement in *magnitude* of effects $\theta_1 = \theta_2$	*Confidence interval overlap:* $1\{(T_1 - T_2)/\sqrt{v_2} > 1.96\}$
	Prediction interval: $1\{(T_1 - T_2)/\sqrt{v_1 + v_2} > 1.96\}$

- The statistical significance procedure concludes study 2 failed to replicate if it disagrees in sign or statistical significance compared to study 1 (e.g., T_1 and T_2 are both positive, but T_1 is statistically significant and T_2 is not). Typically, statistical significance is set to the $\alpha = 0.05$ level for this procedure and two-sided tests are used where appropriate.
- The CIO method concludes a study failed to replicate if T_1 is not contained in a 95% confidence interval for θ_2 in study 2. Note that the confidence interval for study 2 only accounts for estimation error variance in study 2, but not in study 1.
- To adjust for that, the PI approach concludes a study failed to replicate if T_1 is not contained in a 95% *prediction* interval for θ_2, where the prediction interval takes into account estimation error variance in both study 1 and study 2. This is equivalent to concluding study 2 failed to replicate study 1 if 95% confidence intervals from the two studies do not overlap.

Types of Agreement

An important consideration for defining replication is what we mean by "the same" results; that is, what type of agreement do we expect or desire out of our replication studies? The example above described two possible types of agreement: agreement in sign/direction and agreement in magnitude. These appear to be among the most commonly accepted types of agreement in replication research. However, they are not the only possible type of agreement. For instance, we might consider studies to agree qualitatively if their effects are all large enough to be considered clinically relevant, so that $\theta_i > q$ for some threshold value q that corresponds to a clinically relevant effect (see Mathur & VanderWeele, 2020).

In this chapter, I will focus on agreement in magnitude, which can be seen as a finer, more restrictive definition of replication. For instance, if $\theta_1 = 0.2$ in Cohen's d units and $\theta_2 = 20$, this would characterize a "successful" replication if our preferred definition involved the direction of effects, yet few social scientists would consider these to be similar in size given that they differ by two orders of magnitude. In that sense, agreement in direction is a coarser definition of replication. In a clinical setting, agreement in magnitude can provide greater confidence about the stability and predictability of effects and can potentially better inform decisions about implementing an intervention that must weigh potential benefits against anticipated costs or side effects.

Exact Versus Approximate Replication

Agreement in magnitude of effects can also be specified in different ways. An obvious way would be to require effects to be identical, so that $\theta_1 = \ldots = \theta_k$, a scenario referred to as *exact replication* (Hedges & Schauer, 2019b). However, one might expect findings to vary slightly across repeated studies due to sampling subjects from slightly different populations or minor—often unknown—deviations in study implementation. If such differences produce small, but negligible variation between effects, that could still be seen as successful replication so long as the resulting effect parameters would warrant the same scientific or clinical interpretation. For instance, if $\theta_1 = 0.2$ (Cohen's *d*) and $\theta_2 = 0.201$, many social and psychological researchers might consider those to be about the same. Thus, we may also define *approximate replication* as when differences between effect parameters are negligibly small (for further discussion, see Schauer, 2018; Hedges & Schauer, 2019a,b). I will formalize ways to operationalize "negligibly small" in subsequent sections.

Falsification Versus Consistency

When $k > 2$ studies are involved in replication research (e.g., an original study and multiple replications), there are at least two ways to orient an analysis of replication. First, the focus could be on singling out a single study (or group of studies) and comparing it to the others. In such analyses, the original study is typically compared to subsequent replication studies as a means of falsifying the original study. If multiple replication studies have been conducted, analyses of replication may aggregate their results, including via a meta-analysis, so that the analysis compares the effect from the original study and the average effect found in the replication studies. We refer to this type of orientation as a *falsification* approach. Note that analyses based on a *falsification* approach to replication need not result in yes/no conclusions about the original study and could instead focus on continuous metrics, such as the size of the difference between the original study and subsequent replications.

Because falsification definitions contrast the original effect parameter θ_1 to an average of the replication study estimates, it can be seen as treating multiple replication studies (study 2, ..., study *k*) as a single large study. Hence, analyses of falsification definitions of replication are statistically analogous to analyses for $k = 2$ studies, even when $k > 2$ studies (i.e., multiple replication studies) have been conducted.

Rather than singling out one specific study, a definition can focus on whether there is agreement across all studies. In this definition, the focus is on variation across all effects, rather than a comparison of one study versus an average of several others. If there is little or no variation among effects, then we might conclude that the finding is relatively consistent across all studies, and hence we refer to this framing as a *consistency* approach.

When there are only $k = 2$ studies, consistency and falsification analyses are identical. A comparison between an original study and a single replication is equivalent to an analysis that examines differences (i.e., variation) between the two study effects.

Fixed Versus Random Studies

Another consideration in defining and making inferences about replication is whether the studies and their resulting effect parameters are fixed or random (Hedges & Schauer, 2019b). The *fixed effects* model assumes that the studies and effect parameters are the only studies of interest. Inferences will pertain only to the studies that we observe and their corresponding effect parameters θ_i (i.e., did these specific studies successfully replicate?). The *random effects* model assumes that the studies observed are only a sample from a population of replication studies that could be observed. In this model, the θ_i are treated as draws from some distribution or putative population of effect parameters. Inferences about replication pertain to the population of studies, including those not observed. The distinction between fixed and random studies models is analogous to the fixed and random effects models in meta-analysis (Laird & Mosteller, 1990; Hedges & Vevea, 1998).

Putting It All Together: Defining Replication

All of the considerations above are necessary for defining "replication" (i.e., defining results "being the same") as a quantity on which we can conduct inference. Table 14.2 shows different ways we might define replication when we have different views of these considerations. These are not the only possible ways to define replication and other quantities related to replication may possibly be of interest to researchers.

Table 14.2 also demonstrates that the estimand that corresponds to "replication" will depend heavily on the considerations listed in this section. In practice, the distinction between fixed and random effects definitions and analyses is minor, and leads to parameters that differ in their precise statistical interpretation, but have roughly the same scale (see Schauer, 2018; Hedges & Schauer, 2019b). Yet, the framing of the definition of replication (consistency versus falsification) and the type of agreement (magnitude versus direction) can lead to markedly different parameters corresponding to "replication." It is therefore imperative that researchers identify the relevant framing and agreement type in advance.

Once again, in this chapter, I highlight definitions of replication that correspond to consistency across all effects in magnitude. This definition of replication seeks to identify if (and to what degree) effects of an intervention may change over repeated trials. Defining replication in this way can be seen as consistent with moves toward evidence-based practices, as well as with conventional notions regarding the role of replication in the scientific method. Furthermore, it can provide researchers and practitioners with a clearer picture of how stable an intervention's impact is, and potential conditions under which it may change.

Table 14.2 Some possible definitions/parametrizations of replication

		Studies fixed	Studies random
Falsification	Agreement in magnitude	Difference between original study and replication: $\lvert\theta_1 - \theta_2\rvert$ for $k = 2$ $‡\, \theta_1 - \bar{\theta}_{\cdot}\#$ for $k > 2$	Comparison of original study to distribution of effects in replications: $P_{\text{orig}} = P[\theta_1 > \theta_i]$, $i > 1$ $d_{\text{orig}} = (\theta_1 - \mu)/\tau$
	Agreement in direction	Replication effect parameters are in the same direction as the original: sign (θ_1) = sign (θ_2) for $k = 2$ $sign(\theta_1) = sign(\bar{\theta}_{\cdot})$ for $k > 2$	Probability that replication effects from population are in same direction as original: $P[\text{sign}(\theta_i) = \text{sign}(\theta_1)]$, $i > 1$
Consistency	Agreement in magnitude	Variation across effects from observed studies: $\lvert\theta_1 - \theta_2\rvert$ for $k = 2$ $\tau_F^2 = \sum_{i=1}^{k} \frac{(\theta_i - \bar{\theta}_{\cdot})^2}{k-1}$ $\lambda = \sum_{i=1}^{k} \frac{(\theta_i - \bar{\theta}_{\cdot})^2}{v_i}$	Variation across population of effect parameters: $Var[\theta_i] = \tau^2$
	Agreement in sign	Proportion of effects that are positive: $\sum_{i=1}^{k} \frac{1\{\theta_i > 0\}}{k}$	Probability effects are positive: $P_{>0} = P[\theta_i > 0]$

Considerations for Analyses of Replication

When we can precisely state a definition of replication, our next challenge involves formalizing analyses for replication. We can conduct statistical analyses in at least two ways: analyses that lead to binary conclusions about replication and analyses that quantify continuous metrics that correspond with a definition for replication. In addition, any analysis of replication that includes an extant published study must consider the potential impact of publication bias on the analysis (see below). In this section, I will describe some potential choices about how to frame an analysis for replication and discuss possible publication bias adjustments.

Categorical Decisions About Replication: Did the Finding Fail to or Successfully Replicate?

One class of statistical analyses is one that supports qualitative conclusions about replication (e.g., "the replication(s) failed") via some decision procedure. The analysis methods discussed above (Significance, CIO, PI) can be seen as part of this class, as they all result in a success/failure conclusion. More broadly, this type of analysis is common in the null hypothesis test (NHT) framework, wherein we test a null hypothesis about replication and draw conclusions about replication success or failure based on a test of that null hypothesis. For example, each of the Significance, CIO, and PI methods can be seen as tests of a null hypothesis that the replication succeeded, and we would reject that null hypothesis and conclude a finding failed to replicate if a criterion was not met (see Schauer & Hedges, 2021; Schauer et al., 2021).

The Burden of Proof

If using NHTs, an important consideration is the burden of proof, which dictates how to form the null hypothesis. The burden of proof for an NHT about replication can either be on replication or nonreplication (i.e., replication success or failure, respectively). If the burden of proof is on replication, then we would form a null hypothesis that corresponds to replication failure, and we would require evidence to reject that hypothesis and conclude replication success. Conversely, if the burden of proof is on nonreplication, the null hypothesis should correspond with replication success, and we would require evidence to reject that.

As an example, suppose we were interested in exact replication for $k = 2$ studies (i.e., $\theta_1 = \theta_2$). If the burden of proof was on nonreplication, we would form the null hypothesis H_0: $\theta_1 = \theta_2$ and we would only conclude replication failure if our analysis rejected H_0 (e.g., if the PI method indicated replication failure). Conversely, if the burden of proof is on replication, then we would need to form H_0: $\theta_1 \neq \theta_2$ and only conclude that $\theta_1 = \theta_2$ if we reject H_0. Forming a null hypothesis of replication failure that is testable can be done in a manner analogous to equivalence testing, which involves setting H_0: $|\theta_1 - \theta_2| > \varepsilon$ for some constant $\varepsilon > 0$ (see Wellak et al., 2002; Hedges & Schauer, 2019a,b). This null hypothesis contends that the replication failed and the difference between θ_1 and θ_2 is at least as big as some nonzero value ε that characterizes the smallest non-negligible difference between effects consistent with replication failure. We would reject that hypothesis and conclude replication success—that the difference in effects is less than the smallest non-negligible difference between effects ε (i.e., the difference between effects is negligible)—if the data provided evidence to do so.

Decision-Theoretic Properties/Error Rates

As with any NHT, analytic methods that produce qualitative inferences about replication can result in erroneous conclusions about replication. The rate at which these types of errors occur are crucial for interpreting their results. For example, if the burden of proof is on nonreplication, then a Type I error indicates that we conclude replication failure when the replication(s) was successful. If an analytic method has a high Type I error rate, then it has a high probability of labeling successful replications as failures.

The meaning of Type I and Type II errors depends on the burden of proof. A Type I error when the burden of proof is on nonreplication is the same as a Type II error when the burden of proof is on replication; in both cases successful replications are labeled as failures (or at the very least lacking evidence of success). To unify this nomenclature, we use the terms *false failure* and *false success* determinations (Schauer & Hedges, 2021). A false failure occurs when the replication succeeded but the analytic method does not indicate success. Conversely, a false success occurs when a replication failed but the analytic method does not indicate failure (for further discussion, see Schauer & Hedges, 2021).

Continuous Measures and Estimation

Rather than resulting in success/failure decisions about replication, analyses can involve estimating relevant quantities and their related uncertainty. Typically, uncertainty would include standard errors of estimators or confidence or credible intervals. As an example, for $k = 2$ studies, analyses that focus on agreement in magnitude may estimate $\theta_1 - \theta_2$ and report a standard error for that difference. If there are $k > 2$ studies that are treated as random, analyses could involve estimating τ^2, the between-study variation (discussed in subsequent sections) and its standard error. Conversely, if agreement in direction is the preferred definition of replication, there are a variety of alternatives. For example, Mathur and VanderWeele (2020) propose estimating the proportion of effect parameters that exceed some value q, which they denote $P_{>q}$. For agreement in direction, we can specify $q = 0$, and estimate $P_{>0}$ as

$$P_{>0} = 1 - \Phi\left(\frac{\overline{T}_{\cdot}^{*}}{\hat{\tau}_{DL}}\right) \tag{14.12}$$

where $\overline{T}_{\cdot}^{*}$ and $\hat{\tau}_{DL}^{2}$ are given in (14.10) and (14.9), respectively, and Φ is the distribution function for the standard normal distribution. This has an estimated standard error of:

$$\phi\left(\frac{-\overline{T}_{\cdot}^{*}}{\hat{\tau}_{DL}}\right)\sqrt{\frac{Var\left[\overline{T}_{\cdot}^{*}\right]}{\hat{\tau}_{DL}^{2}}+\frac{SE\left[\hat{\tau}_{DL}^{2}\right]^{2}\overline{T}_{\cdot}^{*2}}{4\hat{\tau}_{DL}^{6}}} \tag{14.13}$$

where ϕ is the standard normal density function, the variance of $\hat{\tau}_{DL}^2$, $SE\left[\hat{\tau}_{DL}^2\right]$, is described in (14.26) in later sections, and the variance $Var\left[\overline{T}_{\cdot}^{*}\right]=\left(\sum_{i=1}^{k}\frac{1}{v_i+\hat{\tau}_{DL}^2}\right)^{-1}$.

An alternative approach proposed by Etz and Vandekerckhove (2016) involves examining Bayes factors of original and replication studies. Bayes factors are analogous to hypothesis testing in that they evaluate the relative likelihood of competing hypotheses. In replication research, this often takes the form of a ratio of the likelihood of replication success relative to the likelihood of replication failure. This approach, which is appropriate for $k = 2$ studies, can be seen as examining the evidence provided by the replication study of a nonzero effect under competing models: that the effect in the replication study is $\theta_2 = 0$, and that the effect is equivalent to that estimated in study 1. Though the Etz and Vandekerckhove discuss these as continuous metrics, they also use their value to make qualitative inferences about replication success for failure. For instance, a Bayes factor at least as large as 10 is seen as strong support of a nonzero effect while a Bayes factor of 1/10 or less is seen as strong evidence of a null effect. If such inferences are made, then these methods can be seen as producing qualitative assessments about replication, and their properties should be discussed in terms of false failure and false success error rates rather than standard errors.

Publication Bias

Both empirical and theoretical researches suggest that published findings are subject to a selection process that favors the publication of statistically significant results (see Dickersin, 2005; Rothstein et al., 2005; Francis, 2012). If the probability that a finding is published depends on its statistical significance, this can induce bias in the effect size estimate T_i, and can impact the sampling distribution of T_i so that T_i is no longer normally distributed (see Hedges, 1984; Guan & Vandekerckhove, 2016). In the context of replication research wherein researchers conduct replications of a published study, there may be concern that the estimates reported by studies that were published prior to conducting replication studies may be affected by this process and could therefore be biased. This in turn can impact statistical analyses of replication and their properties.

Analyses can adjust for publication bias if it is suspected (see Rothstein et al., 2005; McShane et al., 2016). Adjustments should be focused on effect estimates for which researchers have good reason to suspect publication bias. This will likely include only a subset of relevant studies in replication research. Because many replications are pre-registered and have not yet been published, it is unlikely that publication selection will bias effect estimates from those studies. However, if extant published findings are to be included in analyses of replication, it would seem more likely that those effect estimates would be biased due to selection.

There are several possible relevant adjustments for publication bias. For example, Hedges (1984) provides a maximum likelihood estimator for unbiased estimation in the face of publication selection. This was one of the first approaches based on selection modeling, in which the process by which findings are selected for publication is based on their effect estimates T_i, variances v_i, or p-values. At their most basic, selection models assume that we only observe a T_i conditional on it being published, which in turn depends on its p-value. For instance, we might expect "statistically significant" T_i with $p_i < 0.05$ to be published with high probability (i.e., near 100%), but "nonsignificant" T_i with $p_i \geq 0.05$ to be published with a lower probability (e.g., near 40–50%). Thus, a published T_i has a conditional distribution affected by the probability of selection. To back out its unconditional distribution (and reduce or eliminate bias), we need to model the probability that T_i is published given its p-value. Selection models typically involve estimating the probability that T_i is published given its p-value and making relevant adjustments based on that probability, but such estimates typically require large numbers of studies subject to publication bias (Hedges & Vevea, 1996).

These models have since been extended to account for increasingly complex relationships between estimators T_i, variance v_i, and the probability of publication (Hedges & Vevea, 2005). Vevea and Woods (2005) propose an adaptation to selection model approaches that would seem appropriate for cases where only one or two effect estimates are biased due to publication selection. This approach assumes that the probability that a significant T_i gets published and a nonsignificant T_i goes unpublished are known a priori and need not be estimated from the data. A Bayesian method that makes (more or less) the same set of assumptions was applied to replication studies by Etz and Vandekerckhove (2016), and a hybrid model was presented by van Aert and van Assen (2017).

Finally, for analyses of $k > 2$ studies that focus on consistency, if published effect estimates are suspected to have severe publication bias, they can be omitted from analyses. Omitting biased effect estimates may make sense if the assumptions made by publication bias adjustments (model specification and parameter values) are untenable or difficult to justify. However, excluding the original study limits the scope and sample size of the analysis.

Perhaps a more principled approach would be to conduct a series of analyses each based on different publication bias adjustments (including no adjustment at all). Results of each analysis would then be presented and interpreted in light of the plausibility and strength of relevant assumptions.

Some Limitations of Statistical Significance and Confidence Interval Overlap

At the time of this writing, conducting a single replication study ($k = 2$ designs) remains a popular approach to studying replication (see Camerer et al., 2018). Moreover, the Significance, CIO, and PI approaches are still in common use. However, this approach to studying replication and these methods have some serious flaws. First, all three analysis methods are really only appropriate for $k = 2$ studies: the original study (study 1) and a replication study (study 2; Schauer, 2018; Schauer & Hedges, 2021). When multiple replication studies are conducted, these methods proceed by aggregating their effect estimates via meta-analysis. Thus, these methods are limited to designs with $k = 2$ studies, or if the framing of replication is falsifiability. Note that this is true of many proposed Bayesian analyses (for discussion see Hedges & Schauer, 2019a; Schauer & Hedges, 2021).

In addition, the statistical properties of these approaches can result in erroneous conclusions about replication with high probability. Previous research examined the false failure and false success rates of the Significance, CIO, and PI methods (Schauer & Hedges, 2021). For example, Schauer and Hedges (2021) found that the error rates of the Significance criterion depend on the power of study 1 and study 2 to detect effects θ_1 and θ_2, respectively. Unless both studies have very high power (i.e., >90% power), the false failure rate can range from 30% to over 70%, while the false success rate is likely between 15 and 30%.

The error rates of the CIO method are largely a function of the ratio of v_1/v_2 (Schauer & Hedges, 2021; Schauer et al., 2021). When v_1/v_2 is high (which occurs if study 2 has a larger sample size than study 1), the false failure rate can be as large as 20–40%. However, when v_1/v_2 is small, this can inflate the false success rate, which could be as large as 80%. In short, depending on the sample sizes and effect sizes of the studies involved, both Significance and CIO may be more likely to result in an error than in an accurate conclusion about replication.

It is worth noting that in addition to potentially high error rates, neither CIO nor the Significance criteria control error rates. In traditional NHTs, the procedures used should (and often do) limit the probability of a Type I error to be no greater than some a priori threshold α. The benefit of controlling the Type I error rate is that (so long as assumptions are met) the probability of a Type I error is (more or less) known and independent of other factors, such as sample size. Because of this, rejection of the null hypothesis can be seen as conclusive, since the probability that it is rejected in error is known (or at least bounded). However, as shown by Schauer and Hedges (2021), the false failure and false success rates of the CIO and Significance methods are functions of the v_i, and hence functions of sample size, as well as the θ_i. In sum, these methods control neither the false failure nor false success rates. Conclusions about replication generated by these procedures may be false with unknown (and possibly large) probability, and therefore it is difficult to view the results of these methods as particularly conclusive in many settings. On a related note, by contrast, Schauer and Hedges (2021) show that the PI criterion is equivalent to z-test for a difference in means, and therefore controls the false failure rate, a result we will point out in later sections of this chapter.

Defining Replication as Consistency of Effects

As argued above, agreement in magnitude may be a more informative definition of replication when it comes to clinical decision-making. To that end, there may be interest in the clinical psychology research community to emphasize definitions of replication that correspond to consistency of effects across studies. Table 14.2 characterizes ways we can parametrize such definitions for the fixed and random studies framework. If we treat the studies as random, we can quantify their consistency in terms of the variance of the distribution from which they were drawn, denoted τ^2. If $\tau^2 = 0$, then all of the effect parameters drawn from that distribution will be identical, and replication will be exact. If $\tau^2 > 0$ but is small, then effect parameters drawn from that distribution will be similar in size and may be seen to replicate approximately (Hedges & Schauer, 2019b).

When the studies are fixed, there are at least two ways to define agreement in magnitude for consistency analyses. One is with their "variance" τ_F^2:

$$\tau_F^2 = \sum_{i=1}^{k} \frac{\left(\theta_i - \bar{\theta}\right)^2}{k-1} \qquad (14.14)$$

Here, $\bar{\theta}$ is the mean of the θ_i given in Eq. (14.4). The parameter τ_F^2 is a summary statistic of the θ_i akin to a sample variance, as opposed to a property of a distribution like τ^2 (Schauer, 2018).

Alternatively, Hedges and Schauer (2019b) suggest that replication be parametrized by

$$\lambda = \sum_{i=1}^{k} \frac{\left(\theta_i - \bar{\theta}_{\cdot}\right)^2}{v_i} \qquad (14.15)$$

where $\bar{\theta}_{\cdot}$ is the precision weighted mean of the θ_i given in Eq. (14.3). The parameter λ is analogous to τ_F^2, but differs in that it accounts for the within-study estimation error variance v_i. When all of the studies have the same estimation error variance (e.g., the same sample size), then $v_1 = \ldots = v_k = v$, and the expression for λ reduces to

$$\lambda = \left(k-1\right)\frac{\tau_F^2}{v} \qquad (14.16)$$

Thus, the primary difference between τ_F^2 and λ is that τ_F^2 is on the scale of the individual θ_i, while λ is a ratio of between-to-within-study variance, a scale commonly used in meta-analysis (Schauer, 2018). This becomes evident when $k = 2$, so that τ_F^2 and λ reduce to expressions that depend on the magnitude of the difference $|\theta_1 - \theta_2|$:

$$\tau_F^2 = \frac{|\theta_1 - \theta_2|^2}{2}; \lambda = \frac{|\theta_1 - \theta_2|^2}{2\bar{v}} \tag{14.17}$$

where $\bar{v} = (v_1 + v_2)/2$ is the mean within-study variance.

The difference between τ_F^2 and λ is analogous to methods for quantifying heterogeneity in a random effects meta-analysis. The parameter τ^2 is on the scale of the individual θ_i. However, meta-analysts more often make judgments about between-study heterogeneity on the scale of between-to-within study variance τ^2/v, where v is some "typical" estimation error variance of the studies observed. Common meta-analytic statistics, such as I^2 or H^2, can be seen as depending on the scale of τ^2/v (see Higgins & Thompson, 2002).

Taken together, regardless of whether the studies are treated as fixed or random, a definition of replication that focuses on consistency of effects can be described as the variation between effect parameters. This variation can be computed on the scale of the θ_i, as with τ_F^2 and τ^2. Alternatively, it can be quantified on the scale of between-versus-within study variance τ^2/v (for random effects models) or λ (for fixed effects models). We note that while parameters like τ_F^2 and τ^2 refer to different quantities, in practice their scales can be interpreted in largely the same way (for discussion, see Hedges & Schauer, 2019b).

Evaluating the Amount of Heterogeneity Among "Consistent" Effects

Specifying an amount of heterogeneity among effects that corresponds with replication success or failure requires we set specific values of τ^2, τ^2/v, τ_F^2, or λ. Moreover, the following sections show that properties of analyses for replication will often depend on these values. Thus, to understand definitions of and analyses for replication, we need to understand the different scales of heterogeneity described above. What is a small or negligible value of λ or τ^2/v? What is a large value of τ^2 or τ^2/v?

The answer to such questions will be subject to scientific and clinical judgment. However, most researchers are used to intuiting the scale of individual effects, rather than variation across effects. In this section, we provide some insight into approaches for quantifying heterogeneity, as well as some conventions for negligible heterogeneity used in various scientific fields.

Hedges and Schauer (2019b) provide several ways to interpret τ^2 or τ_F^2 as a function of differences between pairs of effect parameters $\theta_i - \theta_j$ for $i \neq j$. Since $2\tau^2$ and $2\tau_F^2$ are equal to the mean pairwise squared difference between effects $E[(\theta_i - \theta_j)^2]$, it may be easier to describe replication or replication in terms of a meaningful value of $\theta_i - \theta_j$ and back out a value of τ_F^2 or τ^2. As an example, if the θ_i are standardized mean differences, so that we consider a difference of $|\theta_i - \theta_j| > 0.2$ to be non-negligible, this suggests that $\tau^2 < 0.02$ could be seen as negligible. Alternatively, if the studies are treated as random, we might specify a form of the distribution of the θ_i (e.g., normal) and identify a value of τ^2 that renders large pairwise differences unlikely:

$$\mathrm{P}\left[\left|\theta_i - \theta_j\right| < \varepsilon\right] < \gamma$$

That is, the probability of a large difference between two effect parameters occurs with probability less than some desired level γ.

Using these approaches, Schauer (2018) argues that values of $\tau^2 \leq 0.035$ would result in effect parameters on the scale of Cohen's d that could be characterized as roughly the same size; it would characterize a distribution of parameters such that deviations from the mean of that distribution greater than $d = 0.2$ occur with probability less than 20%. See Hedges and Schauer (2019a,b) and Schauer (2018) for further detail.

As a matter of sound statistical practice, analyses for replication should focus on the parameters τ^2 or $\tau_F{}^2$ rather than on τ^2/v or λ. Since τ^2/v and λ depend on the within-study variance v_i, and hence the sample size within studies n_i, they are not (strictly speaking) parameters. However, the scale of τ^2/v has been easier to work with, with traditional metrics of between-study variation in meta-analysis depending on that scale. For illustration, suppose that the k studies have roughly the same sample size so that $v_1 \approx \ldots v_k \approx v$. Meta-analytic metrics typically used to quantify heterogeneity, such as I^2 or H^2, depend onτ^2/v (Higgins & Thompson, 2002). If the v_i are not similar in value, Higgins and Thompson (2002) provide an expression for the "typical" value v of the v_i (see Eq. 14.9 in their article), and variation can be described on the same τ^2/v scale. When studies are treated as fixed, the parameter λ can be thought of in the same terms, as seen in Eq. (14.16). Whether we treat the studies as fixed or random, a potentially useful scale for heterogeneity is τ^2/v, so long as it is clear what a typical or normative value of v is.

Since τ^2/v is a common scale in meta-analysis, it can be easier to work with. Several different fields that conduct meta-analyses have generated conventions for negligible heterogeneity on that scale. Such conventions may be of use when approaching design and analysis of replication studies. Hedges and Schauer (2019a,b) note that in high-energy physics, the Particle Data Group characterizes minor or unimportant variation in ways that suggest $\tau^2/v < 1/4$ would be seen as negligible (see Olive, et al., 2014). In personnel psychology, Hunter and Schmidt (1990) describe and propose $v/(\tau^2 + v) < 0.75$ as negligible, which means $\tau^2/v < 1/3$. In medicine, a value of $I^2 = 100\% \times \tau^2/(v + \tau^2) < 40\%$ or $\tau^2/v < 2/3$ is seen as "not important" (see Higgins & Green, 2008). These are far from the *only* conventions of negligible heterogeneity (see Pigott, 2012), but they reflect ideas about heterogeneity that guide and inform research in these fields.

Since $\tau^2/v \in [1/4, 2/3]$ could be seen as a range of negligible between-study variation, researchers would likely want to detect differences between studies at least as large as these values; possibly two or three times larger. This would suggest that meaningful values of variation worth studying might range from $\tau^2/v = 1/4$ to $\tau^2/v = 3$ ($I^2 = 0.2$–0.75).

The benefit of using the scale of τ^2/v is that it does not depend on the scale of the θ_i. However, if we have a good idea of the scale of the θ_i, it makes more sense to focus on τ^2 or τ_F^2. Based on the results above, if θ_i are on the scale of Cohen's d, we may consider values of $\tau^2 \in [0, 0.035]$ to be negligible and may be interested in studying values in the range of [0.005, 0.1] (see Schauer, 2018).

A General Approach for Studying Replication: Magnitude of Effects

In this section, I will outline a framework for studying replication when the preferred definition is agreement in magnitude, and the preferred framing involves the consistency of effects. Though this is far from the only way to define replication, it is consistent with research that seeks to understand the conditions under which effects are stable and is in line with the type of knowledge refinement prioritized in various scientific fields including physics, chemistry, and medicine.

I first present results for fixed studies when $k = 2$, and then assume studies are random for $k \geq 3$. The fixed effects analog of the random effects analyses presented here can be found in Hedges and Schauer (2019b). The key distinction between the properties of the fixed and random effects analyses (beyond their scope of inference) is that the random effects analyses will be slightly less powerful and efficient than the fixed effects studies, though the difference in power is relatively small.

A Note aAbout Publication Bias

The analysis methods and their properties that are presented in this section do not include explicit adjustments for publication bias in published effect estimates. Such adjustments could be included in these methods, which would presumably impact their sensitivity. To understand the extent to which they do, note the following two aspects about publication bias adjustments. First, adjusting effect estimate T_i for bias can result in a corrected effect estimate T_i^* with variance v_i^*, where $v_i^* \geq v_i$ (i.e., corrected estimates tend to have greater variance). Second, corrections such as Hedges' (1984) maximum likelihood approach can result in adjusted effect estimates that are asymptotically normally distributed. Because the statistical results that follow depend on the normality of effect estimates, analyses can proceed with T_i and v_i if publication bias is not suspected, and T_i^* and v_i^* if publication bias is likely. Subsequent sections will detail that the sensitivity (statistical power or standard errors) will be worse when the v_i are larger. As a result, one impact of publication bias adjustments is that they reduce the power of tests for replication or increase standard errors or estimates for relevant quantities.

Fixed Effects for Two Studies

When there are $k = 2$ studies, the focus of analysis is on θ_1 and θ_2 being (about) the same value. Thus, analyses of replication can be viewed in terms of the difference $\theta_1 - \theta_2$. When $\theta_1 - \theta_2$ is large in magnitude, then there is a large difference between study results, but when $\theta_1 - \theta_2$ is small, then study results are more similar. We can directly estimate $\theta_1 - \theta_2$ with $T_1 - T_2$. Under the model it is the estimator with the smallest variance, and that variance is simply $v_1 + v_2$. When effect sizes are on the scale of standardized mean differences, the variance of the estimated difference can be written as approximately $4/n_1 + 4/n_2$, where n_i is the total sample size of study i; the standard error can be written as approximately $2\sqrt{\frac{1}{n_1}+\frac{1}{n_2}}$.

Example. Recall the Payne et al. study replicated by the RPP. The original study estimated an effect of $T_1 = 0.75$ (Cohen's d) with a variance $v_1 = 0.066$. The RPP replication study found an estimated effect of $T_2 = 0.30$ with variance of $v_2 = 0.23$. The estimated difference between effects is $T_1 - T_2 = 0.45$, which has a standard error of $\sqrt{v_1 + v_2} = 0.30$. A 95% confidence interval for the difference in effects is [−0.14, 1.04].

Alternatively, there are two different ways we can test null hypotheses about replication for $k = 2$ studies.

Tests When the Burden of Proof is on Nonreplication

If the burden of proof is on nonreplication, then we can structure a null hypothesis

$$H_0 : |\theta_1 - \theta_2| \leq \varepsilon \tag{14.18}$$

for some $\varepsilon \geq 0$. Here, ε corresponds to the largest difference between effect parameters that would be considered negligible. When $\varepsilon = 0$, H_0 corresponds to exact replication, but when $\varepsilon > 0$ (but is still small), H_0 corresponds to approximate replication. To test H_0 in Eq. (14.18), we compute the test statistic

$$Q_2 = \frac{(T_1 - T_2)^2}{v_1 + v_2} \tag{14.19}$$

Note that Q_2 is equivalent to the Q statistic in Eq. (14.6) for $k = 2$ studies. Under H_0, Q_2 follows a chi-square distribution with one degree of freedom and noncentrality parameter

$$\lambda_{02} = \frac{\varepsilon^2}{v_1 + v_2} \tag{14.20}$$

An α-level test would involve rejecting H_0 when $Q_2 > c_{1-\alpha}(1, \lambda_{02})$ where $c_{1-\alpha}(\nu, \lambda)$ is the $1 - \alpha$ percentile of the chi-square distribution with ν degrees of freedom and noncentrality parameter λ. Note that when $\varepsilon = 0$, so that the test concerns exact replication, $\lambda_{02} = 0$ and Q_2 follows a central chi-square distribution under H_0.

The power of this test to detect a difference $|\theta_1 - \theta_2| > \varepsilon$ is given by

$$1 - F\left(c_{1-\alpha}\left(1, \lambda_{02}\right) | , 1 | , \lambda_{12}\right) \tag{14.21}$$

where $c_{1-\alpha}(1, \lambda_{02})$ is defined as above, and $F(x\ |\nu, \lambda)$ is the distribution function of a chi-square random variable with ν degrees of freedom and noncentrality parameter λ. The power depends on a number of quantities:

- It is decreasing in ε, so that tests of exact replication will be more powerful than tests of approximate replication. Tests of increasingly looser notions of approximate replications with larger ε (and hence larger differences between effects that are seen as negligible) will be less powerful.
- It is increasing as a function of the true difference between effects $|\theta_1 - \theta_2|$. If $|\theta_1 - \theta_2|$ is larger, then tests will have more power.
- It is increasing as a function of the variance of each effect estimate $v_1 + v_2$. If v_1 and v_2 are smaller (so that sample sizes in each study are larger), the test will have greater power.
- In practice, the power of this test for $k = 2$ studies is bound to be low unless both studies have uncommonly large sample size. Assuming effects on the scale of Cohen's d and an $\alpha = 0.05$ level test for exact replication, to detect a difference $|\theta_1 - \theta_2| = 0.2$ (Cohen's d) with 80% power would imply $v_1 + v_2 \leq 0.0013$, which is consistent with both studies having sample sizes of at least 1569 given Eq. (14.2). Detecting a difference of $|\theta_1 - \theta_2| = 0.5$ with 80% power would require $v_1 + v_2 \leq 0.008$, which is consistent with both studies having sample sizes of at least 251. Detecting a difference of $|\theta_1 - \theta_2| = 0.5$ with 80% power would require $v_1 + v_2 \leq 0.02$, which is consistent with both studies having sample sizes of at least 98.

It is worth noting that when $\varepsilon = 0$, so that the test concerns exact replication, then this procedure is statistically equivalent to the PI criterion. Because of this, the PI criterion can be seen as a test of the null hypothesis that the studies replicated exactly. In that case, the false failure rate is simply α, and is controlled.

Example. Consider the Payne et al.'s memory studies referenced above ($T_1 = 0.753$, $v_1 = 0.0662$, $T_2 = 0.304$, $v_2 = 0.0229$, converted to Cohen's d). A test for exact replication ($\varepsilon = 0$) would fail to reject the null hypothesis that the studies replicated. The power of this test to detect a difference as large as $|\theta_1 - \theta_2| = 0.5$ is 38%. If instead we consider a difference of $|\theta_1 - \theta_2| < 0.1$ to be negligible, then we might test for approximate replication ($\varepsilon = 0.1$). In doing so we would again fail to reject the null hypothesis that the studies replicated. The power of this test to detect a difference as large as $|\theta_1 - \theta_2| = 0.5$ is 35%. Note that the test for approximate replication is less powerful than the test of exact replication.

Tests When the Burden of Proof Is on Replication

If the burden of proof is on replication, then the null hypothesis can be formed to correspond to nonreplication. As discussed previously in this chapter, forming a testable null hypothesis of replication can be done via approaches used in equivalence testing. Concretely, let ε denote the smallest difference between effects that would be considered non-negligible. Then we form a null hypothesis

$$H_0 : |\theta_1 - \theta_2| \geq \varepsilon \tag{14.22}$$

To test H_0, we compute Q_2. Under the null hypothesis, Q_2 follows a chi-square distribution with one degree of freedom and noncentrality parameter λ_{02} in Eq. (14.20). Since we need conclusive evidence that the studies successfully replicate, an α-level test involves rejecting H_0 if Q_2 is less than $c_\alpha(1, \lambda_{02})$, the α-percentile of the chi-square distribution with one degree of freedom and noncentrality parameter λ_{02}.

The power of this test is given by

$$F\left(c_\alpha\left(1,\lambda_{02}\right)|,1\,|,\lambda_{12}\right) \tag{14.23}$$

where F is as in Eq. (14.21) and $\lambda_{12} = |\theta_1 - \theta_2|^2/(v_1 + v_2) \leq \lambda_{02}$. Note that if the studies replicate exactly, then $\lambda_{12} = 0$. However, if the studies replicate approximately, so that $|\theta_1 - \theta_2| < \varepsilon$, then $0 < \lambda_{12} < \lambda_{02}$.

The power of this test depends on a few quantities:

- It is increasing as function of ε. The bigger the difference between effects that is considered negligible, the greater the power of the test.
- It is increasing as a function of $v_1 + v_2$, so that when the variances for each study decrease (sample sizes within studies increase), the power of the test for replication will increase.
- It is decreasing as a function of $|\theta_1 - \theta_2|$. The smaller the actual difference between effects is, the greater the power. In fact, the power is greatest when the studies replicate exactly, so that $\theta_1 = \theta_2$, and $\lambda_{12} = 0$.
- In practice, unless we consider extremely large differences between effects to be negligible, the test when the burden of proof is on replication is lower than the power of the test when the burden of proof is on nonreplication. For example,

consider effects on the scale of Cohen's d and a design such that $v_1 + v_2 = 0.02$ (i.e., sample size of 98 per study). The power of the test of exact replication when the burden of proof is on nonreplication (H_0: $\theta_1 = \theta_2$) has 80% power to detect a difference of $|\theta_1 - \theta_2| = 0.8$. However, a test for nonreplication assuming $|\theta_1 - \theta_2|$ is at least as large as 0.8 (H_0: $|\theta_1 - \theta_2| \geq 0.8$) has maximum power of 75%, which occurs when the studies replicate exactly (i.e., when $\theta_1 = \theta_2$).

- Note that neither of these tests will be particularly well powered with $k = 2$ studies, as argued in the following section.

Example. Suppose we deem $\varepsilon = 0.2$ to be the smallest non-negligible difference between effects. Then with the replication of Payne et al.'s memory study, we fail to reject the null hypothesis that the studies failed to replicate. The power of this test will be greatest if $\theta_1 = \theta_2$, so that the studies replicate exactly. In that case, the power would be 30%.

More than One Study Is Likely Necessary for Conclusive Results About Replication

Two key questions about the design of replication studies involve how many replications should be conducted and how large the sample size should be for each study in order to ensure sufficiently powerful analyses. If the design a priori sets $k = 2$, as has been common in some social science research, and if the original study has already been conducted and published, then the question of design involves how large the sample size ought to be in the replication study in order to ensure sufficiently powerful analyses.

Hedges and Schauer (2019a) showed that a design with $k = 2$ studies where the original study had already been conducted will almost never support sufficiently sensitive analyses, and in fact the power of tests for replication will typically be bounded by the power of the original study to detect an effect. To see this, note that the power of the original study (study 1) to detect an effect as large as θ_1 is given by

$$1 - F\left(c_{1-\alpha}(1,0) \mid \theta_1^2 / v_1\right) \quad (14.24)$$

where F and $c_{1-\alpha}$ are given in Eq. (14.21). The power of this test depends largely on θ_1^2/v_1.

Now consider the test for replication when the burden of proof is on nonreplication. This test has power given in Eq. (14.21). Note that the power for both the test for an effect in study 1, Eq. (14.24), and the test for replication in Eq. (14.21) depend on the chi-square distribution function with one degree of freedom, and that both are decreasing as the critical value $c_{1-\alpha}$ increases. Further, $c_{1-\alpha}(1, \lambda_{02}) \geq c_{1-\alpha}(1, 0)$ with equality holding only if $\lambda_{02} = 0$, so that the test involves exact replication.

The test for replication is an increasing function in $|\theta_1 - \theta_2|$. Yet, an upper bound on differences between effects we might want to detect likely occurs when

$|\theta_1 - \theta_2| \leq \theta_1$. The reasoning behind this is that if $|\theta_1 - \theta_2| = \theta_1$, this would involve a scenario where θ_1 is in one direction (e.g., $\theta_1 > 0$) and θ_2 is in another direction ($\theta_2 \leq 0$), which would constitute qualitative disagreement in effects (e.g., the effect in study 1 helps patients, the effect in study 2 does nothing for them) and run contrary to agreement of magnitude or direction of effects. Thus, the difference we may wish to detect in a test of replication is bounded above by $|\theta_1|$.

Finally, the power of the test for replication increases as v_2 decreases. The smallest v_2 could possibly be is 0, which would occur if study 2 had an infinite sample size. Putting these pieces together, it follows that the power of study 1 to detect an effect is an upper bound for the test of replication:

$$1 - F\left(c_{1-\alpha}(1,0), \frac{\theta_1^2}{v_1}\right) \geq 1 - F\left(c_{1-\alpha}(1,0), \frac{|\theta_1 - \theta_2|^2}{v_1}\right) \geq$$

$$1 - F\left(c_{1-\alpha}(1,0), \frac{|\theta_1 - \theta_2|^2}{v_1 + v_2}\right) \geq 1 - F\left(c_{1-\alpha}(1,\lambda_{02}), \frac{|\theta_1 - \theta_2|^2}{v_1 + v_2}\right) \quad (14.25)$$

The power of the test for replication will likely be much smaller than the power of study 1, since:

(a) We may wish to detect differences between effects that are themselves smaller than θ_1.
(b) Study 2 will not have an infinite sample size, and so $v_2 > 0$.
(c) We may have to adjust T_1 for publication bias, which will decrease the power of the test for replication.

Further, Hedges and Schauer (2019a) show that similar inequalities hold for tests when the burden of proof is on replication, for parameter estimation, and for Bayesian parameter estimation.

In practice, unless both studies have very high power (>99% power to detect effects θ_1 and θ_2, respectively), the power of the test for replication will be low. This result holds even if we conduct multiple replication studies and aggregate their results via a meta-analysis. Hence, Hedges and Schauer (2019a) argue that analyses framed in terms of falsifiability are likely to be underpowered, and that analyses framed in terms of consistency require more than one replication to ensure high power.

In the absence of conclusive designs and analyses about replication based on $k = 2$ studies, a series of methods have been proposed to better make sense of the evidentiary value of $k = 2$ studies regarding replicability. Maxwell, Lau, and Howard (2015) propose using an equivalence test to analyze a replication study when the original study finds a statistically significant effect. The idea behind this is if the original effect estimate is statistically different from zero, then one way to falsify that is if the effect estimate in the replication study is conclusively close to zero. Held's (2020) skeptical p-value is based on a Bayesian approach to prior skepticism about the original effect estimate and evaluates whether that skepticism is consistent with the replication study effect estimate. Simonsohn's (2015) small telescopes approach involves estimating the statistical power of the original study relative to the effect found in the replication study. All three approaches allow for prospective sample size calculations to ensure that their conclusions are reasonably precise. However, none of these methods concerns definitions of replication focused on similarity of effect size: the equivalence test focuses on how negligibly small the effect is in the replication study, the skeptical p-value concerns the prior beliefs required to doubt the original study results, and small telescopes largely focuses on the sensitivity of the original and replication studies. Because of this, none support inferences about the agreement of effects explicitly, but rather give insight into the evidentiary value of $k = 2$ studies regarding the existence of effect. Though they support conclusions about more diffuse notions of replication, either approach may prove useful when replication research designs cannot include more than two studies (e.g., due to practical or resource constraints).

Random Effects Analyses for Replication ($k > 2$)

Estimation

If the framing of analysis is on consistency of effects, and studies are assumed to be random, then the relevant parameter to estimate is τ^2. There are a variety of possible estimators of τ^2 (see Veroniki et al., 2016, for a review). A common estimator due to DerSimonian and Laird (1986) is given in Eq. (14.9).

The standard error of this estimator is:

$$SE\left[\hat{\tau}^2_{DL}\right] = \sqrt{\frac{2(k-1)}{S^2} + \frac{4\tau^2}{S} + \frac{2\tau^4}{S^2}\left[S_2 - 2\frac{S_3}{S_1} + \frac{S_2^2}{S_1^2}\right]} \tag{14.26}$$

where S_j is defined in Eq. (14.7) and S in Eq. (14.8). To estimate this standard error, we can substitute the estimated variance component $\hat{\tau}^2_{DL}$ for τ^2 in the equation above. Note that the standard error will decrease as the v_i decrease (i.e., with large sample sizes within studies) and as k increases, but will increase as the amount of variation between studies τ^2 increases.

Statistical methodologists have argued that an alternative estimator due to Paule and Mandel (1982) tends to have slightly better properties in certain scenarios (see Veroniki et al., 2016; van Aert & Jackson, 2018). The Paule–Mandel estimator is based on the statistic $Q^*(\tau^2)$:

$$Q^*\left(\tau^2\right) = \sum_{i=1}^{k} \frac{\left(T_i - T_{\cdot}^*\right)^2}{v_i + \tau^2} \tag{14.27}$$

where

$$T_{\cdot}^{*} = \left. \sum_{i=1}^{k} \frac{T_i}{v_i + \tau^2} \middle/ \sum_{i=1}^{k} \frac{1}{v_i + \tau^2} \right. \quad (14.28)$$

The statistic $Q^*(\tau^2)$ is written this way because it is a function of τ^2. Note that $Q^*(\tau^2)$ and $T_{\cdot}^*$ differ from Q and $\overline{T}_{\cdot}$ in that they involve sums weighted by $1/(v_i + \tau^2)$ as opposed to $1/v_i$. Moreover, $T_{\cdot}^*$ differs from $\overline{T}_{\cdot}^*$ in Eq. (14.10) in that $T_{\cdot}^*$ uses weights that depend on the true value of τ^2, while $\overline{T}_{\cdot}^*$ uses weights that depend on an estimate $\hat{\tau}_{DL}^2$.

It can be shown that the expected value of $Q^*(\tau^2)$ is $k - 1$. The Paule–Mandel estimator is thus obtained by using an iterative program to solve the equation $Q^*(\tau^2) = k - 1$ for τ^2:

$$\hat{\tau}_{PM}^2 = \tau^2 : \sum_{i=1}^{k} \frac{\left(T_i - T_{\cdot}^*\right)^2}{v_i + \tau^2} = k - 1 \quad (14.29)$$

The Paule–Mandel estimator can be used in conjunction with a method for constructing confidence intervals for τ^2 called the Q-profile method (Viechtbauer, 2007). If the θ_i are normally distributed, then $Q^*(\tau^2)$ follows a chi-square distribution with $k - 1$ degrees of freedom. A $1 - \alpha$ confidence interval for τ^2 can be obtained by using an iterative program to solve two equations for τ^2: one equation is used to obtain the lower bound ($L_{1-\alpha}$) and one to obtain the upper bound ($U_{1-\alpha}$) of the confidence interval. These equations set $Q^*(\tau^2)$ equal to $c_{1-\alpha/2}(k - 1, 0)$, the $1 - \alpha/2$ percentile of the chi-square distribution with $k - 1$ degrees of freedom, and $c_{\alpha/2}(k - 1, 0)$, the $\alpha/2$ percentile, respectively:

$$L_{1-\alpha} = \tau^2 : \sum_{i=1}^{k} \frac{\left(T_i - \overline{T}_{\cdot}^*\right)^2}{v_i + \tau^2} = c_{1-\frac{\alpha}{2}}\left(k - 1, 0\right) \quad (14.30)$$

$$U_{1-\alpha} = \tau^2 : \sum_{i=1}^{k} \frac{\left(T_i - \overline{T}_{\cdot}^*\right)^2}{v_i + \tau^2} = c_{\frac{\alpha}{2}}\left(k - 1, 0\right) \quad (14.31)$$

In addition to reporting point estimates and uncertainty on the scale of τ^2, researchers may also report statistics such as the H^2 or I^2 values. Both of these statistics can be seen as depending on the ratio of τ^2/v. The statistic H^2 is an estimate of $1 + \tau^2/v$, while I^2 is an estimate of $\tau^2/(v + \tau^2)$ (Higgins & Thompson, 2002). Note that the precise value of H^2 and I^2 depends on an estimated variance τ^2, and hence will possibly differ between the DerSimonian–Laird and Paule–Mandel estimators.

Example. Consider the reverse gambler's fallacy experiment replicated by the Many Labs project described in previous sections. The effect sizes (on the scale of Cohen's d) and their variances from these replications are reported in Table 14.3. Based on the $k = 36$ replication studies, the DerSimonian and Laird estimator is $\hat{\tau}^2_{DL} = 0.013$, which has standard error 0.010 ($H^2 = 1.46$, $I^2 = 31.73\%$). The Paule–Mandel estimator is $\hat{\tau}^2_{PM} = 0.018$ ($H^2 = 1.66$, $I^2 = 39.91\%$), with 95% confidence interval [0.000, 0.060]. Thus, there is some evidence of variation between studies that could be considered modest to moderate (τ^2/v ranging from 0.46 to 0.66, depending on the estimator). The uncertainty in this estimate is such that the variation between studies could possibly be very near zero or as large as 0.06 ($\tau^2/v = 2.28$).

NHT: Burden of Proof on Nonreplication

A test of consistency when the burden of proof is on nonreplication would form a null hypothesis that the studies replicate successfully. Since the variance component τ^2 is the parameter that characterizes "replication," a relevant null hypothesis is

$$\mathrm{H}_0 : \tau^2 \le \tau_0^{\,2} \tag{14.32}$$

where $\tau_0^{\,2}$ constitutes the smallest amount of variation between studies considered non-negligible that would characterize replication failure. Note that if $\tau_0^{\,2} = 0$, this is a test of exact replication, but when $\tau_0^{\,2} > 0$, this is a test of approximate replication.

To test H_0, we can compute the Q statistic in Eq. (14.6), which follows a somewhat complex distribution that can be expressed as a linear combination of chi-square random variables. A reasonable approximation for that distribution is as follows. Denote the following moments of Q that are functions of τ^2

$$\mu_Q\left(\tau^2\right) = k - 1 + S\tau^2 \tag{14.33}$$

is the mean of Q where S is as in Eq. (14.8) and

$$\sigma_Q^{\,2}\left(\tau^2\right) = S^2\left(SE\left[\tau_{DL}^{\,2}\right]\right)^2 \tag{14.34}$$

is the variance of Q where $\mathrm{SE}[\tau_{\mathrm{DL}}^{\,2}]$ is as in Eq. (14.26) and S is in Eq. (14.9) (for further details, see Hedges & Pigott, 2001).

Given these functions, $2\mu_Q(\tau^2)\ Q/\sigma_Q^{\,2}(\tau^2)$ follows a chi-square distribution with $2\mu_Q^{\,2}(\tau^2)/\sigma_Q^{\,2}(\tau^2)$ degrees of freedom. Thus, under H_0, we can use the approximation that $2\mu_Q(\tau_0^{\,2})\ Q/\sigma_Q^{\,2}(\tau_0^{\,2})$ follows a chi-square distribution with $2\mu_Q^{\,2}(\tau_0^{\,2})/\sigma_Q^{\,2}(\tau_0^{\,2})$ degrees of freedom. When all studies have the same estimation error variance $v_i = v$ (i.e., all studies have the same sample size), then this approximation reduces to a much simpler expression that can be written as a constant $(1 + \tau^2/v)$ times a chi-square distribution with $k - 1$ degrees of freedom:

$$Q \sim \left(1+\tau 2/v\right)\chi^2_{k-1} \tag{14.35}$$

Moreover, when $\tau^2 = 0$, as in a null hypothesis of exact replication, then $\mu_Q(0) = k - 1$ and $\sigma_Q^2(0) = 2(k - 1)$ and Q follows a central chi-square distribution with $k - 1$ degrees of freedom.

An α-level test involves rejecting H_0 if $2\mu_Q(\tau_0^2)\, Q/\sigma_Q^2(\tau_0^2)$ exceeds $c_{1-\alpha}(2\mu_Q^2(\tau_0^2)/\sigma_Q^2(\tau_0^2), 0)$, the $1 - \alpha$ percentile of the chi-square distribution with $2\mu_Q^2(\tau_0^2)/\sigma_Q^2(\tau_0^2)$ degrees of freedom. For brevity, we will write

$$C\left(1-\alpha,\tau_0^2\right)=\frac{c_{1-\alpha}\left(\frac{2\mu_Q^2\left(\tau_0^2\right)}{\sigma_Q^2\left(\tau_0^2\right)},0\right)}{\frac{2\mu_Q\left(\tau_0^2\right)}{\sigma_Q^2\left(\tau_0^2\right)}} \tag{14.36}$$

to refer to this percentile/critical value, such that we reject H_0 when Q exceeds $C(1 - \alpha, \tau_0^2)$. The notation $C(1 - \alpha, \tau_0^2)$ denotes that it is a function of α and τ_0^2. When all of the v_i are equal, the test reduces to rejecting H_0 when Q exceeds $C(1 - \alpha, \tau_0^2) = c_{1-\alpha}(k - 1, 0)(1 + \tau_0^2/v)$; that is, when Q exceeds a critical value from the central chi-squared distribution multiplied by $1 + \tau_0^2/v$. In tests of exact replication, $\tau_0^2 = 0$, so the critical value is simply the $1 - \alpha$ percentile of the chi-squared distribution with $k - 1$ degrees of freedom (akin to a traditional Q-test in meta-analysis).

The power of this test to detect some value $\tau^2 > \tau_0^2$ is given by

$$1-F\left(\frac{2\mu_Q\left(\tau^2\right)C\left(1-\alpha,\tau_0^2\right)}{\sigma_Q^2\left(\tau^2\right)}\,\middle|\,\frac{2\left[\mu_Q\left(\tau^2\right)\right]^2}{\sigma_Q^2\left(\tau^2\right)},0\right) \tag{14.37}$$

where F is the chi-square distribution function in Eq. (14.21); since the noncentrality parameter in this function is set to 0, this is a central chi-square distribution function. When all of the $v_i = v$, so that each study has the same estimation error variance, then the power reduces to:

$$1-F\left(\frac{C\left(1-\alpha,\tau_0^2\right)}{\left(1+\tau^2/v\right)}\,\middle|\,k-1,0\right) \tag{14.38}$$

The power of the test for replication is increasing as a function of the number of studies k, as well as in τ^2 in the metric of τ^2/v. Because of this, the power is higher when τ^2/v is larger, which occurs when τ^2 is larger or when v is smaller (i.e., sample sizes within studies are larger). In addition, power is a decreasing function of τ_0^2, which means that tests of approximate replication have lower power than tests of exact replication. Power is discussed further at the end of this section.

Example. The reverse gambler's fallacy example comprises the effect sizes of the $k = 36$ replication studies. Based on the effect estimates and variances in Table 14.3, the Q statistic is 51.27. For an $\alpha = 0.05$ level test of exact replication (H_0: $\tau^2 = 0$), the relevant critical value is $c_{1-\alpha}(35, 0) = 49.8$. Because $Q > 49.8$, we reject H_0 and conclude the studies fail to replicate exactly ($p = 0.04$). This test had 85% power to detect variation on the order of $\tau^2 = 0.027$, or $\tau^2/v = 1$.

Consider a test of approximate replication such that we consider $\tau_0^2 = 0.01$ to be the largest amount of variation considered negligible. Note this would be consistent with roughly $\tau_0^2/v \approx 1/3$, which is roughly the convention specified by Hunter and Schmidt. The relevant critical value is $C(0.95, 0.01) = 69.17$. Since $Q < 61.17$, we do not reject H_0 and so do not conclude the studies failed to replicate approximately ($p = 0.35$). This test had 48% power to detect variation on the order of $\tau^2 = 0.027$, or $\tau^2/v = 1$.

NHT: Burden of Proof on Replication

When the burden of proof is on replication, then, as in the $k = 2$ case, we must form a null hypothesis that the studies fail to replicate. With our focus on between-study variation, our test will involve the following null hypothesis:

$$H_0 : \tau^2 \geq \tau_0^2 \tag{14.39}$$

This can be tested with the Q statistic in Eq. (14.6). Under H_0, we can use the approximation that $2\mu_Q(\tau_0^2)\, Q/\sigma_Q^2(\tau_0^2)$ follows a chi-square distribution with $2\mu_Q^2(\tau_0^2)/\sigma_Q^2(\tau_0^2)$ degrees of freedom, as derived above. An α-level test involves rejecting H_0 if $2\mu_Q(\tau_0^2)\, Q/\sigma_Q^2(\tau_0^2)$ is less than the α-percentile of that distribution $C(\alpha, \tau_0^2)$, where $C(\alpha, \tau_0^2)$ is described in Eq. (14.36). In other words, the test when the burden of proof is on replication proceeds in a similar manner as when the burden of proof is on nonreplication, except that: (1) the critical value now involves the α-percentile of the chi-square approximation, and (2) we reject H_0 if $2\mu_Q(\tau_0^2)\, Q/\sigma_Q^2(\tau_0^2)$ is less than that critical value.

The power of this test to detect some value $\tau^2 < \tau_0^2$ is given by

$$F\left(\frac{2\mu_Q(\tau^2)C(\alpha,\tau_0^2)}{\sigma_Q^2(\tau^2)} \,\middle|\, \frac{2\left[\mu_Q(\tau^2)\right]^2}{\sigma_Q^2(\tau^2)}, 0\right) \tag{14.40}$$

Table 14.3 Data from the Many Labs Replication Project replications of the reverse gambler's fallacy experiment

Site	Effect size	Variance
abington	0.590	0.051
brasilia	0.355	0.036
charles	0.886	0.063
conncoll	0.622	0.050
csun	0.517	0.046
help	0.516	0.043
ithaca	0.782	0.053
jmu	0.715	0.026
ku	0.527	0.039
laurier	0.961	0.042
lse	0.645	0.016
luc	0.528	0.029
mcdaniel	0.510	0.046
msvu	0.340	0.054
mturk	0.620	0.004
osu	0.111	0.038
oxy	1.188	0.048
pi	0.724	0.004
psu	0.605	0.048
qccuny	0.419	0.044
qccuny2	0.338	0.050
sdsu	0.616	0.026
swps	0.114	0.050
swpson	0.593	0.027
tamu	0.747	0.024
tamuc	0.749	0.054
tamuon	0.592	0.020
tilburg	0.687	0.059
ufl	0.378	0.034
unipd	0.765	0.035
uva	1.108	0.059
vcu	0.712	0.040
wisc	0.785	0.045
wku	0.441	0.044
wl	0.072	0.046
wpi	0.978	0.053

Source: Open Science Framework

where F is the chi-square distribution function in Eq. (14.21). When all of the $v_i = v$, so that each study has the same estimation error variance, then the power reduces to:

$$F\left(\frac{C\left(\alpha,\tau_0^2\right)}{1+\tau^2/v} \mid k-1,0\right) \tag{14.41}$$

The power of this test will increase as we test looser notions of replication failure; that is, when τ_0^2 is larger. It will increase when τ^2/v is smaller, and will be greatest when $\tau^2 = 0$, so that the studies replicate exactly. As discussed in the following section, the power of the test when the burden of proof is on replication will typically be lower than when the test puts the burden of proof on nonreplication.

Example. Suppose we wish to test a null hypothesis that the gambler's fallacy replications failed such that $\tau^2 = 0.027$, which would characterize $\tau_0^2/v = 1$. Recall that the Q statistic is 51.27. The relevant critical value is $C(0.95, 0.027) = 41.13$. Because $Q > 41.13$, we fail to reject H_0 and do not conclude the studies replicated successfully ($p = 0.17$). The power of this test will be greatest when $\tau^2 = 0$, and the studies actually replicate exactly, in which case the power would be 78%.

Power and Precision of Analyses

Note that the conclusions of the test for replication can depend on how the null hypothesis is formed. In the gambler's fallacy example, we rejected a null hypothesis of exact replication, but failed to reject a null hypothesis of approximate replication, nor did we reject a null hypothesis of replication failure. Thus, the ultimate conclusions reached about replication will be sensitive to the framing of the null hypothesis.

The sensitivity of analytic methods is key for both planning and analyzing replication studies. Proper interpretation of these tests, however, does not mean we conclude H_0 is true when we fail to reject it. In general, a failure to reject a null hypothesis is ambiguous, and must be interpreted in light of statistical power (or the Type II error rate). So too must interpretations of estimated variance be considered in light of their uncertainty. Moreover, because the power of these tests are known, as are the standard errors of estimators, researchers can plan replication studies prospectively to make sure that relevant analyses are sufficiently sensitive (i.e., have high power or precision). Note that the previous sections demonstrated that the sensitivity analysis will depend on v_i (and hence the within-study sample size n_i), as well as the total number of studies k. Thus, prospective planning of replication studies to ensure sensitivity analyses can be seen as choosing k and n_i or some minimum n for each study.

Understanding the sensitivity of analyses requires some idea of values of τ^2 considered large and what values might be considered negligible. For instance, the power of tests for replication is a function of a value of variation τ_0^2 that distinguishes between negligible and non-negligible variations, as well as the variation one wishes to detect τ^2. The standard error of variance component estimates is a function of τ^2.

In a previous section, I argued that if the scale of the θ_i is understood, it is best practice to derive meaningful values of τ^2 in a manner that is consistent with both the scale of the θ_i, and knowledge regarding the finding under consideration and its scientific and clinical implications. Alternatively, a scale-free approach might conceive of these analyses and their properties as depending on τ^2/v. We noted that when effects are on the scale of standardized mean differences, the negligible values of τ^2 may range from 0 to as large as 0.035, and the values worth studying might range from 0.005 to 0.1.

Figure 14.1 shows the power of replication tests assuming $v_1 = v_2 = 4/100$, analogous to each study having a total sample size of 100 (on the scale of Cohen's *d*) and balanced two-arm designs. The first panel shows the power of a test (*y*-axis) for exact replication (H_0: $\tau^2 = 0$) to detect a given value of τ^2 (*x*-axis) for different numbers of studies $k = 2, 5, 10, 20$, and 40. The second panel shows the power of the test for nonreplication (H_0: $\tau^2 \geq \tau_0^2$) to detect exact replication ($\tau^2 = 0$) as a function of τ_0^2 (*x*-axis) for different numbers of *k*. In addition, the third panel shows the relative standard error (RSE) SE[$\hat{\tau}_{DL}^2$]/τ^2 for the DerSimonian–Laird estimator as a function of τ^2 (*x*-axis) for different numbers of *k*. The figure shows that unless there are a large number of studies, a design with $v = 4/100$ will likely be underpowered when analyzing moderate levels of heterogeneity.

To get a better sense of designs with a given number of studies *k* and sample size per study *n* (assuming balanced designs within studies) that can give sufficiently sensitive analyses, consider Table 14.4. Table 14.4 shows the required sample size per study necessary to obtain 80% power or a relative standard error less than 50%. The sample sizes presented assume that each study has at least *n* participants, each study is a balanced two-armed study, results are the scale of standardized mean differences, and the large sample approximation for $v_i = v \approx 4/n$ in Eq. (14.2) is valid.

The first panel in Table 14.4 gives the requisite sample size per study to ensure a test of exact replication (H_0: $\tau^2 = 0$) has 80% power for an $\alpha = 0.05$ level test.

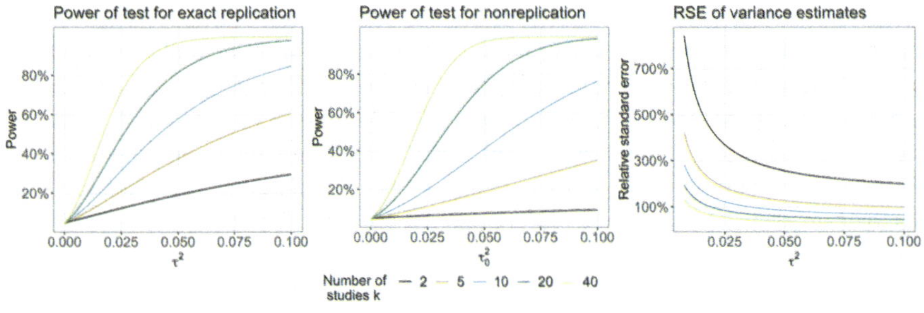

Fig. 14.1 Sensitivity of analyses for replication assuming $v_1 = v_2 = 4/100$

Table 14.4 Required within-study sample sizes to ensure 80% power in tests of exact replication, nonreplication, and estimates with a relative standard error (RSE) less than 50%

	Test of exact replication H_0: $\tau^2 = 0$ (80% power)				Test of nonreplication H_0: $\tau^2 \geq \tau_0^2$ (with $\tau^2 = 0$, 80% power)				Estimation of τ^2 (RSE $\leq$ 50%)			
	$\tau^2 = 0.03$	$\tau^2 = 0.05$	$\tau^2 = 0.07$	$\tau^2 = 0.1$	$\tau_0^2 = 0.03$	$\tau_0^2 = 0.05$	$\tau_0^2 = 0.07$	$\tau_0^2 = 0.1$	$\tau^2 = 0.03$	$\tau^2 = 0.05$	$\tau^2 = 0.07$	$\tau^2 = 0.1$
$k = 5$	635	381	272	191	991	595	425	298	—	—	—	—
$k = 10$	287	172	123	86	359	216	154	108	2199	1319	943	660
$k = 20$	161	97	69	49	183	110	79	55	247	148	106	74
$k = 40$	99	59	43	30	109	65	47	33	111	67	48	34

Note: — means no sample size is possible

Within-study sample sizes are reported for detecting various values of $\tau^2 > 0$ assuming a given number of studies k. The second panel in Table 14.4 gives requisite sample sizes for tests of nonreplication (H_0: $\tau^2 \geq \tau_0^2$) to detect exact replications ($\tau^2 = 0$) with 80% power ($\alpha = 0.05$) for various values of τ_0^2. The third panel gives the requisite sample size to ensure a relative standard error (RSE) SE[$\hat{\tau}_{DL}^2$]/$\tau^2 \leq 50\%$ for a given number of studies k and values of τ^2. Note that cells with "—" indicate that no sample size large enough could generate an RSE below 50%.

For reference, a test of exact replication with $k = 10$ studies would need a sample size of $n \geq 172$ per study to detect $\tau^2 = 0.05$ with 80% power. If $\tau^2 = 0.05$, and $k = 10$ studies are conducted, then we would need a sample size of $n \geq 1319$ to ensure a relative standard error less than 50% for an estimate of τ^2. Large sample sizes will be required for small RSE when τ^2 itself is small because of the relationship between τ^2 and the RSE. Relaxing the RSE slightly in such cases may not result in vastly larger standard errors. For reference, a relative standard error of 50% when $\tau^2 = 0.05$ would ensure a standard error SE[$\hat{\tau}_{DL}^2$] ≤ 0.025. If instead we desire a standard error SE[$\hat{\tau}_{DL}^2$] ≤ 0.03, then for $k = 10$ studies and $\tau^2 = 0.05$, we would require just $n = 294$ subjects per study, less than a quarter of the sample size indicated in the table.

Discussion

Questions about how to define replication as an estimand, analyze replication studies to make inferences on that estimand, and how to design replication studies to support sensitive analyses are intrinsically linked. In this chapter, I have showed that there are myriad approaches to defining replication that are functions of effect *parameters*, and hence there are a variety of analysis methods (functions of effect *estimates*) that are relevant for replication.

Different definitions of replication can and will be preferred in different settings and fields, and for different types of findings. Determining which definition is most relevant is subject to scientific and clinical judgment. Here, I have focused on definitions of replication that involve the consistency of effects (i.e., effects in replication studies are about the same size). This is particularly relevant to approaches for enhancing evidence-based practices, which support inferences about the stability of an effect.

Analyses for replication can support qualitative conclusions about the replicability of scientific findings in a manner consistent with NHT. Indeed, this chapter has presented a series of hypothesis tests for replication. However, recent moves by the American Statistical Association have urged researchers to focus on reporting point estimates and relevant uncertainty over *p*-values (see Wasserstein & Lazar, 2016). In keeping with those developments, I would suggest researchers studying replication to focus on estimating relevant parameters, though qualitative conclusions about replication may still be desired or warranted.

Designing replication studies requires some determination of the requisite number of studies k and sample size per study n to ensure sufficiently sensitive analyses. If one of the studies to be included in analyses is already published, a design with $k = 2$ studies (i.e., conducting only a single replication study) is unlikely to be sufficiently powered. If such $k = 2$ designs are unavoidable due, for instance, to budget or resource constraints, this chapter outlined some relevant methods (small telescopes, skeptical p-value) to make sense of the evidentiary value of $k = 2$ studies.

That a design involving a single replication of a published study is unlikely to be well powered suggests a few considerations for both primary research and replication research. The most obvious is that research seeking to replicate existing findings should (in most cases) have $k \geq 3$ studies. However, there are two possible ways around the statistical limitations of the $k = 2$ design that do not involve increasing the number of studies conducted. First, the research community could prioritize conducting primary studies with larger sample sizes. Sound statistical practice dictates that experiments be devised so that they are well powered. This typically means a power of at least 80%. However, even studies with 80% power will limit the power of analyses for replication. Thus, seeking designs of primary studies with higher power (e.g., at least 90%) may reduce the resources required to replicate them in the future.

Second, improving transparency of primary studies and their publication can improve statistical analyses for replication (see Collins & Tabak, 2014; Bollen et al., 2015; Schauer & Hedges, 2020). Recall that adjustments for publication bias will only reduce the sensitivity of replication analyses. Therefore, pre-registering studies and analysis plans, reporting all relevant effects, and reporting regardless of statistical significance may reduce the impact of selection and hence reduce bias in extant findings.

An alternative approach is to conduct replications prior to publishing any single study (e.g., Schweinsberg et al., 2016). Rather than designing a single study, the focus can be on designing an ensemble of replication studies. The results of these studies (including their consistency) can be reported as part of a single article or series of articles. This is analogous to the type of work done by the PPIR and to efforts possible under the Psychological Science Accelerator. Embedding replication into the process of primary inquiry can help improve our understanding about a phenomenon and the conditions under which it is studied.

Further Reading

For discussion regarding meta-analytic approaches to studying replication, Valentine et al. (2011) describe a general framework that was later refined by Hedges and Schauer (2019b). Finer points about fixed effects analyses were identified and discussed by Schauer (2018), including various ways for intuiting the scale of between-study variation. A more detailed demonstration of the methods discussed in this chapter was done by Schauer and Hedges (2020). Hedges and Schauer (2021) proposed methods for identifying cost-effective or otherwise optimal designs of replication studies that support powerful analyses. In addition, Schauer (2018) derives some corrections to many of the analyses discussed in this chapter to account for small sample sizes in studies that could lead to violations of the assumption that the v_i are known and not estimated.

Though the focus of this chapter was on the replicability of results, particularly as operationalized as variation between effect parameters, replication research programs can provide insight into other parameters. For instance, meta-analytic methods can support estimation of mean effects across studies, as well as prediction intervals of effects (see Cooper, Hedges, & Valentine, 2019). Estimates of relevant parameters, including variance components, are possible with most meta-analytic software, including with the metafor library in the R computing language (Viechtbauer, 2010). Similarly, the Replicate library in R can conduct inference on the $P_{>q}$ and P_{orig} statistics (Mathur & VanderWeele, 2020).

The analyses presented here are primarily for direct replications, where studies are devised to be as similar as possible. Contrast that with conceptual replications, which may systematically vary aspects of a study to examine the potential impact of those variations on study results. This can be conceived of in a meta-analytic analysis of variance (ANOVA) framework, where studies can be grouped according to how they were conducted; if we denote a variable X that is systematically varied across studies, then we can group studies according to their value of X. Relevant analyses are discussed by Schauer (2018) and Schauer and Hedges (2020), which also includes empirical demonstrations.

As an alternative to the frequentist analysis methods that this chapter focused on, there are several possible Bayesian analyses of replications. Schauer (2018) describes Bayesian approaches to estimating λ and τ_F^2, and outlines various considerations for Bayesian estimation of τ^2. These latter discussions have been considered thoroughly in the statistical literature (for a good discussion, see Gelman et al., 2014). Alternatively, there have been approaches devised for $k = 2$ studies including those by Etz and Vandekerckhove (2016), van Aert and van Assen (2017), or Held (2020).

This chapter relied on meta-analytic notation as a matter of simplicity. This approach can be used even if data on individual participant are not available to the analyst. In programs of research regarding replication, this may not be the case; analysts may have access to individual participant data in all or a portion of relevant studies. In such cases, analyses can use multilevel models (also referred to individual participant data meta-analysis), which are analogous to the models presented in this chapter (see Raudenbush & Bryk, 1992; Riley et al., 2010; Tierney et al., 2015).

Finally, this chapter demonstrated that there are configurations of replication research programs wherein enough studies are conducted each with large enough sample sizes such that analyses are sufficiently powered. These are key design choices that should be made prior to collecting data. It is worth noting that there may be multiple configurations of study sample sizes n and number of studies k that provide sufficient power. Choosing between these can be as simple as what configurations can be implemented. Alternatively, Schauer (2018) and Hedges and Schauer (2021) provide an approach for optimally allocating sample sizes and numbers of studies.

References

Alogna, V. K., Attaya, M. K., Aucoin, P., Bahník, Š., Birch, S., Birt, A. R., ... Zwaan, R. A. (2014). Registered replication report: Schooler and Engstler-Schooler (1990). *Perspectives on Psychological Science, 9*(5), 556–578.

Bollen, K., Cacioppo, J. T., Kaplan, R. M., Krosnick, J. A., & Olds, J. L. (2015). Reproducibility, replicability, and generalization in the social, behavioral, and economic sciences. In *Report of the Subcommittee on Replicability in Science Advisory Committee to the National Science Foundation Directorate for Social, Behavioral, and Economic Sciences*. National Science Foundation.

Borenstein, M., Hedges, L. V., Higgins, J. P. T., & Rothstein, H. R. (2009). *Introduction to meta-analysis*. Wiley.

Bouwmeester, S., Verkoeijen, P. P. J. L., Aczel, B., Barbosa, F., Bègue, L., Brañas-Garza, P., ... Wollbrant, C. E. (2017). Registered replication report: Rand, Greene, and Nowak (2012). *Perspectives on Psychological Science, 12*(3), 527–542.

Brandt, M. J., IJzerman, H., Dijksterhuis, A., Farach, F. J., Geller, J., Giner-Sorolla, R., et al. (2014). The replication recipe: What makes for a convincing replication? *Journal of Experimental Social Psychology, 50*, 217–224.

Camerer, C. F., et al. (2016). Evaluating the reproducibility of laboratory experiments in economics. *Science, 351*, 1433–1436.

Camerer, C. F., Dreber, A., Holzmeister, F., et al. (2018). Evaluating the replicability of social science experiments in *Nature* and *Science* between 2010 and 2015. *Nature Human Behavior, 2*, 637–644.

Cheung, I., Campbell, L., LeBel, E. P., Ackerman, R. A., Aykutoğlu, B., Bahník, Š., ... Yong, J. C. (2016). Registered replication report: Study 1 from Finkel, Rusbult, Kumashiro, & Hannon (2002). *Perspectives on Psychological Science, 11*(5), 750–764.

Collins, H. M. (1992). *Changing order: Replication and induction in scientific practice*. University of Chicago Press.

Collins, F. S., & Tabak, L. A. (2014). NIH plans to enhance reproducibility. *Nature, 505*, 612–613.

Cooper, H. M., Hedges, L. V., & Valentine, J. (2019). *The handbook of research synthesis and meta-analysis* (3rd ed.). The Russell Sage Foundation.

DerSimonian, R., & Laird, N. M. (1986). Meta-analysis in clinical trials. *Controlled Clinical Trials, 7*(3), 177–188.

Dickersin, K. (2005). Publication bias: Recognizing the problem, understanding its origins and scope, and preventing harm. In H. R. Rothstein, A. J. Sutton, & M. Borenstein (Eds.), *Publication bias in meta-analysis: Prevention, assessment, and adjustments* (pp. 11–33). Wiley.

Ebersole, C. R., Atherton, O. E., Belanger, A. L., Skulborstad, H. M., Allen, J. M., Banks, J. B., Baranski, E., Bernstein, M. J., Bonfiglio, D. B. V., Boucher, L., Brown, E. R., Budiman, N. I., Cairo, A. H., Capaldi, C. A., Chartier, C. R., Chung, J. M., Cicero, D. C., Coleman, J. A., Conway, J. G., ... Nosek, B. A. (2016). Many labs 3: Evaluating participant pool quality across the academic semester via replication. *Journal of Experimental Social Psychology, 67*, 68–82.

Eerland, A., Sherrill, A. M., Magliano, J. P., Zwaan, R. A., Arnal, J. D., Aucoin, P., ... Prenoveau, J. M. (2016). Registered replication report: Hart & Albarracín (2011). *Perspectives on Psychological Science, 11*(1), 158–171.

Etz, A., & Vandekerckhove, J. (2016). A Bayesian perspective on the reproducibility project: Psychology. *PLoS ONE, 11*(2), e0149794. https://doi.org/10.1371/journal.pone.0149794

Francis, G. (2012). Publication bias and the failure of replication in experimental psychology. *Psychonomic Bulletin & Review, 19*(6), 975–991.

Gelman, A., Carlin, J. B., Stern, H. S., Dunson, D. B., Vehtari, A., & Rubin, D. B. (2014). *Bayesian data analysis* (3rd ed.). CRC Press.

Guan, M., & Vandekerckhove, J. (2016). A Bayesian approach to mitigation of publication bias. *Psychonomic Bulletin and Review, 23*, 74–86. https://doi.org/10.3758/s13423-015-0868-6

Hagger, M. S., Chatzisarantis, N. L. D., Alberts, H., Anggono, C. O., Batailler, C., Birt, A. R., … Zwienenberg, M. (2016). A multilab preregistered replication of the ego-depletion effect. *Perspectives on Psychological Science, 11*(4), 546–573.

Head, M. L., Holman, L., Lanfear, R., Kahn, A. T., & Jennions, M. D. (2015). The extent and consequences of *p*-hacking in science. *PLoS Biology, 13*(3), e1002106. https://doi.org/10.1371/journal.pbio.1002106

Hedges, L. V. (1984). Estimation of effect size under nonrandom sampling: The effects of censoring studies yielding statistically insignificant mean differences. *Journal of Educational Statistics, 9*, 61–85.

Hedges, W.L.V. (1982). Estimating effect size from a series of independent experiments. *Psychological Bulletin*, 92, 490–499.

Hedges, L. V., & Olkin, I. (1985). *Statistical methods for meta-analysis*. Academic Press.

Hedges, L. V., & Pigott, T. D. (2001). The power of statistical tests in meta-analysis. *Psychological Methods, 6*, 203–217.

Hedges, L. V., & Schauer, J. M. (2019a). More than one replication study is needed for unambiguous tests of replication. *Journal of Educational and Behavioral Statistics, 44*(5), 543–570.

Hedges, L. V., & Schauer, J. M. (2019b). Statistical analyses for studying replication: Meta-analytic perspectives. *Psychological Methods, 24*(5), 557–570.

Hedges, L. V., & Schauer, J. M. (2021). The design of replication studies. *Journal of the Royal Statistical Society, Series A, 184*, 868–886.

Hedges, L. V., & Vevea, J. L. (1996). Estimating effect size under publication bias: Small sample properties and robustness of a random effects selection model. *Journal of Educational and Behavioral Statistics, 21*(4), 299–332. https://doi.org/10.3102/10769986021004299

Hedges, L. V., & Vevea, J. L. (1998). Fixed- and random-effects models in meta-analysis. *Psychological Methods, 3*(4), 486–504.

Hedges, L. V., & Vevea, J. L. (2005). Selection method approaches. In H. R. Rothstein, A. J. Sutton, & M. Borenstein (Eds.), *Publication bias in meta-analysis: Prevention, assessment, and adjustments* (pp. 145–174). Wiley.

Held, L. (2020). A new standard for the analysis and design of replication studies. *Journal of the Royal Statistical Society, Series A, 183*, 431–448. https://doi.org/10.1111/rssa.12493

Higgins, J. P. T., & Green, S. (2008). *The Cochrane handbook for systematic reviews of interventions*. John Wiley.

Higgins, J. P. T., & Thompson, S. G. (2002). Quantifying heterogeneity in meta-analysis. *Statistics in Medicine, 21*, 1539–1558.

Hunter, J. E., & Schmidt, F. L. (1990). *Methods of meta-analysis: Correcting error and bias in research findings*. Sage.

Ioannidis, J. P. A. (2005). Contradicted and initially stronger effects in highly cited clinical research. *Journal of the American Medical Association, 294*(2), 218–228.

Klein, R. A., Ratliff, K. A., Vianello, M., Adams, R. B., Bahník, Š., Bernstein, M. J., … Nosek, B. A. (2014). Investigating variation in replicability: A "many labs" replication project. *Social Psychology, 45*(3), 142–152.

Klein, R. A., Vianello, M., Hasselman, F., Adams, B. G., Adams, R. B., Alper, S., … Nosek, B. A. (2018). Many Labs 2: Investigating variation in replicability across samples and settings. *Advances in Methods and Practices in Psychological Science, 1*(4), 443–490.

Klein, R. A., Cook, C. L., Ebersole, C. R., Vitiello, C. A., Nosek, B. A., Chartier, C. R., … Ratliff, K. A. (2019). *Many Labs 4: Failure to replicate mortality salience effect with and without original author involvement*. Retrieved from: https://psyarxiv.com/vef2c

Laird, N. M., & Mosteller, F. (1990). Some statistical methods for combining experimental results. *International Journal of Technology Assessment in Health Care, 6*(1), 5–30.

Lawrance, R., Degtyarev, E., Griffiths, P., Trask, P., Lau, H., D'Alessio, D., Griebsch, I., Wallenstein, G., Cocks, K., & Rufibach, K. (2020). What is an estimand & how does it relate to quantifying the effect of treatment on patient-reported quality of life outcomes in clinical trials? *Journal of Patient-Reported Outcomes, 4*(1), 68. https://doi.org/10.1186/s41687-020-00218-5

Mathur, M., & VanderWeele, T. J. (2020). New statistical metrics for multisite replication projects. *Journal of the Royal Statistical Society, Series A, 183*, 1145–1166.

Maxwell, S. E., Lau, M. Y., & Howard, G. S. (2015). Is psychology suffering from a replication crisis? What does "failure to replicate" really mean? *American Psychologist, 70*(6), 487–498.

McShane, B. B., Böckenholt, U., & Hansen, K. T. (2016). Adjusting for publication bias in meta-analysis: An evaluation of selection methods and some cautionary notes. *Perspectives on Psychological Science, 11*(5), 730–749.

Moshontz, H., Campbell, L., Ebersole, C. R., IJzerman, H., Urry, H. L., Forscher, P. S., … Chartier, C. R. (2018). The psychological science accelerator: Advancing psychology through a distributed collaborative network. *Advances in Methods and Practices in Psychological Science, 1*(4), 501–515. https://doi.org/10.1177/2515245918797607

Olive, K. A., et al. (2014). Review of particle properties. *Chinese Physics Journal C, 38*, 090001. http://iopscience.iop.org/issue/1674-1137/38/9

Open Science Collaboration. (2012). An open, large-scale, collaborative effort to estimate the reproducibility of psychological science. *Perspectives on Psychological Science, 7*(6), 657–660.

Open Science Collaboration. (2015). Estimating the reproducibility of psychological science. *Science, 349*, 943–951.

Oppenheimer, D. M., & Monin, B. (2009). Investigations in spontaneous discounting. *Memory & Cognition, 37*(5), 608–614. https://doi.org/10.3758/MC.37.5.608

Payne, J. D., Stickgold, R., Swanberg, K., & Kensinger, E. A. (2008). Sleep preferentially enhances memory for emotional components of scenes. *Psychological Science, 19*(8), 781–788. https://doi.org/10.1111/j.1467-9280.2008.02157.x

Perrin, S. (2014). Make mouse studies work. *Nature, 507*, 423–425.

Pashler, H., & Harris, C. R. (2012). Is the replicability crisis overblown? Three arguments examined. *Psychological Science, 7*, 531–536.

Patil, P., Peng, R. D., & Leek, J. T. (2016). What should researchers expect when they replicate studies? A statistical view of replicability in psychological science. *Perspectives on Psychological Science, 11*(4), 539–544.

Paule, R., & Mandel, J. (1982). Consensus values and weighting factors. *Journal of Research of the National Bureau of Standards, 87*(5), 377–385. https://doi.org/10.6028/jres.087.022

Pigott, T. (2012). *Advances in meta-analysis*. Springer.

Raudenbush, S. W., & Bryk, A. S. (1992). *Hierarchical linear models: Applications and data analysis methods*. Sage Publications.

Riley, R. D., Lambert, P. C., & Abo-Zaid, G. (2010). Meta-analysis of individual participant data: Rationale, conduct, and reporting. *BMJ, 340*, c221. https://doi.org/10.1136/bmj.c221

Rothstein, H. R., Sutton, A. J., & Borenstein, M. (Eds.). (2005). *Publication bias in meta-analysis: Prevention, assessment, and adjustments*. Wiley.

Schauer, J. M. (2018). *Statistical methods for assessing replication: A meta-analytic framework*. (Doctoral thesis). Retrieved from https://search.proquest.com/docview/2164811196?accountid=12861

Schauer, J. M., Fitzgerald, K. G., Peko-Spicer, S., Whalen, M. C. R., Zejnullahi, R., & Hedges, L. V. (2021). An evaluation of statistical methods for aggregate patterns of replication failure. *Annals of Applied Statistics, 15*(1), 208–229. https://doi.org/10.1214/20-AOAS1387

Schauer, J. M., & Hedges, L. V. (2020). Assessing heterogeneity and power in replications of psychological experiments. *Psychological Bulletin, 146*(8), 701–719.

Schauer, J. M., & Hedges, L. V. (2021). Reconsidering statistical methods for assessing replication. *Psychological Methods, 26*(1), 127–139. https://doi.org/10.1037/met0000302

Schmidt, S. (2009). Shall we really do it again? The powerful concept of replication is neglected in the social sciences. *Review of General Psychology, 13*, 90–100.

Schweinsberg, M., Madan, N., Vianello, M., Sommer, S. A., Jordan, J., Tierney, W., … Uhlmann, E. L. (2016). The pipeline project: Pre-publication independent replications of a single laboratory's research pipeline. *Journal of Experimental Social Psychology, 66*, 55–67.

Simons, D. J., Holcombe, A. O., & Spellman, B. A. (2014). An introduction to registered replication reports at Perspectives on Psychological Science. *Perspectives on Psychological Science, 9*(5), 552–555.

Simonsohn, U. (2015). Small telescopes: Detectability and the evaluation of replication results. *Psychological Science, 26*(5), 559–569.

Tierney, J. F., Vale, C., Riley, R., Smith, C. T., Stewart, L., Clarke, M., & Rovers, M. (2015). Individual Participant Data (IPD) meta-analyses of randomised controlled trials: Guidance on their use. *PLoS Medicine, 12*(7), e1001855. https://doi.org/10.1371/journal.pmed.1001855

Valentine, J. C., Biglan, A., Boruch, R. F., Castro, F. G., Collins, L. M., Flay, B. R., Kellam, S., Mościcki, E. K., & Schinke, S. P. (2011). Replication in prevention science. *Prevention Science, 12*(2), 103–117. https://doi.org/10.1007/s11121-011-0217-6

van Aert, R., & Jackson, D. (2018). Multistep estimators of the between-study variance: The relationship with the Paule-Mandel estimator. *Statistics in Medicine, 37*(17), 2616–2629. https://doi.org/10.1002/sim.7665

van Aert, R. C., & Van Assen, M. A. (2017). Bayesian evaluation of effect size after replicating an original study. *PLoS One, 12*(4), e0175302.

Veroniki, A. A., Jackson, D., Viechtbauer, W., Bender, R., Bowden, J., Knapp, G., Kuss, O., Higgins, J. P., Langan, D., & Salanti, G. (2016). Methods to estimate the between-study variance and its uncertainty in meta-analysis. *Research Synthesis Methods, 7*(1), 55–79. https://doi.org/10.1002/jrsm.1164

Vevea, J. L., & Woods, C. M. (2005). Publication bias in research synthesis: Sensitivity analysis using a priori weight functions. *Psychological Methods, 10*, 428–443.

Viechtbauer, W. (2007). Confidence intervals for the amount of heterogeneity in meta-analysis. *Statistics in Medicine, 26*(1), 37–52. https://doi.org/10.1002/sim.2514

Viechtbauer, W. (2010). Conducting meta-analysis in R with the metafor package. *Journal of Statistical Software, 36*(3), 1–48.

Wasserstein, R. L., & Lazar, N. A. (2016). The ASA statement on p-values: Context, process, and purpose. *The American Statistician, 70*(2), 129–133.

Wagenmakers, E.-J., Beek, T., Dijkhoff, L., Gronau, Q. F., Acosta, A., Adams, R. B., Jr., ... Zwaan, R. A. (2016). Registered Replication Report: Strack, Martin, & Stepper (1988). *Perspectives on Psychological Science*, 11, 917–928.

Wellak, S. (2002). *Testing statistical hypotheses of equivalence*. CRC Press.

Chapter 15
Preregistration: Definition, Advantages, Disadvantages, and How It Can Help Against Questionable Research Practices

Angelos-Miltiadis Krypotos, Gaetan Mertens, Irene Klugkist, and Iris M. Engelhard

Abstract Questionable research practices (QRPs), such as p-hacking (i.e., the inappropriate manipulation of data analysis to find statistical significance) and post hoc hypothesizing, are threats to the replicability of research findings. One key solution to the problem of QRPs is preregistration. This refers to time-stamped documentation that describes the methodology and statistical analyses of a study before the data are collected or inspected. As such, readers of the study's report can evaluate whether the described research is in line with the planned methods and analyses or whether there are deviations from these (e.g., analyses performed so that the research hypotheses is confirmed). Here, we aim to describe what preregistration entails and why it is useful for psychology research. In this vein, we present the key elements of a sufficient preregistration file, its advantages as well as its disadvantages, and why preregistration is a key, yet partially insufficient, solution against QRPs. By the end of this chapter, we hope that readers are convinced that there is little reason not to preregister their research.

Keywords Questionable research practices · Preregistration · Psychological science · Clinical science

---

A.-M. Krypotos (✉)
Department of Clinical Psychology, Utrecht University, Utrecht, the Netherlands

Group of Health Psychology, KU Leuven, Leuven, Belgium

G. Mertens
Medical and Clinical Psychology, Tilburg University, Tilburg, the Netherlands

I. Klugkist
Methodology and Statistics, Utrecht University, Utrecht, the Netherlands

I. M. Engelhard
Department of Clinical Psychology, Utrecht University, Utrecht, the Netherlands

W. O'Donohue et al. (eds.), *Avoiding Questionable Research Practices in Applied Psychology*, https://doi.org/10.1007/978-3-031-04968-2_15

Introduction

Credible psychological science implies that research is reproducible or replicable. Concerns about whether the psychology literature is reliable have been raised a long time ago (Babbage, 1830; Rosenthal, 1979; Stroebe et al., 2012). However, this past decade, a crisis in the confidence of psychology as well as other scientific fields has risen (Camerer et al., 2016; National Academies of Sciences & Medicine, 2019; Pashler & Wagenmakers, 2012). This was primarily due to the publication of studies showing poor replicability (the repetition of a study's findings with new data) of many important psychological findings (Open Science Collaboration, 2015; Ritchie, 2020), as well as poor reproducibility (finding of identical results when performing the original analyses on the same data) of research (Hardwicke et al., 2019, 2020).

One of the proposed reasons for the low replicability and reproducibility in psychology is questionable research practices (QRPs). QRPs include the formation of a research hypothesis after the results are known (HARKing; Kerr, 1998), the flexible use of data analyses to obtain evidence for a hypothesis, even when it is not supported by the data (Simmons et al., 2011), and the collection of data until the null hypothesis is rejected in Null-Hypothesis Significance Testing (NHST; Strube, 2006). The reported high prevalence of QRPs in psychology (John et al., 2012) demands immediate changes in our research practices and the establishment of ways that prevent researchers from using these.

Diverse methods for eliminating QRPs have been proposed. The first method is to change the incentive structures in science (Bruton et al., 2020; Chambers et al., 2015). In particular, academic success is commonly evaluated by the number of articles scientists have published in journals with high impact factors. Given that the report of significant results increases the chances that a paper will be published (Fanelli, 2012; Rosenthal, 1979), QRPs bias the results towards this direction (Fanelli, 2010). These QRPs include the deviation of data collection procedures so that the results would support the predictions made by the authors, the removal of data without a justifiable reason, or even data fabrication. Different proposals have been made to solve this problem, such as a training in ethics in science (Bruton et al., 2020) or emphasizing science quality as an indicator of academic success.

A second step towards tackling QRPs is the open sharing of data and materials as well as the replication of past findings. Reproduction, however, is often challenging given that researchers typically do not share their data and materials broadly (Alsheikh-Ali et al., 2011; Hardwicke et al., 2019, 2020; Vanpaemel et al., 2015; Vines et al., 2014). Replication of past findings had also been done sparingly, given that replication studies are traditionally harder to publish compared to original findings. To date, however, more and more journals (e.g., *Journal of Experimental Psychology: General*) and funding agencies (e.g., the Netherlands Organization for Scientific Research) call for such studies, giving hope that this will be a way to reduce QRPs (Zwaan et al., 2018).

A third way to battle QRPs is the *preregistration* of a study before data analysis. Preregistration of studies is not new: the first registries were introduced in the 1960s (see Wiseman et al., 2019 for a full historical review). To date, preregistration is routinely done in some scientific fields (e.g., for clinical trials; see clinicaltrials.gov in the United States of America and eudract.ema.europa.eu in Europe). However, there is a call for extensive preregistration in psychology for all experimental studies, meta-analyses, and literature reviews. Preregistration is increasingly used (Lindsay et al., 2016; Nosek & Lindsay, 2018; Simmons et al., 2021b). It is promoted by journals, for instance, by providing a badge to a published article if the reported study was preregistered (Kidwell et al., 2016) or by not allowing a paper to be published unless the authors preregistered the research or explain why they did not. Also, more and more journals are requiring (e.g., *The Journal of Politics*) or at least encouraging (e.g., *PAIN*) the preregistration of experimental studies. Similar strategies are also encouraged for researchers and graduate students (e.g., in the Behavioural Science Institute of Radboud University), and the same goes for some grant agencies (e.g., the Dutch organization for health research and healthcare innovation, ZonMw, in the Netherlands).

Preregistration is the topic of this chapter. Specifically, we aim to explain what preregistration is, why it is useful, what its shortcomings are, as well as why it can provide a shield against some QRPs.

The structure of this chapter is as follows: We first describe what preregistration is and the key distinctions between the types of preregistrations. Then, we discuss the advantages and challenges of preregistration and end by providing alternatives to preregistration. Furthermore, we provide key sections that are typically included in a preregistration document. At the end of this chapter, we hope that readers will be convinced that it is imperative to preregister their research.

What Is Preregistration?

Preregistration consists of a collection of time stamped documents that typically describe a study's research questions, hypotheses, methodology, and statistical analyses. Although a study should be preregistered before the beginning of data collection, this is not always possible (e.g., see below about preregistration of pre-existing data). As such, as a general rule, a study should be preregistered at least *before* the research data are inspected (Nosek et al., 2018).

Multiple preregistration templates have been introduced (e.g., Johnson & Cook, 2019; Krypotos et al., 2019; Mertens & Krypotos, 2019), with different criteria to be fulfilled depending on the type of study (e.g., meta-analysis, laboratory studies, single-case designs) or when the study was preregistered (e.g., prior to data collection or to the data analysis). As such, researchers should first define clearly under which category their study falls.

A widely accepted distinction in preregistration is between studies in which original data are collected (e.g., laboratory research or a randomized controlled trial) and studies that use pre-existing data (e.g., re-analysis of an available data set). In the former case, researchers should define their research questions, hypotheses (if any), methods, and analyses. In the latter case, the preregistration of methods is more limited as the data have been already collected. In the case of pre-existing data, researchers should acknowledge that they already have, at least partial, knowledge of the data set to be analyzed, which could influence their analytic choices. Another commonly used distinction is whether a study aims to confirm a hypothesis (i.e., *confirmatory research*) or whether its goal is to explore different data patterns, without having an a priori hypothesis or a computational model to confirm (i.e., *exploratory research*) (De Groot, 2014; Dirnagl, 2020). Notably, the distinction between the different types of studies is not qualitative, as the different types of studies serve different purposes. For example, exploratory studies may enable the development of novel computational models, whereas confirmatory studies are needed to provide supportive evidence or gain more confidence in the structure of a particular pre-specified model.

Despite different preregistration templates for different studies, there are many commonalities in preregistration templates. Below, we will present the common aspects of preregistration, and we will describe which deviations are needed for different types of studies. Before that, though, we present the advantages and disadvantages of preregistration.

Advantages of Preregistration

There are plenty of reasons to preregister a study. Here we provide some of the key advantages of preregistration, before moving to a series of challenges in the next section.

First, preregistration allows researchers to take full credit for making an accurate prediction. Think, for example, of someone pulling random numbers from a bag. She picks one number, sees it, and claims, "It is number 4 as I had predicted." Without her having mentioned her prediction publicly in advance, her claim is not strong enough. Preregistration allows scientists to take full credit for the accuracy of their predictions by providing clear evidence that these were made in advance.

Second, in line with open science (Allen & Mehler, 2019), preregistration is a way to show that you are conducting transparent research, with results that are not based on post hoc reasoning and analyses (see QRPs above) but with concrete predictions made in advance. Increasingly, science funders and journals require researchers to demonstrate that their research practices are in line with open science principles. Preregistration is one way to achieve this.

Third, from a philosophy of science standpoint, preregistration allows researchers to transparently evaluate the *severity of their tests* (Mayo, 2018). Dating back to the time of Sir Karl Popper (e.g., Popper, 2005), a test is argued to be severe when it is strong enough to falsify a theory. In this line of reasoning, a preregistration allows others to evaluate whether a performed test was capable of falsifying a tested theory (Hitzig & Stegenga, 2020; Lakens, 2020; O'Donohue, 2021; Vanpaemel, 2019).

From a practical point of view, preregistration allows researchers to wrap up a project faster compared to when they have to decide on all analytic options after data collection. Also, in the case of results that do not confirm someone's hypothesis, there is the temptation to abandon a project altogether. By preregistrating the study, researchers already have the basic material for writing their methods/analysis section and a specific plan for carrying out all the analyses. Lastly, preregistration may also help researchers in protecting themselves against unwarranted requests for additional data analyses by reviewers, which can delay the publication of their results.

Despite the advantages of preregistrating a study, there are also arguments against preregistration of (some) studies. We turn to these below.

Disadvantages of Preregistration (and How to Counter Them)

No scientific practice is without its shortcomings, and as such, preregistration is not without its shortcomings (e.g., Rubin, 2020). We will discuss eight disadvantages below and will show that they are less important than the relative advantages.

First, it is not uncommon that by the end of data collection, researchers have thought of a different and better way to analyze their data than the way they mentioned in the preregistration. Also, a new statistical method may have been introduced between the preregistration and the end of data collection. As mentioned above, in these cases, the authors may update their preregistration by providing the reasons for applying a new analysis and mentioning the reasons why such an analysis is superior to the preregistered ones.

The second disadvantage relates to the limits of preregistration although it makes research design and analysis plans transparent but not necessarily correct or relevant. To illustrate, even when a researcher preregisters that she is going to perform a paired-samples *t*-test for group comparisons, this does not mean that such an analysis is correct. In this specific example, a paired-samples *t*-test would not be the right option as one of its basic assumptions is that each pair of observations comes from the same participant/group, making the between group comparisons impossible. A solution to this disadvantage could be given in the form of registered reports (Chambers, 2013). This type of article includes an evaluation of the study's introduction and methods *prior* to the beginning of data collection. The reviewers can evaluate the soundness of the methodology, the statistical analyses, as well as the relevance of the study to the specific journal. After the paper has been accepted as a registered report, the authors can collect the data and resubmit the article. The registered report format also protects researchers from reviewers' critiques after the results are known (Wagenmakers & Dutilh, 2016). Registered reports are currently adopted by almost 300 journals (see https://www.cos.io/initiatives/registered-reports for the full list of journals).

Third, preregistration calls for a change in the workflow of doing research, which could be particularly difficult, especially for seasoned researchers. In order to ensure that researchers use preregistration in their work, many relevant user-friendly programs have been introduced (e.g., see Krypotos et al., 2019).

Fourth, there is an ongoing discussion as to whether preregistration is worthwhile in the first place (Nosek et al., 2019; Szollosi et al., 2019). This point relates to the idea that preregistration improves the diagnostic value of the statistical tests (see also severity tests above). For example, preregistration is argued to enable an accurate familywise error rate (i.e., the probability of making at least one false discovery when running multiple statistical tests) and to force people to think deeply about their theories (Nosek et al., 2019). Still, these ideas have been challenged (Olken, 2015; Szollosi et al., 2019), and a call for better theories has been made instead of the ubiquitous adoption of preregistration.

Fifth, it has been argued that preregistration cannot really limit QRPs, as it may give them a different form, such as preregistration after the study has been completed (Yamada, 2018). Related to that, there has been a misuse of the badges awarded to some studies, with some articles reporting multiple studies and gaining badges for only preregistering part of the studies (Claesen, Gomes, Tuerlinckx, & Vanpaemel, 2019). Still, such disadvantages do not relate to the limitations of preregistration as a tool, but to its misuse by researchers.

Sixth, exploratory research is sometimes considered to be less strong compared to confirmatory research, so a concern could be that "safer" research will be promoted that is focused on the confirmation of the largest effects and that exploratory research is put in second place (Pham & Oh, 2020). This argument relates to the faulty misconception that exploratory research is a second-tier research, although it is equally important as confirmatory research (Scheel et al., 2020). Preregistration just helps researchers to better separate these two types of research, but it does not value one as better than the other (Simmons et al., 2021a, b).

Seventh, the preregistration does not just concern the authors but also the reviewers and editors. It is important that the reviewers and editors carefully confirm that the preregistration plan has been followed and, if not, that the deviations are reported in the manuscript. Although this may seem like a lot of work (Pham & Oh, 2020), in principle, it will result in less work as the reviewers or editors do not have to question whether the results were p-hacked as preregistration plan has been shown (Wagenmakers & Dutilh, 2016).

Lastly, it is tempting not to follow the preregistration plan, especially when the research results go against the hypotheses of the study. Sadly, there is evidence that often the published results differ from the preregistration plan (Claesen et al., 2019). However, this is not a disadvantage of preregistration per se, but of the current incentive system in science.

What to Include in a Study's Preregistration

Research Hypotheses

Confirmatory research is conducted to prove or falsify a hypothesis (see O'Donohue, 2021 for a discussion of this issue). As such, specific hypotheses should be determined explicitly in advance in the preregistration. General research hypotheses may leave too much room for flexible data analyses. In contrast, exploratory research does not require explicit or specific hypotheses.

Methodology

Following the research hypotheses, the methodology for testing the hypotheses should be described. Although the methodology is more extensive when original data are collected (see below), even studies with preexisting data should include a methodology section, including how the data were acquired or, in case of a meta-analysis, how these will be retrieved and extracted from the literature. Notably, in case the data have already been published, a link to the previous research should be included, and prior information about the data should be disclosed that could influence the analytic decisions (see below).

The methodology of a study includes the definition of (if applicable) stimuli that will be used, questionnaires and answering scales, procedures, blinding of the experimenters, and randomization. In line with open-science practices, it would be desirable if all relevant materials are uploaded into a repository, so other researchers would have access to all original materials in case they want to replicate the study.

Sample

The characteristics of the (intended) sample should be described in the preregistration document. Although sex and age descriptions are standard in psychology research, other sample characteristics that are relevant to the research questions should be included as well. For example, a study regarding anxiety disorders may also include the anxiety levels of the sample. Notably, characteristics of the sample can influence the generalizability of results to other samples. This is particularly important, because current samples in psychology mostly include Western, educated, industrial, rich, and democratic (WEIRD) samples (Henrich et al. (2010); Muthukrishna et al. (2020), which often limits the generalizability of the findings to other populations.

An important decision that has to be made before the beginning of the study is the size of the sample. There are different ways to justify the sample size of a study.

For example, the researcher could run a power analysis (Cohen, 1992) based on the effect sizes previously reported in the literature or by defining the effect size that is minimally interesting for a study (i.e., *the minimal statistically detectable effect*) (Albers & Lakens, 2018; Lakens, 2020). This is the minimum effect that, if present, would be statistically significant given the sample size of the study and the chosen α level (Cook et al., 2014). Importantly, the size of the sample has an important influence on the direction of the results, especially when the analyses are run within a Null-Hypothesis Significance Testing (NHST) Framework. Within a NHST framework, p-values (i.e., the probability of observing the current or more extreme data given that the null hypothesis is true; Wagenmakers, 2007) will almost always turn out to be statistically significant given that enough sample data are collected. This is also the case when the tested effects come from the population correctly as described by the null-hypothesis.

As an alternative to defining the sample size in advance, it is also possible to use adaptive procedures. In these procedures, data collection is completed when adequate evidence has been accumulated for or against a hypothesis, or it is completed based on other objective criteria, such as the time the lab is available. For example, an investigator could use sequential analyses, where the α level is divided by the times a test is planned to be performed (Lakens, 2014), or use a Bayesian data planning procedure (Schönbrodt & Wagenmakers, 2018). Our goal here is to make it explicit that no matter which stopping rule is used in the study, this should be mentioned clearly in advance in a preregistration document.

In the case of pre-existing data sets, the sample characteristics that need to be reported depend on the research question. To illustrate, in the case of a genome-wide association study, which is conducted to test whether specific genes predict the development of psychopathology, the preregistration should include only the characteristics of the subset of the sample. When a new model will be tested, there may be a distinction between the data that are used for tuning the model parameters and validating the model (also referred to as the *training* and *validating* data set in machine learning; Dwyer et al. (2018)). If this is the case, then the preregistration should mention how the two separate data sets will be determined.

Data Preprocessing

Before the data analyses, scientists often transform their data, reject outliers, and aggregate values into sums or mean scores. For example, in the case of reaction time (RT) analyses, extreme values are typically removed, and the distribution of RTs is log transformed (Heathcote et al., 1991). It is important that all data transformations be also included in the preregistration given that different transformations may change the direction of the results. Whether each choice is defendable or not is up to the researcher and the scientific community. Still, the modification of the data may determine the direction of results, and non-specification of data transformation/reduction processes leaves room for QRPs. In some research fields, however, exact predefining of the data reduction/transformation procedures is almost impossible, given that such procedures are often dictated by the data per se (e.g., normalize distribution of data only if they show that they are distributed normally). In such cases, researchers are advised to list the sensitivity analyses they will perform to ensure that the direction of the findings is not the result of the data reduction procedure.

Statistical Analysis

Statistical analyses follow from the theoretical background of the study and the research questions. In cases of concrete formal theories, the statistical analyses follow such models, and researchers have reduced flexibility in choosing which analysis they should perform (van Rooij & Blokpoel, 2020). However, such formal models are rare in psychology, and usually generic statistical models for drawing inferences are selected, such as regression, *t*-tests and analysis of variance (ANOVA). Nonetheless, and due to the absence of a formal model (van Rooij & Baggio, 2021), the same research question can be answered with different analyses. To illustrate, during a fear conditioning task, in which initially neutral stimuli are paired with unpleasant stimuli across multiple trials, someone could run a repeated measures ANOVA or a multilevel model. Given that different analyses can yield different results, flexibility in such statistical analyses may inflate false-positive rates (Simmons et al., 2011).

To convince readers that the analyses were free from biases stemming from data inspection, a clear description of the planned statistical analyses should be included in the preregistration document. This description includes the inferential framework (e.g., NHST, Bayesian analyses) and the statistical models that will be used, with a clear definition of the variables in the model. Arguably, many decisions cannot be taken without inspecting the data, and some data reduction procedures cannot be predicted (e.g., a data pattern against expectations). There are at least two ways to solve problems with decisions that need to be made before data inspection. First, researchers can create a flow chart of how the data could inform the statistical decisions. For instance, someone could argue that when the assumption of a normality in the variables is violated, a Welch's *t*-test will be used instead of a Student's *t*-test. Accounting for each possible data pattern will be daunting, especially when complex statistical models are used. An alternative strategy would be to update the preregistration file and to argue why the newly proposed analyses are a better approach to the data analyses. Preregistration should be viewed as a plan, not a prison (DeHaven, 2017), that can be updated. Such updates should be shared timely and transparently with the rest of the community so they can be judged accordingly.

Remaining Sections

Above, we described common characteristics of current preregistration templates (e.g., Crüwell & Evans, 2019; Kirtley et al., 2020; Krypotos et al., 2019; Mertens & Krypotos, 2019; Van den Akker et al., 2019; van't Veer & Giner-Sorolla, 2016). As mentioned previously, however, different types of studies need different preregistration elements, and more templates are being introduced depending on the field of study. We suggest that authors first inspect available templates and choose the template that best fits their study (see https://osf.io/zab38/ for an overview).

Where to Preregister

The completed preregistration document should be submitted to an official repository. To date, most repositories are online. The type of repository that will be used also depends on the type of study. For example, clinical trials are most commonly registered on clinicaltrials.gov in the United States and eudract.ema.europa.eu for Europe. For experimental and modelling work within psychology, two databases are commonly used. The first one is *aspredicted* (aspredicted.org). It enables researchers to preregister a study by answering nine simple questions relating to the study's research design and analyses. The second one is *osf* (osf.io) where researchers have the option to select templates in which they answer many more questions compared to AsPredicted and go much more in depth in their study. We urge authors to prefer including enough information in their study compared to vague specifications, something that will leave less room for misinterpretations as well as flexible data analyses. A limitation of the AsPredicted website is that although the preregistration is quite easy to complete, the website is *not* a formal registry, given that preregistrations could be kept private forever. In contrast, on the osf, preregistrations are released online after a maximum of 4 years. Allowing researchers to keep their preregistration private could result in preregistrating multiple hypotheses and releasing only the ones that support the preregistration file that supports their study.

Alternatives to Preregistration

Preregistration is only one way to counter QRPs. In this section, we will suggest additional tools that could be used in combination with preregistration, although these are not integral parts of preregistration.

The first one is crowdsourcing analyses. This includes the sharing of a dataset with different groups that are allowed to analyze the data set in any desired way (Dutilh et al., 2019a, b; Silberzahn et al., 2018). To illustrate, in Dutilh, Annis, et al. (2019a), the first author shared the same set of data with different experts of the *diffusion model*, a computational model used for decomposing reaction time performance into different model parameters (Ratcliff & McKoon, 2008). The different groups then had to fit the diffusion model and report back the model parameter values. This allowed the different groups to use any version of the diffusion model they wanted, with no two groups selecting the same model, without knowing explicitly what the research question was. This approach can reduce the bias towards presenting results supporting an effect.

The second alternative is a multiverse analysis (Steegen et al., 2016). The rationale for multiverse analyses is that in the absence of a concrete background, more than one type of analyses seems reasonable. Let us return to the reaction time example. Reaction time distributions are typically skewed. Typically, a summary value is used for a reaction time distribution (e.g., the mean). It could be argued that a given researcher prefers computing the median of the distribution, because it is less influenced by extreme values compared to the mean. Another approach would be to normalize the distribution and then compute the mean. In the absence of a theory about the best option, both options are reasonable. In multiverse analyses, researchers need to conduct all reasonable analyses. Then, the distribution of results is plotted. Multiverse analyses can be specified not only on the level of data reduction procedure (e.g., different data transformation) but also on the level of the selected statistical models (e.g., multilevel analyses or analyses of variance).

Concluding Remarks

To date, preregistration of a study constitutes one of the most important tools towards battling QRPs. As shown above, however, it does not provide absolute immunity against them. Nevertheless, given the advantages presented above, there is little reason not to preregister a study. It is likely that in the next few years, preregistration of a study will become the norm, rather than an exception, and it is possible that over the next decade, there will be hardly any experimental study in our field that is not preregistered. This norm, however, can have exceptions, and as such, researchers can always simply argue as to why they did not preregister their study.

In order to achieve the goal of science being open, transparent, and replicable, we will need to move towards adopting better practices, such as the open sharing of data and materials. Ultimately, the goal of science is the collection of reliable information that is useful for science itself and for the whole society. QRPs do not serve that goal and should be maximally eliminated, such as by the adoption of study preregistration.

Funding AMK is supported by a senior postdoctoral grant from FWO (Reg. # 12X5320N) and a replication grant from the Netherlands Organization for Scientific Research (NWO) (Reg. # 401.18.056). Iris M. Engelhard is supported with a Vici innovational research grant by NWO (grant number: 453-15-005).

References

Albers, C., & Lakens, D. (2018). When power analyses based on pilot data are biased: Inaccurate effect size estimators and follow-up bias. *Journal of Experimental Social Psychology, 74*, 187–195.

Allen, C., & Mehler, D. M. (2019). Open science challenges, benefits and tips in early career and beyond. *PLoS Biology, 17*, e3000246.

Alsheikh-Ali, A. A., Qureshi, W., Al-Mallah, M. H., & Ioannidis, J. P. (2011). Public availability of published research data in high-impact journals. *PLoS One, 6*, e24357.

Babbage, C. (1830). *Reflections on the decline of science in England: And on some of its causes, by charles babbage (1830)* (Vol. 1). B. Fellowes.

Bruton, S. V., Medlin, M., Brown, M., & Sacco, D. F. (2020). Personal motivations and systemic incentives: Scientists on questionable research practices. *Science and Engineering Ethics, 26*, 1–17.

Camerer, C. F., Dreber, A., Forsell, E., Ho, T.-H., Huber, J., Johannesson, M., … others. (2016). Evaluating replicability of laboratory experiments in economics. *Science, 351*, 1433–1436.

Chambers, C. D. (2013). Registered reports: A new publishing initiative at cortex. *Cortex, 49*, 609–610.

Chambers, C. D., Dienes, Z., McIntosh, R. D., Rotshtein, P., & Willmes, K. (2015). Registered reports: Realigning incentives in scientific publishing. *Cortex, 66*, A1–A2.

Claesen, A., Gomes, S. L. B. T., Tuerlinckx, F., & Vanpaemel, W. (2019). Preregistration: Comparing dream to reality.

Cohen, J. (1992). Statistical power analysis. *Current Directions in Psychological Science, 1*, 98–101.

Cook, J. A., Hislop, J., Adewuyi, T. E., Harrild, K., Altman, D. G., Ramsay, C. R., … others. (2014). Assessing methods to specify the target difference for a randomised controlled trial: DELTA (difference elicitation in trials) review. *Health Technology Assessment (Winchester, England), 18*.

Crüwell, S., & Evans, N. J. (2019). Preregistration in complex contexts: A preregistration template for the application of cognitive models. *Preprint Available at PsyArXiv*.

De Groot, A. (2014). The meaning of "significance" for different types of research [translated and annotated by Eric-Jan Wagenmakers, Denny Borsboom, Josine Verhagen, Rogier Kievit, Marjan Bakker, Angelique Cramer, Dora Matzke, Don Mellenbergh, and Han LJ van der Maas]. *Acta Psychologica, 148*, 188–194.

DeHaven, A. (2017). *Preregistration: A plan, not a prison*. Retrieved October 29, 2019.

Dirnagl, U. (2020). Preregistration of exploratory research: Learning from the golden age of discovery. *PLoS Biology, 18*, e3000690.

Dutilh, G., Annis, J., Brown, S. D., Cassey, P., Evans, N. J., Grasman, R. P., … others. (2019a). The quality of response time data inference: A blinded, collaborative assessment of the validity of cognitive models. *Psychonomic Bulletin & Review, 26*(4), 1051–1069.

Dutilh, G., Sarafoglou, A., & Wagenmakers, E.-J. (2019b). Flexible yet fair: Blinding analyses in experimental psychology. *Synthese, 198*, 1–28.

Dwyer, D. B., Falkai, P., & Koutsouleris, N. (2018). Machine learning approaches for clinical psychology and psychiatry. *Annual Review of Clinical Psychology, 14*, 91–118.

Fanelli, D. (2010). Do pressures to publish increase scientists' bias? An empirical support from US States Data. *PLoS One, 5*, e10271.

Fanelli, D. (2012). Negative results are disappearing from most disciplines and countries. *Scientometrics, 90*, 891–904.

Hardwicke, T. E., Bohn, M., MacDonald, K., Hembacher, E., Nuijten, M. B., Peloquin, B., … Frank, M. C. (2020). Analytic reproducibility in articles receiving open data badges at psychological science: An observational study. *Royal Society Open Science, 8*, 1–9.

Hardwicke, T. E., Wallach, J. D., Kidwell, M. C., Bendixen, T., Crüwell, S., & Ioannidis, J. P. (2019). An empirical assessment of transparency and reproducibility-related research practices in the social sciences (2014–2017). *Royal Society Open Science, 7*, 190806.

Heathcote, A., Popiel, S. J., & Mewhort, D. (1991). Analysis of response time distributions: An example using the Stroop task. *Psychological Bulletin, 109*, 340.

Henrich, J., Heine, S. J., & Norenzayan, A. (2010). Most people are not WEIRD. *Nature, 466*, 29–29.

Hitzig, Z., & Stegenga, J. (2020). The problem of new evidence: P-hacking and pre-analysis plans. *Diametros, 17*, 10–33.

John, L. K., Loewenstein, G., & Prelec, D. (2012). Measuring the prevalence of questionable research practices with incentives for truth telling. *Psychological Science, 23*, 524–532.

Johnson, A. H., & Cook, B. G. (2019). Preregistration in single-case design research. *Exceptional Children, 86*, 95–112.

Kerr, N. L. (1998). HARKing: Hypothesizing after the results are known. *Personality and Social Psychology Review, 2*, 196–217.

Kidwell, M. C., Lazarević, L. B., Baranski, E., Hardwicke, T. E., Piechowski, S., Falkenberg, L.-S., … others. (2016). Badges to acknowledge open practices: A simple, low-cost, effective method for increasing transparency. *PLoS Biology, 14*, e1002456.

Kirtley, O., Lafit, G., Achterhof, R., Hiekkaranta, A., & Germeys, I. (2020). Making the black box transparent: A template and tutorial for (pre-) registration of studies using experience sampling methods (ESM). *Advances in Methods and Practices in Psychological Science*.

Krypotos, A.-M., Klugkist, I., Mertens, G., & Engelhard, I. M. (2019). A step-by-step guide on preregistration and effective data sharing for psychopathology research. *Journal of Abnormal Psychology, 128*, 517–527.

Lakens, D. (2014). Performing high-powered studies efficiently with sequential analyses. *European Journal of Social Psychology, 44*, 701–710.

Lakens, D. (2020). Sample size justification. *Preprint Available at PsyArXiv*.

Lindsay, D. S., Simons, D. J., & Lilienfeld, S. O. (2016). Research preregistration 101. *APS Observer, 29*.

Mayo, D. G. (2018). *Statistical inference as severe testing*. Cambridge University Press.

Mertens, G., & Krypotos, A.-M. (2019). Preregistration of analyses of preexisting data. *Psychologica Belgica, 59*, 338.

Muthukrishna, M., Bell, A. V., Henrich, J., Curtin, C. M., Gedranovich, A., McInerney, J., & Thue, B. (2020). Beyond western, educated, industrial, rich, and democratic (weird) psychology: Measuring and mapping scales of cultural and psychological distance. *Psychological Science*, 0956797620916782.

National Academies of Sciences, Engineering, & Medicine. (2019). *Reproducibility and replicability in science*. National Academies Press.

Nosek, B. A., Beck, E. D., Campbell, L., Flake, J. K., Hardwicke, T. E., Mellor, D. T., … Vazire, S. (2019). Preregistration is hard, and worthwhile. *Trends in Cognitive Sciences, 23*, 815–818.

Nosek, B. A., Ebersole, C. R., DeHaven, A. C., & Mellor, D. T. (2018). The preregistration revolution. *Proceedings of the National Academy of Sciences, 115*, 2600–2606.

Nosek, B. A., & Lindsay, D. S. (2018). Preregistration becoming the norm in psychological science. *APS Observer, 31*(3).

O'Donohue, W. (2021). Some Popperian notes regarding replication failures in psychology.

Olken, B. A. (2015). Promises and perils of pre-analysis plans. *Journal of Economic Perspectives, 29*, 61–80.

Open Science Collaboration. (2015). Estimating the reproducibility of psychological science. *Science, 349*.

Pashler, H., & Wagenmakers, E. (2012). Editors' introduction to the special section on replicability in psychological science: A crisis of confidence? *Perspectives on Psychological Science, 7*, 528–530.

Pham, M. T., & Oh, T. T. (2020). Preregistration is neither sufficient nor necessary for good science. *Journal of Consumer Psychology*.

Popper, K. (2005). *The logic of scientific discovery*. Routledge.

Ratcliff, R., & McKoon, G. (2008). The diffusion decision model: Theory and data for two-choice decision tasks. *Neural Computation, 20*, 873–922.

Ritchie, S. (2020). *Science fictions: How fraud, bias, negligence, and hype undermine the search for truth*. The Bodley Head.

van Rooij, I., & Baggio, G. (2021). Theory before the test: How to build high-verisimilitude explanatory theories in psychological science. *Perspectives on Psychological Science*.

van Rooij, I., & Blokpoel, M. (2020). Formalizing verbal theories. *Social Psychology, 51*, 285–298.

Rosenthal, R. (1979). The file drawer problem and tolerance for null results. *Psychological Bulletin, 86*, 638–641.

Rubin, M. (2020). Does preregistration improve the credibility of research findings? *The Quantitative Methods in Psychology, 16*, 376–390.

Scheel, A. M., Tiokhin, L., Isager, P. M., & Lakens, D. (2020). Why hypothesis testers should spend less time testing hypotheses. *Perspectives on Psychological Science, 16*, 744–755.

Schönbrodt, F. D., & Wagenmakers, E.-J. (2018). Bayes factor design analysis: Planning for compelling evidence. *Psychonomic Bulletin & Review, 25*, 128–142.

Silberzahn, R., Uhlmann, E. L., Martin, D. P., Anselmi, P., Aust, F., Awtrey, E., … others. (2018). Many analysts, one data set: Making transparent how variations in analytic choices affect results. *Advances in Methods and Practices in Psychological Science, 1*, 337–356.

Simmons, J., Nelson, L., & Simonsohn, U. (2021a). Pre-registration is a game changer. But, like random assignment, it is neither necessary nor sufficient for credible science. *Journal of Consumer Psychology, 31*, 177–180.

Simmons, J., Nelson, L., & Simonsohn, U. (2021b). Pre-registration: Why and how. *Journal of Consumer Psychology, 31*, 151–162.

Simmons, J. P., Nelson, L. D., & Simonsohn, U. (2011). False-positive psychology: Undisclosed flexibility in data collection and analysis allows presenting anything as significant. *Psychological Science, 22*, 1359–1366.

Steegen, S., Tuerlinckx, F., Gelman, A., & Vanpaemel, W. (2016). Increasing transparency through a multiverse analysis. *Perspectives on Psychological Science, 11*, 702–712.

Stroebe, W., Postmes, T., & Spears, R. (2012). Scientific misconduct and the myth of self-correction in science. *Perspectives on Psychological Science, 7*, 670–688.

Strube, M. J. (2006). SNOOP: A program for demonstrating the consequences of premature and repeated null hypothesis testing. *Behavior Research Methods, 38*, 24–27.

Szollosi, A., Kellen, D., Navarro, D. J., Shiffrin, R., van Rooij, I., Van Zandt, T., & Donkin, C. (2019). Is preregistration worthwhile. *Trends in Cognitive Sciences, 24*, 94–95.

Van den Akker, O., Weston, S. J., Campbell, L., Chopik, W. J., Damian, R. I., Davis-Kean, P., … others. (2019). Preregistration of secondary data analysis: A template and tutorial.

Vanpaemel, W. (2019). The really risky registered modeling report: Incentivizing strong tests and HONEST modeling in cognitive science. *Computational Brain & Behavior, 2*, 218–222.

Vanpaemel, W., Vermorgen, M., Deriemaecker, L., & Storms, G. (2015). Are we wasting a good crisis? The availability of psychological research data after the storm. *Collabra: Psychology, 1*.

van't Veer, A. E., & Giner-Sorolla, R. (2016). Pre-registration in social psychology—A discussion and suggested template. *Journal of Experimental Social Psychology, 67*, 2–12.

Vines, T. H., Albert, A. Y., Andrew, R. L., Débarre, F., Bock, D. G., Franklin, M. T., … Rennison, D. J. (2014). The availability of research data declines rapidly with article age. *Current Biology, 24*, 94–97.

Wagenmakers, E.-J. (2007). A practical solution to the pervasive problems of p-values. *Psychonomic Bulletin & Review, 14*, 779–804.

Wagenmakers, E.-J., & Dutilh, G. (2016). Seven selfish reasons for preregistration. *APS Observer, 29*.

Wiseman, R., Watt, C., & Kornbrot, D. (2019). Registered reports: An early example and analysis. *PeerJ, 7*, e6232.
Yamada, Y. (2018). How to crack pre-registration: Toward transparent and open science. *Frontiers in Psychology, 9*, 1831.
Zwaan, R. A., Etz, A., Lucas, R. E., & Donnellan, M. B. (2018). Making replication mainstream. *Behavioral and Brain Sciences, 41*.

Chapter 16
Adversarial Collaboration

Tim Rakow

Abstract Adversarial collaboration is an approach to resolving scientific disputes, wherein researchers who have different positions on the issue at hand collaborate with the aim of making progress on their disputed research question. As an approach to research, adversarial collaboration sits squarely within the open science framework because it puts a premium on transparency in hypothesis specification, study design, data analysis, study interpretation and reporting, and supplies a framework that can encourage rigour in these components of the research process. It is, however, far less common than many other open science innovations such as open materials and data sharing or study preregistration. Therefore, this chapter will begin by familiarising readers with adversarial collaboration, outlining some of its key features, and identifying potential benefits of the approach. The chapter ends with a discussion of what the approach can offer.

Keywords Adversarial collaboration · Open science · Remedies for questionable research practices

What is Adversarial Collaboration?

> Iron sharpens iron. – Proverbs 27:17.

> It was science at its best. – Latham et al. (1988, p. 767).

Adversarial collaboration is an approach to resolving scientific disputes, wherein researchers who have different positions on the issue at hand collaborate with the aim of making progress on their disputed research question. As an approach to research, adversarial collaboration sits squarely within the open science framework because it puts a premium on transparency in hypothesis specification, study design,

T. Rakow (✉)
King's College London, Institute of Psychiatry, Psychology and Neuroscience, Department of Psychology, London, UK
e-mail: tim.rakow@kcl.ac.uk

W. O'Donohue et al. (eds.), *Avoiding Questionable Research Practices in Applied Psychology*, https://doi.org/10.1007/978-3-031-04968-2_16

data analysis, study interpretation and reporting, and supplies a framework that can encourage rigour in these components of the research process. It is, however, far less common than many other open science innovations such as open materials and data sharing or study preregistration. Therefore, this chapter will begin by familiarising readers with adversarial collaboration, outlining some of its key features, and identifying potential benefits of the approach. The chapter ends with a discussion of what the approach can offer. As the chapter unfolds, you will see that adversarial collaborations vary somewhat in their form and function and are occasionally known by other names such as proponent-skeptic collaborations (Matzke et al., 2015) – or indeed may not be labelled as any specific type of investigation. To help you as I endeavour to build a fuller appreciation of the approach, the following brief quotes can function as outline definitions for adversarial collaboration:

> ...a project carried out by two individuals or research groups who, having proposed conflicting hypotheses, seek to resolve their dispute. (Bateman et al., 2005, p.1561).

> ...a cooperative research effort that is undertaken by two (groups of) investigators who hold different views on a particular empirical question....The goal of an adversarial collaboration is to reach consensus on an experimental design and the corresponding testable hypotheses. (Matzke et al., 2015, p .e1).

> The approach requires both parties to agree on empirical tests for resolving a dispute and to conduct these tests with the help of an arbiter. (Mellers et al., 2001, p. 269).

Even from such brief definitions, you will likely appreciate that adversarial collaboration adds an additional level of oversight to a research study because the work of one researcher is open to scrutiny by another (the other investigator in the adversarial collaborator). Moreover, this scrutiny is a mutual – each collaborating investigator has oversight of the other's work. As with several other open science initiatives, this serves to reduce the researcher degrees of freedom (Simmons et al., 2011) which might otherwise make the design, implementation, conduct, analysis, or reporting of a piece of research more likely to favour one researcher's preferred views. I hope that you will also see in this chapter that the benefits of adversarial collaboration extend beyond those afforded by additional scrutiny. Collaborative design of research that involves researchers with opposing views facilitates the design of more severe tests of a hypothesis, thereby creating a valuable opportunity to move research closer to the Popperian ideal of critical testing of theory (Popper, 1974).

Case Studies in Adversarial Collaboration

Through a series of case studies, presented in approximate chronological order, this section will illustrate the various forms that adversarial collaborations take and will share some of the reflections of the "adversaries" who have participated in them. You will see that only a few of these examples fall within the field of clinical psychology. Nonetheless, the approach is applicable to any field that uses behavioural research to generate new knowledge, and therefore the wisdom and lessons that can be gleaned from these case studies are applicable to clinical psychology research.

A Trail Blazing Collaboration Applying the Joint Design of Crucial Experiments *to Resolve a Dispute – Latham et al. (1988)*

In this paper, Latham and Erez describe how – with the assistance of Locke serving as their mediator – they jointly designed a series of experiments which sought to resolve a scientific dispute on which they held different positions. This is a gem of a paper, partly because it might represent the first published adversarial collaboration in the psychological sciences but also because it raises several of the concerns about research practices that the open science movement brought to prominence some 15–20 years later. But more than that, it offers constructive guidance on how some of those concerns can be addressed.

Latham et al. (1988) reported four studies, for which the design of each was agreed upon between themselves – the two "antagonists" – and their mediator. Two studies were each directed by one of the antagonists and were run by research assistants who were blind to the hypotheses. Work by Latham and others had indicated that active participation in goal setting makes little difference to goal commitment or task performance. However, Erez had found higher goal acceptance with participation (group discussion) and that the subsequent goal commitment predicted performance. From brainstorming with the mediator present, Locke and Erez identified five sets of differences between their experiments, each of which generated candidate hypotheses that might explain the variance in their findings: task importance, group discussion, instructions to participants, timing of the goal setting, and cultural differences in personal values (e.g. collectivism). Each study used a factorial design to examine two or three of the candidate hypotheses for the differences in previous findings. The main conclusion was that participant instructions were an important source of difference between the antagonists' previous sets of findings.

The *Discussion* included (named) comments from each author – something that is not unusual in adversarial collaborations. In his comments, Latham described the extent of their collaboration, which extended to "…systematically reviewing one another's studies, formulating hypotheses, arguing over proper procedures for testing hypotheses, implementing the procedures, re-implementing the procedures, analyzing the data, and reanalyzing the data…" (p.767). Erez's comments focussed on some of the benefits from the process required by this kind of collaborative investigation. These included establishing boundary conditions for predictions, illuminating how contextual factors can affect research findings – yet do so in ways that a researcher working within one context may be blind to the effects of context. For example, Latham et al. (1988) found that manipulating the instructions used to assign goals moderated the effect of goal participation and that goal commitment was affected by manipulating goal difficulty and goal importance. However, in their prior investigations, these factors of participant instructions and task importance had not been manipulated by Latham or Erez but, rather, had been features that differed *between* the studies run by each researcher. Consequently, these factors provided some explanation for the differences in the effects that Latham and Erez had previously observed. More generally, this highlights how methods can be important for what is found and what is concluded. Locke's comments also focussed on what he had learned from the collaboration about the importance of methods. The antagonists had identified at least nine differences in the methods of their previous investigations – even though they were "allegedly studying the same phenomenon" (p. 769). Importantly, not all of these would be evident from the published *Method* sections, leading to the recommendation that fuller reporting be encouraged. The paper's final set of concluding comments emphasise that for this approach to the joint design of experiments to be fruitful, the antagonists must be sufficiently aligned at the philosophical level that they can agree on how variables can be operationalised and that antagonists require the scientific curiosity necessary to invest them in the process. The authors also concluded that the role of the third-party mediator was important and that while there had been no need for them to "become heavy-handed" (p.770), the mediator had made important interventions to ask the antagonists to reconsider their opinions or conclusions. For this reason, it was felt necessary that the mediator had the "trust and respect" (p.770) of both antagonists.

An Adversarial Collaboration on the Nature of Regret – Gilovich et al. (1998)

In three jointly designed survey studies, Gilovich et al. (1998) sought to distinguish between two accounts of regret, each of which aims to explain the observation that people tend to regret actions more than inactions in the recent past but regret inactions more than actions when considering the distant past. Gilovich and Medvec argued that differential patterns of regret for actions and inactions arise because the regret associated with actions dissipates more quickly than the regret that is associated with inaction. Kahneman offered an alternative explanation. He argued that there are two distinct forms of regret: "hot" regret, which typically arises soon after the event to which it is attached, and "wistful" regret, which arises much later because it is often associated with consequences that occur long after the decisions and actions that might cause regret. Upon completing their studies, the collaborators agreed that – in line with Kahneman's view – their new data point to two distinct forms of regret. However, Kahneman also conceded that the data also point to a degree of pain that can be associated with long-term regrets that he had not previously acknowledged. Thus, the data appears to have moved each author towards some shared interpretations of the phenomenon under examination. An additional benefit that the article points to is the opportunity for Gilovich and Medvec to articulate their theoretical position in a more precise fashion, reducing the opportunity for misinterpretation of their position and clarifying which of their claims have only weak evidential support. Therefore, the collaboration seems to have had value in identifying next steps for empirical investigation of the research question. All authors agreed (p. 605) "…that the exercise was worthwhile and that joint research is often a better way to deal with scholarly disagreement than are critiques and rejoinders."

The Best-Known Adversarial Collaboration – Mellers et al. (2001)

A search of the *APA Psych Info* database conducted in late 2020 generated fewer than 300 hits for the term "adversarial collaboration". Only a minority of the papers identified by this search actually report an adversarial collaboration. Rather, many papers that came up in this search did so because they cited this paper by Mellers et al. (2001) which includes the term 'adversarial collaboration' in its title. Most commonly, their paper was cited because the topic of the research was related to the paper's findings – though often Mellers et al. (2001) is cited when authors define or discuss adversarial collaboration as an approach to research. Thus, Mellers et al. (2001) is a good candidate for the most prominent example of adversarial collaboration. It may also be the first paper to apply the term "adversarial collaboration" to a joint investigation by antagonists, though – as demonstrated by the preceding case studies – it is not the first such collaboration.

Mellers et al. (2001) report three studies that were jointly designed by Ralph Hertwig and Daniel Kahneman, with the data collected by an independent arbiter, Barbara Mellers. The paper also includes valuable guidance on how an adversarial collaboration that includes an arbiter might be approached and conducted (reproduced in the Appendices of this chapter). A key feature of these suggestions is an emphasis on agreeing and recording the rationale and conduct of the collaboration *before* embarking upon data collection. This includes recording what results from the initial study would lead each researcher to change their mind, agreeing on the principles by which any subsequent follow-up studies are to be planned and implemented, and allowing the arbiter to set, *in advance*, how the resulting paper will be structured and co-authored. These suggested guidelines do not propose that an arbiter is essential to an adversarial collaboration, but rather a *possibility* to consider in cases where the differences in theoretical position or research methodology between the adversaries are substantial. The article itself gives no explanation for why Barbara Mellers was invited to be the arbiter, though the fact that she collected the data and was the first author – together with the focus of the paper's

Introduction – suggests that, likely, both methodological and theoretical differences between Hertwig and Kahneman underpinned the decision to involve her as arbiter.

The investigation focussed on the conjunction fallacy, which is when the conjunction of two events is judged more probable than one (or either) of the constituent events (e.g. Tversky & Kahneman, 1983). Hertwig and Kahneman held opposing positions regarding the reason why the fallacy is less common when events are described in a frequency format (e.g. "Of 200 instances …") than when described using probabilities (e.g. "which is more likely?"). A critical point of difference was that Hertwig claimed that frequency formats disambiguate the term "and", making it less likely that participants interpret this as a union of two sets, rather than as a conjunction of two sets. The experiments therefore compared probability estimates between event conjunctions and individual events, each of which were made in a frequency format, and varied the terms used to signal the conjunction of events (e.g. "and" vs. "and are" vs. "who are"). The paper describes each of the second and third studies as having been proposed by one author (Hertwig Study 2, Kahneman Study 3) in order to clarify findings from the first study. The main difference between the studies was the presence/absence of filler items. The results were in line with Kahneman's predictions for Studies 1 and 3, while the results of Study 2 aligned with Hertwig's predictions. Most of the *Discussion* section is given over to separate interpretations of the results, first by Hertwig and then by Kahneman. Despite these separate discussions that reflect non-shared interpretations of the findings, the paper includes points of agreement, some of which include shared reflections on the value of adversarial collaboration. A key one of these was the conclusion that the collaborating authors each came to a fuller appreciation of the limitations of their own claims. In line with these acknowledged limitations, each collaborator included explicit suggestions for the new lines of enquiry that would further test those claims.

An Adversarial Collaboration That Advocates Eloquently for the Approach – Bateman et al. (2005)

Adversarial collaborations are relatively rare events, and what little prominence the approach has could be attributed to the fact that it has been championed by one of the world's most prominent behavioural scientists, Nobel Prize winner, Daniel Kahneman. This is the third (and last) of the case studies in this section to involve Kahneman as one of the antagonists in an adversarial collaboration. The paper by Bateman et al. (2005) that reports this collaboration is worth reading, even if the scientific content holds no interest for you. This is because the paper includes a notable discussion of the features and potential benefits of adversarial collaboration and how these relate to the task of "doing science". One such benefit is that adversarial collaboration forces researchers to understand the arguments that oppose their preferred theoretical position. This, together with the requirement for collaborative design, should increase the chances that studies provide stringent tests of hypotheses, thereby reducing the chance that weak tests of hypotheses serve to strengthen the researcher's prior beliefs via an apparent (though perhaps trivial) confirmation of their hypothesis. The paper also includes this definition of adversarial collaboration, which explained how Bateman et al. approached their collaboration and sets out a potential roadmap for others seeking to set up an adversarial investigation:

> In an adversarial collaboration, the two parties agree on the design of an experiment which they will conduct jointly. Before knowing what the experiment will find, they accept its validity as a test of their respective hypotheses. Each party anticipates its interpretation of possible outcomes of the experiment, particularly those that it does not predict. The two parties agree that particular outcomes of the experiment would support one hypothesis, and particular other outcomes would support the other. Both parties commit to publishing the results, whatever they may be. (p.1563).

Bateman et al. (2005) report a single experiment with a design for which the two collaborating parties proposed different patterns of predictions across the conditions. The key idea under test relates to the status of monetary outlays (i.e. expenditures) in theories of reference-dependent choice, specifically, whether or not monetary outlays are regarded as losses. The paper discusses several general benefits of adversarial collaboration, sets out the two opposing theories in a formal fashion (as is standard in experimental economics), and discusses the design of the experiment at length. Notably, the parties favoured different approaches to valuation, and in the spirit of an adversarial collaboration, they settled on a composite design, which was acceptable to all, and that included both sets of methods. This resulted in eight treatment conditions. An additional two conditions were added to test an explanation for one surprising result from the original experiment. Both parties agreed that the data more strongly supported the predictions of one party (the "Norwich group" of researchers) than those of the other party (Kahneman). However, the parties had some differences on the precise theoretical interpretation of the patterns of data that were found, both of which were presented in the paper. One benefit that the authors point to is the sharpening of ideas as a result of the discipline of designing and reporting the experiment. This arose because the need for an agreed design required one party (Kahneman) to formulate their hypothesis with greater precision.

In their discussion of the features of adversarial collaboration, Bateman et al. (2005) highlight that the approach requires researchers to pay particular attention to understanding the other side of the debate. Experiments should therefore be more balanced tests of opposing theories, with less chance that the design has been biased to favour one theory. This should produce tests that can be genuinely decisive. When parties disagree on the interpretation of the findings, both sides of the argument are given a fair airing, without one side being downplayed. This should give the readers greater opportunity to decide the interpretation for themselves. It also circumvents some of the problems of selective publication because both parties agree to publish irrespective of the findings.

Allegiance Effects in Psychotherapy: Fertile Ground for Adversarial Collaboration – e.g. DeRubeis et al. (2005)

Leykin and DeRubeis (2009) outline that various meta-analyses point to a relationship between researcher allegiance and effect size in studies of psychotherapy, where allegiance is defined as belief in the superiority of an outcome. Put simply, clinician-researchers who favour a given therapy tend to report stronger evidence of that therapy's superiority over other therapies. In the absence of experimental evidence, this could be an association or a causal bias. For instance, it is possible that causality runs from observed outcome to belief. Thus, the clinician's allegiance may follow from the treatment efficacies that they have observed in their own research and therefore could simply be a consequence of the variation in treatment effect sizes across studies that arises from sampling variability. There are, however, several less mundane possibilities for why treatment effects increase with allegiance. Allegiances may generate unrepresentative results because those with allegiance are more likely to be experts in delivering their preferred therapy in its most effective form, and therefore the effect arises from "honest differences" in expertise. Alternatively, allegiances may motivate researchers to select inferior comparison treatments or could prompt selective reporting (i.e. a file drawer problem; Rosenthal, 1979). Outlining the potential for adversarial collaboration on this question, Leykin and DeRubeis (2009) point to four papers from a special issue of the journal *Clinical Psychology: Science and Practice* in 1999 – all of which advocate for researchers with opposing allegiances and complementary expertise, to collaborate when therapies are compared. Berman and Reich (2010) join this call, arguing that such adversarial collaborations involving researchers with different allegiances and clinical competence should increase the chance that treatment delivery is comparable, that analyses are unbiased, and that treatment expectations bias findings or their interpretation.

Importantly, such adversarial collaborations have occurred (though they have not necessarily been labelled as such) with several multisite comparisons between psychotherapy and pharmacotherapy taking place from the 1990s onwards (e.g. Heimberg et al., 1998). Hollon (1999) describes one such collaboration between psychiatrists with expertise in drug treatments (including Amsterdam and Shelton) and cognitive therapists (including Hollon and DeRubeis). Hollon explains how collaboration between those with opposing allegiances can create "...a fair "horse race," one in which each modality will have a reasonable chance of showing what it can do and one that incorporates the necessary controls to interpret "tie scores" should they occur" (p.108). Hollon's assumption is that allegiance effects most likely arise from honest differences in expertise in treatment delivery. Therefore, if each "side" looks after the effective delivery of the treatment for which they have expertise, this balances out allegiance effects. The fruits of this particular adversarial collaboration are reported in DeRubeis et al. (2005).

Adversarial Collaboration to Resolve Replication Failures – Matzke et al. (2015) and Kerr et al. (2018)

Matzke et al. (2015) describe their research as a "proponent-skeptic collaboration" (p. e1). They report the outcome of a single preregistered experiment with the design agreed by both parties after a referee (i.e. mediator/arbiter, van der Molen) had set up a collaboration agreement. The issue under test was whether horizontal eye-movement improves episodic memory, which the "proponents" (Nieuwenhuis and Slagter) had previously found to be the case, while the "skeptics" (Matzke, van Rijk and Wagenmakers) had failed to replicate this effect in pilot work, as well as noting inconsistent findings in the literature. The study compared horizontal movement vs. vertical movement vs. no movement, with the proponents expecting that horizonal movement would generate better recall than the other conditions. Using Bayesian analyses, Matzke et al. established that the evidence more strongly favoured the (skeptic's) null predictions over the alternative predictions of the proponents. The paper's *Discussion* includes separate discussions by the proponents, the skeptics, and the referee. In their discussion, the proponents still held to the validity of the basic result, but less strongly than before. All those involved in the collaboration reflected positively on the outcome, emphasising that not only did the collaboration bring the parties somewhat closer together but also generated new ideas for future testing. They also felt that their design had benefitted from bringing together the expertise and knowledge of both parties and that the public disclosure inherent in this kind of collaboration ensured that the research was confirmatory with respect to the hypothesis under test.

This latter benefit of adversarial collaboration – keeping researchers honest and precise about their predictions, thereby mitigating against the questionable practice of HARK-ing (**h**ypothesising **a**fter **r**esults are **k**nown; Kerr, 1998; see Chap. 5 in this volume) – has been touted elsewhere as a positive feature of adversarial collaboration. This precision in hypothesis specification is, of course, one of the primary drivers behind the practice of study preregistration. Therefore, it is notable that Matzke et al. (2015) doubly bound themselves in this regard by preregistering their hypotheses, which would already have been known to the other party in the collaboration as well as to the independent referee. Other adversarial collaborations have undertaken this "wide open science" approach of preregistering their adversarial collaboration (e.g. Van Dessel et al., 2017). This can be regarded as a safety net for adversarial collaborations, which might be particularly useful if undertaking a collaboration without an independent referee/arbiter. Matzke et al. (2015) propose some guidance for running a preregistered adversarial collaboration (which is reproduced in the Appendices of this chapter).

The adversarial collaboration reported by Kerr et al. (2018) also arose from failures to replicate. They frame their investigation in relation to social psychology's replication crisis, presenting adversarial collaboration as an additional tool that can address some difficulties associated with replicability and the reluctance to publish replication studies. The agonists examined ingroup favouritism in minimal groups, where a frequently replicated finding is that study participants disproportionately allocate rewards to or positively evaluate performance by members of their own group, even though group membership is based on trivial or arbitrary characteristics (e.g. Mullen et al., 1992). The impetus for the adversarial collaboration was conversations between Hogg and Kerr – both active researchers in this field with different preferred accounts of the phenomenon – after Hogg had reported several failures to replicate the basic effect which had been reported by Kerr.

Hogg and Kerr identified around a dozen differences between their experimental protocols, then set themselves the task of identifying whether one or more of these (e.g. the proximity of fellow study participants, delivery mode of instructions) might account for the non-replications. They describe this as a "bottom-up approach" (p. 68): determining whether a methodological feature can moderate the effect and using that to revise the theory, rather than testing competing predictions derived from different theories. Their large, collaboratively run experiment not only replicated the in-group favouritism effect but also identified new moderators of the effect. First, in-group favouritism was reduced with greater social distancing from other study participants, a finding that either of the agonists' preferred theories could accommodate. Second, behaviour differed between cultures, with Australians allocating resources more fairly than Americans, and a number of related differences in attitude being apparent. Third, there was a greater pull towards fairness when participant instructions were given orally rather than in writing. Each agonist's preferred theory would require amendment (e.g. additional post hoc assumptions) to account for these second and third findings. Moreover, the direction of the cultural differences ran counter to what would have been expected based on the previous pattern of replications and non-replications. This, therefore, opened up a new line of inquiry. The authors' reflections point to the specific value of adversarial collaboration when replicability is an issue, in particular because it addresses some of the difficulty of communicating every detail of the method to support replication. They also point to an increase in collegiality, creating a better appreciation for each other's viewpoints.

An Adversarial Collaboration on a Topic of Public Concern – Lindner et al. (2020)

This is a single-experiment paper on the possible effect of character sexualisation in video games, which the authors describe in a footnote as "intended as an adversarial collaboration" because "the authors come from different traditions and beliefs regarding the potential impact of media on self-objectification" (p.553). The experimenters manipulated the representation of the female character in a video game that participants played for 30 minutes and assessed the effect of this character's sexualisation on a number of measures such as self-objectification, body image, and attitudes towards women. No significant effects were found, and Bayesian analyses suggested meaningful support for the null hypotheses of no effect on the dependent variables. Lindner et al. conclude that public and academic concerns about the effects of sexualized video games may be greater than warranted by the data. The authors list adversarial collaboration as one of several recommendations for improving methodological rigour in their research field, and – helpfully – the lead authors describe some specific benefits of their adversarial collaboration in a separate article (Lindner & Trible, 2020). These include bringing together expertise in experimental methods and objectification theory and collaboration on the best way to operationalise variables.

A Programmatic Adversarial Collaboration Between Three Research Groups (e.g. Cowan et al., 2020)

This final case study in this section, arguably, represents the state of the art in adversarial collaboration because the collaboration extends over a series of investigations. It brings together three research groups, each working with a different model of working memory. The multicomponent model of memory (MCM, favoured by Logie and colleagues at the University of Edinburgh) posits separate storage and processing components and codes specific to different modalities (visual, semantic). The time-based resource sharing model (TBRS, favoured by Barrouillet and colleagues at the Universities of Geneva and Fribourg) assumes that processing and storage share the same attention limited resources. The embedded processes model (EP, favoured by Cowan and colleagues at the University of Missouri) assumes that features are temporarily activated in long-term memory and that memory is governed by a limited capacity domain-general controller. Importantly for the purposes of an extended adversarial collaboration, the three models make different sets of predictions across a range of memory phenomena, and the research groups can agree on what kinds of experiments represent legitimate tests of those predictions. Together, the three groups secured grant funding for their collaboration (see https://womaac.psy.ed.ac.uk).

In one of the papers from their collaboration, Doherty et al. (2019) discuss a frequently asked question in adversarial collaborations: *why do findings differ between labs?* One possibility for the kind of dual-task paradigms that these researchers use is that, if participant characteristics vary across labs, a given level of cognitive load could be "low" load for high-capacity individuals but "high" load for low-capacity individuals. This, therefore, confounds any experimental prediction that is contingent on the memory load that the research participant placed under. Therefore, in this and their other investigations, the parties favoured collecting data in more than one lab. It is notable that a suite of predictions is made for the four experiments in this paper, with each model making around a dozen predictions. These predictions vary not only according to the presence of effects across different main effects and interactions but also as to the predicted size (or relative size) of those effects (Table 1, p. 1536). This is important because it means that a researcher cannot simply focus on the presence of one or two effects that were predicted. The accuracy of each model's predictions must be examined in the round – using the full set of predictions – as indeed must the set of competing predictions when the models are compared. Perhaps unsurprisingly, given the number and precision of these predictions, none of the models predicted every effect that was observed across the four experiments. One of the paper's conclusions emphasises the value of this as an impetus to theory revision:

> There was mixed success by each framework in predicting trends in the data, but all missed large trends in the data. Each theory requires some reconsideration of its core assumptions, or at least under what circumstances expected effects should be observed. (p. 1547).

Related to this, the authors highlight that their investigation generated valuable new hypotheses to be tested in subsequent experiments. They also reflected on the difficulty of designing experiments that generate fully contrasting predictions – a particular challenge, presumably, when three models are being compared.

Cowan et al. (2020) review and discuss their programmatic adversarial collaboration in detail and outline some of the benefits of adversarial collaboration – several of which have already been identified in this chapter. Of particular note among these benefits are (1) collaborative design of research increasing trust in the results, (2) collaborative analysis and writing ensuring balanced reporting of the data and conclusions, (3) enhanced theory development as *different* theories are amended to align with shared datasets, and (4) collaboratively published research providing a more fruitful platform for other researchers to build on because it exposits a range of theoretical positions. Based on their specific experiences, Cowan, Belletier, Doherty et al. outline benefits from adversarial collaboration for clarifying theory, formulating hypotheses more precisely, and generating appropriate study designs. They also emphasise the benefits of the parties being forced to reach consensus on the *General Discussion* section when no one model accounts for all effects, acknowledging their suspicion that "…if these same results were collected by any one of our groups, that group's discussion would be tilted much more in favour of the group's theory" (p. 1019). Cowan Belletier, Doherty et al. also acknowledge some challenges and the probable limits of what can be achieved by adversarial collaboration. For example, they conclude that the approach will not lead senior researchers to abandon their view but should prompt "varieties" of these views to emerge to account for the new data.

In addition to outlining what they see as the important characteristics of their adversarial collaboration, Cowan, Belletier, Doherty et al. also offer advice to others seeking to initiate an adversarial collaboration. This advice includes working hard at what the other party is meaning when they set out their position. For practical reasons, they recommend that two to four labs provide the optimal scope for fruitful discussion and planning. They argue that it is important not only to preregister predictions but also to allow time for this in order to clarify what is crucial and what is not critical to the theory being tested. Cowan, Belletier, Doherty et al. also counsel that it is valuable to have some collaborators who are not strongly committed to one position. In their adversarial collaboration, this role was often taken by postdoctoral researchers who – perhaps unsurprisingly – had less commitment than the senior researchers who had developed their respective positions over many years of research. The benefits of this "light grip" on a particular theoretical position for achieving consensus are similar to the function served by including an independent referee (Matzke et al., 2015) arbiter (Mellers et al., 2001) or mediator (Latham et al., 1988) in some adversarial collaborations.

Approaches Related to Adversarial Collaboration

There are notable examples of research collaborations that share aspects of the adversarial collaborations reviewed above (such as bringing together researchers with opposing positions) but that lack one or more of the key features of adversarial collaboration (such as the joint design of research studies). For example, Alempaki et al. (2019) describe their series of 14 experiments on context effects in decision-making as a "quasi-adversarial collaboration". The investigation shares aspects of adversarial collaboration in that the researchers came from different disciplines with somewhat different research perspectives (psychology and economics) and had different levels of prior commitment to the competing theoretical frameworks that their experiments investigated. However, there was no formal collaboration agreement, nor was there collaboration on the entire set of experiments. Rather, the initial experiments were designed and conducted independently before the separate research groups learned of each other's endeavours and began to consult with each other on experimental design.

Other collaborations seek consensus or some more modest degree of restoration between researchers with opposing positions, but involve no new empirical research. Finkel et al. (2015) described one such collaboration between "erstwhile adversaries" (p. 3). The "Norton group" opposed the position of the "Reis group" that familiarity increases interpersonal attraction. This dispute had been pursued in several papers from each group, objecting to the methods, analysis, or findings of the other group. Finkel, Norton, Reis et al. did not describe any crucial experiments arising from their investigation. Instead, they presented a new framework to incorporate the findings of the two groups and existing findings in the literature. Two major figures in the field of decision research, Daniel Kahneman and Gary Klein, report a somewhat similar collaboration in which they explored whether their different interpretations of research on expertise and intuition could be aligned (Kahneman & Klein, 2009). Their article describes some alignment in their respective positions that might not have been apparent from their separate work, which they summarise as a "failure to disagree" (p. 515).

One reflection that is often made when adversarial collaborations are reported is the importance and benefits of seeking to understand the other party's position (Bateman et al., 2005; Cowan et al., 2020; Kerr et al., 2018). Allied to this is the view that approaching disputes via critiques and rejoinders rarely resolves disputes or advances science to a meaningful degree (Gilovich et al., 1998). In this spirit, many have resolved to conduct academic debate in a more measured, respectful manner with an element of collaboration in how the debate is structured and presented. One example of this is a special issue of the *International Journal of Transpersonal Studies* in which two scholars (Taylor and Hartelius) with distinct positions on the psychology of spirituality undertook an open debate via three pairs of articles. The editor and commentators described this as an adversarial collaboration (Lancaster & Friedman, 2017; Thouin-Savard, 2017). And while these articles did not involve a shared programme of empirical research, as would be expected for adversarial collaborations as defined for this chapter, it does appear that the structure imposed by this particular special issue format made for a constructure sharing and advancement of ideas.

Discussion and Conclusions

In his concluding remarks of the first case study of this chapter, Latham declared that his experience of adversarial collaboration was both "exciting" and "illuminating" and that the collaboration with Erez and Locke represented "science at its best" (Latham et al., 1988, p. 767). I think we should be forgiving of this outpouring of immodesty – not least because Latham and Erez had engaged in something bold and innovative, which stood to challenge their own strongly held ideas and to open them up to the potential for a form of public failure. I also think that adversarial collaboration can indeed embody important aspects of what science aims for. For example, because it pushes the antagonists to formulate precise hypotheses that distinguish effectively between their positions, adversarial collaborations can move researchers closer to the Popperian ideal of critical testing of falsifiable hypotheses by severe tests (Popper, 1974). It does so by avoiding weak confirmations of hypotheses in which theories compete only against straw man alternatives – possibly with no real substance to such competition. With its emphasis on the collaborative identification of methods for a study, adversarial collaboration also sits well with other accounts of what process science should follow. Having a set of methods that are held in common by both parties in the collaboration can help to create a shared language and a set of rules by which their science is conducted. When researchers work in this way according to a shared paradigm (Kuhn, 1962), they should be less inclined to ignore or misunderstand each other's viewpoint and better disposed to learn from each other's data. One hopes that this then encourages research programmes that generate new findings and better explanations for those findings, in keeping with the notion of a progressive research programme (Lakatos, 1970). This certainly seems to be the sentiment expressed by many who have themselves been involved in adversarial collaborations.

One point of diversity across adversarial collaborations, which is illustrated in this chapter's case studies, is the presence or absence of an impartial third-party mediator/arbiter/referee. The mediator's agreed duties can include chairing discussions (e.g. Latham et al., 1988), collecting data (e.g. Mellers et al., 2001), trouble-shooting issues for which the resolution had not been specified in advance (e.g. Matzke et al., 2015), and carrying the responsibility to resolve disputes by prompting antagonists to reconsider their position or by making a binding adjudication. The reports of adversarial collaborations that have included a mediator are explicit that the mediator's role was substantial and imply that they were therefore important to the success of the endeavour. Nonetheless, the inclusion of a mediator is not a defining feature of adversarial collaboration. Indeed, most published adversarial collaborations have not included one, seemingly without regret that no mediator was involved. The reports of these "unmediated" collaborations often point to the importance of a mix of formal agreement (e.g. preregistration) and healthy attitudes (e.g. openness, curiosity) and are therefore implicit that the functions carried out by the mediator can be achieved by other means. Of course, it may be that there are adversarial collaborations that have collapsed for lack of a mediator – or for other reasons – of which we are unaware. We simply do not know whether a file drawer of uncompleted adversarial collaborations exists – and therefore can only hope that the advice we can glean from apparently successful adversarial collaborations provides a reliable roadmap to a successful collaboration.

I hope that the case studies reviewed above make clear the links that adversarial collaboration has with other open science initiatives. Sharing in the design of critical experiments and planning in advance for how hypotheses will be tested ensures precision in the specification of hypotheses and predictions. This embodies the ancient wisdom that "iron sharpens iron" – a researcher's theoretical ideas are sharpened when they must be honed alongside those of another researcher who takes a different theoretical position. And this sharpening of ideas can occur for both researchers. Sharing in the task of specifying hypotheses and predictions also makes these specifications public *in advance* of running the experiment(s). In this regard, adversarial collaboration is a form of advance public disclosure of study details akin to study preregistration or registered replication (van 't Veer & Giner-Sorolla, 2016) – albeit one for which the information disclosed might have a limited distribution (i.e. only among the collaborating parties).

The joint design of research studies places a premium on full disclosure of methods and procedures, because only if both parties understand each other's methods can they design and implement a suitable plan of research. Indeed, several of the collaborations reviewed above relied heavily on bringing methodological differences into the open, as a means of designing suitable studies for the collaboration (e.g. Kerr et al., 2018; Latham et al., 1988). And many of these differences could not easily have been identified from even a close reading of the parties' previous publications. Such transparency about methods – and supporting its ensuing benefits – is, of course, the main goal of open materials.

Because both parties engaged in an adversarial collaboration hold the data in common, data are necessarily shared beyond a single research group who might have a restricted set of expectations about what the data will show. This should reduce the chance that questionable practices in the analyses of data arise, both in the case where "honest expectations" inadvertently bias data analysis decisions and where more sinister motives are at work (Simmons et al., 2011). In either case, there is another party on hand (and sometimes an independent referee) to call out such practices.

There have been many calls for adversarial collaborations to be more common than they are at present (e.g. Hobson, 2019; Nier & Campbell, 2013). And these calls include invitations to submit manuscripts that report adversarial collaborations to journals (e.g. *Judgment and Decision Making; Thinking & Reasoning*) which include offers of practical support for the approach from journals, in terms of both suggesting guidelines for adversarial collaboration and supporting authors with the process of publishing their collaboration (Rakow et al., 2014). Indeed, sometimes it seems that the number of papers that call for adversarial collaboration might exceed the number of papers that report an adversarial collaboration. This apparent under-utilisation of the approach may reflect that the costs of initiating and implementing an adversarial collaboration are quite high, in terms of time, effort, emotional energy, and perhaps even the fear of loss of reputation (if the results do not fit with one party's position). However, what I hope to have achieved in this chapter is to show that there are benefits from adversarial collaboration which can outweigh those costs because the approach can improve the quality and impact of the science that researchers' conduct. So let me encourage you, if you see an opportunity for adversarial collaboration, grasp that opportunity – and go do *science at its best*.

Appendices: Guidance for Conducting an Adversarial Collaboration

Appendix A Guidance for Conducting Adversarial Collaboration, Including Such That Includes an Independent Arbiter (Mellers et al., 2001)

Reproduced verbatim from Mellers et al. (2001, Table 1, p.270).

Suggestions for adversarial collaboration

1. When tempted to write a critique or to run an experimental refutation of a recent publication, consider the possibility of proposing joint research under an agreed protocol. We call the scholars engaged in such an effort participants. If theoretical differences are deep or if there are large differences in experimental routines between the laboratories, consider the possibility of asking a trusted colleague to coordinate the effort, referee disagreements, and collect the data. We call that person an arbiter.
2. Agree on the details of an initial study, designed to subject the opposing claims to an informative empirical test. The participants should seek to identify results that would change their mind, at least to some extent, and should explicitly anticipate their interpretations of outcomes that would be inconsistent with their theoretical expectations. These predictions should be recorded by the arbiter to prevent future disagreements about remembered interpretations.

3. If there are disagreements about unpublished data, a replication that is agreed to by both participants should be included in the initial study.
4. Accept in advance that the initial study will be inconclusive. Allow each side to propose an additional experiment to exploit the fount of hindsight wisdom that commonly becomes available when disliked results are obtained. Additional studies should be planned jointly, with the arbiter resolving disagreements as they occur.
5. Agree in advance to produce an article with all participants as authors. The arbiter can take responsibility for several parts of the article: an introduction to the debate, the report of experimental results, and a statement of agreed-upon conclusions. If significant disagreements remain, the participants should write individual discussions. The length of these discussions should be determined in advance and monitored by the arbiter. An author who has more to say than the arbiter allows should indicate this fact in a footnote and provide readers with a way to obtain the added material.
6. The data should be under the control of the arbiter, who should be free to publish with only one of the original participants if the other refuses to cooperate. Naturally, the circumstances of such an event should be part of the report.
7. All experimentation and writing should be done quickly, within deadlines agreed to in advance. Delay is likely to breed discord.
8. The arbiter should have the casting vote in selecting a venue for publication, and editors should be informed that requests for major revisions are likely to create impossible problems for the participants in the exercise.

Appendix B Guidance for Conducting a Preregistered Adversarial Collaboration (Matzke et al., 2015)

Reproduced verbatim from Matzke et al. (2015, Table 1, p. e2) with punctuation amended.

Proposed guidelines for a preregistered proponent-skeptic collaboration

First, the adversaries reach consensus on an optimal research design. This precaution eliminates the possibility of later disputes regarding the execution of the study.

Second, the two parties formulate their hypotheses and expectations in advance. This precaution decreases the probability of the investigators falling prey to various cognitive biases, such as hindsight bias (i.e. judging an event as more predictable after it has occurred; Roese & Vohs, 2012) and confirmation bias (i.e. favouring information that confirms one's own hypotheses; Nickerson, 1998).

Third, the adversaries agree to write a joint article and submit it to an academic journal regardless of the outcome of the study. This precaution may in the long term counteract publication bias and the file drawer problem (Greenwald, 1975; Rosenthal, 1979).

Last, as the novel but crucial ingredient, the two parties set up an adversarial collaboration agreement. The agreement describes the proposed research design and all foreseeable aspects of the preprocessing and analysis of the data. This precaution secures the purely confirmatory nature of the investigation and increases the transparency of scientific communication.

References

Alempaki, D., Canic, E., Mullett, T. L., Skylark, W. J., Starmer, C., Stewart, N., & Tufano, F. (2019). Reexamining how utility and weighting functions get their shapes: A quasi-adversarial collaboration providing a new interpretation. *Management Science, 65*(10), 4841–4862. https://doi.org/10.1287/mnsc.2018.3170

Bateman, I., Kahneman, D., Munro, A., Starmer, C., & Sugden, R. (2005). Testing competing models of loss aversion: An adversarial collaboration. *Journal of Public Economics, 89*, 1561–1580.

Berman, J. S., & Reich, C.M. (2010). Investigator allegiance and the evaluation of psychotherapy outcome research, *European Journal of Psychotherapy and Counselling, 12*(1), 11–21. https://doi.org/10.1080/13642531003637775

Cowan, N., Belletier, C., Doherty, J. M., Jaroslawska, A. J., Rhodes, S., Forsberg, A., Naveh-Benjamin, M., Barrouillet, P., Camos, V., & Logie, R. H. (2020). How do scientific views change? Notes from an extended adversarial collaboration. *Perspectives on Psychological Science, 15*(4), 1011–1025.

DeRubeis, R. J., Hollon, S. D., Amsterdam, J. D., Shelton, R. C., Young, P. R., Salomon, R. M., et al. (2005). Cognitive therapy vs medications in the treatment of moderate to severe depression. *Archives of General Psychiatry, 62*, 409–416.

Doherty, J. M., Belletier, C., Rhodes, S., Jaroslawska, A. J., Barrouillet, P., Camos, V., Cowan, N., Neveh-Benjamin, M., & Logie, R. H. (2019). Dual-task costs in working memory: An adversarial collaboration. *Journal of Experimental Psychology: Learning, Memory, and Cognition, 45*, 1529–1551.

Finkel, E. J., Norton, M. I., Reis, H. T., Ariely, D., Caprariello, P. A., Eastwick, P. W., Frost, J. H., & Maniac, M. R. (2015). When does familiarity promote versus undermine interpersonal attraction? A proposed integrative model from erstwhile adversaries. *Perspectives on Psychological Science, 10*(1), 3–19.

Gilovich, T., Medvec, V. H., & Kahneman, D. (1998). Varieties of regret: A debate and partial resolution. *Psychological Review, 105*, 602–605.

Greenwald, A. G. (1975). Consequences of prejudice against the null hypothesis. *Psychological Bulletin, 82*, 1–20.

Heimberg, R. G., Liebowitz, M. R., Hope, D. A., Schneier, F. R., Holt, C. S., Welkowitz, L. A., et al. (1998). Cognitive behavioral group therapy vs phenelzine therapy for social phobia: 12-week outcome. *Archives of General Psychiatry, 55*, 1133–1141.

Hobson, H. (2019). Must replication attempts be battlegrounds? *Cortex, 113*, 355–356.

Hollon, S. D. (1999). Allegiance effects in treatment research: A commentary. *Clinical Psychology: Science and Practice, 6*, 107–112.

Kahneman, D., & Klein, G. (2009). Conditions for intuitive expertise: A failure to disagree. *American Psychologist, 64*, 515–526.

Kerr, N. L., Xiang, A., Hogg, M. A., & Zhang, J. (2018). Addressing replicability concerns via adversarial collaboration: Discovering hidden moderators of the minimal intergroup discrimination effect. *Journal of Experimental Social Psychology, 78*, 66–76.

Kerr, R. L. (1998). HARKing: Hypothesizing after the results are known. *Personality and Social Psychology Bulletin, 2*(3), 196–217. https://doi.org/10.1207/s15327957pspr0203_4

Kuhn, T. (1962). *The structure of scientific revolutions* (1st ed.). The University of Chicago Press.

Lakatos, I. (1970). Falsification and the methodology of scientific research programmes. In I. Lakatos & A. Musgrave (Eds.), *Criticism and the growth of knowledge* (pp. 91–195). Cambridge University Press.

Latham, G. P., Erez, M., & Locke, E. A. (1988). Resolving scientific disputes by the joint design of crucial experiments by the antagonists: Application to the Erez-Latham dispute regarding participation in goal setting. *Journal of Applied Psychology, 73*(4), 753–772.

Leykin, Y., & DeRubeis, R. J. (2009). Allegiance in psychotherapy outcome research: Separating association from bias. *Clinical Psychology: Science and Practice, 16*(1), 54–65.

Lindner, D. & Trible, M. (2020, October). Keep your friends close, your adversaries closer. *The Psychologist*, 34–35.

Lindner, D., Trible, M., Pilato, I., & Ferguson, C. J. (2020). Examining the effects of exposure to a sexualized female video game protagonist on women's body image. *Psychology of Popular Media, 9*(4), 553–560. https://doi.org/10.1037/ppm0000251

Matzke, D., Nieuwenhuis, S., van Rijn, H., Slagter, H. A., van der Molen, M. W., & Wagenmakers, E.-J. (2015). The effect of horizontal eye movements on free recall: A preregistered adversarial collaboration. *Journal of Experimental Psychology: General, 144*(1), e1–e15.

Mellers, B., Hertwig, R., & Kahneman, D. (2001). Do frequency representations eliminate conjunction effects? *Psychological Science, 12*, 269–275.

Mullen, B., Brown, R., & Smith, C. (1992). Ingroup bias as a function of salience, relevance, and status: An integration. *European Journal of Social Psychology, 22*, 103–122. https://doi.org/10.1002/ejsp.2420220202

Nickerson, R. S. (1998). Confirmation bias: A ubiquitous phenomenon in many guises. *Review of General Psychology, 2*, 175–220.

Nier, J. A., & Campbell, S. D. (2013). Two outsiders' view on feminism and evolutionary psychology: An opportune time for adversarial collaboration. *Sex Roles, 69*, 503–506. https://doi.org/10.1007/s11199-012-0154-2

Popper, K. R. (1974). *Conjectures and refutations: The growth of scientific knowledge* (5th ed.). Routledge and Kegan Paul.

Rakow, T., Thompson, V., Ball, L., & Markovits, H. (2014). Rationale and guidelines for empirical adversarial collaboration: A *Thinking & Reasoning* initiative. *Thinking & Reasoning, 21*(2), 167–175. https://doi.org/10.1080/13546783.2015.975405

Roese, N. J., & Vohs, K. D. (2012). *Hindsight bias. Perspectives on Psychological Science, 7*, 411–426.

Rosenthal, R. (1979). The file drawer problem and tolerance for null results. *Psychological Bulletin, 86*, 638–641.

Simmons, J. P., Nelson, L. D., & Simonsohn, U. (2011). False-Positive Psychology: Undisclosed flexibility in data collection and analysis allows presenting anything as significant. *Psychological Science, 22*, 1359–1366.

Thouin-Savard, M. (2017). Adversarial collaboration: How free and open debate leads to better transpersonal ideas (Editor's Introduction). *International Journal of Transpersonal Studies, 36*(2), iii–iv. https://doi.org/10.24972/ijts.2017.36.2.iii

Lancaster, B., & Friedman, H. (2017). Introduction to the special topic section: The Taylor-Hartelius debate on psychology and spirituality. *International Journal of Transpersonal Studies, 36*(2), 72–74. https://doi.org/10.24972/ijts.2017.36.2.72

Tversky, A., & Kahneman, D. (1983). Extensional versus intuitive reasoning: The conjunction fallacy in probability judgment. *Psychological Review, 90*, 293–315.

Van Dessel, P., Gawronski, B., Smith, C. T., & De Houwer, J. (2017). Mechanisms underlying approach-avoidance instruction effects on implicit evaluation: Results of a preregistered adversarial collaboration. *Journal of Experimental Social Psychology, 69*, 23–32.

van 't Veer, A. E., & Giner-Sorolla, R. (2016). Pre-registration in social psychology—A discussion and suggested template. *Journal of Experimental Social Psychology, 67*, 2–12. http://dx.doi.org/10.1016/j.jesp.2016.03.004

Chapter 17
Assessing and Improving Robustness of Psychological Research Findings in Four Steps

Michèle B. Nuijten

Abstract Increasing evidence indicates that many published findings in psychology may be overestimated or even false. An often-heard response to this "replication crisis" is to replicate more: replication studies should weed out false positives over time and increase the robustness of psychological science. However, replications take time and money – resources that are often scarce. In this chapter, I propose an efficient alternative strategy: a four-step robustness check that first focuses on verifying reported numbers through reanalysis before replicating studies in a new sample.

Keywords Robustness of psychological research findings · Four-step robustness check · Replication crisis

Introduction

The Replication Crisis

Around 2012, scientists started speaking of a "replication crisis" in psychology (Pashler & Harris, 2012; Pashler & Wagenmakers, 2012). A growing number of published psychological findings did not seem to hold up when the study was done again in a new sample (e.g., Chabris et al., 2012; Doyen et al., 2012; LeBel & Campbell, 2013; Matthews, 2012; Pashler et al., 2013). Since the 1950s, research consistently shows that over 90% of published psychology papers find support for their main hypothesis (Fanelli, 2010; Sterling, 1959; Sterling et al., 1995), whereas this is virtually impossible given the generally low statistical power in the field (Bakker et al., 2012; Francis et al., 2014). In other words, it seems that many published findings in psychology are too good to be true.

M. B. Nuijten (✉)
Tilburg University, Tilburg, Netherlands
e-mail: M.B.Nuijten@tilburguniversity.edu

W. O'Donohue et al. (eds.), *Avoiding Questionable Research Practices in Applied Psychology*, https://doi.org/10.1007/978-3-031-04968-2_17

A possible explanation for the excess of positive results in psychology is a combination of publication bias, publication pressure, and questionable research practices (QRPs). Publication bias occurs when articles with statistically significant results have a higher probability of being published than articles with nonsignificant results (Greenwald, 1975). When researchers experience the pressure to publish (and they often do; van Dalen & Henkens, 2012; Tijdink et al., 2014), publication bias provides a direct incentive to only report positive, significant results. This may cause researchers to (consciously or unconsciously) exploit the inherent flexibility in data collection, processing, and analysis until they find the desired result. When only the "successful" strategy is then reported, one could speak of QRPs (Gelman & Loken, 2013; Kerr, 1998; Simmons et al., 2011). Evidence from surveys and from comparisons of research plans with the resulting publications seemed to indicate a high prevalence of such QRPs in the psychological literature (Franco et al., 2016; Agnoli et al., 2017; John et al., 2012; but see Fiedler & Schwarz, 2016).

Suggested Solution: More Replications

One reaction to the replication crisis is the call to perform more replication studies. See, for example, the suggestion to require researchers to perform replication studies in their research area in proportion to the number of original studies they conduct (LeBel, 2015) or to require undergraduate, graduate, or PhD students to perform replication studies (Frank & Saxe, 2012; Kochari & Ostarek, 2018). Journals have also taken up this sentiment by actively encouraging authors to publish replication studies (Registered Replication Reports; Association for Psychological Science, n.d.; Jonas et al., 2017; Nosek & Lakens, 2014), and even funders have set funds aside specifically for replication research (The Dutch Research Council, n.d.). We have also seen an increase in the number of large-scale, multi-lab replication attempts, in which sometimes dozens of labs across the world set out to replicate the same study (e.g., Alogna et al., 2014; Klein et al., 2014; ManyBabies Consortium, 2020; Moshontz et al., 2018) in order to give a (somewhat) definitive answer to a certain research question.

One of the most well-known examples of a multi-lab replication project is the Reproducibility Project: Psychology, a project with 270 contributing authors led by Prof. Brian Nosek (Open Science Collaboration, 2012, 2015). The goal of the project was to estimate the replicability of psychology by systematically replicating a selection of 100 psychology studies published in prominent journals. They found that the mean effect size in the replication studies was much lower than the mean effect size of the original studies (replication: $r = 0.197$, SD = 0.257; original: $r = 0.403$, SD = 0.188). Furthermore, where the original studies found statistically significant results in 97% of the cases, only 36% of the replications did. There is some debate about how exactly these results should be interpreted (Anderson et al., 2016; see, e.g., Bavel et al., 2016; Etz & Vandekerckhove, 2016; Gilbert et al.,

2016), but the results are generally taken as a sign that the replicability rate in psychology might be low.

The call to replicate more is perfectly sensible and in line with the notion that replication is a cornerstone of science (Lakatos & Musgrave, 1970; Meehl, 1990) . Already in introductory research methods classes, it is generally taught that a single study cannot provide definitive answers to a research question (Morling, 2020, p. 14). Instead, multiple studies need to convincingly show an effect before we continue to build on it any further. Unfortunately, the psychological literature does not seem to reflect this notion. Even though the literature contains many conceptual replications that investigate the boundaries and generalizability of a theory (Neuliep & Crandall, 1993), direct replication studies that are mainly aimed at checking whether an effect is robust in the first place seem to be rare (Makel et al., 2012). Without direct replications, researchers would have to rely on individual studies to build their work upon. Unfortunately, the replication crisis has shown us that the findings in many of these individual studies may be false positives. This means that many researchers may be trying to build research lines based on dead ends. By encouraging more replication studies, we could ideally weed out many of the false positives, and the foundation to build upon would be stronger.

Even though encouraging replications could in theory help in assessing and improving the robustness of published findings, there is a significant disadvantage: performing a replication study takes considerable time and money – resources that are usually scarce. In this chapter, I would like to suggest a more efficient way to assess whether a published result is robust through a *four-step robustness check* that first focuses on verification of reported numbers through reanalysis before replicating a study in a new sample.

Statistical Reproducibility Is a Prerequisite for Replication

When replicating a study, researchers are often interested in comparing the results of their replication with those of the original study. There is no consensus on the best way to decide whether or not the original finding has been replicated (Open Science Collaboration, 2015; Zwaan et al., 2017), but in general, the main statistical results from the original study are compared to the main statistical results from the replication. If the two sets of results are similar enough, one could conclude that the original study has successfully been replicated, and if the two sets of results lie too far apart, one could conclude that the original study was not successfully replicated. One key point to keep in mind here is that, for this comparison to be meaningful, the reported numbers have to be *correct*: the reported results should not contain typos, calculation errors, or other mistakes. In other words, the reported results should be *statistically reproducible*: reanalysis of the original data following the reported procedures should result in the same numbers as those reported in the paper (Nuijten et al., 2018).

Roughly, there can be two reasons why a result is not statistically reproducible (Nosek et al., 2021). First, a *process reproducibility failure* occurs when it is not possible to repeat the steps of the original analysis, for example, because of unavailable data or code, unclear description of the analysis steps, or unavailable software or tools. In psychology, raw data have been notoriously unavailable (Wicherts et al., 2006), although there has been some improvement in recent years (Hardwicke et al., 2018; Kidwell et al., 2016; Nuijten et al., 2017a). Furthermore, even when raw data are available, there are often insufficient details reported to redo the original analysis (e.g., Kidwell et al., 2016). For example, the data are insufficiently documented (e.g., instead of informative variable names, variables still have SPSS' default labels VAR0001, VAR0002 etc.), or the paper only states that "an ANOVA" has been done, without elaborating on any data preprocessing steps (e.g., the removal of outliers) or specific details about the analysis.

Second, an *outcome reproducibility failure* occurs when the original analysis can be repeated but leads to a different result than the one that is reported (Nosek et al., 2021). In general, it is plausible to assume that reanalyzing the same data according to the same methods leads to the same results. Unfortunately, this is not always the case. For example, two recent studies reran the original analyses on the original data of a set of psychology studies and compared the outcomes with the reported results. Both studies found numerical discrepancies in over 60% of the reanalyzed studies (Hardwicke et al., 2018, 2020). Furthermore, evidence from over 16,000 psychology papers showed that roughly half of the papers contained at least one p-value that was not consistent with the reported test statistic and degrees of freedom. In roughly one in eight articles, the recomputed p-value was not significant, whereas the reported p-value was, or vice versa (Nuijten et al., 2016).

Statistical reproducibility of results is a basic, necessary requirement for scientific quality (Chambers, 2020; Peng, 2011). If a reported result cannot be linked back to the underlying data, it is extremely difficult (if not impossible) to meaningfully interpret that result. As such, if an investigator wants to know whether a certain result is robust or not, that investigator may not need to perform a full replication study: if the result is not statistically reproducible, it is not robust. I would therefore like to argue that verifying reported results should be the first step in assessing the robustness of a result (see also LeBel et al., 2018; Nuijten et al., 2018; Stark, 2018). To facilitate such statistical reproducibility checks, I suggest a practical four-step approach.

Checking Robustness of a Finding in Four Steps

Step 1: Check for Internal Inconsistencies in Reported Statistics

A first step when checking if a result is robust is to check if the reported statistics are internally consistent. To complete this step, you do not need access to the raw data; you only need the reported statistical results. A statistical reporting inconsistency occurs when a set of numbers that belong together does not match. For example, consider the following sentence: "70% of patients recovered within three months after the first diagnosis (65/100)". Only by looking at the reported numbers in this sentence, we can already see that something is wrong: 65/100 is 65%, not 70%. At this point, it is unclear *which* of the reported numbers is incorrect, but what *is* clear is that this set of numbers presents an impossible combination and can therefore not have come from the underlying raw data. In other words: this result is not statistically reproducible.

Internal inconsistencies could be detected in a wide range of statistics. Other than percentages that have to match the accompanying fractions, examples are:

- Reported total sample sizes should match the subgroup sizes.
- Reported effect estimates should fall within the bounds of the accompanying confidence interval.
- Reported odds ratios should match the accompanying frequency table.
- Reported sensitivity of a diagnostic test should match the true/false positive and true/false negative rates.
- Reported t- and F-values should match the reported means and standard deviations.
- Reported p-values should match the test statistic and degrees of freedom.

Currently, several tools and algorithms are being developed to automatically (or semi-automatically) detect statistical reporting inconsistencies. One of these tools is *statcheck*: a free R package (Epskamp & Nuijten, 2014) and accompanying web app (http://statcheck.io; Rife et al., 2016) that automatically extracts Null Hypothesis Significant Tests (NHST) results from articles and recomputes p-values based on the reported test statistic and degrees of freedom. For example, say that an article reports the following sentence: "*We found that the treatment group scored significantly higher on well-being than the control group, $t(28) = 1.46$, $p < .05$.*" If you scan this article with *statcheck*, it would recognize the reported t-test and use the reported test statistic (1.46) and degrees of freedom (28) to recompute the p-value. In this case, the recomputed p-value would be 0.155. This is not consistent with the reported p-value of < .05. What is more in this case is that the reported result would be flagged as a *decision inconsistency* (also sometimes called a *gross inconsistency* or *gross error*): based on the recomputed p-value, one would draw a different conclusion (i.e., the difference between groups is *not* significant). *Statcheck* currently only recognizes statistics reported in APA style (American Psychological Association, 2019), which makes the tool primarily useful for psychology papers. Also please see Nuijten et al. (2017b) for a full analysis of *statcheck*'s accuracy in spotting (decision) inconsistencies.

Another example of a tool to spot statistical reporting inconsistencies is the tool *GRIM* (granularity-related inconsistency of means; Brown & Heathers, 2017). *GRIM* can spot whether reported means that are based on integer data (e.g., from Likert-type scales) are possible in combination with a certain sample size and total number of items. Other examples include an algorithm to check whether reported effect sizes match their confidence intervals and p-values (Georgescu & Wren,

2018) and a semiautomated protocol to assess inconsistencies in a wide range of statistics (van Aert et al., 2021).

In sum, regardless of whether one can automate the process of looking for inconsistencies in statistical reporting or not, the abovementioned tools offer relatively quick procedures that do not require access to anything else but the paper itself. This makes it an efficient first "sanity check" in assessing whether or not a reported result is robust.

Step 2: Reanalysis of Original Data

A second step when checking if a result is robust is to reanalyze the original data according to the reported procedure to see if one can find the same results as reported. As opposed to step 1, one now does need access to the original data and information about the original analysis.

There is not one way to approach a reanalysis, but there are some general steps one could follow. First, determine whether the raw data underlying the finding of interest are available. A quick way to do so is to search for a data availability statement (a standardized short statement about whether data are available, requested by an increasing number of journals, incl. e.g., *PLOS* journals) or an Open Data Badge (a badge printed at the top of the paper that signals that data are available, used by an increasing number of psychology journals, incl. APS and APA journals; Center for Open Science, n.d.-a). If data are stated to be available, it is unfortunately still not a guarantee that they actually are. For example, 25–30% of articles published in *Frontiers in Psychology* and in several *PLOS* journals that stated that data were available did not contain or link to the raw data (Chambers, 2017, p. 86; Nuijten et al., 2017a). Of course, it is also possible to contact the original authors to ask for the data and other relevant materials, although historically, this is often not very successful (Wicherts et al., 2006).

Next, download the data and try to open the file. Ideally, raw data files are in a (relatively) standard format, such as .csv, .txt, .xlsx, or .sav, and one can open the data without needing expensive software. Once one can open the data, it is important to skim through the file to see whether the data seem to be understandable and complete (i.e., are all variables mentioned in the paper also mentioned in the data? Does the number of rows correspond to the reported number of participants?). This step includes determining whether the authors also shared a codebook that (among other things) explains all variables in the data and their values, how missing data are coded, and whether any data preprocessing steps have already taken place (e.g., reverse coding of contraindicative items).

Beside the data, one also needs as much information as possible about the original data analysis. Therefore, a good next step, once one has access to the data and understands the file, is to determine whether the authors also shared their analysis script. An analysis script is preferable to the analysis description in the methods or results section of the paper, because a script usually contains more explicit and detailed information about the subsequent steps in the analysis. If an analysis script is not available, extract as much detail about the data preprocessing and analysis from the paper (and supplementary files) as possible.

Once an investigator has all the available data and information about the analysis, the reanalysis phase itself can begin. Here, the investigator has to decide which reported values you want to try and reproduce: this could be all the reported numbers in the paper or just some key values related to the main conclusion, or something in between. In this reanalysis phase, follow the original analysis steps themselves as closely as possible, where possible using the same software, version, and operating system.

How much time it takes to reanalyze original data according to the original procedure depends for a large part on the complexity of the data and the analysis and on the clarity with which the original procedure was reported. For example, in one reanalysis project, the researchers spent between 1 and 30 hours (median = 7) on each reproducibility check (Hardwicke et al., 2020). Roughly speaking, a reanalysis will likely take more time than a check for statistical reporting inconsistencies in the paper, but less time than a full replication study.

Step 3: Sensitivity Checks

Even if one were able to reproduce the same results through reanalysis of the original data, there is no guarantee that the result is robust. Therefore, a third step when checking if a result is robust is to reanalyze the original data using slightly different but still justifiable preprocessing and analysis steps than the ones reported. This can shed light on whether the result is robust to alternative choices.

Say that an investigator followed the reported analytical procedures and that she was able to reproduce the main result from the original data. However, it could be the case that the original authors removed an outlier in their analysis. It may also be the case that if she does not remove this outlier, the result changes substantively. Similarly, it may happen that the removal of a seemingly arbitrary covariate can make the result disappear. And what if other ways of constructing the final variables of interest lead to different results? There are many choices involved in data preprocessing and statistical analysis, and if only a very specific combination of analytical steps leads to a significant result, one may question its robustness (Gelman & Loken, 2013).

It is difficult to provide general instructions for how to do such sensitivity checks, because the set of justifiable analytical choices is highly dependent on the specific research question, type of study, available variables, and other contextual factors. That said, there are several general questions one could ask that could guide the sensitivity checks (see also Patel et al., 2015; Steegen et al., 2016).

First, one could consider the data preprocessing steps that led to the final data on which the analyses were performed. For example, if the original authors used questionnaire data, how did they summarize the scores on individual items to a score on the construct of interest? In case not all items were included, what happens if one does include them all? What if instead of calculating a sum score over the items, one calculates a factor score? If the authors turned a continuous variable into a categorical one (e.g., classifying BMI into underweight, normal weight, overweight, or obese), would other cutoffs to determine the categories also be justifiable? Yet other questions related to data preprocessing could concern inclusion/exclusion criteria; were participants excluded from the final analysis? What happens if one does include them or use slightly different exclusion criteria?

A second type of question one could ask is to what extent a finding is sensitive to different choices in the analysis itself. First, check if the authors did the correct analysis in the first place (e.g., including an interaction effect in the analysis, instead of erroneously comparing the p-values of two different effects; Nieuwenhuis et al., 2011). But also within a correct analysis, different choices can be made. This could include questions, such as: how did the authors deal with missing values, and would another strategy be justifiable as well? Which (if any) covariates were included in the analysis, and would another selection be equally justifiable? In the case of frequentist hypothesis testing, one could consider adding/removing corrections for multiple testing, trying other cutoffs (such as the often arbitrary a significance level of $p < .05$), choices for one-tailed testing, or comparisons with the outcome of Bayesian hypothesis testing.

Step 4: Replication in a New Sample

A fourth and final step when checking if a result is robust could be to perform a replication study in a new sample (Nuijten et al., 2018). Replications come in many shapes and sizes, but they are usually classified along a continuum ranging from a direct replication (also known as exact replication or close replication) to a conceptual replication (LeBel et al., 2017). In a direct replication, the methods of the original study are followed as closely as possible. The results of a direct replication (or preferably multiple direct replications) can be used to assess the *reliability* of a result: will the effect (or lack thereof) show up again if we repeat a study? Conceptual replications, on the other hand, aim to test the *generalizability* of a result by testing the original hypothesis using different methods than the original study (e.g., in a different setting, in a different population, or using different operationalizations). Arguably, it makes most sense to start with direct replications to first rule out (at least to some extent) that an original result is not a false positive or false negative, before setting out to check the generalizability of a result that might turn out to be a fluke (Zwaan et al., 2017).

A lot has been written about ways to conduct (or evaluate) a good direct replication study (e.g., Brandt et al., 2014; LeBel et al., 2019), and in this chapter, I mainly want to focus on the less-discussed statistical reproducibility checks. However, there are some general guidelines when doing a direct replication study that one can take into account. Generally speaking, a direct replication aims to study the same effect as the original study and should follow the original study's procedures as closely as possible, barring some inevitable differences (e.g., the actual participants or the point in time that the studies take place). It is also advised to have high statistical power, which often means significantly increasing the sample size as compared to the original study (Anderson & Maxwell, 2017; Open Science Collaboration, 2015). Furthermore, several methods have been proposed to interpret the results of the replication study compared to the original study, including subjective evaluation, comparing *p*-values, effect sizes, confidence intervals, Bayes factors, and more (Open Science Collaboration, 2015; Simonsohn, 2015; Verhagen & Wagenmakers, 2014; Zwaan et al., 2017). Note that several recommendations to improve the quality of replications also hold for original studies, such as transparent reporting, high statistical power, robust statistical methods, and sharing data and materials (Benjamin et al., 2018; Brandt et al., 2014; Lakens et al., 2018; Lakens & Evers, 2014; LeBel et al., 2019; Munafò et al., 2017; Nosek et al., 2012; Simmons et al., 2011).

Interpreting the Outcomes of Reproducibility Checks

Strictly speaking, *any* failure to reproduce a reported result in any of the first three steps of the robustness check would allow one to conclude that a result – as reported – is not robust. However, to what extent this is problematic for the overall conclusion and whether it is still useful to follow subsequent steps of the robustness checks is for a large part context-dependent. For example, it matters (a) if the original authors can help clear up any discrepancies, (b) how big any discrepancies are between the reported and recalculated numbers, (c) how important the result is for the overall conclusion, (d) if a failure to reproduce was due to a process or outcome reproducibility failure, and (e) the overall goal, for which one is doing a robustness check.

Contacting the Original Authors

When encountering a reproducibility failure (process or outcome), one can consider contacting the original authors to ask for help and/or clarification. In previous reanalysis studies, original authors were often able to help resolve reproducibility issues (Hardwicke et al., 2018, 2020). Even though it is a positive sign that they were helpful in resolving these specific issues, it is far from ideal if reported results can only be reproduced with the help of the original authors. Especially if one wants to assess the robustness and reproducibility of results published several years ago, the authors may not be able to help anymore: they may not have access to the data or scripts themselves anymore, or it may not even be possible to contact them at all. Full statistical reproducibility can only be achieved when the published paper and materials contain sufficient information to independently and successfully redo the original analysis.

Size and Importance of Analytical Discrepancies

The size and context of any outcome reproducibility failure matter. In most cases, a rounding error in the fourth decimal of a *p*-value mentioned in a footnote is probably less consequential than a major discrepancy in the reported and recalculated key outcome of a paper.

There are several ways to judge the size and importance of a reproducibility failure. One option is to look at the difference between the reported and recalculated numbers. For example, if a reported correlation is .80, a recalculated correlation of .60 presents a larger discrepancy than a recalculated correlation of .78 (see e.g., Petrocelli et al., 2013). Instead of using absolute differences, you could also look at relative differences, expressed in percentages. If we stick to the same example, the percentage error in the first scenario is equal to $(|.60 - .80|) / .80 * 100\% = 25\%$, and in the second scenario, it is equal to $(|.78 - .80|) / .80 * 100\% = 2.5\%$. Earlier research defined a percentage error larger than 10% as a major numerical error (Hardwicke et al., 2018, 2020).

Another option to classify the size and/or importance of a discrepancy is to look at the statistical decision based on the reported numbers. For example, most of the research in psychology retains a significance level of .05, meaning that a *p*-value smaller than .05 is considered statistically significant. Several studies that recalculated *p*-values based on the reported test statistic and degrees of freedom used this cutoff to distinguish between inconsistencies and gross inconsistencies (or errors and decision errors, respectively). If the recalculated *p*-value did not match the reported one, but both were on the same side of the .05 threshold, e.g., reported $p = .03$ vs. recalculated $p = .04$, this was classified as an inconsistency. If a reported *p*-value was statistically significant, but the recalculated *p*-value was not (or vice versa), e.g., reported $p = .03$ vs. recalculated $p = .07$, this was classified as a gross inconsistency (see, e.g., Bakker & Wicherts, 2011; Nuijten et al., 2016).

The location of a statistical result and its importance for the main conclusion can also help in determining the seriousness of a reproducibility failure. With respect to location, results reported in the abstract of a paper can be assumed to be of more importance for a conclusion than results in a footnote or appendix (Georgescu & Wren, 2018).

Finally, it is also possible to look at the implications of a discrepancy in more depth. In one reanalysis study, the authors concluded that for the studies that contained reproducibility problems (errors in the code and small discrepancies in the number of participants included), the overall conclusions did not change (Naudet et al., 2018).

Process Reproducibility Failure

When interpreting a reproducibility failure in one of the first three steps above (checking internal inconsistencies, reanalysis, or sensitivity analyses), it matters whether one encountered a process or outcome reproducibility failure. Remember that a process reproducibility failure occurs when not all steps could be followed to redo the original analysis, whereas an outcome reproducibility failure occurs when the outcome of the reanalysis shows a different result than the one originally reported (Nosek et al., 2021). An outcome reproducibility failure is a more clear-cut outcome than a process reproducibility failure: in the former case, one can conclude that a reported result is not robust, whereas in the latter case, one cannot assess the robustness of a result at all.

In case of an outcome reproducibility failure, the reported numbers are not in line with the underlying data and reported analytical method. In such a case, trust in the reported results and possibly also the conclusion decreases. However, in case of a process reproducibility failure, it is not possible to verify the reported results. This is problematic, because this means that one just has to "trust" that all reported numbers are correct when interpreting the conclusion, and unfortunately, we know from previous research that this may not be the case (Hardwicke et al., 2020; Nuijten et al., 2016). Furthermore, if the process reproducibility failure is caused by a lack of access to the raw data (as opposed to unclear analytical steps), it is also not possible to assess to what extent a result is sensitive to alternative analytical choices. If earlier steps of the four-step robustness check cannot be completed, it may be risky to proceed to step 4 and perform a replication anyway. After all, it is hard to meaningfully compare replication results to the original results, if one does not know if the numbers in the original study are correct in the first place. If and how a process reproducibility failure should affect one's decision whether or not to do a replication study will depend on the reason why one wanted to assess the robustness of a finding in the first place.

Reason to Assess Robustness

The four-step robustness check could help you decide whether it is worth investing the time and money in performing a replication study. In some cases, one may conclude that if the numbers in the original paper already do not add up, conducting a replication would not be useful. To what extent this holds depends on one's reason for assessing the robustness of an original result.

One's goal could be to say something about the robustness of a specific original finding (as opposed to a phenomenon in general). This is often the goal in multi-lab replication projects, such as the Reproducibility Project: Psychology (Open Science Collaboration, 2015). In these cases, the replication studies are often high-powered and pre-registered, arguably enhancing their evidential value compared to the original study they are replicating (Nosek et al., 2021). They therefore attempt to provide a more or less definitive conclusion about the robustness of the original result.

The strict quality controls in these large-scale replication projects can make them very costly, so it is important that their results can be meaningfully compared to the original study. Here, I would argue that following the steps of the four-step robustness check can be very valuable to first check the statistical reproducibility of the original study and avoid "wasting" resources on a large-scale replication project.

Contrarily, one's goal could also be to get a step closer towards learning the truth about the underlying phenomenon studied in a particular paper. If one then encounters a reproducibility failure in the original study, it does not necessarily have to mean that it is useless to do a replication. It can still be valuable to gather more empirical evidence to answer the original question. However, in such a case, one should be careful when comparing the replication results to the original results. More specifically, one may even want to consider discarding the original study entirely, depending on the severity of the reproducibility failure, and only take the results of the replication study into account to answer the research question.

Successful Robustness Check

The sections above are mainly considering scenarios in which the four-step robustness check fails. However, it could, of course, also be the case that all steps can be completed successfully. First and foremost: this is good news. It means that the reported results are consistently reported, can be traced back to the underlying data, are robust to different analytical choices, and are replicable in new samples. However, passing the four-step robustness check is not sufficient to definitively conclude that a result is robust. As with anything in science, it is hard to draw such a black-and-white conclusion at all. Instead, it makes more sense to talk about the *degree* of robustness or the *strength of the evidence* that a result is robust.

Some potential problems remain unchecked after following the four-step robustness check. First, the four-step check assumes that the raw data are correct. In other words, the procedure does not take into account errors (or fraud) in data entry. Similarly, these steps do not say anything about the theoretical or methodological quality of a study. For example, if a study uses a biased design and non-validated questionnaires to measure the main constructs, it could still pass all steps in the four-step robustness check. Finally, due to sampling error and the probabilistic nature of psychological research, it is possible that two studies find the same results (i.e., the study is successfully replicated), but in both cases, the result is a false positive or false negative.

In the end, answering a research question will likely require a long research line comprising of multiple independent studies (incl. direct and conceptual replications) consisting of severe tests (Mayo, 2018), and even then the answer will likely remain tentative (Popper, 1959; O'Donohue, 2021). The decision whether to invest in such a research line as opposed to another, however, can be informed by the success rates of previous robustness checks.

Improving Robustness in Your Own Manuscripts

The framework of the four-step robustness check is not only useful to *assess* robustness but also to *improve* robustness of a result. Below, I will outline some concrete actions researchers can take that are in line with the logic of the four-step robustness check.

Step 1: Report All Relevant Statistical Information

It would greatly facilitate robustness checks if relevant statistics were reported in full and with sufficient detail to be able to assess their internal consistency. To illustrate, consider the following conclusion: "All planned contrasts showed support for our hypotheses (all $ps < .05$)." There is insufficient information here to check whether the reported statistical results are internally consistent. In addition, in this specific case, reporting only p-values (and not even the exact p-values) also omits important information about effect size and uncertainty in the estimate.

Luckily, many psychology journals require that authors to follow the reporting guidelines of the American Psychological Association (APA; American Psychological Association, 2019), which contains specific guidelines on how to report statistical results. For example, they require the following information concerning inferential statistics: "*Results of all inferential tests conducted, including exact p-values if null hypothesis statistical testing (NHST) methods were employed, including reporting the minimally sufficient set of statistics (e.g., dfs, mean square [MS] effect, MSerror) needed to construct the tests*" (Appelbaum et al., 2018; Table 1). In addition, effect size estimates, confidence intervals, and other relevant details concerning data preprocessing and analysis should be reported.

Step 2: Provide Raw Data and Analysis Scripts

Sharing data has many benefits (Wicherts, 2013; Wicherts et al., 2012). If raw data are available, others could reanalyze the data to detect potential errors, check the robustness of the conclusions to different analytical choices, or even answer entirely new research questions. Ideally, not only the data but also the original analysis scripts are shared. Methods and results sections in scientific articles often do not contain sufficient detail to rerun an analysis exactly according to the original procedure, whereas analysis code does.

Ideally, data are fully and freely available and well-documented. In other words, the data should be shared according to the FAIR principles (Findable, Accessible, Interoperable, Reusable; Wilkinson et al., 2016). In order to achieve this, authors should share not only the data but also a detailed codebook that includes information on the variables and other important metadata, such as where and when the data were collected and under which license the data are shared (for instructions on how to do this, see, e.g., Horstmann et al., 2020; Klein et al., 2018; Stodden, 2010). Finally, it is important to take privacy legislation into account and protect the confidentiality and/or anonymity of the participants.

It is also important to provide an explicit and detailed explanation of the data preprocessing and analysis steps, ideally (again) in the form of an analysis script. Even better would be to share the data and analysis script in a "container" that allows others to rerun the analysis using the same software and operating system as used in the original study (see e.g., Klein et al., 2018).

There can be very good reasons why sharing data is not possible. For example, one may not be allowed to share data because of privacy reasons or because the data are not yours. In such cases, there may be intermediate solutions that *are* possible. It may, for instance, be possible to remove identifying data and share only the anonymized part of the data. Or it may be possible to share data with someone else as long as certain agreements not to share the data any further are signed. A more advanced solution includes simulating data that have the same properties as the real data (see also Sweeney, 2002). Running analyses on the simulated data should lead to the same general conclusions as the ones reported (although here a one-on-one comparison of the specific numbers may be difficult). If this is not possible, sharing only part of the data or even only the analysis script without the accompanying data is useful. In sum, when considering the possibilities of sharing data, I encourage authors to focus on what *is* possible instead of what is not (see also Klein et al., 2018).

Step 3: Perform Sensitivity Analyses

Step 3 in *assessing* robustness entailed running alternative, justifiable analyses to see whether a reported result would still hold up (see above). Such alternative analyses are also known as sensitivity analyses: how *sensitive* is the result to different analytical approaches? In some scientific fields, running sensitivity analyses is a standard part of the research process (e.g., in economics). In psychology, however, this is not yet the case. There are roughly two options to fulfill this step. The first one is to actually perform relevant and justifiable alternative data preprocessing steps and statistical analyses oneself and, importantly, to report *all* outcomes. Such a multiverse analysis can give insight in how easily one's effect "breaks" under different circumstances and, conversely, how robust it is (Silberzahn et al., 2018; Simonsohn et al., 2019; Steegen et al., 2016). The guidelines in Step 3 of *assessing* robustness can also be used to list reasonable alternative preprocessing and analytical choices for your own case.

A second option is to explicitly state that no sensitivity analyses have taken place, for example, when an investigator is confident that one's own analysis is the only sensible analysis to perform. It can also help readers evaluate the analyses that the investigator conducted and conclusions that the investigator reached if the investigator clearly justifies certain analytical choices (e.g., why include a certain covariate but not another?). Such a strategy gains in strength when the analytic plan was preregistered. In a preregistration, the hypotheses, methods, and analysis plans are publicly registered *before* any data collection has taken place. It ensures a clear division between planned, confirmatory analyses and ad hoc exploratory analyses that may have a higher chance of finding a false positive (Kerr, 1998; Munafò et al., 2017; Simmons et al., 2011; Wagenmakers et al., 2012).

Whether an investigator performs sensitivity analyses or not, being transparent and explicit about analytical choices and the reasoning behind them will help readers to evaluate the results and guide them in any reanalysis they might wish to do.

Step 4: Share Study Materials

To facilitate replication of their studies, it is important for researchers to share as many study materials as possible. In theory, a Methods section in a paper should contain sufficient detail to allow for a replication of the study, but in practice, this is often not feasible. In order for other researchers to conduct a direct replication, they need to know the exact instruments that were used (e.g., the specific questionnaire), the exact procedure followed (What were the instructions that participants received? In what kind of setting did the study take place?), the exact population that the tested sample was drawn from, etc. This level of detail is usually not accepted in a manuscript (nor beneficial for the readability of a paper), but that does not mean that this information cannot be shared at all. I encourage authors to create extensive supplemental materials for their studies, in which the specific instruments, stimuli, procedural videos, and additional methodological details are shared.

Concluding Remarks

In this chapter, I have argued for a four-step robustness check to both assess and improve robustness of psychological findings by first focusing on verifying reported numbers before replicating in a new sample. Statistical reproducibility is a necessary requirement for scientific quality and deserves a place in the spotlight in the current discussions on how to improve psychological science.

Making the four-step robustness check common practice would require both bottom-up and top-down actions. Researchers themselves can use the steps described in this chapter to improve the robustness of their own work. This is in the best interest of science but can also have direct benefits for the scientists themselves. For example, there is evidence that sharing data is associated with an increased citation rate (Christensen et al., 2019; Piwowar et al., 2007). Researchers can also play a role in studying whether and how the four-step robustness check and other interventions affect the robustness of practices and results, ideally by performing randomized controlled trials. With respect to top-down actions: several journals already perform checks for internal consistency of submitted manuscripts (e.g., *Psychological Science* and *the Journal of Experimental Social Psychology* use the tool *statcheck* to scan submitted manuscripts for inconsistent *p*-values). Furthermore, an increasing number of journals require open data (see e.g., the current signatories to the Transparency and Openness Promotion Guidelines; Center for Open Science, n.d.-b; Chambers, 2018) and accept new article formats focused on rigor and transparency, including registered reports (Chambers, 2013), registered replication reports (Association for Psychological Science, n.d.), or verification reports (Chambers, 2020).

It is important to keep in mind that while statistical reproducibility is *necessary*, it is not *sufficient* for robustness. When determining and/or improving the robustness of a finding, other factors play a role as well. These factors include (but are not limited to) strong theory (Eronen & Bringmann, 2021), valid measurement (Flake & Fried, 2020), high statistical power (Button et al., 2013), robust statistics (Benjamin et al., 2018; Cumming, 2014; Marsman & Wagenmakers, 2017), severe tests (Mayo, 2018), and transparent reporting (Aczel et al., 2020).

Improving robustness of scientific results is complex and hard work. That said, the scores of initiatives aimed at improving psychological science we have seen in recent years stem hopeful. I would like to close this chapter by encouraging researchers not to become overwhelmed by all these initiatives and practices. If you pick one initiative at a time (e.g., scan a paper with *statcheck*, share your data, or try out a Bayesian analysis in addition to a traditional frequentist analysis), the robustness of our field will improve. One step at a time.

Acknowledgements The preparation of this chapter was supported by a Veni grant (no. 11507) from the Dutch Research Council.

References

Aczel, B., Szaszi, B., Sarafoglou, A., Kekecs, Z., Kucharský, Š., Benjamin, D., Chambers, C. D., Fisher, A., Gelman, A., Gernsbacher, M. A., Ioannidis, J. P. A., Johnson, E., Jonas, K., Kousta, S., Lilienfeld, S. O., Lindsay, D. S., Morey, C. C., Munafò, M., Newell, B. R., … Wagenmakers, E.-J. (2020). A consensus-based transparency checklist. *Nature Human Behaviour, 4*(1), 4–6. https://doi.org/10.1038/s41562-019-0772-6

Agnoli, F., Wicherts, J. M., Veldkamp, C. L. S., Albiero, P., & Cubelli, R. (2017). Questionable research practices among Italian research psychologists. *PLoS One, 12*(3), e0172792. https://doi.org/10.1371/journal.pone.0172792

Alogna, V. K., Attaya, M. K., Aucoin, P., Bahník, Š., Birch, S., Birt, A. R., Bornstein, B. H., Bouwmeester, S., Brandimonte, M. A., Brown, C., Buswell, K., Carlson, C., Carlson, M., Chu, S., Cislak, A., Colarusso, M., Colloff, M. F., Dellapaolera, K. S., Delvenne, J.-F., … Zwaan, R. A. (2014). Registered replication report: Schooler and Engstler-Schooler (1990). *Perspectives on Psychological Science, 9*(5), 556–578. https://doi.org/10.1177/1745691614545653

American Psychological Association. (2019). *Publication manual of the American Psychological Association* (7th ed.). American Psychological Association.

Anderson, C. J., Bahník, Š., Barnett-Cowan, M., Bosco, F. A., Chandler, J., Chartier, C. R., Cheung, F., Christopherson, C. D., Cordes, A., Cremata, E. J., Penna, N. D., Estel, V., Fedor, A., Fitneva, S. A., Frank, M. C., Grange, J. A., Hartshorne, J. K., Hasselman, F., Henninger, F., ... Zuni, K. (2016). Response to comment on "Estimating the reproducibility of psychological science". *Science, 351*(6277), 1037–1037. https://doi.org/10.1126/science.aad9163

Anderson, S. F., & Maxwell, S. E. (2017). Addressing the "replication crisis": Using original studies to design replication studies with appropriate statistical power. *Multivariate Behavioral Research, 52*(3), 305–324. https://doi.org/10.1080/00273171.2017.1289361

Appelbaum, M., Cooper, H., Kline, R. B., Mayo-Wilson, E., Nezu, A. M., & Rao, S. M. (2018). Journal article reporting standards for quantitative research in psychology: The APA Publications and Communications Board task force report. *American Psychologist, 73*(1), 3. https://doi.org/10.1037/amp0000191

Association for Psychological Science. (n.d.). *Registered replication reports*. Association for Psychological Science – APS. Retrieved 27 Feb 2021, from https://www.psychologicalscience.org/publications/replication

Bakker, M., van Dijk, A., & Wicherts, J. M. (2012). The rules of the game called psychological science. *Perspectives on Psychological Science, 7*(6), 543–554. https://doi.org/10.1177/1745691612459060

Bakker, M., & Wicherts, J. M. (2011). The (mis)reporting of statistical results in psychology journals. *Behavior Research Methods, 43*(3), 666–678. https://doi.org/10.3758/s13428-011-0089-5

Bavel, J. J. V., Mende-Siedlecki, P., Brady, W. J., & Reinero, D. A. (2016). Contextual sensitivity in scientific reproducibility. *Proceedings of the National Academy of Sciences, 113*(23), 6454–6459. https://doi.org/10.1073/pnas.1521897113

Benjamin, D. J., Berger, J. O., Johannesson, M., Nosek, B. A., Wagenmakers, E.-J., Berk, R., Bollen, K. A., Brembs, B., Brown, L., Camerer, C., Cesarini, D., Chambers, C. D., Clyde, M., Cook, T. D., De Boeck, P., Dienes, Z., Dreber, A., Easwaran, K., Efferson, C., ... Johnson, V. E. (2018). Redefine statistical significance. *Nature Human Behaviour, 2*(1), 6–10. https://doi.org/10.1038/s41562-017-0189-z

Brandt, M. J., IJzerman, H., Dijksterhuis, A., Farach, F. J., Geller, J., Giner-Sorolla, R., Grange, J. A., Perugini, M., Spies, J. R., & van't Veer, A. (2014). The replication recipe: What makes for a convincing replication? *Journal of Experimental Social Psychology, 50*, 217–224. https://doi.org/10.1016/j.jesp.2013.10.005

Brown, N. J. L., & Heathers, J. A. J. (2017). The GRIM test: A simple technique detects numerous anomalies in the reporting of results in psychology. *Social Psychological and Personality Science, 8*(4), 363–369. https://doi.org/10.1177/1948550616673876

Button, K. S., Ioannidis, J. P. A., Mokrysz, C., Nosek, B. A., Flint, J., Robinson, E. S., & Munafò, M. R. (2013). Power failure: Why small sample size undermines the reliability of neuroscience. *Nature Reviews Neuroscience, 14*(5), 365–376. https://doi.org/10.1038/nrn3475

Center for Open Science. (n.d.-a). *Open Science Badges*. Retrieved 23 Feb 2021, from https://www.cos.io/initiatives/badges

Center for Open Science. (n.d.-b). *TOP guidelines*. Retrieved 28 Feb 2021, from https://www.cos.io/initiatives/top-guidelines

Chabris, C. F., Hebert, B. M., Benjamin, D. J., Beauchamp, J., Cesarini, D., van der Loos, M., Johannesson, M., Magnusson, P. K. E., Lichtenstein, P., Atwood, C. S., Freese, J., Hauser, T. S., Hauser, R. M., Christakis, N., & Laibson, D. (2012). Most reported genetic associations with general intelligence are probably false positives. *Psychological Science, 23*(11), 1314–1323. https://doi.org/10.1177/0956797611435528

Chambers, C. D. (2013). Registered reports: A new publishing initiative at Cortex. *Cortex, 49*(3), 609–610. https://doi.org/10.1016/j.cortex.2012.12.016

Chambers, C. D. (2017). *The seven deadly sins of psychology: A manifesto for reforming the culture of scientific practice*. Princeton University Press. https://doi.org/10.1515/9781400884940

Chambers, C. D. (2018). Introducing the transparency and openness promotion (TOP) guidelines and badges for open practices at Cortex. *Cortex, 106*, 316–318. https://doi.org/10.1016/j.cortex.2018.08.001

Chambers, C. D. (2020). Verification reports: A new article type at Cortex. *Cortex*. https://doi.org/10.1016/j.cortex.2020.04.020

Christensen, G., Dafoe, A., Miguel, E., Moore, D. A., & Rose, A. K. (2019). A study of the impact of data sharing on article citations using journal policies as a natural experiment. *PLoS One, 14*(12), e0225883. https://doi.org/10.1371/journal.pone.0225883

Cumming, G. (2014). The new statistics: Why and how. *Psychological Science, 25*(1), 7–29. https://doi.org/10.1177/0956797613504966

Doyen, S., Klein, O., Pichon, C.-L., & Cleeremans, A. (2012). Behavioral priming: It's all in the mind, but whose mind? *PLoS One, 7*(1), e29081. https://doi.org/10.1371/journal.pone.0029081

Epskamp, S., & Nuijten, M. B. (2014). *statcheck: Extract statistics from articles and recompute p-values*. Retrieved from http://CRAN.Rproject.org/package=statcheck. (R package version 1.0.0)

Eronen, M. I., & Bringmann, L. F. (2021). The theory crisis in psychology: How to move forward. *Perspectives on Psychological Science: A Journal of the Association for Psychological Science*, 1745691620970586. https://doi.org/10.1177/1745691620970586

Etz, A., & Vandekerckhove, J. (2016). A Bayesian perspective on the reproducibility project: Psychology. *PLoS One, 11*(2), e0149794.

Fanelli, D. (2010). "Positive" results increase down the hierarchy of the sciences. *PLoS One, 5*(4), e10068. https://doi.org/10.1371/journal.pone.0010068

Fiedler, K., & Schwarz, N. (2016). Questionable research practices revisited. *Social Psychological and Personality Science, 7*(1), 45–52. https://doi.org/10.1177/1948550615612150

Flake, J. K., & Fried, E. I. (2020). Measurement schmeasurement: Questionable measurement practices and how to avoid them. *Advances in Methods and Practices in Psychological Science*, 2515245920952393. https://doi.org/10.1177/2515245920952393

Francis, G., Tanzman, J., & Matthews, W. J. (2014). Excess success for psychology articles in the journal *Science*. *PLoS One, 9*(12), e114255.

Franco, A., Malhotra, N., & Simonovits, G. (2016). Underreporting in psychology experiments: Evidence from a study registry. *Social Psychological and Personality Science, 7*(1), 8–12. https://doi.org/10.1177/1948550615598377

Frank, M. C., & Saxe, R. (2012). Teaching replication. *Perspectives on Psychological Science, 7*(6), 600–604. https://doi.org/10.1177/1745691612460686

Gelman, A., & Loken, E. (2013). *The garden of forking paths: Why multiple comparisons can be a problem, even when there is no "fishing expedition" or "p-hacking" and the research hypothesis was posited ahead of time* (p. 348). Department of Statistics, Columbia University.

Georgescu, C., & Wren, J. D. (2018). Algorithmic identification of discrepancies between published ratios and their reported confidence intervals and P-values. *Bioinformatics, 34*(10), 1758–1766. https://doi.org/10.1093/bioinformatics/btx811

Gilbert, D. T., King, G., Pettigrew, S., & Wilson, T. D. (2016). Comment on "Estimating the reproducibility of psychological science". *Science, 351*(6277), 1037–1037. https://doi.org/10.1126/science.aad7243

Greenwald, A. G. (1975). Consequences of prejudice against the null hypothesis. *Psychological Bulletin, 82*(1), 1–20.

Hardwicke, T. E., Bohn, M., MacDonald, K., Hembacher, E., Nuijten, M. B., Peloquin, B. N., DeMayo, B. E., Long, B., Yoon, E. J., & Frank, M. C. (2020). Analytic reproducibility in articles receiving open data badges at Psychological Science: An observational study. Preprint Retrieved from https://Osf.Io/Preprints/Metaarxiv/H35wt/. https://doi.org/10.31222/osf.io/h35wt.

Hardwicke, T. E., Mathur, M. B., MacDonald, K., Nilsonne, G., Banks, G. C., Kidwell, M. C., Hofelich Mohr, A., Clayton, E., Yoon, E. J., & Henry Tessler, M. (2018). Data availability, reusability, and analytic reproducibility: Evaluating the impact of a mandatory open data policy at the journal *Cognition*. *Royal Society Open Science, 5*(8), 180448. https://doi.org/10.1098/rsos.180448

Horstmann, K. T., Arslan, R. C., & Greiff, S. (2020). Generating codebooks to ensure the independent use of research data: Some guidelines. *European Journal of Psychological Assessment, 36*(5), 721–729. https://doi.org/10.1027/1015-5759/a000620

John, L. K., Loewenstein, G., & Prelec, D. (2012). Measuring the prevalence of questionable research practices with incentives for truth telling. *Psychological Science, 23*(5), 524–532. https://doi.org/10.1177/0956797611430953

Jonas, K. J., Cesario, J., Alger, M., Bailey, A. H., Bombari, D., Carney, D., Dovidio, J. F., Duffy, S., Harder, J. A., van Huistee, D., Jackson, B., Johnson, D. J., Keller, V. N., Klaschinski, L., LaBelle, O., LaFrance, M., Latu, I. M., Morssinkhoff, M., Nault, K., ... Tybur, J. M. (2017). Power poses – Where do we stand? *Comprehensive Results in Social Psychology, 2*(1), 139–141. https://doi.org/10.1080/23743603.2017.1342447

Kerr, N. L. (1998). HARKing: Hypothesizing after the results are known. *Personality and Social Psychology Review, 2*(3), 196–217. https://doi.org/10.1207/s15327957pspr0203_4

Kidwell, M. C., Lazarević, L. B., Baranski, E., Hardwicke, T. E., Piechowski, S., Falkenberg, L.-S., Kennett, C., Slowik, A., Sonnleitner, C., & Hess-Holden, C. (2016). Badges to acknowledge open practices: A simple, low-cost, effective method for increasing transparency. *PLoS Biology, 14*(5), e1002456. https://doi.org/10.1371/journal.pbio.1002456

Klein, O., Hardwicke, T. E., Aust, F., Breuer, J., Danielsson, H., Hofelich Mohr, A., IJzerman, H., Nilsonne, G., Vanpaemel, W., & Frank, M. C. (2018). A practical guide for transparency in psychological science. *Collabra: Psychology, 4*(1), 1–15. https://doi.org/10.1525/collabra.158

Klein, R. A., Ratliff, K. A., Vianello, M., Adams, R. B., Jr., Bahník, Š., Bernstein, M. J., Bocian, K., Brandt, M. J., Brooks, B., & Brumbaugh, C. C. (2014). Investigating variation in replicability: A "Many Labs" replication project. *Social Psychology, 45*(3), 142–152.

Kochari, A. R., & Ostarek, M. (2018). Introducing a replication-first rule for PhD projects (commentary on Zwaan et al., 'Making replication mainstream'). *Behavioral and Brain Sciences, 41*. https://doi.org/10.1017/S0140525X18000730

Lakatos, I., & Musgrave, A. (1970). *Criticism and the growth of knowledge*. Cambridge University Press.

Lakens, D., Adolfi, F. G., Albers, C. J., Anvari, F., Apps, M. A. J., Argamon, S. E., Baguley, T., Becker, R. B., Benning, S. D., Bradford, D. E., Buchanan, E. M., Caldwell, A. R., Van Calster, B., Carlsson, R., Chen, S.-C., Chung, B., Colling, L. J., Collins, G. S., Crook, Z., ... Zwaan, R. A. (2018). Justify your alpha. *Nature Human Behaviour, 2*(3), 168–171. https://doi.org/10.1038/s41562-018-0311-x

Lakens, D., & Evers, E. R. K. (2014). Sailing from the seas of chaos into the corridor of stability: Practical recommendations to increase the informational value of studies. *Perspectives on Psychological Science, 9*(3), 278–292. https://doi.org/10.1177/1745691614528520

LeBel, E. P. (2015). A new replication norm for psychology. *Collabra, 1*(4). https://doi.org/10.1525/collabra.23

LeBel, E. P., Berger, D., Campbell, L., & Loving, T. J. (2017). Falsifiability is not optional. *Journal of Personality and Social Psychology, 113*(2), 254–261. https://doi.org/10.1037/pspi0000106

LeBel, E. P., & Campbell, L. (2013). Heightened sensitivity to temperature cues in individuals with high anxious attachment: Real or elusive phenomenon? *Psychological Science, 24*(10), 2128–2130. https://doi.org/10.1177/0956797613486983

LeBel, E. P., McCarthy, R. J., Earp, B. D., Elson, M., & Vanpaemel, W. (2018). A unified framework to quantify the credibility of scientific findings. *Advances in Methods and Practices in Psychological Science, 1*(3), 389–402. https://doi.org/10.1177/2515245918787489

LeBel, E. P., Vanpaemel, W., Cheung, I., & Campbell, L. (2019). A brief guide to evaluate replications. *Meta-Psychology, 3*. https://doi.org/10.15626/MP.2018.843

Makel, M. C., Plucker, J. A., & Hegarty, B. (2012). Replications in psychology research: How often do they really occur? *Perspectives on Psychological Science, 7*(6), 537–542. https://doi.org/10.1177/1745691612460688

ManyBabies Consortium. (2020). Quantifying sources of variability in infancy research using the infant-directed-speech preference. *Advances in Methods and Practices in Psychological Science, 3*(1), 24–52. https://doi.org/10.1177/2515245919900809

Marsman, M., & Wagenmakers, E.-J. (2017). Bayesian benefits with JASP. *European Journal of Developmental Psychology, 14*(5), 545–555. https://doi.org/10.1080/17405629.2016.1259614

Matthews, W. J. (2012). How much do incidental values affect the judgment of time? *Psychological Science, 23*(11), 1432–1434. https://doi.org/10.1177/0956797612441609

Mayo, D. G. (2018). Statistical inference as severe testing. https://www.cambridge.org/core/books/statistical-inference-as-severe-testing/copyright-page/55AF1D228E1401D0912B1D59E7400BD3

Meehl, P. E. (1990). Appraising and amending theories: The strategy of Lakatosian defense and two principles that warrant it. *Psychological Inquiry, 1*(2), 108–141. https://doi.org/10.1207/s15327965pli0102_1

Morling, B. (2020). *Research methods in psychology* (4th ed.). W. W. Norton & Company.

Moshontz, H., Campbell, L., Ebersole, C. R., IJzerman, H., Urry, H. L., Forscher, P. S., Grahe, J. E., McCarthy, R. J., Musser, E. D., Antfolk, J., Castille, C. M., Evans, T. R., Fiedler, S., Flake, J. K., Forero, D. A., Janssen, S. M. J., Keene, J. R., Protzko, J., Aczel, B., … Chartier, C. R. (2018). The psychological science accelerator: Advancing psychology through a distributed collaborative network. *Advances in Methods and Practices in Psychological Science, 1*(4), 501–515. https://doi.org/10.1177/2515245918797607

Munafò, M. R., Nosek, B. A., Bishop, D. V., Button, K. S., Chambers, C. D., Du Sert, N. P., Simonsohn, U., Wagenmakers, E.-J., Ware, J. J., & Ioannidis, J. P. A. (2017). A manifesto for reproducible science. *Nature Human Behaviour, 1*(1), 1–9. https://doi.org/10.1038/s41562-016-0021

Naudet, F., Sakarovitch, C., Janiaud, P., Cristea, I., Fanelli, D., Moher, D., & Ioannidis, J. P. A. (2018). Data sharing and reanalysis of randomized controlled trials in leading biomedical journals with a full data sharing policy: Survey of studies published in The BMJ and PLOS Medicine. *British Medical Journal, 360*, k400. https://doi.org/10.1136/bmj.k400

Neuliep, J. W., & Crandall, R. (1993). Everyone was wrong: There are lots of replications out there. *Journal of Social Behavior and Personality, 8*(6), 1–8.

Nieuwenhuis, S., Forstmann, B. U., & Wagenmakers, E.-J. (2011). Erroneous analyses of interactions in neuroscience: A problem of significance. *Nature Neuroscience, 14*(9), 1105–1107. https://doi.org/10.1038/nn.2886

Nosek, B. A., Hardwicke, T. E., Moshontz, H., Allard, A., Corker, K. S., Almenberg, A. D., Fidler, F., Hilgard, J., Kline, M., Nuijten, M. B., Rohrer, J. M., Romero, F., Scheel, A. M., Scherer, L., Schönbrodt, F., & Vazire, S. (2021). Replicability, robustness, and reproducibility in psychological science. PsyArXiv. https://doi.org/10.31234/osf.io/ksfvq

Nosek, B. A., & Lakens, D. (2014). Registered reports: A method to increase the credibility of published results. *Social Psychology, 45*(3), 137–141. https://doi.org/10.1027/1864-9335/a000192

Nosek, B. A., Spies, J. R., & Motyl, M. (2012). Scientific Utopia: II. Restructuring incentives and practices to promote truth over publishability. *Perspectives on Psychological Science, 7*(6), 615–631. https://doi.org/10.1177/1745691612459058

Nuijten, M. B., Bakker, M., Maassen, E., & Wicherts, J. M. (2018). Verify original results through reanalysis before replicating. *Behavioral and Brain Sciences, 41*, e143. https://doi.org/10.1017/S0140525X18000791

Nuijten, M. B., Borghuis, J., Veldkamp, C. L., Dominguez-Alvarez, L., Van Assen, M. A., & Wicherts, J. M. (2017a). Journal data sharing policies and statistical reporting inconsistencies in psychology. *Collabra: Psychology, 3*(1). https://doi.org/10.1525/collabra.102

Nuijten, M. B., Hartgerink, C. H. J., van Assen, M. A. L. M., Epskamp, S., & Wicherts, J. M. (2016). The prevalence of statistical reporting errors in psychology (1985–2013). *Behavior Research Methods, 48*(4), 1205–1226. https://doi.org/10.3758/s13428-015-0664-2

Nuijten, M. B., Van Assen, M. A. L. M., Hartgerink, C. H. J., Epskamp, S., & Wicherts, J. (2017b). The validity of the tool "statcheck" in discovering statistical reporting inconsistencies. PsyArXiv. https://doi.org/10.31234/osf.io/tcxaj

O'Donohue, W. (2021). Are psychologists appraising research properly? Some Popperian notes regarding replication failures in psychology. *Journal of Theoretical and Philosophical Psychology, 41*(4), 233–247. https://doi.org/10.1037/teo0000179

Open Science Collaboration. (2012). An open, large-scale, collaborative effort to estimate the reproducibility of psychological science. *Perspectives on Psychological Science, 7*(6), 657–660. https://doi.org/10.1177/1745691612462588

Open Science Collaboration. (2015). Estimating the reproducibility of psychological science. *Science, 349*(6251), aac4716. https://doi.org/10.1126/science.aac4716

Pashler, H., & Harris, C. R. (2012). Is the replicability crisis overblown? Three arguments examined. *Perspectives on Psychological Science, 7*(6), 531–536. https://doi.org/10.1177/1745691612463401

Pashler, H., Rohrer, D., & Harris, C. R. (2013). Can the goal of honesty be primed? *Journal of Experimental Social Psychology, 49*(6), 959–964. https://doi.org/10.1016/j.jesp.2013.05.011

Pashler, H., & Wagenmakers, E. (2012). Editors' introduction to the special section on replicability in psychological science: A crisis of confidence? *Perspectives on Psychological Science, 7*(6), 528–530. https://doi.org/10.1177/1745691612465253

Patel, C. J., Burford, B., & Ioannidis, J. P. A. (2015). Assessment of vibration of effects due to model specification can demonstrate the instability of observational associations. *Journal of Clinical Epidemiology, 68*(9), 1046–1058.

Peng, R. D. (2011). Reproducible research in computational science. *Science, 334*(6060), 1226–1227. https://doi.org/10.1126/science.1213847

Petrocelli, J. V., Clarkson, J. J., Whitmire, M. B., & Moon, P. E. (2013). When $ab \neq c - c'$: Published errors in the reports of single-mediator models. *Behavior Research Methods, 45*(2), 595–601. https://doi.org/10.3758/s13428-012-0262-5

Piwowar, H. A., Day, R. S., & Fridsma, D. B. (2007). Sharing detailed research data is associated with increased citation rate. *PLoS One, 2*(3), e308. https://doi.org/10.1371/journal.pone.0000308

Popper, K. R. (1959). *The logic of scientific discovery*. University Press.

Rife, S. C., Nuijten, M. B., & Epskamp, S. (2016). *statcheck: Extract statistics from articles and recompute p-values* [*web application*]. http://statcheck.io

Silberzahn, R., Uhlmann, E. L., Martin, D. P., Anselmi, P., Aust, F., Awtrey, E., Bahník, Š., Bai, F., Bannard, C., Bonnier, E., Carlsson, R., Cheung, F., Christensen, G., Clay, R., Craig, M. A., Dalla Rosa, A., Dam, L., Evans, M. H., Flores Cervantes, I., … Nosek, B. A. (2018). Many analysts, one data set: Making transparent how variations in analytic choices affect results. *Advances in Methods and Practices in Psychological Science, 1*(3), 337–356. https://doi.org/10.1177/2515245917747646

Simmons, J. P., Nelson, L. D., & Simonsohn, U. (2011). False-positive psychology: Undisclosed flexibility in data collection and analysis allows presenting anything as significant. *Psychological Science, 22*(11), 1359–1366. https://doi.org/10.1177/0956797611417632

Simonsohn, U. (2015). Small telescopes: Detectability and the evaluation of replication results. *Psychological Science, 26*(5), 559–569. https://doi.org/10.1177/0956797614567341

Simonsohn, U., Simmons, J. P., & Nelson, L. D. (2019). Specification curve: Descriptive and inferential statistics on all reasonable specifications (SSRN Scholarly Paper ID 2694998). *Social Science Research Network.* https://doi.org/10.2139/ssrn.2694998

Stark, P. B. (2018). Before reproducibility must come preproducibility. *Nature, 557*(7707), 613–613. https://doi.org/10.1038/d41586-018-05256-0

Steegen, S., Tuerlinckx, F., Gelman, A., & Vanpaemel, W. (2016). Increasing transparency through a multiverse analysis. *Perspectives on Psychological Science, 11*(5), 702–712. https://doi.org/10.1177/1745691616658637

Sterling, T. D. (1959). Publication decisions and their possible effects on inferences drawn from tests of significance—Or vice versa. *Journal of the American Statistical Association, 54*(285), 30–34. https://doi.org/10.1080/01621459.1959.10501497

Sterling, T. D., Rosenbaum, W. L., & Weinkam, J. J. (1995). Publication decisions revisited: The effect of the outcome of statistical tests on the decision to publish and vice versa. *The American Statistician, 49*(1), 108–112. https://doi.org/10.1080/00031305.1995.10476125

Stodden, V. C. (2010). Reproducible research: Addressing the need for data and code sharing in computational science. *Computing in Science & Engineering, 5*, 8–12.

Sweeney, L. (2002). K-anonymity: A model for protecting privacy. *International Journal of Uncertainty, Fuzziness and Knowledge-Based Systems, 10*(05), 557–570. https://doi.org/10.1142/S0218488502001648

The Dutch Research Council. (n.d.). *Replication studies | NWO*. Retrieved 24 Feb 2021, from https://www.nwo.nl/en/researchprogrammes/replication-studies

Tijdink, J. K., Verbeke, R., & Smulders, Y. M. (2014). Publication pressure and scientific misconduct in medical scientists. *Journal of Empirical Research on Human Research Ethics, 9*(5), 64–71. https://doi.org/10.1177/1556264614552421

van Aert, R. C. M., Nuijten, M. B., Olsson-Collentine, A., Stoevenbelt, A. H., Van den Akker, O. R., & Wicherts, J. M. (2021). Comparing the prevalence of statistical reporting inconsistencies in COVID-19 preprints and matched controls: A registered report. *Royal Society Open Science*. https://doi.org/10.17605/OSF.IO/WCND4

van Dalen, H. P., & Henkens, K. (2012). Intended and unintended consequences of a publish-or-perish culture: A worldwide survey. *Journal of the American Society for Information Science and Technology, 63*(7), 1282–1293. https://doi.org/10.1002/asi.22636

Verhagen, J., & Wagenmakers, E.-J. (2014). Bayesian tests to quantify the result of a replication attempt. *Journal of Experimental Psychology: General, 143*(4), 1457.

Wagenmakers, E.-J., Wetzels, R., Borsboom, D., van der Maas, H. L. J., & Kievit, R. A. (2012). An agenda for purely confirmatory research. *Perspectives on Psychological Science, 7*(6), 632–638. https://doi.org/10.1177/1745691612463078

Wicherts, J. M. (2013). Science revolves around the data. *Journal of Open Psychology Data, 1*(1), e1. https://doi.org/10.5334/jopd.e1

Wicherts, J. M., Borsboom, D., Kats, J., & Molenaar, D. (2006). The poor availability of psychological research data for reanalysis. *American Psychologist, 61*(7), 726. https://doi.org/10.1037/0003-066X.61.7.726

Wicherts, J. M., Kievit, R. A., Bakker, M., & Borsboom, D. (2012). Letting the daylight in: Reviewing the reviewers and other ways to maximize transparency in science. *Frontiers in Computational Neuroscience, 6*. https://doi.org/10.3389/fncom.2012.00020

Wilkinson, M. D., Dumontier, M., Aalbersberg, I. J., Appleton, G., Axton, M., Baak, A., Blomberg, N., Boiten, J.-W., da Silva Santos, L. B., Bourne, P. E., Bouwman, J., Brookes, A. J., Clark, T., Crosas, M., Dillo, I., Dumon, O., Edmunds, S., Evelo, C. T., Finkers, R., … Mons, B. (2016). The FAIR guiding principles for scientific data management and stewardship. *Scientific Data, 3*(1), 160018. https://doi.org/10.1038/sdata.2016.18

Zwaan, R., Etz, A., Lucas, R., & Donnellan, B. (2017). Making replication mainstream. *Behavioral and Brain Sciences: An International Journal of Current Research and Theory with Open Peer Commentary*, 1–50. https://doi.org/10.1017/S0140525X17001972

Chapter 18
Reflections on the Reproducibility Project in Psychology and the Insights It Offers for Clinical Psychology

Elizabeth W. Chan, Johnny Wong, Christian S. Chan, and Felix Cheung

Abstract The overarching goal of this chapter is to discuss the Reproducibility Project in Psychology (RP:P), the resulting credibility movement, and its implications for scientific practices in clinical psychology and beyond. We start by introducing the RP:P and then describe how the RP:P led to other replication projects and the development of improved research practices. We conclude with a discussion of the replicability in clinical psychology, with attention to the challenges and opportunities of replicating clinical psychological findings in different places, with different people, and at different times.

Keywords Reproducibility project in psychology · Clinical psychology

In the past decade, a range of scientific disciplines have witnessed a rise in recognizing the importance of replicability. Replicability is fundamental to the verification of empirical research. It entails repeating the same study procedure in a different location with a different population at a different time to examine if the results are consistent with the original study (Barba, 2018; Peng et al., 2006). Despite the importance of replicability for science, until recently, replication studies are rarely done. Prior to any large-scale replications, Makel et al. (2012) documented that only 1.07% of 500 randomly chosen articles from 100 psychology journals with high impact factors were replications. The replicability of research findings has long been speculated to be very low, particularly when researchers have flexibility in their methodological and analytical decisions (Ioannidis, 2005). They may be

Elizabeth W. Chan and Johnny Wong contributed equally with all other contributors.

E. W. Chan · J. Wong · F. Cheung (✉)
Department of Psychology, University of Toronto, Toronto, ON, Canada
e-mail: f.cheung@utoronto.ca

C. S. Chan
Department of Psychology, University of Hong Kong, Hong Kong, SAR China

W. O'Donohue et al. (eds.), *Avoiding Questionable Research Practices in Applied Psychology*, https://doi.org/10.1007/978-3-031-04968-2_18

incentivized to find statistically significant effects in order to increase their chances of getting studies published, which often influence job prospects and lead to financial gains. Furthermore, the publication of a series of parapsychological experiments (Bem, 2011), its subsequent failed replications (e.g., Wagenmakers et al., 2011), and cases of scientific misconduct (e.g., Crocker, 2011) compelled the scientific community to reflect on their research practices.

The Reproducibility Project: Psychology (RP:P; Open Science Collaboration, 2015) was the first major replication effort that aimed to provide an empirical estimate of the replicability in psychological research. The RP:P marked the beginning of the "replication crisis" (also known as the "credibility movement"), which led to growing efforts to promote open science, transparency, and improved scientific practices. Since then, there have been many other large-scale replication projects in psychology like *Many Labs* (Klein et al., 2014) and *ManyBabies* (Frank et al., 2017).

The overarching goal of this chapter is to discuss the RP:P, the resulting credibility movement, and its implications for scientific practices in clinical psychology and beyond. We start by introducing the RP:P and then describe how the RP:P led to other replication projects and the development of improved research practices. We conclude with a discussion of the replicability in clinical psychology, with attention to the challenges and opportunities of replicating clinical psychological findings in different places, with different people, and at different times.

Introducing the Reproducibility Project: Psychology

The Open Science Collaboration, led by Dr. Brian Nosek, consisted of a network of 270 psychological researchers who collectively replicated 100 studies from three highly reputable psychology journals as part of the Reproducibility Project: Psychology (Open Science Collaboration, 2015). Participating research teams conducted replications of studies from *Psychological Science*, the *Journal of Personality and Social Psychology*, and the *Journal of Experimental Psychology: Learning, Memory, and Cognition*. The first journal covers a wide range of articles from across the fields of psychology; the latter two focus on articles from social, personality, and cognitive psychology, respectively. Of note is that this project did not focus on articles directly related to clinical psychology. The research teams outlined a specific research protocol that largely followed the original study design, after having contacted the original authors for study materials where possible. Each replication was preregistered and internally reviewed, meaning that the data collection plan and analytical approaches were pre-specified prior to actual data collection and such plans were verified to be reasonable tests of the original studies by other team member(s) of the Open Science Collaboration. The advantage of a preregistered approach is that the preregistration document makes transparent any changes one makes before and after conducting the analyses.

Of the 97 studies that originally found evidence for statistically significant results, the replication studies achieved 92% of the statistical power needed to detect an effect size similar to the original studies. In other words, based on the high statistical power, one would expect around 91.2% (89) of the 97 replication studies to replicate the original studies. The actual results, however, were striking. Whereas 97% of the original studies revealed statistically significant effects ($p < .05$), the RP:P found that only 36% of studies replicated with $p < .05$. To complement this analysis and reduce the focus on null hypothesis significance testing, the RP:P further evaluated the replication and original studies by effect sizes. The effect sizes in the original studies doubled those found in the replications. In sum, the large-scale replication effort found that fewer than half of the original studies were successfully replicated despite high statistical power and the use of original study materials, and the effect sizes substantially declined in the replications.

There were also marked differences across subdisciplines, such that 50% of cognitive psychology studies replicated in contrast to only 25% of social psychology studies. Original studies with smaller p-values were also more likely to be replicated than those with larger p-values. Specifically, 63% of original studies that achieved $p < .001$ were later replicated with $p < .05$; 41% of original studies with $p < .02$ were replicated with $p < .05$. In contrast, only 26% of studies that had a p-value between .02 to .04 were replicated with $p < .05$. As such, a smaller p-value in the original study was related to a greater likelihood that the findings would replicate with $p < .05$.

The RP:P marked the beginning of a movement that would shape the field of psychology and other scientific fields. It provides the first systematic estimate of reproducibility in psychology, demonstrating that even among articles published in some of the most reputable journals, just over one-third of studies replicated with significant effects, and the rate of reproducibility differed by subdiscipline. Although the RP:P focused on social and cognitive psychology studies, the findings raise important questions for psychological science as a whole, as many research practices share commonality across subfields.

Developments Since the Reproducibility Project: Psychology

Replications Become More Mainstream

Since the RP:P, large-scale replication efforts that involved multiple labs have become more common, and they can be broadly categorized into three groups. First, the ManyLab format, which involves many labs replicating multiple original studies simultaneously, typically seeks to answer metascientific questions such as whether the quality of data obtained from college student samples is consistent throughout the semester (Ebersole et al., 2016). Second, studies that take a Registered Replication Reports format are conducted with the targeted goal of verifying a specific phenomenon, especially those that are documented in frequently cited papers. Finally, similar to the RP:P itself, the Reproducibility Project: Cancer Biology,

Experimental Economics Replication Project, and a series of replications of *Nature* and *Science* articles aim to quantify the level of reproducibility of studies conducted within a specific discipline or published in specific outlets.

Many Labs The initial Many Labs project consisted of 36 independent research teams who tried to replicate 13 well-known effects in psychology with over 6300 participants altogether (Klein et al., 2014). All studies had a confirmatory analysis plan prior to data collection and were conducted through Project Implicit (www.projectimplicit.com). The replication studies were conducted in different settings (27 in-lab studies and nine online studies) and geographical regions (25 in the US and 11 outside of the US). Aggregated across the research teams, 10 of the 13 effects were replicated, and there were little to no differences depending on the setting or sample. Thus, replicability appeared to rely more on whether an effect truly exists rather than the context in which the effect was tested. A subsequent Many Labs 2.0 project found that 54% of classic psychology effects replicated with $p < .05$ and 50% replicated with $p < .0001$ (Klein et al., 2018). Moreover, 75% of the replication studies found smaller effect sizes than those seen in the original studies. In the subsequent Many Labs projects, 14 of 28 effects were replicated in Many Labs 2.0, and 3 of 10 effects replicated in Many Labs 3.0 (Ebersole et al., 2016). Similar large-scale replications have taken place in developmental science as part of the ManyBabies project (https://manybabies.github.io/) (Frank et al., 2017). In general, developmental psychology research with infants has small sample sizes and thus lower statistical power. They often rely on nonverbal measures (e.g., habituation) when working with infants, who have limited verbal communication abilities. ManyBabies strives to replicate landmark developmental psychology findings in order to determine their robustness and replicability. The first ManyBabies replication project looked at a preference for infant-directed speech over adult-directed speech and successfully replicated this classic finding. Ongoing projects aim to replicate findings related to theory of mind, rule learning, and social evaluations.

Registered Replication Reports Registered Replication Reports (RRR) were introduced as a new article type for disseminating findings from replications (Simons et al., 2014). Unlike the RP:P, participating research teams in RRR attempt to replicate the same effects. When applying for the RRR, researchers first submit their proposed replication effort directly to the journal. A team of editors then reviews it, and if deemed appropriate, the original authors (or qualified researchers recommended by the original authors) will be invited to review the proposed study plan. Once the plan is approved by the editor, the resulting replication study will be published regardless of the statistical significance of the results. The first issue was released in *Perspectives on Psychological Science*, and this article type is also available in *Advances in Methods and Practices in Psychological Science*. An example of a RRR is a 17-team replication of the facial feedback hypothesis, or the idea that incidental facial expressions shape emotional experiences. Results from each individual lab and a meta-analysis synthesizing the available evidence ($N = 1894$; Wagenmakers et al., 2016) failed to replicate the original study (Strack et al., 1988).

Replication Projects Across the Sciences Fields outside of psychology are also attempting to replicate these findings. Beginning in 2013, the Reproducibility Project: Cancer Biology aimed to replicate 50 cancer biology studies from high impact-factor journals, including *Science*, *Nature*, and *Cell*, and there have been mixed results thus far (Baker & Dolgin, 2017; Morrison, 2014). Additionally, the results of a survey with researchers at a cancer research center revealed that half of them were unable to replicate a finding at least once (Mobley et al., 2013). Altogether, the preliminary data points to potentially low reproducibility in cancer biology research, which, given the potential implications of these studies, warrants careful consideration, especially concerning questions about whether it is related to statistical power, incentive structures, or study design (Begley & Ellis, 2012; Nosek & Errington, 2017).

In addition, the Experimental Economics Replication Project attempted to replicate 18 studies from two top-tier economic journals, the *American Economic Review* and the *Quarterly Journal of Economics* (Camerer et al., 2016). Only 61% of the studies replicated with $p < .05$, and 66% achieved the same effect size, which was quite promising compared to RP:P. A team of researchers also attempted to replicate 21 behavioral economic studies from *Nature* and *Science* (Camerer et al., 2018). Research teams were in touch with the original study authors for materials and comments on their Registered Replication Reports, and they conducted the same statistical tests as those used in the original studies. Of the 21 original studies, the findings of 61.9% (i.e., 13 studies) were replicated. In other words, the findings of 38.1% of these studies that were published in two of the most highly regarded and influential scientific journals failed to be replicated.

Shaping Research Practices

Many researchers have also introduced a range of open, transparent, and rigorous scientific practices related to conducting and evaluating empirical research. For practices related to producing empirical research, this includes adversarial collaboration (see Chap. 16 in this volume), increased sample size, "Big Science" (i.e., research involving large teams of researchers), preregistration, open data, open materials, publications of null results, etc. Proposed practices for improving the evaluation of research include signed reviews, p-curve analysis, replication indices, and more. A p-curve analysis involves plotting the distribution of p-values for study findings to check for selective reporting (Simonsohn et al., 2014), and the replication index measures the likelihood that a study will replicate (R-Index; Schimmack, 2016). We commend such efforts and refer interested readers to other chapters in this volume for more detailed discussions of these practices.

There is emerging evidence that these improved research practices indeed contribute to greater replicability. For example, Mullinix et al. (2015) conducted a replication of 20 studies on the Time-sharing Experiments for the Social Sciences

(TESS) platform and successfully replicated over 80% of the studies. The TESS is a platform where researchers can submit proposals (including experimental stimuli) for conducting online experiments and, if accepted, a large sample of participants will be recruited to complete the experiments. When conducting the replications, the authors (Mullinix et al., 2015) were able to sample from existing studies from TESS, which, unlike the published literature, does not filter studies based on statistical significance (i.e., publication bias). All in all, the replications of studies on the TESS platform present a scenario where the original studies were adequately powered, the exact experimental materials were shared, and the selection of studies was not based on p-values. The high replicability demonstrated in this study highlights the promising possibility that when we do science the right way, science replicates.

A Sociological Model of the Philosophy of Science

A key feature of the credibility movement following the RP:P is the recognition that the context in which research is conducted matters. This sociological model of the philosophy of science suggests that scientific practices are ultimately embedded within the context of different institutions (i.e., universities, journals, publishers, funding bodies, professional organizations). Researchers, journals, and institutions all have different goals and incentives that do not necessarily align with each other or with the broader goal of describing, understanding, explaining, and changing human thoughts and behaviors. In other words, the responsibility to produce a replicable, reliable, and relevant science does not solely lie with individual researchers, but rather it relies also on a healthy scientific community that prioritizes and incentivizes rigor and openness.

Academic Institutions Most rankings of academic institutions put a strong emphasis on the number of publications by their researchers. As such, many institutions reward their staff based on their productivity in the form of publications. The "publish or perish" ethos inevitably pressures many to publish more in order to secure promotion and tenure. These incentive structures may inadvertently cultivate a culture where publishing is prioritized over scientific rigor, leading to questionable research practices (QRPs) as potential shortcuts for publishing more studies (John et al., 2012). Moreover, in recruiting new faculty members or in tenure and promotion, novelty and creative thinking are often highly valued characteristics, and efforts invested in conducting replications may not be viewed as evidence that supports these traits. Realigning the goal of such endeavors towards the betterment of science will likely require giving a stronger emphasis to open, transparent, and rigorous scientific practices. One relatively simple change towards this goal is to value replication efforts, especially when they are done in accordance with the aforementioned principles.

Journals Given researchers' pressure to "publish or perish," journals (and their publishers) play a sizeable role in the incentive structure. For example, publication bias describes a tendency for researchers to *file away* null results because journals may prioritize significant results. This assumption is not entirely unfounded as journals tend to prefer significant results because these attract readership (Duyx et al., 2017) and media attention. Both researchers and journals are responsible for this vicious cycle. Since the RP:P, a growing number of journals are tackling publication bias by either stating a preference for preregistered research (Kravitz et al., 2020) or offering a registered report format. For example, *Nature Human Behaviour* accepts submissions of registered reports (Editorial., 2018). The researcher first submits an initial manuscript that includes an introduction, preliminary data or study, methods, disclosures, and planned analyses. Any manuscripts that pass this screening phase then undergo expert peer review. Authors whose manuscripts are provisionally accepted after the initial review need to register the study protocol. The journal will then publish the final manuscript regardless of the significance of the results as long as the authors followed the study protocol. Furthermore, the *British Medical Journal Open* prioritizes open and transparent scientific practices in medical research. They publish stand-alone study protocols, which encourage more in-depth descriptions of complicated procedures to facilitate replication (Munro & Prendergast, 2019).

Professional Organizations Professional organizations can also help lower the barriers for researchers to improve their scientific practices. For example, the Center for Open Science hosts a website (https://osf.io/) for researchers to preregister their study plan, upload and share data or other files, and share their findings in pre- and post-prints. All files and preregistrations are time-stamped. Authors can upload new preregistrations to reflect unexpected changes that occur throughout the research process, but individuals can still view the original, frozen preregistrations. Furthermore, the International Standard Randomised Controlled Trials Number (ISRCTN) registry is used for clinical trials and is recognized by the World Health Organization and the International Committee of Medical Journal Editors. This registry assigns a unique code for each clinical trial and is required to publish the paper. The ISRCTN is committed to transparency and makes study information publicly and freely available. These infrastructures provide standardized platforms for engaging in preregistrations and open science.

Moreover, the Society for the Improvement of Psychological Sciences (SIPS) is a professional organization aimed at addressing issues surrounding replicability and striving towards conducting better science (http://improvingpsych.org/). They held their inaugural meeting at the Center for Open Science in 2016 and continue to meet annually. Their values include enhancing transparency and open sharing to further encourage criticism and mutual respect. The key activities of SIPS include refining institutional policies to advocate for practicing proper science, evaluating current practices and reforms, and reaching out to parties within and beyond psychology. Through their outreach efforts, they may learn more about cultural or field-specific practices and ultimately promote better practices across the sciences. Their official journal, *Collabra: Psychology*, is an open-access journal that welcomes articles across the subdisciplines of psychology and has a section dedicated to methodological practices. Altogether, the SIPS initiatives help build a scientific community aimed at making science more open and transparent, and thus more valid and reliable.

What Does the Reproducibility Project: Psychology and the Credibility Movement Mean for Clinical Psychology?

To some extent, clinical psychology concerns the testing and assessment of mental health as well as prevention and intervention strategies to ameliorate psychological abnormality or enhance psychological well-being. Replicability is central to clinical psychology. The prevailing approach of evidence-based practice relies on the synthesis of scientific evidence to determine the reliability and validity of its tools, including the efficacy of psychological treatments. Much like medical interventions prior to receiving approval for dissemination, psychological interventions are expected to undergo multiple clinical trials before they are adopted by practitioners. An initial trial showing evidence for a new type of psychotherapy would require further independent replications. Specifically, it is critical to encourage openness and transparency in both the basic and applied sides of clinical psychology. Basic research attempts to understand the basic processes of psychopathology and its possible interventions. Replication studies can help determine whether the extant research findings are well-supported. These can strengthen the clinical psychological research base and clarify which pieces of evidence have the strongest support. Furthermore, basic research often acts as the foundation for applied research which is geared towards practical applications (e.g., treatment). For example, evidence showing that a treatment is efficacious should be replicated with different patients and by different researchers. If the goal is to develop treatments that can be used beyond a highly controlled lab setting, then the treatments should be tested across contexts. Open science principles are crucial to both basic and applied research. Below, we discuss the opportunities and challenges posed by the replication crisis in clinical psychology.

Empirically supported treatments (ESTs) are considered the gold standard for determining which treatments should be used for different diagnoses (Kendall, 1998). The American Psychological Association originally used criteria from Chambless and Hollon (1998) to indicate the strength of a treatment's empirical support for a certain psychiatric diagnosis or mental disorder. These criteria would lead to a categorization of strong (well-established), modest (probably efficacious), or controversial research support. Treatments were deemed well-established if they had empirical support from at least two well-designed RCTs conducted by two independent research teams. Probably efficacious treatments were those that had empirical support from at least one well-designed study or a few decently designed studies.

More recently, a new classification system was recommended by Tolin et al. (2015), which labels treatments as very strong, strong, weak, or having insufficient evidence. To be classified as very strong, there should be high-quality evidence showing a clinically meaningful effect on functional outcomes and symptom reduction, and at least one of these effects should continue for no less than 3 months. In addition, there must be a minimum of one well-designed study supporting the effectiveness of the treatment outside of the research context. As a result, many treatments have a pending status according to these recommendations and thus rely on the older criteria.

The Tolin et al. (2015) criteria, however, remains reliant on statistical significance to evaluate whether a treatment should be labeled an EST despite controversies related to focusing on statistical significance alone (Open Science Collaboration, 2015). Equipped with additional evaluation tools developed since the RP:P, a meta-analysis (Sakaluk et al., 2019) sought to evaluate ESTs using a range of metrics such as statistical power, Bayes Factors, Replicability-Index (Schimmack, 2016), and the degree to which researchers misreported inferential statistics. While there were generally few misreported statistics, the authors found that nearly all ESTs were underpowered and had low R-indices, indicating that the evidence supporting ESTs may not be as strong as previously thought (Sakaluk et al., 2019). A considerable number of ESTs that were labeled as Strong received low scores on the aforementioned metrics, and several ESTs had too little information available to obtain scores on these metrics. The authors further noted that studies have been increasingly well-powered over time, and researchers should continue to strive for highly powered designs.

Similar to the strong push for replicability in other domains, recently, clinical psychological researchers have also been joining the open science movement (Tackett et al., 2019). A metascientific study (Nutu et al., 2019) reviewed 201 articles published across 60 clinical psychology journals that have policies in favor of at least 4 of 5 best scientific practices: (a) preprints, (b) preregistration, (c) open data, (d) reporting guidelines, and (e) conflict of interest (COI) disclosure statement. The study found that most of these practices were not mandated by the journals, aside from the COI disclosure statement, which has long been seen as important for transparency. According to the editorial policies of these 60 journals, 15 allowed preprints, 15 mentioned preregistrations, 40 mentioned open data, 28 had reporting guidelines, and 52 required COI disclosure statements. However, when reviewing the sampled articles published in these journals, only 3% were preregistered, 2% had open data, and one article had a preprint. Another study reviewed 165 randomized controlled trials (RCTs) published in 2013 in the 25 highest-impact clinical psychology journals and found that just 15% of them were preregistered (Cybulski et al., 2016). Of the articles that were preregistered, 58% included their registration information in the published article. Additionally, only 1% of them were both preregistered and fully outlined their main outcome variables. Mentioning the outcome variables of interest prior to conducting analyses increases transparency, ensures that the results presented reflect the researchers' initial analytic plan, and reduces the likelihood of Type I errors. Moreover, Grant et al. (2013) found that only 27.5% (11 of 40) of high-impact journals that publish psychological intervention trials (n = 239) provide reporting guidelines for their authors. Among the reviewed articles, only 20% mentioned that they conducted an RCT in the article name, and 55% included this in the abstract. Approximately 42% of the articles followed the journals' recommended reporting guidelines. Overall, these studies suggest that journals should consider moving beyond merely recommending these practices and take concrete steps to further encourage, if not mandate, open, transparent, and rigorous practices.

A welcoming sight is that *Clinical Psychological Science* and *Collabra: Clinical Psychology* now offer a badge system in which researchers can earn badges for open data, open materials, and/or preregistrations. This can be considered an evidence-based approach as Kidwell et al. (2016) found that data sharing substantially increased from 3% to 40% after journals implemented badge systems, though researchers may be motivated to share their data for other reasons, for example, to allow other research teams to replicate their work. Moreover, in August of 2019, the *Journal of Abnormal Psychology* published a special issue titled Increasing Replicability, Transparency, and Openness in Clinical Psychological Research. These efforts are conducive to facilitating more widespread awareness and adoption of open science practices.

Given the resource-intensive nature of clinical research, replication studies may not always be feasible. Recall that replicability refers to conducting an identical procedure in a different place with a different population at a different time to test whether the results match the results of the initial study (Barba, 2018; Peng et al., 2006). A starting point for clinical psychological researchers aiming to make their studies more replicable is to consider how replicability can vary depending on the context and timing in which studies are conducted, as well as the individuals involved. The goal of the following paragraphs is to caution against the direct application of insights gained from social-personality and cognitive psychology to clinical psychological research and call for a more careful field-specific adaptation.

Different Place

Soon after the publication of the RP:P, a metascientific question emerges related to whether the replicability of studies is context-dependent. Van Bavel et al. (2016) reanalyzed the RP:P data and tested whether contextual factors like time, culture, and location were related to replication success for the 100 replication attempts. The authors found that contextual factors were related to replication success ($r = -0.23$, $p = .024$), even when controlling for methodological features like effect sizes. However, Inbar (2016) argued against this finding by pointing out how Van Bavel et al. (2016) failed to consider a third variable: social-personality vs. cognitive psychology as a confounder. By reanalyzing the RP:P data and distinguishing the cognitive psychology studies from the social-personality psychology studies, Inbar (2016) yielded contradicting results about the context-dependent nature of replication success. He found no significant relationship between the replicability of a study and its context-dependence when looking within subdisciplines. Both social-personality psychology studies ($r = -0.08$, $p = .54$) and cognitive psychology studies ($r = -0.04$, $p = .79$) did not find an association between contextual factors and replication success. The context-dependent hypothesis was further informed by ManyLab 2.0 (Klein et al., 2018), which found low heterogeneity across replication studies conducted in Western, educated, industrialized, rich, and democratic (WEIRD) societies and non-WEIRD societies. This suggests that replication success may be less impacted by the setting or sample than by the actual effect itself.

Therefore, at least within the social-personality and cognitive psychology literature, there is some evidence that replicability may not heavily depend on context. However, whether this holds true in clinical psychology remains an open empirical question. Indeed, with the substantial cross-cultural clinical psychology literature questioning even the core foundation of the science, including the presumed universality of psychopathology (e.g., on the cultural differences in the manifestation of depressive symptoms among Chinese and Americans; Kleinman, 2004; Ryder et al., 2008; Yen et al., 2000), replication efforts in clinical psychology should adequately and explicitly account for the potential influence of cultural context. Notably, the pursuit of cross-cultural replicability of clinical psychology should not come at the cost of downplaying the importance of indigenous research. Indigenous practices are not presumed to work outside of the cultural context in which they were developed, but they are still a valuable area of study that can inform culture-specific interventions on psychopathology. Ultimately, more empirical studies are required to evaluate the extent to which context-dependence is associated with replication success in clinical psychology, while being mindful that not all findings (and practices) within clinical psychology are presumably universal.

Another aspect distinguishing social and cognitive psychology studies from clinical psychology studies that may impact replication attempts is the centrality of self-reported measures. While social and cognitive psychology studies can employ reasonably standardized lab-based experimental procedures, clinical psychology studies rely heavily on self-reported measures of psychological phenomena (e.g., mental health symptoms). A major issue that results from the dominant use of self-reports regards the translation of instruments. A poorly translated instrument in both linguistic and cultural terms can lead to inaccurate assessments of the construct in question. Even instruments that are written in the same language (e.g., Chinese) can use wording that has different meanings in different cultures (e.g., mainland China vs. Taiwan vs. Hong Kong). As such, cross-cultural replication attempts in clinical psychology should pay careful attention to the instruments being used. This may entail consulting with local experts to ensure that the linguistic translation is adequate, as well as testing for factorial invariance to examine the cross-cultural comparability of a measurement scale across populations.

Different People

Individuals who are involved in clinical psychological studies also warrant attention when considering the replicability of studies. This includes the resource-intensive nature of data collection in clinical psychology. In addition, many studies point to the critical roles of health care providers in the success of a treatment, e.g., *the therapist effect* (Kim et al., 2006; McKay et al., 2006).

Participants Replication studies may have become more common in cognitive and social-personality psychology partly because of easily accessible convenience samples (e.g., participant pool in psychology departments or the Mturk platform). The cost-effectiveness of these samples facilitates increasing the sample size to achieve greater statistical power. Given that clinical studies require substantial resources, including time and money, they tend to consist of smaller sample sizes. This may contribute to low statistical power, making it hard to detect meaningful effects (Button et al., 2013). In fact, Cuijpers et al. (2016) found that only up to 40% of the sample size needed to detect a clinically meaningful effect was achieved in a review of trials for depression. Moreover, when reviewing the sample sizes of studies published in the *Journal of Abnormal Psychology* and the *Journal of Consulting and Clinical Psychology* from 2000 to 2015, Reardon and her colleagues (Reardon et al., 2019) found limited evidence for increases in statistical power over time. Although it may be challenging to obtain larger clinical sample sizes, researchers in clinical psychology can potentially benefit from the Big Science approach pioneered by the RP:P and collaborate on pooling resources to achieve greater sample sizes.

Health Care Providers Unlike social-personality and cognitive psychology, where studies can often be self-administered or computerized, clinical studies tend to involve therapeutic or otherwise specialized procedures delivered by trained professionals. This feature adds another layer of complexity because the estimation of treatment effects could be biased if the therapist effect is not properly accounted for. Studies have examined whether therapist effects (Kim et al., 2006; McKay et al., 2006) play a role in treatment outcomes in clinical trials. For example, the therapist effect has been tested by reanalyzing data from the landmark Treatment of Depression Collaborative Research Program, commissioned by the National Institute of Mental Health in 1985 (McKay et al., 2006). An implicit assumption in the original analyses was that health care providers do not differ in terms of their effectiveness. When therapist effects were modeled in a multilevel model, the reanalysis found that although the actual treatments impacted clients' outcomes, health care providers played an even larger role. In particular, 9.1% of the variance in clients' depression scores was attributable to the psychiatrists (versus 3.4% attributable to the medication; McKay et al., 2006). Corroborating this finding, another study showed that 8% of the variance in clients' outcomes was due to the therapist (versus 0% due to the specific treatment that the client received; Kim et al., 2006). Multilevel models have the potential to disentangle therapy and therapist effects and to identify the characteristics and actions of therapists that account for therapist differences (Kim et al., 2006). A more accurate estimation of treatment effects is necessary to plan for an adequate sample size in replication studies.

Different Time

The COVID-19 pandemic may present opportunities and challenges for the replicability of clinical psychology. Although COVID-19-related studies may not be readily replicable, they are still important to conduct in order to advance the research based on timely and pressing topics and to understand how COVID-19 is impacting mental health. In the future, replications may reveal the cross-temporal generalizability of findings and help capture whether certain treatments remain or are more efficacious during a highly unusual time, such as during a pandemic. Do well-documented clinical psychology phenomena look substantially different in the COVID-19 world or post-COVID-19 world? The COVID-19 pandemic has increased both the physical and mental health burden in global communities. For example, 41% of U.S. adults have reported anxiety and/or depression symptoms in January of 2021, compared to 11% between January and June of 2019, according to the Household Pulse Survey (U.S. Census Bureau). This was more prevalent among 18- to 29-year-olds and women. It would be of interest to examine whether current clinical psychology findings still hold (e.g., basic research) and whether treatments remain effective (e.g., applied research) in a global crisis such as the COVID-19 pandemic.

Under the challenges of COVID-19, telehealth platforms, including emails, videoconferencing, smartphone applications, texting, and web forums, saw exponential growth in usage. Studies have shown that telehealth can be effective (Zhou et al., 2020) when working with clients who have depression (García-Lizana & Muñoz-Mayorga, 2010), anxiety (Rees & Maclaine, 2015), or post-traumatic stress disorder (Turgoose et al., 2018). It is critical to understand whether treatments from the pre-COVID era will replicate today or whether they will be less effective in light of the mental health challenges posed by COVID-19. If treatments are less effective than they were in past studies, then it can be informative for the development of future treatments. Future replication and meta-analytic efforts should comprehensively consider different measures of evidential support (Anderson & Maxwell, 2016; Simonsohn, 2015) and take into consideration how heterogeneity in findings could be potentially attributable to data collected during COVID-19. Furthermore, even assuming a replicable treatment effect of individual therapies, given the unforeseen mental health burden at the population level, further reflections and a stronger emphasis on a public health approach towards mental health are warranted (e.g., Kazdin & Blase, 2011).

Conclusion

Situated in the historical context that featured few replication studies, fantastical extrasensory findings (Bem, 2011), and well-known fraud cases, the RP:P marks a turning point that prompts reflection on and revision in research practices. The resulting credibility movement contributes to making replication studies more mainstream, shaping transparent and verifiable research practices, and leading to a systemic view towards a healthy science. In this chapter, we outlined some opportunities and challenges in implementing the various open, transparent, and rigorous scientific practices in clinical psychological science. Perhaps more important than its findings, the RP:P serves as a reminder that grassroots movements can chisel away long-standing perverse vested interests and create lasting changes for the better. Time would tell.

References

Anderson, S. F., & Maxwell, S. E. (2016). There's more than one way to conduct a replication study: Beyond statistical significance. *Psychological Methods, 21*(1), 1–12. https://doi.org/10.1037/met0000051

Baker, M., & Dolgin, E. (2017). Cancer reproducibility project releases first results. *Nature News, 541*(7637), 269.

Barba, L. A. (2018). Terminologies for reproducible research. *arXiv* preprint arXiv:1802.03311.

Begley, C. G., & Ellis, L. M. (2012). Raise standards for preclinical cancer research. *Nature, 483*(7391), 531–533.

Bem, D. J. (2011). Feeling the future: Experimental evidence for anomalous retroactive influences on cognition and affect. *Journal of Personality and Social Psychology, 100*, 407–425. https://doi.org/10.1037/a0021524

Button, K. S., Ioannidis, J. P., Mokrysz, C., Nosek, B. A., Flint, J., Robinson, E. S., & Munafò, M. R. (2013). Power failure: Why small sample size undermines the reliability of neuroscience. *Nature Reviews Neuroscience, 14*(5), 365–376.

Camerer, C. F., Dreber, A., Forsell, E., Ho, T. H., Huber, J., Johannesson, M., … Wu, H. (2016). Evaluating replicability of laboratory experiments in economics. *Science, 351*(6280), 1433–1436.

Camerer, C. F., Dreber, A., Holzmeister, F., Ho, T. H., Huber, J., Johannesson, M., … Wu, H. (2018). Evaluating the replicability of social science experiments in nature and science between 2010 and 2015. *Nature Human Behaviour, 2*(9), 637–644.

Chambless, D. L., & Hollon, S. D. (1998). Defining empirically supported therapies. *Journal of Consulting and Clinical Psychology, 66*(1), 7–18. https://doi.org/10.1037/0022-006X.66.1.7

Crocker, J. (2011). The road to fraud starts with a single step. *Nature, 479*, 151.

Cuijpers, P., Cristea, I. A., Karyotaki, E., Reijnders, M., & Huibers, M. J. (2016). How effective are cognitive behavior therapies for major depression and anxiety disorders? A meta-analytic update of the evidence. *World Psychiatry, 15*(3), 245–258.

Cybulski, L., Mayo-Wilson, E., & Grant, S. (2016). Improving transparency and reproducibility through registration: The status of intervention trials published in clinical psychology journals. *Journal of Consulting and Clinical Psychology, 84*(9), 753–767. https://doi.org/10.1037/ccp0000115

Duyx, B., Urlings, M. J., Swaen, G. M., Bouter, L. M., & Zeegers, M. P. (2017). Scientific citations favor positive results: A systematic review and meta-analysis. *Journal of Clinical Epidemiology, 88*, 92–101.

Ebersole, C. R., Atherton, O. E., Belanger, A. L., Skulborstad, H. M., Allen, J. M., Banks, J. B., ... & Nosek, B. A. (2016). Many Labs 3: Evaluating participant pool quality across the academic semester via replication. *Journal of Experimental Social Psychology, 67*, 68–82. https://doi.org/10.1016/j.jesp.2015.10.012

Editorial. (2018). What next for registered reports. *Nature Human Behaviour, 2*, 789–790. https://doi.org/10.1038/s41562-018-0477-2

Frank, M. C., Bergelson, E., Bergmann, C., Cristia, A., Floccia, C., Gervain, J., ... Yurovsky, D. (2017). A collaborative approach to infant research: Promoting reproducibility, best practices, and theory-building. *Infancy, 22*(4), 421–435.

Grant, S. P., Mayo-Wilson, E., Melendez-Torres, G. J., & Montgomery, P. (2013). Reporting quality of social and psychological intervention trials: A systematic review of reporting guidelines and trial publications. *PLoS One, 8*(5), e65442. https://doi.org/10.1371/journal.pone.0065442

García-Lizana, F., & Muñoz-Mayorga, I. (2010). What about telepsychiatry? A systematic review. *The Primary Care Companion for CNS Disorders, 12*(2), 26919. https://doi.org/10.4088/PCC.09m00831whi

Ioannidis, J. P. (2005). Why most published research findings are false. *PLoS Medicine, 2*(8), e124. https://doi.org/10.1371/journal.pmed.0020124

Inbar, Y. (2016). Association between contextual dependence and replicability in psychology may be spurious. *Proceedings of the National Academy of Sciences of the United States of America, 113*(34), E4933–E4934. https://doi.org/10.1073/pnas.1608676113

John, L. K., Loewenstein, G., & Prelec, D. (2012). Measuring the prevalence of questionable research practices with incentives for truth telling. *Psychological Science, 23*(5), 524–532. https://doi.org/10.1177/0956797611430953

Kazdin, A. E., & Blase, S. L. (2011). Rebooting psychotherapy research and practice to reduce the burden of mental illness. *Perspectives on Psychological Science, 6*(1), 21–37.

Kendall, P. C. (1998). Empirically supported psychological therapies. *Journal of Consulting and Clinical Psychology, 66*(1), 3–6. https://doi.org/10.1037/0022-006X.66.1.3

Kim, D. M., Wampold, B. E., & Bolt, D. M. (2006). Therapist effects in psychotherapy: A random-effects modeling of the National Institute of Mental Health Treatment of Depression Collaborative Research Program data. *Psychotherapy Research, 16*(02), 161–172. https://doi.org/10.1080/10503300500264911

Kidwell, M. C., Lazarević, L. B., Baranski, E., Hardwicke, T. E., Piechowski, S., Falkenberg, L. S., ... & Nosek, B. A. (2016). Badges to acknowledge open practices: A simple, low-cost, effective method for increasing transparency. *PLoS Biology, 14*(5), e1002456. https://doi.org/10.1371/journal.pbio.1002456

Kleinman, A. (2004). Culture and depression. *New England Journal of Medicine, 351*(10), 951–953. https://doi.org/10.1056/NEJMp048078

Klein, R. A., Ratliff, K. A., Vianello, M., Adams, R. B., Jr., Bahník, Š., Bernstein, M. J., Bocian, K., Brandt, M. J., Brooks, B., Brumbaugh, C. C., Cemalcilar, Z., Chandler, J., Cheong, W., Davis, W. E., Devos, T., Eisner, M., Frankowska, N., Furrow, D., Galliani, E. M., . . . Nosek, B. A. (2014). Investigating variation in replicability: A "many labs" replication project. *Social Psychology, 45*(3), 142–152. https://doi.org/10.1027/1864-9335/a000178

Klein, R. A., Vianello, M., Hasselman, F., Adams, B. G., Adams, R. B., Jr., Alper, S., ... Sowden, W. (2018). Many labs 2: Investigating variation in replicability across samples and settings. *Advances in Methods and Practices in Psychological Science, 1*(4), 443–490.

Kravitz, D. J., Mitroff, S. R., & Bauer, P. J. (2020). Practicing good laboratory hygiene, even in a pandemic. *Psychological Science, 31*, 483.

Makel, M. C., Plucker, J. A., & Hegarty, B. (2012). Replications in psychology research: How often do they really occur? *Perspectives on Psychological Science, 7*(6), 537–542.

McKay, K. M., Imel, Z. E., & Wampold, B. E. (2006). Psychiatrist effects in the psychopharmacological treatment of depression. *Journal of Affective Disorders, 92*(2–3), 287–290. https://doi.org/10.1016/j.jad.2006.01.020
Morrison, S. J. (2014). Reproducibility project: Cancer biology: Time to do something about reproducibility. *eLife, 3*, e03981.
Mobley, A., Linder, S. K., Braeuer, R., Ellis, L. M., & Zwelling, L. (2013). A survey on data reproducibility in cancer research provides insights into our limited ability to translate findings from the laboratory to the clinic. *PLoS One, 8*(5), e63221. https://doi.org/10.1371/journal.pone.0063221
Mullinix, K., Leeper, T., Druckman, J., & Freese, J. (2015). The generalizability of survey experiments. *Journal of Experimental Political Science, 2*(2), 109–138. https://doi.org/10.1017/XPS.2015.19
Munro, K. J., & Prendergast, G. (2019). Encouraging pre-registration of research studies. *International Journal of Audiology, 58*(3), 123–124. https://doi.org/10.1080/14992027.2019.1574405
Nutu, D., Gentili, C., Naudet, F., & Cristea, I. A. (2019). Open science practices in clinical psychology journals: An audit study. *Journal of Abnormal Psychology, 128*(6), 510–516. https://doi.org/10.1037/abn0000414
Nosek, B. A., & Errington, T. M. (2017). Reproducibility in cancer biology: Making sense of replications. *eLife, 6*, e23383.
Open Science Collaboration. (2015). Estimating the reproducibility of psychological science. *Science, 349*, 6251.
Peng, R. D., Dominici, F., & Zeger, S. L. (2006). Reproducible epidemiologic research. *American Journal of Epidemiology, 163*(9), 783–789.
Reardon, K. W., Smack, A. J., Herzhoff, K., & Tackett, J. L. (2019). An N-pact factor for clinical psychological research. *Journal of Abnormal Psychology, 128*(6), 493–499.
Rees, C. S., & Maclaine, E. (2015). A systematic review of videoconference-delivered psychological treatment for anxiety disorders. *Australian Psychologist, 50*(4), 259–264. https://doi.org/10.1111/ap.12122
Ryder, A. G., Yang, J., Zhu, X., Yao, S., Yi, J., Heine, S. J., & Bagby, R. M. (2008). The cultural shaping of depression: Somatic symptoms in China, psychological symptoms in North America? *Journal of Abnormal Psychology, 117*(2), 300–313.
Sakaluk, J. K., Williams, A. J., Kilshaw, R. E., & Rhyner, K. T. (2019). Evaluating the evidential value of empirically supported psychological treatments (ESTs): A meta-scientific review. *Journal of Abnormal Psychology, 128*(6), 500–509.
Schimmack, U. (2016). The Replicability-Index: Quantifying statistical research integrity. https://wordpress.com/post/replication-index.wordpress.com/920.
Simons, D. J., Holcombe, A. O., & Spellman, B. A. (2014). An introduction to registered replication reports at perspectives on psychological science. *Perspectives on Psychological Science, 9*(5), 552–555.
Simonsohn, U. (2015). Small telescopes: Detectability and the evaluation of replication results. *Psychological Science, 26*(5), 559–569. https://doi.org/10.1177/0956797614567341
Simonsohn, U., Nelson, L. D., & Simmons, J. P. (2014). P-curve: A key to the file-drawer. *Journal of Experimental Psychology: General, 143*(2), 534–547. https://doi.org/10.1037/a0033242
Strack, F., Martin, L. L., & Stepper, S. (1988). Inhibiting and facilitating conditions of the human smile: A nonobtrusive test of the facial feedback hypothesis. *Journal of Personality and Social Psychology, 54*(5), 768–777. https://doi.org/10.1037/0022-3514.54.5.768
Tackett, J. L., Brandes, C. M., King, K. M., & Markon, K. E. (2019). Psychology's replication crisis and clinical psychological science. *Annual Review of Clinical Psychology, 15*, 579–604.
Tolin, D. F., McKay, D., Forman, E. M., Klonsky, E. D., & Thombs, B. D. (2015). Empirically supported treatment: Recommendations for a new model. *Clinical Psychology: Science and Practice, 22*(4), 317–338.

Turgoose, D., Ashwick, R., & Murphy, D. (2018). Systematic review of lessons learned from delivering tele-therapy to veterans with post-traumatic stress disorder. *Journal of Telemedicine and Telecare, 24*(9), 575–585. https://doi.org/10.1177/1357633X17730443

Van Bavel, J. J., Mende-Siedlecki, P., Brady, W. J., & Reinero, D. A. (2016). Contextual sensitivity in scientific reproducibility. *Proceedings of the National Academy of Sciences, 113*(23), 6454–6459.

Wagenmakers, E. J., Beek, T., Dijkhoff, L., Gronau, Q. F., Acosta, A., Adams, R. B., Jr., … Zwaan, R. A. (2016). Registered replication report: Strack, martin, & stepper (1988). *Perspectives on Psychological Science, 11*(6), 917–928.

Wagenmakers, E.-J., Wetzels, R., Borsboom, D., & van der Maas, H. L. J. (2011). Why psychologists must change the way they analyze their data: The case of psi: Comment on Bem (2011). *Journal of Personality and Social Psychology, 100*, 426–432. https://doi.org/10.1037/a0022790

Yen, S., Robins, C. J., & Lin, N. (2000). A cross-cultural comparison of depressive symptom manifestation: China and the United States. *Journal of Consulting and Clinical Psychology, 68*(6), 993–999.

Zhou, X., Snoswell, C. L., Harding, L. E., Bambling, M., Edirippulige, S., Bai, X., & Smith, A. C. (2020). The role of telehealth in reducing the mental health burden from COVID-19. *Telemedicine and e-Health, 26*(4), 377–379. https://doi.org/10.1089/tmj.2020.0068

Chapter 19
Psychological Science Accelerator: A Promising Resource for Clinical Psychological Science

Julie Beshears, Biljana Gjoneska, Kathleen Schmidt, Gerit Pfuhl, Toni Saari, William H. B. McAuliffe, Crystal N. Steltenpohl, Sandersan Onie, Christopher R. Chartier, and Hannah Moshontz

Abstract The Psychological Science Accelerator (PSA) is an international collaborative network of psychological scientists that facilitates rigorous and generalizable research. In this chapter, we describe how the PSA can help clinical

---

J. Beshears
University of Southern Indiana, Evansville, IN, USA
e-mail: jebeshears@eagles.usi.edu

B. Gjoneska
Macedonian Academy of Sciences and Arts, Skopje, North Macedonia
e-mail: biljanagjoneska@manu.edu.mk

K. Schmidt
Southern Illinois University, Carbondale, IL, USA

G. Pfuhl
UiT The Arctic University of Norway, Tromsø, Norway
e-mail: gerit.pfuhl@uit.no

T. Saari
University of Eastern Finland, Kuopio, Finland
e-mail: toni.saari@uef.fi

W. H. B. McAuliffe
Cambridge Health Alliance, Cambridge, MA, USA

C. N. Steltenpohl
University of Southern Indiana, Evansville, IN, USA
e-mail: cnsteltenp@usi.edu

S. Onie
Black Dog Institute, UNSW Sydney, Sydney, Australia

C. R. Chartier
Ashland University, Ashland, OH, USA
e-mail: cchartie@aashland.edu

H. Moshontz (✉)
Department of Psychology, University of Madison-Wisconsin, Madison, WI, USA

W. O'Donohue et al. (eds.), *Avoiding Questionable Research Practices in Applied Psychology*, https://doi.org/10.1007/978-3-031-04968-2_19

psychologists and clinical psychological science more broadly. We first describe the PSA and outline how individual clinical psychologists can use the PSA as a helpful resource in numerous capacities: leading or contributing to clinical research or research with clinical relevance, building collaborative relationships, obtaining experience and expertise, and learning about systems and tools, particularly those related to open science practices, that they can adapt to their own research. We then describe how the PSA supports rigor and transparency at each stage of the research process. Finally, we discuss the challenges of the PSA's large, collaborative approach to research.

Keywords Psychological science accelerator · Clinical psychology · Clinical psychological science

Clinical Psychology and the PSA

Psychological science benefits society to the extent that it produces reliable and generalizable knowledge about human behavior and mental processes. Valid and broadly generalizable empirical evidence for a claim must come from large, geographically broad, and culturally diverse samples (Simons et al., 2017). Yet, even in high-impact journals, researchers often make universal claims based on convenience samples from Western, educated, industrialized, rich, and democratic (WEIRD; Henrich et al., 2010; Rad et al., 2018) populations. Typically, participants are White American (Cheon et al., 2020) undergraduate students (Sears, 1986).

This general tendency for psychology research to rely on samples from a single geographic and cultural context particularly characterizes the clinical specialty. Many clinical psychologists study phenomena and experiences that are uncommon and must recruit participants from small populations. Obtaining large, appropriately diverse samples from these hard-to-reach populations can be further challenging because participation in clinical psychology research may require people in target populations (e.g., people in crisis or experiencing clinical depression) to share personal, stigmatized information or engage in other tasks they may find uncomfortable. Additionally, inclusion criteria for clinical psychology research (e.g., participants cannot have certain comorbid diagnoses) further restrict the pool of potential participants in order to achieve internal validity or diminish the potential harm to participants. Because of these factors, sample sizes in clinical psychology research are often small (e.g., the median sample size in top clinical psychology journals is 179; Reardon et al., 2019). Thus, most clinical psychology studies are unable to provide sufficiently precise estimates of correlation coefficients (e.g., accurately estimating an $r = .10$ requires a sample of about 250 participants; Schönbrodt & Perugini, 2013), let alone produce sufficiently precise estimates of the low-probability outcomes that clinical psychology research often seeks to predict and understand (Davison & Lazarus, 2006).

Recent methodological reforms have succeeded in improving the rigor, accessibility, and transparency of psychological science (Christensen et al., 2020; Nelson et al., 2018), but these advances have not successfully proliferated certain subfields, including clinical psychology (Hopwood & Vazire, 2020; Nutu et al., 2019; Tackett et al., 2019; Tackett & Miller, 2019). The relative lack of methodological reform can have detrimental downstream effects on clinical practice and, ultimately, negatively affect mental health outcomes (Suliman et al., 2019; Tackett et al., 2017). For example, insufficient description of study procedures and the use of study materials that are not or cannot be publicly shared prevent other researchers from building on or appropriately applying interventions (Premachandra & Lewis, 2020). Questionable research practices, like failing to report all tested outcomes, can produce false positive findings (Simmons et al., 2011) which can cause harm when they motivate the implementation of less effective treatments (Sakaluk et al., 2019; Tajika et al., 2015). Practical constraints explain much of the slow progress towards improved methodology. For example, in clinical psychology research with sampling constraints, obtaining samples that provide 95% power to detect hypothesized effects (e.g., as is currently required for Registered Reports at Nature Human Behavior) can take an impractically long time for small research groups using even simple research designs.

Large-scale, crowdsourced collaborations offer clinical psychological scientists a way to conduct rigorous research on a scale not otherwise accessible to most researchers (Uhlmann et al., 2019). Individual research teams wanting to conduct a study in a sample that generalizes beyond a single context might not have the knowledge or resources to conduct language or cultural translation of study materials and measures, know how and where to recruit participants at every research site, or know how best to model the resulting data (Leong & Kalibatseva, 2013). By pooling research resources together, clinical psychologists can accomplish what no single research group could do alone without significant outside grant funding.

The Psychological Science Accelerator (PSA) is an international collaborative network of psychological scientists that facilitates rigorous and generalizable research (Moshontz et al., 2018). In this chapter, we describe how the PSA can help clinical psychologists and clinical psychological science more broadly. We first describe the PSA and outline how individual clinical psychologists can use the PSA as a helpful resource in numerous capacities: leading or contributing to clinical research or research with clinical relevance, building collaborative relationships, obtaining experience and expertise, and learning about systems and tools, particularly those related to open science practices, that they can adapt to their own research. We then describe how the PSA supports rigor and transparency at each stage of the research process. Finally, we discuss the challenges of the PSA's large, collaborative approach to research.

About the PSA

The PSA was formed in 2017 as a proactive response to critical issues facing psychological science such as replicability and generalizability (John et al., 2012; Nelson et al., 2018; Open Science Collaboration, 2015; Simons, 2014; Simons et al., 2017; Uhlmann et al., 2019). The PSA's strategy of pooling the resources of individual labs together in order to conduct sufficiently powered, geographically distributed research was inspired by crowdsourced collaborations, including the Emerging Adulthood Measured at Multiple Institutions project (EAMMI; Reifman & Grahe, 2016) and the Reproducibility Project: Psychology (Open Science Collaboration, 2015). Outside of psychology, the European Organization for Nuclear Research (CERN) inspired the conception of the PSA as a standing collaborative network of researchers from different nations committed to conducting ambitious, novel research rather than specific projects. Within weeks of a blog post inviting psychology researchers to join a standing collaborative network that would later become the PSA, dozens around the world had joined (Chartier, 2017a). These early members began formalizing an organizational structure and procedures that were later detailed in a paper introducing the network (Moshontz et al., 2018).

As of December 2020, the PSA is a large, active organization. The network contains over 1400 individual researchers, including undergraduate students, graduate students, professors of all ranks, staff scientists, and people in nonacademic roles (e.g., in industry or government). PSA members are based in over 70 countries spread across all six populated continents. Just under 25% of researchers in the PSA network are based in North America, and about 40% are based in Western Europe (Paris et al., 2020). Currently, clinical psychology is the reported specialty for 145 members (~6%), relatively fewer than those who specialize in social and personality psychology (~20%), experimental psychology (~14%), cognitive psychology (~14%), and quantitative psychology (~10%).

The members of the PSA network collaboratively and transparently select, design, and conduct research as guided by five core principles: *diversity and inclusion*, *decentralized authority*, *transparency*, *rigor*, and *openness to criticism*. These principles shape the policies and procedures of the PSA. *Diversity and inclusion* are reflected in both the collaborating researchers and the studied participants and are central to the plans for the future of the PSA. The network members who help propose, select, design, translate, and conduct research represent a diverse collection of geographic regions, research institutions, academic positions, and training areas. Additionally, the PSA recruits socioculturally and geographically diverse research samples. Although member labs in the network are globally distributed, they mostly have access to already well-represented samples, like undergraduate university students and people who live in densely populated areas. With funding, the PSA can better promote the principle of diversity and inclusion by supporting labs to broaden their sampling approach into local communities and more rural areas.

The *decentralized authority* principle is reflected in the governance structure of the PSA; specifically, stages of the research process are managed by different committees and decisions are made democratically, either by the entire network or by committee. *Transparency* finds expression with respect to both the internal workings of the PSA (e.g., network members can view all committee meeting notes) and its research products. The PSA shares policy documents (e.g., Forscher, Aczel, et al., 2020) and the materials, analysis code, and data from all studies that it conducts to the extent allowable (e.g., by ethics considerations; Meyer, 2018).

The core principle of *rigor* shapes the PSA research process. Proposed studies are selected on the basis of their rigor, and the primary purpose of a key PSA committee, the Data and Methods Committee, is to ensure the quality of study protocols and analyses. Finally, the PSA strives to function with an *openness to criticism.* PSA procedures involve soliciting and incorporating critical feedback from within and outside the network on aspects of both research projects and the PSA's processes for selecting and conducting research.

Although the PSA produces research projects similar to other crowdsourced, large-scale collaborations in psychology (Ebersole et al., 2016; Klein et al., 2018), it differs from these efforts in several key ways. First, rather than existing for the purpose of completing a particular project, the PSA is an ongoing network that runs multiple projects simultaneously. Second, anyone can contribute to research at the PSA. Membership in the network is not contingent upon professional connections, training, background, job title, or geographic location. Third, the PSA is flexible; rather than conducting research in a specific content area or population, the PSA selects studies that range in their focus and population of interest. Studies are not selected on the basis of their psychological research area or the prestige of the study proposers, whose identities are concealed during the review and selection process. However, resource availability does constrain what projects are feasible. As described in calls for study submissions, feasibility constraints have resulted in a preference for studies with samples that are fairly small and easy to reach (e.g., requiring fewer than 150 participants per collection site), protocols that are rather short (e.g., less than 90 minutes per session), and equipment that is readily available (e.g., using open source software and no specialized hardware) and does not pose a risk to participant health. Such parameters have changed over time, and, given the growing membership and resources of the PSA, research that targets harder-to-reach populations or uses longer, more complex procedures may soon be feasible.

The ten completed and ongoing PSA studies use large and often international samples to investigate a broad range of research questions. For example, the first completed PSA study assessed the global generalizability of a model of face perception in 11,570 participants in 41 countries and 11 world regions (Jones et al., in press). One study that has yet to begin data collection will assess different operationalizations of stereotype threat among Black college students in the United States with an anticipated sample of 2700 students across 27 geographically distributed schools (Forscher, Taylor, et al., 2020). In 2021, the PSA anticipates collecting data from a minimum of 20,000 participants in total (Paris et al., 2020). Although most PSA studies follow a standard process and were proposed in response to open calls, special-topic projects with different foci have also been implemented; for example, a teaching-focused replication project invited undergraduate students in member labs to collect data, conduct analyses, and contribute to the final manuscript (Hall et al., 2020; Wagge et al., 2019). The PSA has also successfully collected data for three particularly accelerated projects related to COVID-19 (Dorison et al., 2020; Legate et al., 2020; Wang et al., 2020). For these projects, which were run in a bundled protocol, PSA members selected and revised studies, translated materials into 43 languages and dialects, and collected data from over 44,000 participants around the world, all within only 8 months. PSA studies have been led by people at different career stages, including graduate students (i.e., Hall et al., 2020; Wang et al., 2020).

How Individual Clinical Psychological Scientists Can Benefit from the PSA

Clinical psychological scientists aiming to produce rigorous and generalizable research can use the PSA in several ways. First, they can lead or contribute to rigorous research with potential clinical relevance; in fact, the PSA just completed a study on the effectiveness of brief cognitive reappraisal interventions for reducing people's negative emotions during the COVID-19 pandemic (Wang et al., 2020). As of December 2020, the relatively small number of PSA network researchers specializing in clinical psychology would restrict clinical psychologists from leading studies at the PSA that require specialized equipment or clinically-trained experimenters. Consequently, clinical psychologists could not currently conduct research involving screening or treating participants with psychopathology as a PSA project, but they could conduct research on more easily recruited populations using widely-available equipment. Within these constraints, the PSA enables researchers without external funding to lead studies that would otherwise require large grants and specialized training (e.g., the ability to translate materials) to conduct. One particularly important area of clinical psychology research that is well-suited to the PSA is research establishing the properties of clinical measures across cultures (i.e., assessing measurement equivalence; Leong & Kalibatseva, 2013). Additionally, clinical psychological scientists could perform secondary analysis on any of the datasets collected by the PSA or use translated materials from completed and ongoing PSA studies for their own research.

Second, by joining the PSA network or contributing to a PSA study, clinical psychological scientists can build collaborative connections with other researchers. The PSA network is a community; members share research, conference and grant calls, and other opportunities. Members have developed collaborative projects beyond the primary studies selected and run by the PSA. For example, several member labs collaborated on a study of the perceived efficacy of COVID-19 restrictions and their effect on mental health that collected data from over 2000 participants in six countries (Mækelæ et al., 2020). PSA projects have also resulted in secondary analysis collaborations (Adkins et al., 2020; Batres, 2020; Chandrashekar, 2020;

Durkee & Ayers, 2020; Hester & Hehman, 2020; Martin et al., 2020; Oh & Todorov, 2020; Xie & Hehman, 2020). Further, the PSA also supports collaborative discussions outside of empirical projects. Network members often discuss and debate issues in psychological science more broadly, in informal and formal outlets (IJzerman et al., 2020; Onie, 2020).

Third, scientists in the PSA can use the PSA to gain experience and develop expertise. Many roles on PSA projects can serve as experiential learning for experienced clinical psychological scientists. For example, membership in the Data and Methodology Committee offers opportunities to work with and learn from methodological experts. Members of the Project Monitoring Committee can learn about how to manage research projects with hundreds of research sites and thousands of participants. By becoming more involved in any capacity, PSA members are given additional opportunities to contribute to and benefit from PSA resources.

Finally, clinical psychologists can use the PSA to learn about systems and tools that support rigorous, transparent, collaborative research. Methodological reforms evolve, and researchers who do not adopt reforms may simply not know about them or know how to implement them (Washburn et al., 2018). For example, some researchers may not know that failing to report all study outcomes can severely undermine a study's evidentiary value (Nelson et al., 2018; Simmons et al., 2011). PSA membership can be a means by which clinical psychologists can learn about the need for particular methodological reforms and ways to implement them. Further, the PSA organizes and collaboratively produces projects using tools (e.g., collaboration agreements, translation protocols, project tracking templates) that can also benefit small groups. Many of the challenges of working in large international collaborations are present in other group contexts. Members can prevent and overcome problems in their outside collaborations by using the solutions that the PSA has devised and tested over time.

How the PSA Supports Rigor and Transparency at Each Stage of the Research Process

By design, the PSA supports rigor and transparency at every stage of the research process. Some challenges of conducting rigorous, transparent clinical psychological research are inherent to a clinical research question (e.g., researchers cannot randomly assign participants to experience trauma to see what factors predict who develops PTSD). The PSA's practices and procedures cannot eliminate these challenges, although they can, in some cases, lessen the impact of unavoidable challenges on the quality of the final research product. In this section, we describe the PSA's current research processes, which are similar to those described in Moshontz et al. (2018), but reflect improvements made as the PSA has grown in membership and experience.

Selecting a Research Project

Research questions at the PSA are selected from a pool of masked protocols submitted in response to an open call for proposals from all areas of psychology. Proposed studies may be confirmatory or exploratory, test a novel research question or propose a replication, or explore the validity of measures or stimuli. When study proposals are submitted for consideration, authors are asked to explicitly address feasibility, implementation, and ethics concerns. Submissions requiring specialized samples are asked to explain and justify this requirement and elaborate on risk mitigation steps taken for any vulnerable populations. The study selection committee reviews masked submissions for quality (e.g., whether the proposal is complete and well-considered), feasibility (e.g., if the PSA has the capacity and resources to support the research), and appropriateness (e.g., whether the project necessitates a lab network). Submissions that pass this initial phase are sent to expert reviewers both in and outside of the PSA. Reviewers evaluate study-specific threats to inference (e.g., a confound unique to the paradigm), and threats to inference common to all cross-cultural research (e.g., measurement invariance). Submissions that reach the second round are also made available to the full network for members to evaluate. After feedback from reviewers and the network are compiled and synthesized, the committee votes and decides whether to provisionally accept the proposal, request proposal revisions, or reject the submission.

Identifying Project Needs

Accepted proposals enter a needs assessment process, which identifies the lead (i.e., proposing) authors' needs with respect to all major aspects of conducting the study: methodology, data management, ethics, translation, logistics, adhering to PSA policies, and writing. During this process, the lead authors provide information about their study that will determine which committee members and labs they will be paired with. In addition, they meet with the PSA Director and members of PSA committees that focus on each aspect of the study to ensure that they understand how studies are run at the PSA. For example, lead authors are asked to describe any special requirements for some or all participants, whether data collection teams need specialized knowledge or equipment, and whether the submitting authors will clean and analyze the data, and if so, in what programming language. The questions asked at this stage are designed to ensure that each project is conducted rigorously and transparently and that the lead authors and network contributors have appropriate and clearly defined roles. For example, if lead authors propose a design that requires complex analyses and do not have an analytic expert on their team, after the needs assessment process, the lead authors would be matched with a PSA collaborator with relevant expertise, who would most likely join the lead author team. During this stage, the lead authors and members of the expert committees identify aspects of the PSA's standard practices and procedures that need to be adjusted for that particular project. For example, sharing data publicly is standard practice at the PSA, but if a team of clinical psychological scientists led a study that involved collecting sensitive data, the standard data sharing plan would be adjusted at this stage. Once the lead author team describes their needs, the project is matched with a member of each expert committee accordingly to further develop the proposal before Registered Report submission and study implementation.

All submitting authors on Psychological Science Accelerator projects are held to ten expectations by default if they agree to lead a collaborative project. These ten expectations align with core principles and formal policies (Moshontz et al., 2018):

1. Work collaboratively with PSA member labs and committee personnel.
2. Create a collaboration agreement that describes authorship criteria.
3. Obtain a demonstration video for every data collection site.
4. Preregister methods, materials, and analyses.
5. Obtain ethics approval (or equivalent) at every data collection site.
6. Make all study materials open access (unless prohibited by copyright).
7. Make all data open access in accordance with the PSA-approved data management plan.
8. Make all analysis scripts openly accessible.
9. Make any final report openly accessible.
10. Adhere to a code of conduct.

Refining the Study Design and Analytic Approach

Regardless of the lead authors' expertise, all accepted proposals are assigned to one or more members of the Data and Methods Committee. The primary goal of the Data and Methods Committee is to provide expertise and oversight on the methodological components of PSA projects. The process of ensuring the rigor of PSA studies begins even before a study is accepted; every submitted proposal is reviewed by at least one Data and Methods Committee member or external reviewer appointed by the committee. The Data and Methods Committee member on each project collaborates with the lead authors to develop an analysis plan and write statistical analysis scripts. The committee also appoints a data manager to each project to ensure that researchers comply with the analysis plan, archive data in a public or private repository in a timely manner, and correct any analytic errors that are found in manuscripts. More informally, the committee provides technical support as needed. The committee's secondary goal is to organize or implement projects of methodological and meta-scientific interest. For example, the committee might examine the performance of a new analytic tool that has only been evaluated via simulations using PSA data; implementing a new tool will likely include many complications that routinely arise in collaborative projects but are glossed over in the initial vetting of a method.

Submitting a Registered Report and Preregistering the Study

Once the study design and analysis plan has been refined, that plan is submitted as part of a Registered Report or preregistration. Registered Reports, which share similarities with registered clinical trials, allow studies to be considered for publication before they have been conducted. Registered Reports support rigorous methodology (Scheel et al., 2020; Soderberg et al., 2020) and help ensure researcher resources and participant time are well-spent. Such concerns are critical in clinical research settings given the high opportunity cost of resources and the potential to influence practitioner behavior (Tackett et al., 2017; Cristea & Naudet, 2019). Because Registered Reports are submitted before data collection or analysis, studies are reviewed on the basis of their rationale, writing, methods, measures, analysis plans, and contingent conclusions.

The evaluation of studies at this stage, rather than after results are known, protects against publication biases that result in the overrepresentation of positive results in the published literature and the inflation of effect size estimates (Fanelli, 2010; Ferguson & Heene, 2012; Kühberger et al., 2014; Simonsohn et al., 2014). Once a Registered Report is accepted in principle (i.e., as a Stage 1 Registered Report), the described methods and analytic approach are preregistered. Accepted Stage 1 Registered Reports can provide lead authors, data collection labs, and other contributors peace of mind, knowing that as long as they execute the project as described, they will be rewarded with what is still the most important professional incentive for most psychological scientists: a publication.

The PSA process is particularly well-aligned with the Registered Report format. Lead authors submit proposals in the format of Registered Reports. Further, many requirements for PSA proposals are also requirements of Registered Reports, including the requirement to conduct a priori power analyses (Moshontz et al., 2018). Due to the similarity in formatting and content of PSA proposals and Registered Reports, many completed and ongoing PSA studies have been submitted as Registered Reports (Bago et al., 2020; Chen et al., 2020; Forscher, Taylor, et al., 2020; Jones et al., in press; Wang et al., 2020) or Registered Replication Reports (Hall et al., 2020), which are Registered Reports focused on replication.

Translating the Study Protocol

If an accepted study needs materials to be translated, the Translation and Cultural Diversity Committee provides expertise in and oversight of the translation process. Most PSA studies are conducted in different geographic regions, where participants speak different languages, and the meaning and impact of study procedures might differ as a function of culture. Thus, material translation is a challenging but essential aspect of the research process.

The PSA uses a standard translation protocol, adapted from Brislin (1970), to standardize the translation process for all languages. A translation coordinator oversees the entire process, and a language-wise coordinator oversees the process for each target language. Language-wise coordinators work closely with the translation coordinator to ensure efficient and high-quality translations. To begin the translation process, the source material is first translated into the target language by two independent translators. Then, these translators and language-wise coordinators compare and discuss the translations to create a single forward translation (Version A). Two independent translators then translate Version A back to the source language (i.e., back-translate). The two back-translators and language-wise coordinators discuss discrepancies and create a single back-translation (Version B). The translation coordinator and language-wise coordinators compare Version B and Version A, identifying and discussing discrepancies with input from the lead authors. The language-wise coordinator then creates a new version of the translated materials (Version C), which is sent to at least two external readers who evaluate the wording and clarity.

Language-wise coordinators discuss the need for cultural adjustments with the data collection labs that will use the translated material. These cultural considerations are particularly important in clinical contexts because they help establish *linguistic* (related to translation of words), *functional* (related to translation of behaviors), *conceptual* (related to translation of constructs) and *metric equivalence* (related to psychometric properties of instruments) across cultures (Leong & Kalibatseva, 2013). Psychopathology can be culturally-specific in its expression and effect (Henrich et al., 2010; Patel & Sumathipala, 2001), so clinical research that fails to use culturally heterogeneous samples or account for cultural context cannot ensure clinical relevance or broad generalizability and may be of limited value (Nagayama Hall, 2006). In the final step of the translation process, the language-wise coordinators (and participating labs) construct a final version of the materials with attention to cultural considerations and feedback from the external readers. All the translation materials - including all versions and notes - are stored publicly to allow interested researchers to investigate or otherwise make use of these materials.

Ethics Review

Prior to data collection, the study protocol is subjected to ethics review, first at the PSA and then by ethics review boards. Every data collection site must obtain an ethics review exemption or approval before they begin data collection. The involvement of a local ethics review, when possible, is most appropriate when the risks associated with a particular study procedure may differ as a function of culture. For example, whereas many clinical studies pose some risk to participants (e.g., by collecting sensitive data; Cristea & Naudet, 2019; Meyer, 2018), the risk and data sensitivity of a protocol may also vary by data collection site, as a function of cultural norms and stigma associated with the focal topic. Revisions to the study procedure based on ethics review at a data collection site are not common; thus far, ethics reviews at each data collection site have not resulted in any major revisions to study procedures.

Data Collection

Data collection labs are matched with studies based on expressed interest and match with study needs. Collaboration agreements that are written before data collection begins describe authorship criteria and expectations. After obtaining ethics review exemption or approval at their institution, if applicable, data collection labs practice the study procedure and record demonstration videos. In the videos, one researcher typically plays the role of a participant while another conducts the study procedure. Recording demonstration videos serves multiple purposes. First, demonstration videos help ensure procedural fidelity at every site. Lead authors can review the demonstration videos, identify discrepancies between the protocol as written and the protocol as administered, and give labs feedback as needed before data collection. Second, recording demonstration videos documents aspects of the data collection context that can be examined or otherwise used in the future. Demonstration videos serve as a record of fidelity and of data collection site features (e.g., the physical space where the study was conducted). Demonstration videos can also be used by data collection labs to train research assistants. When data collection sites have completed all the requirements specified in the collaboration agreement (e.g., ethics review exemption or approval, demonstration video submission), they can begin collecting data.

Data Analysis and Final Manuscript Submission

After data collection at all sites is complete, data are cleaned and analyzed in accordance with the study preregistration. The submitting author team then drafts the final manuscript (e.g., a Stage 2 Registered Report) for submission with help from other members of the collaboration team (e.g., the Data and Methods Committee member who works on the project). People who meet the authorship criteria defined in the collaboration agreement provide feedback on the manuscript draft and approve the final version prior to submission.

Challenges

The PSA's introductory paper outlines six challenges the PSA faces in conducting research (Moshontz, et al., 2018). These challenges include resource management, linguistic and cultural translation, inclusivity, research ethics, funding, and crediting contributions. Many of these challenges are particularly relevant to clinical research, and the increasing membership of clinicians in the PSA will proportionally increase the likelihood of finding suitable solutions.

A persistent challenge for the PSA is drawing on and distributing research resources effectively. Not all studies require large, international samples or are equally deserving of the participant hours and researcher time required to conduct a PSA study. The first fully completed PSA project collected data from 11,570 participants and took 3 years to conclude from the initial proposal (in October 2017; Chartier, 2017b) to finalized publication as a Stage 2 Registered Report (in October 2020; Jones et al., in press). The appropriate use of research resources is a challenge that affects individual labs and the PSA as a whole. Individual labs decide on a study-by-study basis what they will contribute to, ensuring no lab will spend their resources on projects that they do not deem valuable. PSA studies are often more efficient and scientifically valuable than the typical single-lab or small collaborative projects to which lead authors and data collection labs might otherwise contribute.

The PSA carefully considers different perspectives in deciding which studies to conduct. Conducting research that addresses trivial scientific questions at the scale of a PSA project would contribute to, rather than detract from, research waste, but the scientific value of research questions is often subjective. The lead authors of study proposals are asked to justify the required resources in their proposals, and data collection labs provide feedback on proposals during the study selection process related to scientific value. In addition, for accepted studies submitted as Registered Reports, editors and peer-reviewers evaluate and improve the evidentiary value of PSA studies and reduce the risk of wasted resources.

A second challenge is of particular relevance to clinical psychology: any international, cross-cultural study requires the translation of stimuli and instructions into dozens of languages, dialects, and cultures. As described previously, PSA procedures aim to address this challenge, but a perfect solution for translation often does not exist (Brislin, 1976). For example, translating questions about mental health and psychopathology requires great care, as do seemingly simple questions, like demographic questions about gender, sexuality, race, and ethnicity that are not defined or publicly discussed in the same way across cultures. The PSA cannot avoid translation challenges for its international studies but is well positioned to make thoughtful and culturally contextualized translation decisions.

A third challenge for the PSA, inclusion, remains the most difficult to address. Although the PSA has been successful in recruiting members across the globe, not all regions are equally represented in the network; over 60% of member labs are based in Europe and the United States (Paris et al., 2020). Because minimum sample sizes are needed to conduct cross-cultural comparisons in PSA studies, undue pressure is placed on contributing labs from areas with low membership to provide sufficient sample sizes.

Sociocultural diversity is likewise lacking among the leadership of the PSA, and a minority of members have been directly involved in formulating the PSA's policies and procedures. Though many decisions are made democratically, participation is not equally accessible to all members (e.g., all formal PSA communication is in English). Further, decision-making processes are most influenced by people who voice their opinions and argue on behalf of their preferences. Paralleling trends in psychological science more broadly, participation in the decision-making processes at the PSA can be inaccessible to certain PSA members, such as members of traditionally marginalized groups and researchers at institutions without research infrastructure or support. Those who would likely benefit the most from participation in the PSA and whose involvement would advance the PSA's mission to accelerate the accumulation of reliable and generalizable evidence in psychological science may be the least likely to join the network or seek out leadership roles. However, the PSA, through the Community Building and Network Expansion Committee and other means, remains focused on identifying and addressing barriers to inclusion (e.g., Chartier, 2020).

Another challenge of conducting research at the PSA is ensuring participant protection. Guidelines for ethical human subject research vary considerably across nations and institutions. The PSA's Ethics Review Committee is well-equipped to help coordinate the ethics review process and ensure compliance with requirements at each data collection site. The PSA has thus far only conducted research that does not involve vulnerable populations or the collection of identified, sensitive data. However, PSA policies were designed to accommodate sensitive data and specify that research will be shared to the extent allowable due to legal (e.g., proprietary measures) or ethical constraints (Moshontz et al., 2018).

The biggest challenge facing the PSA is providing sufficient material and administrative support to research. The PSA largely relies on member volunteerism, which can take a heavy toll on people who carry the biggest load of responsibilities. This reliance on volunteer labor can create project delays and workload asymmetries, inequalities, and tensions between collaborators. However, without outside funding, the PSA has no alternative means of viability. A recent internal report estimated that the administrative support provided by the members of PSA committees alone is equivalent to at least 200,000 US dollars per year (Paris et al., 2020). The administrative support required for complex studies like clinical trials is even greater than for the single session lab studies implemented by the PSA thus far. Complex clinical psychology research conducted at the PSA would likely strain its volunteer workforce and exacerbate the concomitant problems of relying upon volunteers. Fortunately, the PSA's continued growth helps mitigate this challenge. As the PSA grows, the number of people who can share administrative duties grows, and the burden on individual people lessens.

A final challenge of the PSA's crowdsourced approach to research is properly crediting all contributors within an authorship system that is not designed for projects with hundreds of collaborators. Large authorship lists pose logistical challenges. Project leaders must document the contributions and changing affiliations of the hundreds of contributing researchers and maintain communication during the process of writing, revising, and submitting a manuscript. In addition, providing meaningful credit to all of a project's contributors can be difficult when so many people have made contributions. Manuscripts describing PSA projects report how each author contributed to the research, but the labels representing author contributions may inadequately capture the scale or impact of the effort required to conduct a large-scale collaboration.

Conclusion

By joining the PSA, clinical psychologists can help produce rigorous research while gaining experience and expertise, learning about new systems and tools, and advancing the improvement of clinical psychological science more broadly. Clinical psychology is not a common specialization at the PSA (Paris, et al., 2020), and although the PSA has conducted a simple intervention study with clinical relevance (Wang et al., 2020), it has not yet conducted studies proposed by clinical psychological scientists.

The PSA welcomes submissions from clinical psychological scientists. Currently, study submissions using simple protocols (e.g., surveys administered with a computer) are most likely to meet the PSA's feasibility requirements. In addition, the PSA could easily support measurement research that involves translating and assessing the properties of clinical psychological surveys. Such research is both easy to administer and important (Flake & Fried, 2020). Looking ahead, the more researchers with clinical training who join the network, the more able the PSA is to support more complex, resource-intensive clinical research protocols.

Finally, clinical psychological scientists who join the network can shape the PSA to support and improve their field. By voting for studies and in leadership elections, providing feedback on submitted studies, and otherwise taking part in the PSA's decentralized decision-making processes, clinical psychological scientists can broaden the PSA's scope to include the critical questions that drive clinical psychological research. More rigorous, international, collaborative clinical psychological science, both at the PSA and beyond, can accelerate the discovery and refinement of treatments that improve people's lives.

References

Adkins, M. C., Beribisky, N., Bonfield, S., & Farmus, L. (2020). Hierarchical modelling of facial perceptions: A secondary analysis of aggressiveness ratings [Preprint]. *PsyArXiv*. https://doi.org/10.31234/osf.io/5bv2p.

Bago, B., Aczel, B., Kekecs, Z., Protzko, J., Kovacs, M., Nagy, T., Hoekstra, R., Li, M., Musser, E. D., Arvanitis, A., Iones, M. T., Bayrak, F., Papadatou-Pastou, M., Belaus, A., Storage, D., Thomas, A. G., Buchanan, E. M., Becker, B., Baskin, E., … Dutra, N. B. (2020). Moral thinking across the world: Exploring the influence of personal force and intention in moral dilemma judgments (stage 1 registered report). *Nature Human Behavior*. https://doi.org/10.31234/osf.io/9uaqm

Batres, C. (2020). PSA001 secondary analysis: Examining the "attractiveness halo effect" [Preprint]. *PsyArXiv*. https://doi.org/10.31234/osf.io/c7hf3.

Brislin, R. W. (1970). Back-translation for cross-cultural research. *Journal of Cross-Cultural Psychology*, 1(3), 185–216. https://doi.org/10.1177/135910457000100301

Brislin, R. W. (1976). Comparative research methodology: Cross-cultural studies. *International Journal of Psychology, 11*(3), 215–229.

Chandrashekar, S. P. (2020). The facial width-to-height ratio (fWHR) and perceived dominance and trustworthiness: Moderating role of social identity cues (gender and race) and ecological factor (pathogen prevalence) [preprint]. *PsyArXiv*. https://doi.org/10.31234/osf.io/64t9s

Chartier, C. R. (2017a, August 26). Building a CERN for psychological science. *Christopher R. Chartier*. https://christopherchartier.com/2017/08/26/building-a-cern-for-psychological-science/

Chartier, C. R. (2017b, October 12). On the verge of acceleration: The PSA has received its first submissions. *Christopher R. Chartier*. https://christopherchartier.com/2017/10/12/on-the-verge-of-acceleration-the-psa-has-received-its-first-submissions/

Chartier, C. R. (2020, October 28). News from the accelerator- October 2020. *Psychological Science Accelerator*. https://psysciacc.org/2020/10/28/news-from-the-accelerator-october-2020/

Chen, S.-C., Szabelska, A., Chartier, C. R., Kekecs, Z., Lynott, D., Bernabeu, P., Jones, B. C., DeBruine, L. M., Levitan, C., Werner, K. M., Wang, K., Milyavskaya, M., Musser, E. D., Papadatou-Pastou, M., Coles, N. A., Janssen, S., Özdoğru, A. A., Storage, D., Manley, H., … Schmidt, K. (2020). Investigating object orientation effects across 14 languages [preprint]. *PsyArXiv*. https://doi.org/10.31234/osf.io/t2pjv

Cheon, B. K., Melani, I., & Hong, Y. (2020). How USA-centric is psychology? An archival study of implicit assumptions of generalizability of findings to human nature based on origins of study samples. *Social Psychological and Personality Science, 11*(7), 928–937. https://doi.org/10.1177/1948550620927269

Christensen, G., Wang, Z., Levy Paluck, E., Swanson, N., Birke, D., Miguel, E., & Littman, R. (2020). *Open science practices are on the rise: The state of social science (3S) survey*. https://escholarship.org/uc/item/0hx0207r

Cristea, I. A., & Naudet, F. (2019). Increase value and reduce waste in research on psychological therapies. *Behaviour Research and Therapy, 123*, 103–479. https://doi.org/10.1016/j.brat.2019.103479

Davison, G. C., & Lazarus, A. A. (2006). Clinical case studies are important in the science and practice of psychotherapy. In S. O. Lilienfeld & W. T. O'Donohue (Eds.), *The Great Ideas of Clinical Science: 17 Principles that Every Mental Health Professional Should Understand* (pp. 149–162). Routledge. https://www.taylorfrancis.com/chapters/clinical-case-studies-important-science-practice-psychotherapy-gerald-davison-arnold-lazarus/10.4324/9780203942789-14

Dorison, C., Lerner, J. S., Heller, B. H., Rothman, A., Kawachi, I. I., Wang, K., Rees, V. W., Gill, B. P., Gibbs, N., & Coles, N. A. (2020). A global test of message framing on behavioural intentions, policy support, information seeking, and experienced anxiety during the COVID-19 pandemic [Preprint]. PsyArXiv. https://doi.org/10.31234/osf.io/sevkf.

Durkee, P., & Ayers, J. D. (2020). Is facial width-to-height ratio reliably associated with social inferences? A large cross-national examination [preprint]. *PsyArXiv*. https://doi.org/10.31234/osf.io/tpngz

Ebersole, C. R., Atherton, O. E., Belanger, A. L., Skulborstad, H. M., Allen, J. M., Banks, J. B., Baranski, E., Bernstein, M. J., Bonfiglio, D. B. V., Boucher, L., Brown, E. R., Budiman, N. I., Cairo, A. H., Capaldi, C. A., Chartier, C. R., Chung, J. M., Cicero, D. C., Coleman, J. A., Conway, J. G., … Nosek, B. A. (2016). Many labs 3: Evaluating participant pool quality across the academic semester via replication. *Journal of Experimental Social Psychology, 67*, 68–82. https://doi.org/10.1016/j.jesp.2015.10.012

Fanelli, D. (2010). "Positive" results increase down the hierarchy of the sciences. *PLoS One, 5*(4), e10068. https://doi.org/10.1371/journal.pone.0010068

Ferguson, C. J., & Heene, M. (2012). A vast graveyard of undead theories: Publication bias and psychological science's aversion to the null. *Perspectives on Psychological Science, 7*(6), 561–555. https://doi.org/10.1177/1745691612459059

Flake, J. K., & Fried, E. I. (2020). Measurement schmeasurement: Questionable measurement practices and how to avoid them [preprint]. *PsyArXiv*. https://doi.org/10.31234/osf.io/hs7wm

Forscher, P. S., Aczel, B., Chartier, C. R., Musser, E. D., Horstmann, K. T., Grahe, J. E., Kekecs, Z., Corker, K. S., Sirota, M., Vaughn, L. A., Wichman, A. L., Papadatou-Pastou, M., Evans, T. R., Szecsi, P., Storage, D., Pfuhl, G., Borghi, J., Urry, H. L., Isager, P. M., … Flake, J. K. (2020). Psychological science accelerator data management bylaws [preprint]. *PsyArXiv*. https://doi.org/10.31234/osf.io/buqyc

Forscher, P. S., Taylor, V. J., Cavagnaro, D., Lewis, N. A., Buchanan, E. M., Moshontz, H., Mark, A. Y., Appleby, S., Batres, C., Bennett-Day, B., Chopik, W. J., Damian, R. I., Ellis, C. E., Faas, C., Gaither, S., Green, D., Hall, B. F., Hinojosa, B. M., Howell, J. L., … Chartier, C. R. (2020). Stereotype threat in black college students across many operationalizations (stage 1 registered report). *Nature Human Behavior*. https://doi.org/10.31234/osf.io/6hju9

Hall, B. F., Wagge, J. R., Pfuhl, G., Stieger, S., Vergauwe, E., Ijzerman, H., Musser, E. D., Schmidt, K., Weissgerber, S. C., Buchanan, E. M., Chen, S.-C., Werner, K. M., Field, A. P., Meijer, E., Andreychik, M., Storage, D., Voracek, M., Tran, U. S., Hartanto, A., … Lewis, S. (2020). Registered replication report: Turri, Buckwalter, & Blouw (2015). *Advances in Methods and Practices in Psychological Science*. https://doi.org/10.31234/osf.io/zeux9

Henrich, J., Heine, S. J., & Norenzayan, A. (2010). Most people are not WEIRD. *Nature, 466*(7302), 29. https://doi.org/10.1038/466029a

Hester, N., & Hehman, E. (2020). Hester PSA001 preregistration preprint—Region- and language-level ICCs for judgments of faces [preprint]. *PsyArXiv*. https://doi.org/10.31234/osf.io/fz6kr

Hopwood, C. J., & Vazire, S. (2020). Reproducibility in clinical psychology. In A. G. C. Wright & M. N. Hallquist (Eds.), *The Cambridge handbook of research methods in clinical psychology* (1st ed., pp. 371–382). Cambridge University Press. https://doi.org/10.1017/9781316995808.035

Ijzerman, H., Lewis, N. A., Przybylski, A. K., Weinstein, N., DeBruine, L., Ritchie, S. J., Vazire, S., Forscher, P. S., Morey, R. D., Ivory, J. D., & Anvari, F. (2020). Use caution when applying behavioural science to policy. *Nature Human Behaviour, 4*(11), 1092–1094. https://doi.org/10.1038/s41562-020-00990-w

John, L. K., Loewenstein, G., & Prelec, D. (2012). Measuring the prevalence of questionable research practices with incentives for truth telling. *Psychological Science, 23*(5), 524–532. https://doi.org/10.1177/0956797611430953

Jones, B. C., DeBruine, L. M., Flake, J. K., Liuzza, M. T., Antfolk, J., Arinze, N. C., Ndukaihe, I. L. G., Bloxsom, N., Lewis, S., Foroni, F., Willis, M., Cubillas, C. P., Vadillo, M. A., Gilead, Michael, Simchon, A., Saribay, S. A., Owsley, N. C., Calvillo, D. P., Włodarczyk, A., … Coles, N. A. (in press). To which world regions does the valence-dominance model of social perception apply? *Nature Human Behavior*. https://doi.org/10.31234/osf.io/n26dy.

Klein, R. A., Vianello, M., Hasselman, F., Adams, B. G., Adams, R. B., Alper, S., Aveyard, M., Axt, J. R., Babalola, M. T., Bahník, Š., Batra, R., Berkics, M., Bernstein, M. J., Berry, D. R., Bialobrzeska, O., Binan, E. D., Bocian, K., Brandt, M. J., Busching, R., … Nosek, B. A. (2018). Many labs 2: Investigating variation in replicability across samples and settings. *Advances in Methods and Practices in Psychological Science, 1*(4), 443–490. https://doi.org/10.1177/2515245918810225

Kühberger, A., Fritz, A., & Scherndl, T. (2014). Publication bias in psychology: A diagnosis based on the correlation between effect size and sample size. *PLoS One, 9*(9), e105825. https://doi.org/10.1371/journal.pone.0105825

Legate, N., Nguyen, T.-V., Moller, A., Legault, L., Weinstein, N., & Psychological Science Accelerator. (2020). Effect of message framing on motivation to follow vs. defy social distancing guidelines during the COVID 19 pandemic. *PsychArchives*. https://doi.org/10.23668/PSYCHARCHIVES.3014

Leong, F. T. L., & Kalibatseva, Z. (2013). *Clinical research with culturally diverse populations*. Oxford University Press. https://doi.org/10.1093/oxfordhb/9780199793549.013.0021

Mækelæ, M. J., Reggev, N., Dutra, N., Tamayo, R. M., Silva-Sobrinho, R. A., Klevjer, K., & Pfuhl, G. (2020). Perceived efficacy of COVID-19 restrictions, reactions and their impact on mental health during the early phase of the outbreak in six countries. *Royal Society Open Science, 7*(8), 200644. https://doi.org/10.1098/rsos.200644

Martin, J., Wood, A. R., & Oh, D. (2020). Population diversity is associated with trustworthiness impressions from faces [preprint]. *PsyArXiv*. https://doi.org/10.31234/osf.io/sbvkz

Meyer, M. N. (2018). Practical tips for ethical data sharing. *Advances in Methods and Practices in Psychological Science, 1*(1), 131–144. https://doi.org/10.1177/2515245917747656

Moshontz, H., Campbell, L., Ebersole, C. R., Ijzerman, H., Urry, H. L., Forscher, P. S., Grahe, J. E., Mccarthy, R. J., Musser, E. D., Antfolk, J., Castille, C. M., Evans, T. R., Fiedler, S., Flake, J. K., Forero, D. A., Janssen, S. M. J., Keene, J. R., Protzko, J., Aczel, B., … Chartier, C. R. (2018). The psychological science accelerator: Advancing psychology through a distributed collaborative network. *Advances in Methods and Practices in Psychological Science, 1*(4), 501–515. https://doi.org/10.1177/2515245918797607

Nagayama Hall, G. C. (2006). Diversity in clinical psychology. *Clinical Psychology: Science and Practice, 13*(3), 258–262. https://doi.org/10.1111/j.1468-2850.2006.00034.x

Nelson, L. D., Simmons, J., & Simonsohn, U. (2018). Psychology's renaissance. *Annual Review of Psychology, 69*(1), 511–534. https://doi.org/10.1146/annurev-psych-122216-011836

Nutu, D., Gentili, C., Naudet, F., & Cristea, I. A. (2019). Open science practices in clinical psychology journals: An audit study. *Journal of Abnormal Psychology, 128*(6), 510–516. https://doi.org/10.1037/abn0000414

Oh, D., & Todorov, A. (2020). Do regional gender and racial biases predict gender and racial biases in social face judgments? [preprint]. *PsyArXiv*. https://doi.org/10.31234/osf.io/v7hpe

Onie, S. (2020). Redesign open science for asia, africa and latin america. *Nature, 587*(7832), 35–37. https://doi.org/10.1038/d41586-020-03052-3

Open Science Collaboration. (2015). Estimating the reproducibility of psychological science. *Science, 349*(6251), aac4716. https://doi.org/10.1126/science.aac4716

Paris, B., Ijzerman, H., & Forscher, P. S. (2020). PSA 2020-2021 study capacity report [preprint]. *PsyArXiv*. https://doi.org/10.31234/osf.io/v9zma

Patel, V., & Sumathipala, A. (2001). International representation in psychiatric literature: Survey of six leading journals. *British Journal of Psychiatry, 178*(5), 406–409. https://doi.org/10.1192/bjp.178.5.406

Premachandra, B., & Lewis, N. A. (2020). Do we report the information that is necessary to give psychology away? A scoping review of the psychological intervention literature 2000-2018 [preprint]. *PsyArXiv*. https://doi.org/10.31234/osf.io/nr8kh

Rad, M. S., Martingano, A. J., & Ginges, J. (2018). Toward a psychology of Homo sapiens: Making psychological science more representative of the human population. *Proceedings of the National Academy of Sciences, 115*(45), 11401–11405.

Reardon, K. W., Smack, A. J., Herzhoff, K., & Tackett, J. L. (2019). An N-pact factor for clinical psychological research. *Journal of Abnormal Psychology, 128*(6), 493–499. https://doi.org/10.1037/abn0000435

Reifman, A., & Grahe, J. E. (2016). Introduction to the special issue of emerging adulthood. *Emerging Adulthood, 4*(3), 135–141. https://doi.org/10.1177/2167696815588022

Sakaluk, J. K., Williams, A. J., Kilshaw, R. E., & Rhyner, K. T. (2019). Evaluating the evidential value of empirically supported psychological treatments (ESTs): A meta-scientific review. *Journal of Abnormal Psychology, 128*(6), 500–509. https://doi.org/10.1037/abn0000421

Scheel, A. M., Schijen, M., & Lakens, D. (2020). An excess of positive results: Comparing the standard psychology literature with registered reports [preprint]. *PsyArXiv*. https://doi.org/10.31234/osf.io/p6e9c

Schönbrodt, F. D., & Perugini, M. (2013). At what sample size do correlations stabilize? *Journal of Research in Personality, 47*(5), 609–612.

Sears, D. O. (1986). College sophomores in the laboratory: Influences of a narrow data base on social psychology's view of human nature. *Journal of Personality and Social Psychology, 51*(3), 515–530.

Simmons, J. P., Nelson, L. D., & Simonsohn, U. (2011). False-positive psychology: Undisclosed flexibility in data collection and analysis allows presenting anything as significant. *Psychological Science, 22*(11), 1359–1366.

Simons, D. J. (2014). The value of direct replication. *Perspectives on Psychological Science, 9*(1), 76–80. https://doi.org/10.1177/1745691613514755

Simons, D. J., Shoda, Y., & Lindsay, D. S. (2017). Constraints on generality (COG): A proposed addition to all empirical papers. *Perspectives on Psychological Science, 12*(6), 1123–1128. https://doi.org/10.1177/1745691617708630

Simonsohn, U., Nelson, L. D., & Simmons, J. P. (2014). P-curve and effect size: Correcting for publication bias using only significant results. Perspectives on Psychological Science, 9(6), 666–681. https://doi.org/10.1177/1745691614553988.

Soderberg, C. K., Errington, T. M., Schiavone, S. R., Bottesini, J. G., Singleton Thorn, F., Vazire, S., Esterling, K., & Nosek, B. A. (2020). Research quality of registered reports compared to the traditional publishing model [Preprint]. *MetaArXiv*. https://doi.org/10.31222/osf.io/7x9vy.

Suliman, S., van den Heuvel, L., Suryapranata, A., Bisson, J. I., & Seedat, S. (2019). Publication and non-publication of clinical trials in PTSD: An overview. *Research Integrity and Peer Review, 4*(1), 15.

Tackett, J. L., Brandes, C. M., King, K. M., & Markon, K. E. (2019). Psychology's replication crisis and clinical psychological science. *Annual Review of Clinical Psychology, 15*(1), 579–604. https://doi.org/10.1146/annurev-clinpsy-050718-095710

Tackett, J. L., Lilienfeld, S. O., Patrick, C. J., Johnson, S. L., Krueger, R. F., Miller, J. D., Oltmanns, T. F., & Shrout, P. E. (2017). It's time to broaden the replicability conversation: Thoughts for and from clinical psychological science. *Perspectives on Psychological Science, 12*(5), 742–756. https://doi.org/10.1177/1745691617690042

Tackett, J. L., & Miller, J. D. (2019). Introduction to the special section on increasing replicability, transparency, and openness in clinical psychology. *Journal of Abnormal Psychology, 128*(6), 487–492. https://doi.org/10.1037/abn0000455

Tajika, A., Ogawa, Y., Takeshima, N., Hayasaka, Y., & Furukawa, T. A. (2015). Replication and contradiction of highly cited research papers in psychiatry: 10-year follow-up. *British Journal of Psychiatry, 207*(4), 357–362. https://doi.org/10.1192/bjp.bp.113.143701

Uhlmann, E. L., Ebersole, C. R., Chartier, C. R., Errington, T. M., Kidwell, M. C., Lai, C. K., McCarthy, R. J., Riegelman, A., Silberzahn, R., & Nosek, B. A. (2019). Scientific utopia III: Crowdsourcing science. *Perspectives on Psychological Science, 14*(5), 711–733. https://doi.org/10.1177/1745691619850561

Wagge, J. R., Brandt, M. J., Lazarevic, L. B., Legate, N., Christopherson, C., Wiggins, B., & Grahe, J. E. (2019). Publishing research with undergraduate students via replication work: The collaborative replications and education project. *Frontiers in Psychology, 10*, 247. https://doi.org/10.3389/fpsyg.2019.00247

Wang, K., Goldenberg, A., Dorison, C., Miller, J. K., Uusberg, A., Lerner, J. S., & Gross, J. (2020). A global test of brief reappraisal interventions on emotions during the COVID-19 pandemic (stage 1 registered report). *Nature Human Behavior*. https://doi.org/10.31234/osf.io/m4gpq.

Washburn, A. N., Hanson, B. E., Motyl, M., Skitka, L. J., Yantis, C., Wong, K. M., Sun, J., Prims, J. P., Mueller, A. B., Melton, Z. J., & Carsel, T. S. (2018). Why do some psychology researchers resist adopting proposed reforms to research practices? A description of researchers' rationales. Advances in Methods and Practices in Psychological Science, 1(2), 166–173. https://doi.org/10.1177/2515245918757427.

Xie, S. Y., & Hehman, E. (2020). Variance and homogeneity of facial trait space across world regions [Preprint]. *PsyArXiv*. https://doi.org/10.31234/osf.io/d4zma

Index

W. O'Donohue et al. (eds.), *Avoiding Questionable Practices in Applied Psychology*, https://doi.org/10.1007/978-3-031-04968-2

Q

The manufacturer's authorised representative in the EU is Springer Nature Customer Service Centre GmbH, Europaplatz 3, 69115 Heidelberg, Germany. If you have any concerns regarding our products, please contact ProductSafety@springernature.com

Printed and bound by CPI Group (UK) Ltd, Croydon, CR0 4YY
30/09/2026
02225250-0016